Robotic Nondestructive Testing Technology

Robotic Nondestructive Testing Technology

Chunguang Xu

CRC Press
Taylor & Francis Group
Boca Raton London New York

CRC Press is an imprint of the
Taylor & Francis Group, an **informa** business

北京理工大学出版社
BEIJING INSTITUTE OF TECHNOLOGY PRESS

First edition published 2022
by CRC Press
6000 Broken Sound Parkway NW, Suite 300, Boca Raton, FL 33487-2742

and by CRC Press
2 Park Square, Milton Park, Abingdon, Oxon, OX14 4RN

CRC Press is an imprint of Taylor & Francis Group, LLC

Library of Congress Cataloging-in-Publication Data
Names: Xu, Chunguang, 1964- author.
Title: Robotic nondestructive testing technology / Chunguang Xu.
Description: First edition. | Boca Raton, FL : CRC Press, 2022. |
Includes bibliographical references. | Summary: "This book introduces a variety
of non-destructive testing (NDT) methods, including testing and application cases.
New ultrasonic testing technology for complex workpieces is proposed"— Provided by publisher.
Identifiers: LCCN 2021031691 (print) | LCCN 2021031692 (ebook) |
ISBN 9781032079547 (hbk) | ISBN 9781032079561 (pbk) |
ISBN 9781003212232 (ebk)
Subjects: LCSH: Nondestructive testing. | Automatic test equipment. |
Robotics—Industrial applications.
Classification: LCC TA417.2 .X83 2022 (print) | LCC TA417.2 (ebook) |
DDC 620.1/127—dc23
LC record available at https://lccn.loc.gov/2021031691
LC ebook record available at https://lccn.loc.gov/2021031692

ISBN: 978-1-032-07954-7 (hbk)
ISBN: 978-1-032-07956-1 (pbk)
ISBN: 978-1-003-21223-2 (ebk)

DOI: 10.1201/9781003212232

Typeset in Minion
by codeMantra

Contents

Preface

NONDESTRUCTIVE TESTING (NDT) IS a comprehensive engineering application technology using ultrasound, X-ray, electromagnetic wave, penetration and other physical methods to nondestructively test and evaluate the internal discontinuity and material properties of mechanical components. It is also an important basis for evaluation of the integrity and service safety of mechanical components. It plays an important role in controlling and improving the manufacturing quality of mechanical products and ensuring the safe and reliable operation of in-service equipment, thus laying a foundation for the development of modern industry and high-end manufacturing equipment.

It is hard for traditional NDT technology to realize the automatic detection of complex curved components, especially the automatic high-precision NDT of curved-surface components with variable curvature, variable thickness and complex contour, which, in most cases, rely on manual point-selection detection. Traditionally, a constant coupling distance between the detection transducer and the component under test, a stable and effective coupling state as well as flexible adjustment of spatial detection attitude is hard to achieve, resulting in the difficulty in ensuring the NDT precision. In contrast, robotic has been widely used in many industrial fields because of its good flexibility and wide adaptability. The technology of robotic NDT combines the physical principle of nondestructive testing with the flexible motion control of spatial attitude of articulated robotic. With NDT as the constraint, it controls the motion attitude and azimuth angle of a transmitting or receiving transducer. Thus, traditional NDT technique has developed from plane to curved surface, from 2D to many dimensions and from artificial testing to intelligent detection, into the interdisciplinary theory and technology of robotic NDT. Now the robotic NDT has been applied in aerospace, weapon, vehicle and other industrial fields.

This book is divided into 16 chapters. Chapters 1–3 mainly provide the basic knowledge on manipulator, covering the basic theories and algorithms of robotic coordinate system transformation, water-waveguide ultrasonic testing (UT) and manipulator trajectory planning. This part is the theoretical basis of robotic NDT technique. Chapters 4 and 5 introduce the single-robotic NDT and dual-robotic NDT, respectively, describe the composition of single/dual-robotic NDT systems and present the coordinate transformation method of a robotic NDT system in detail. Chapters 6–8 describe the basic theory of error analysis and calibration of single-robotic NDT. Among them, Chapter 6 mainly presents three error sources, Chapter 7 discusses the geometric error of the system and the method to correct it and Chapter 8 describes the motion error of the system and the method to compensate for it. Chapters 9 and 10 present the trajectory planning of dual-robotic UT of semi-closed narrow small components and the calibration method of a special-shaped tool coordinate system. Chapter 11 discusses the robotic X-ray detection technology and describes the composition of a robotic X-ray CT detection system and the coordination control method of X-ray detection data and manipulator attitude data. Chapter 12 discusses the robotic electromagnetism eddy current technology and presents the composition, data processing and imaging method of a robotic electromagnetism eddy current testing system. Chapter 13 discusses the robotic measurement method of ultrasonic transducer sound field and presents the basic theory and composition of measurement system of ultrasonic transducer sound field. Chapter 14 presents the basic theory and composition of the robotic laser measurement system of sound field intensity in solids. Chapters 15 and 16 address the main applications of single/dual-robotic NDT technology, including the applications in revolving components, box components and large composite-material components.

This book is based on the long-term scientific research and teaching experience of Professor Xu Chunguang and his research team in Beijing Institute of Technology, which has been combined with the summary and extraction of the latest NDT research results. I would like to express my heartfelt thanks to Xiao Dingguo, Hao Juan, Zhou Shiyuan, Pan Qinxue, Jia Yuping, Meng Fanwu and many other advisors and graduate students for their contributions to the contents of this book. I would like to express my special thanks to Doctor Ma Pengzhi for his great work on the formation of the first draft of this book and on the editing and page make-up of characters and formulas. Finally, I would like to express my sincere thanks to Lu Zongxing, Xiao Zhen, Zhang Hanming, Li Fei, Guo Chanzhi, Lu Yuren and other doctoral and master students for their hard work in research and their extensive editing and organizing efforts.

This book can be used as a learning and reference material for the theoretical and technical researchers working on automatic NDT as well as for advanced NDT personnel in training courses and relevant majors in colleges and universities. It can also be used as the teaching material publicizing the national standard "Non-destructive Testing – Test Method for Robotic Ultrasonic Testing" (GB/T34892-2017).

Due to the author's limited proficiency, shortcomings and mistakes are inevitable in this book. The readers' criticisms and corrections are invited.

Chunguang Xu

MATLAB® is a registered trademark of The MathWorks, Inc. For product information, please contact:

The MathWorks, Inc.
3 Apple Hill Drive
Natick, MA 01760-2098 USA
Tel: 508-647-7000
Fax: 508-647-7001
E-mail: info@mathworks.com
Web: www.mathworks.com

Author

Chunguang Xu, PhD, is a professor at Beijing Institute of Technology, China. He earned his PhD in mechanical manufacturing from Beijing Institute of Technology in 1995. His current research interests include ultrasonic testing and manipulator control.

CHAPTER 1

Introduction

WITH THE CONTINUOUS PROGRESS and development of social civilization and industry, the safety and reliability of the advanced equipment have raised increasingly high requirements on the automatic nondestructive testing (NDT) technology for the internal damage of mechanical components. Modern mechanized equipment is equipped with a large number of new materials and complex-shaped components that are hard to effectively quantify, test and evaluate by traditional manual NDT. Therefore, it is urgent to study and develop a flexible intelligent automatic NDT technology with adaptive function.

NDT is usually used for observing and evaluating the continuity, integrity, safety and physical properties of the component material under test. It plays an important role in controlling and improving the manufacturing quality of products, ensuring the safe operation of equipment in service, and evaluating the performance and reliability of products and their components. It is an important foundation of modern industry and advanced intelligent manufacturing [1,2].

It is difficult for traditional NDT technology to realize the roboticized testing and quantitative evaluation of complex curved components, especially the high-precision contour tracking and scanning of the curved components with local curvature discontinuity, thickness discontinuity and complex shape. Moreover, the accuracy, repeatability and traceability of manual NDT are hard to effectively guarantee. Currently, the multi-axis-linked NDT technology based on the Cartesian coordinate system is only applicable to planar or regular-shaped components and is hard to meet the automatic NDT requirement of variable-curvature/large complex-shaped components such as aero-engine blades, turbine disks and wheel hubs.

In recent years, the articulated manipulator has been widely used in welding, machining, handling, loading and unloading, painting and other manufacturing processes due to its advantages such as multi-dimensional flexibility, spatial adaptability and attitude motion accuracy. It is only in recent years that the application of the articulated manipulator in industrial NDT has been in full swing [3–7]. When applying the articulated manipulator to NDT, it is necessary to ensure that the update frequency of the manipulator position data meets the requirement for ultrasonic triggering frequency. That is to say, the update

DOI: 10.1201/9781003212232-1

rate of position information should match the trigger repetition rate of NDT, so that the information on the angle of each manipulator joint and on the position and attitude of end-effector and detection probe can be obtained in time. Generally, the basic characteristics of the articulated manipulator used in NDT are as follows:

1. The accuracy of spatial position and attitude and motion trajectory should be high enough. It is necessary to plan the scanning trajectory of the manipulator holding a detection probe according to the geometry profile information of the detected component, and to ensure that the errors in probe position/attitude and trajectory meet the requirements for component profile accuracy and probe position/attitude accuracy;
2. The smoothness and stability of motion process should be good enough. The manipulator holding a probe moves along the planned trajectory for real-time detection. For example, in the process of ultrasonic reflection and transmission detection, the excitation and reception of ultrasonic signals should be completed. In the process of DR (Direct Digit Radiography) ray detection, the X-ray transmitting and receiving units should be highly coordinated and synchronized.
3. The refresh frequency of attitude information and spatial position information should be high enough. The update frequency of multi-degree of freedom (DOF) position information of the manipulator should be not lower than the trigger acquisition frequency of the NDT system. For example, the frequency of position and attitude update should not be lower than the trigger frequency of ultrasonic or electromagnetic signal or the frame rate of DR ray screen, so as to realize the synchronous acquisition and visualization of detection position signal and NDT signal without loss.
4. The control system should be able to memorize and recover the motion trajectory. The robotic NDT is characterized by automation and continuity. It can accurately record the detection position information and the corresponding detection signal characteristics in real time and can repeat the detection and testing. It also has the function of self-locking and self-recovery of abnormal interrupt, which means the detection process can be continued after the abnormal system stop and restart.
5. Good protection. The articulated manipulator can meet the requirements of electromagnetic interference and vibration impact. It is dustproof and waterproof. For immersion-type ultrasonic testing (UT), waterproofing is necessary.
6. Good functional interfaces for utilities. The manipulator can meet the requirements such as the automatic replacement of the pneumatic tooling for roboticized testing, the power supply and signal transmission of sensors, as well as the transportation of water-coupled and air-coupled media.

Based on the research of motion control technology and NDT theory and technology of articulated manipulator, a set of automatic single/dual-manipulator NDT system has been established in this book to realize the automatic NDT of minor defects/cracks in complex components. Meanwhile, in order to realize the high-efficiency high-precision automatic NDT of complex components, the theoretical and experimental research has been carried out for some theories, key technologies and methods that need to be solved in the robotic NDT of complex components, especially for the robotic scanning control modes and detection principles of ultrasonic NDT, such as trajectory planning algorithm, error compensation and calibration as well as 3D data processing. The test results can be displayed and stored in real time to facilitate the observation of spatial location and distribution of defects in the tested component.

Due to the advantages such as high sensitivity to defect identification, wide detection range, easy automatic scanning, low cost and no harm to human body, the UT has become one of the most widely used and researched NDT means. Its application in the NDT field is also increasing day by day. It is not only widely used in manufacturing, aerospace, shipping, petrochemical engineering, electronics, transportation and other industrial fields but also favored in medicine, material, biology and other fields [8,9], where it plays an irreplaceable role. Therefore, this book will highlight the principle of single/dual-manipulator UT technique while shedding light on the principles of robotic X-ray and electromagnetic testing techniques. To sum up, the automatic control technology of the articulated manipulator can be applied to the automatic NDT of all kinds of complex components and has a broad prospect of engineering application.

1.1 BACKGROUND

The traditional UT method is used to obtain the A-scan waveforms of refraction/reflection/transmission ultrasonic signals of the tested component through contact coupling and to acquire the characteristics of ultrasonic echo signal [10–12] (e.g. sound speed, phase, frequency, peak-to-peak value) corresponding to the properties of component material (such as density, stress and discontinuity) after the data processing. The ultrasonic probe moves along the component surface to obtain ultrasonic B-scan and C-scan images. The accuracy of ultrasonic scanning result depends on the condition of coupling between the probe surface and the component to be tested and on the precision of probe profile trajectory and spatial attitude positioning. To solve the problem of coupling between the ultrasonic probe and the component surface to be contacted, water coupling is usually obtained by means of water immersion or spraying to realize ultrasonic water-coupled reflection/transmission detection or air coupling is used to realize ultrasonic air-coupled transmission detection. Although the positioning and motion precision of traditional multi-axis-linked scanning mechanism (including AB axis) is high enough to fully meet the requirements of general NDT, the mechanical interference often occurs when completing the full-size scanning of a complex-shaped component between the scanning track of curved component and the tested component (or the scanning mechanism) due to the limitation of spatial structure and scanning mechanism freedom, so that the high-precision NDT requirement of complex components or narrow spatial structures is hard to meet. For single-sided electromagnetic detection, X-ray transmission detection and single-sided or two-sided infrared thermal wave detection, the motion interference of the scanning mechanism mentioned above also exists to limit the engineering application of traditional multi-axis-linked scanning mechanism.

With a high repeatable positioning accuracy and a high degree of flexible automation within its motion space, the multi-DOF articulated manipulator is suitable for the copying measurement and roboticized testing of complex-shaped samples. Supported by auxiliary tooling and workpiece rotation, it can meet the NDT requirement of complex curved components. This manipulator has a broad prospect of engineering application.

1.1.1 Automatic NDT Methods of Complex Components

According to the way of coupling, the NDT methods of complex-shaped components can be divided into contact testing and non-contact testing [13–15]. Flexible phased-array UT and electromagnetic eddy current flexible array testing are contact testing, while air coupling, water coupling, laser UT, X-ray detection and infrared thermal wave detection are non-contact testing. The way of coupling can be changed to meet the need for testing.

In the UT of a complex component, the flexible phased-array probe can scan the whole component along the blade surface. Through the contact and coupling with the curved surface of the tested component under pressure, the probe can receive stable pulse echo signals. However, the probe wear and the scratch of workpiece surface can be caused easily. When a large curved component is tested with this method, a large amount of coupling agent will be consumed due to the need for decoupling, so that the testing environment will be polluted. Moreover, the flexible probe is hard to clamp and locate, causing the difficulty in realizing fully roboticized testing and the need for manual intervention all the time. As a result, the testing efficiency is low and the equipment is expensive. The electromagnetic eddy current flexible array has a high sensitivity to the surface defects of complex-shaped components, but the eddy current is susceptible to cracked material, electrical and magnetic conductivities, component thickness/shape/size and other factors. At present, mature commercial eddy current detectors are available to test the curved components in real time. However, most of them are hand-held. They can detect the change in the voltage signal of electromagnetic eddy current but can't easily make quantitative judgment on the size and position of defects.

Currently, water-coupled or air-coupled UT, electromagnetic testing, penetration, infrared thermal imaging, X-ray and many other non-contact testing methods are commonly used at home and abroad for the automatic NDT of complex-shaped components [16]. Basically, most of NDT methods can turn into automatic NDT techniques. For example, the fluorescent penetrant testing system combined with the manipulator can become a three-dimensional omnidirectional visual inspection system. With the support of a manipulator, the detection efficiency of an automatic penetration system can be greatly improved. The eddy current system can become a robotic testing system very easily. By combining the trajectory planning with the lift-off effect, the cracking or opening on the surface or superficial surface of a mechanical component can be detected very easily. The infrared thermal imaging system can also become a robotic NDT system very easily. However, the stitching of CCD thermal-wave images requires that the spatial positioning accuracy of a manipulator is high enough to match the pixel spacing of CCD camera. Laser UT is non-contact testing. It can become robotic NDT very easily only if the laser beam is perpendicular to the surface of complex component for excitation. In the laser UT, the receiving probe may be a laser ultrasonic beam or an ultrasonic contact probe. The laser shearography system can also become a robotic testing system very easily due to its non-contact. Especially, it can adapt to the detection of any defects in complex composite components very well.

In addition, several contact and non-contact NDT methods can be combined into an automatic NDT system for complex components provided that the manipulator technique is introduced. For example, UT and laser shearography can be combined. UT is used for the areas where the glass fiber laminates are bonded with adhesive, and laser shearography is used for the sandwich structure layers, so the NDT of the whole composite component can be realized. Pulsed eddy current and thermal imaging can also be combined to detect the layered structures.

Judging from the detection efficiency and result resolution of the above non-contact NDT methods used for complex components, the non-contact ultrasonic NDT is a perfect automatic NDT and evaluation method for complex-shaped components. In particular, modern UT technique has become quite mature with a low equipment cost. The UT-based quantitative analysis and evaluation method for the defects such as pores, inclusions, delamination and cracks in the tested components has been widely applied and verified in the NDT industry. Therefore, it has vast potential for future development.

1.1.2 Development Trend of Robotic NDT Technique

In recent years, with the increasing demand for the roboticized testing of curved components, various kinds of special roboticized equipment in the Cartesian coordinates have been developed and applied one after another to greatly improve the speed, accuracy and repeatability of NDT. For example, the NDT scanning equipment special for a type of blade can track and approximate the profile of curved blade surface. The blade surface is divided into several small planes to adapt to the insufficient flexibility of the testing equipment so as to obtain good detection accuracy. However, this roboticized testing equipment has poor universality and adaptability, and still can't effectively check some components with a small size, a large curvature or a narrow space.

Generally, the six-axis roboticized UT method is limited by the motion range of the testing mechanical mechanism. When testing the specimens with large curvature changes, the detection efficiency is often affected due to the large changes in manipulator position and attitude. Therefore, it is relatively easy for this method to realize the ultrasonic scanning of planar or regular-shaped mechanical components. For the components whose curvature changes greatly (such as blades), higher detection accuracy cannot be guaranteed in this method due to the required change of ultrasonic incidence angle.

With the increasing maturity of manipulator manufacturing and control technology, the manipulators (including bionic robots and industrial robots) have been increasingly applied to the automatic NDT of industrial materials and their products. However, most of the manipulators are only applied to the NDT of large regular curved components, such as petroleum pipelines, high-risk buildings, reservoirs or the nuclear reactor vessels, oil tanks and boiler tubes in dangerous environments. In these applications, manual detection usually doesn't work, and the efficiency of traditional detection device is low. Using the manipulator as the carrier of detection device is an important technical means to achieve flexible roboticized testing.

Biomimetic manipulators, especially mobile manipulators, were first used in the automatic NDT of large specimens. This is mainly because the mobile manipulator can walk a long distance, automatically track the surface of the tested component (usually by using a camera installed on the manipulator) and receive instructions in the detection process to adjust the scanning path. For example, the mobile manipulator is used to detect the defects such as welds in natural gas pipelines. The crawling manipulator is used for NDT by communicating with the manipulator through wireless interconnection and obtaining its real-time position in the detection process. The peristaltic manipulator can make use of the thrust formed by the difference between gas pressure and fluid pressure in a pipeline to axially scan the pipe with an mm-level accuracy. The underwater mobile manipulator can also be used for the automatic NDT of service components such as hulls, naval vessels and underwater oil pipelines.

Industrial manipulators are mainly applied to the NDT of large curved specimens such as aerospace parts and auto parts. The robotic detection system shown in Figure 1.1 is a self-transmit self-received detection system of eddy current type. The mechanical play error of scanning stroke under the six-axis linkage condition can be less than 0.5 mm.

Currently, the industrial manipulators are mainly used for welding and handling. The NDT-specific manipulators are still in the development stage and are mostly modified from other types of manipulators. Their prototypes are mainly welding robots. To realize the automatic NDT of a complex component, the scanning path of the manipulator-based

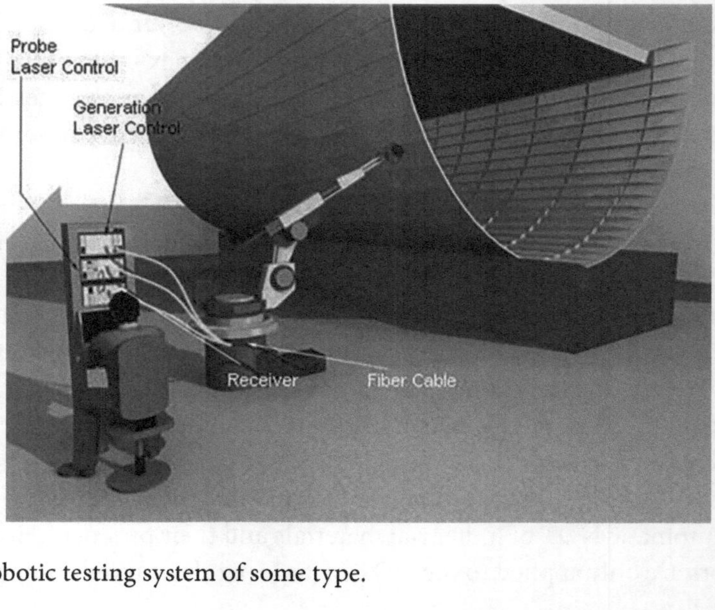

FIGURE 1.1 Robotic testing system of some type.

testing equipment needs to be planned according to the model of the tested component. Meanwhile, the synchronous acquisition of manipulator position information and ultrasonic pulse echo signal has become the key robotic NDT technology. The industrial manipulator products for commercial use can only provide the calling interface of internal position information, whose transmission speed and update time are limited to some extent. This is a technical bottleneck to the high efficiency and speed of a robotic NDT system. In addition, the data processing and real-time imaging of the scanning results is also a key technology. For example, most of the existing UT methods are based on spatial pseudocolor processing. That is to say, the pulse echo peak-to-peak value of ultrasonic A-scan waveform is converted into the gray value representing it according to some rules and then into different RGB color values according to the linear mapping function, so as to obtain the pseudocolor images whose ultrasonic echo information is represented by different color values (ultrasonic C-scan). To realize the real-time display of the detection results in the scanning process, a reasonable worker thread should be arranged to process the collected signals. The programming language in which the control program and the image display program are written is also one of the factors that affect the way of data transmission and communication with the manipulator. Especially, when writing the 3D display program of detection results or calling the OpenGL and Direct3D for display, the transmission and communication of some parameters need to be coordinated with the timing sequence of the computer worker thread.

In conclusion, the NDT tools represented by the manipulator have high positioning accuracy, have high degree of flexible automation and other strong points and are suitable for the automatic NDT of complex components. In particular, they can realize the multi-method multi-station parallel combined detection for the tested components.

With the coordinated motion control technique of a six-DOF manipulator as the core and the principles, technical standards and specifications of NDT methods (such as ultrasonic, ray and electromagnetism) as constraints, this book has constructed a robotic NDT system, explained the construction principle, analyzed the construction technology, described the work flow and discussed the typical applications. It lays an important foundation for robotic NDT technology and its applications.

1.2 BASIS OF MANIPULATOR

The most important parameters for the manipulator used for NDT or measurement are the 3D spatial position coordinates and spatial attitude vector of the probe held by the manipulator end-effector, as well as the position and trajectory relationships between the probe and the tested target object. Usually the coordinates of the manipulator system, the coordinates of the motion control system and the profile coordinates of the tested object are not the same. The coordinate conversion between different reference coordinate systems is needed to obtain the motion control coordinate system of the whole testing system and finally to obtain the spatial positions and attitudes of the detection probe and of the tested object.

For the convenience of subsequent discussion, some basic definitions and terms are given below for the manipulators used in NDT or measurement.

1.2.1 Type and Structure of Manipulators

Almost all the motion arms of articulated manipulators are composed of rigid connecting rods that are connected by the joints possibly in relative motion. A position sensor is mounted on each joint to measure the relative position of adjacent rods. If the joint is rotational, this displacement is called joint angle. Some motion arms contain the sliding (or moving) joints, so the displacement of the two adjacent rods is in rectilinear motion. This displacement is sometimes referred to as joint offset. As shown in Figure 1.2, the spatial position and attitude of each joint can be represented by four structural kinematics parameters: rod offset d_i, joint angle θ_i, rod length a_i and rotation angle a_i. Among them, the rotation angle a_i and the rod length a_i are used to describe the connecting rod itself, while the joint angle θ_i and the rod offset d_i are used to describe the connection relationship between the two rods.

The rod length a_i is the length of the common perpendicular connecting the joint axes on both ends of the connecting rod *i*. The rotation angle a_i is used to define the relative position of the two joint axes. The rod offset d_i describes the distance along the common axis of the two adjacent connecting rods. The joint angle θ_i describes the angle at which the two adjacent rods rotate around the common axis. The rule that uses the rod parameters to describe the mechanism's kinematic relationship is called Denavit–Hartenberg parameter. The spatial position and attitude of the articulated manipulator in any mechanism can be expressed by this parameter.

For a motion arm with *n* DOFs, the position of the connecting rod at its end can be determined by a set of *n* joint variables. The space formed by all the joint vectors is called joint space. When the trajectory planned for the manipulator is established in the Cartesian coordinate system with orthogonal spaces (called task spaces or operating spaces), the vector representation of the manipulator end-effector should undergo the transformation of the coordinate systems for position and attitude.

1.2.2 Working Mode of Manipulators

Depending on the different requirements for detection motion, the testing systems can be divided into two types: the ultrasonic transducer-held system (as shown in Figure 1.3) and

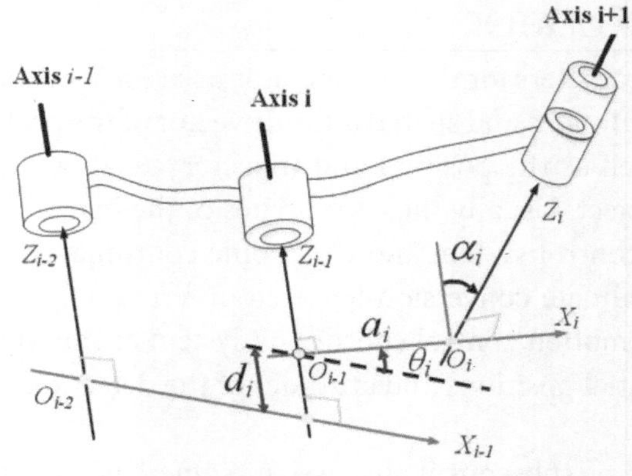

FIGURE 1.2 Structural parameters of a connecting rod.

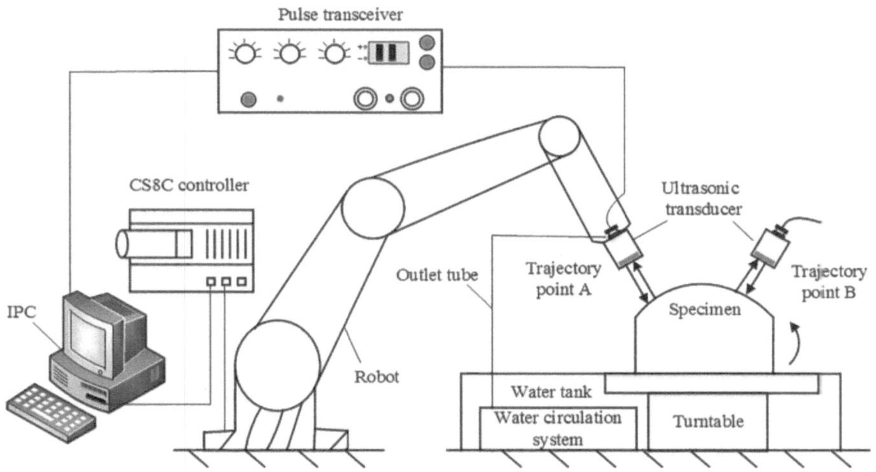

FIGURE 1.3 Structural diagram of a transducer-held ultrasonic reflection detection system.

the specimen-held system (as shown in Figure 1.4). According to the propagation path of ultrasonic wave in the tested sample, the testing systems can be divided into ultrasonic reflection system and ultrasonic transmission system. Therefore, the testing system based on manipulator motion control has four main working modes: (1) the reflection UT mode in which the manipulator with specimen moves relative to the ultrasonic transducer; (2) the reflection UT mode in which the manipulator with ultrasonic transducer moves relative to the specimen; (3) the transmission UT mode in which the manipulator with specimen moves relative to the ultrasonic transducer; (4) the transmission UT mode in which the manipulator with ultrasonic transducer moves relative to the specimen.

The hardware compositions and testing processes of the above testing methods are similar, except for the design requirements of auxiliary tooling. Figures 1.3 and 1.4, respectively, show the reflection UT mode in which the manipulator with specimen moves relative to the ultrasonic transducer and the reflection UT mode in which the manipulator with ultrasonic transducer moves relative to the specimen. The robotic ultrasonic NDT system is mainly composed of the manipulator motion control device, ultrasonic transmitted-received signal acquisition and processing device, industrial personal computer (IPC) and testing software, water injector and other auxiliary equipment.

When the robotic ultrasonic NDT system is adopted, the manipulator will complete the measurement movement copying the specimen surface along the planned trajectory. In the testing process, the ultrasonic transducer is synchronously triggered to transmit ultrasonic waves and the echo signal reflected by the specimen is received. After the data processing, the UT image representing the defect information is obtained for the nondestructive evaluation of the specimen.

The technical solution of the robotic testing system is illustrated in Figure 1.5. The scanning trajectory is composed of pre-planned discrete point clouds covering the specimen surface. The excitation waveform of ultrasonic transducer is triggered synchronously based on the manipulator position information or with equal time intervals. The ultrasonic

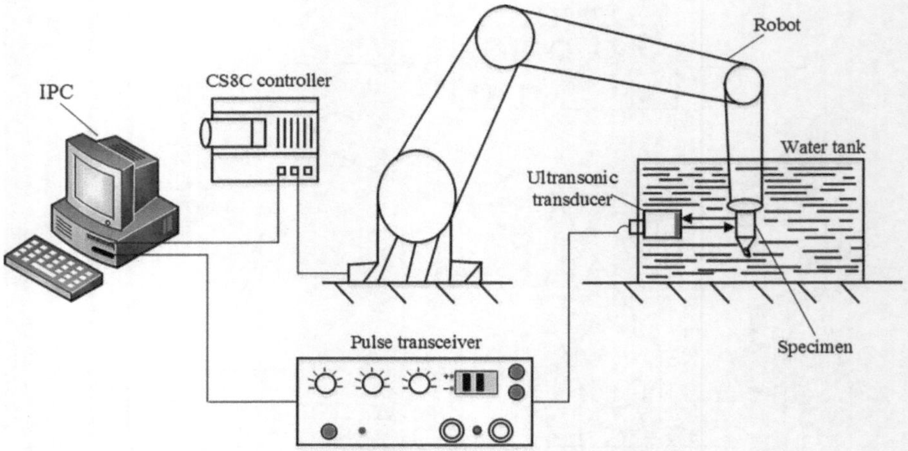

FIGURE 1.4 Structural diagram of a specimen-held ultrasonic reflection detection system.

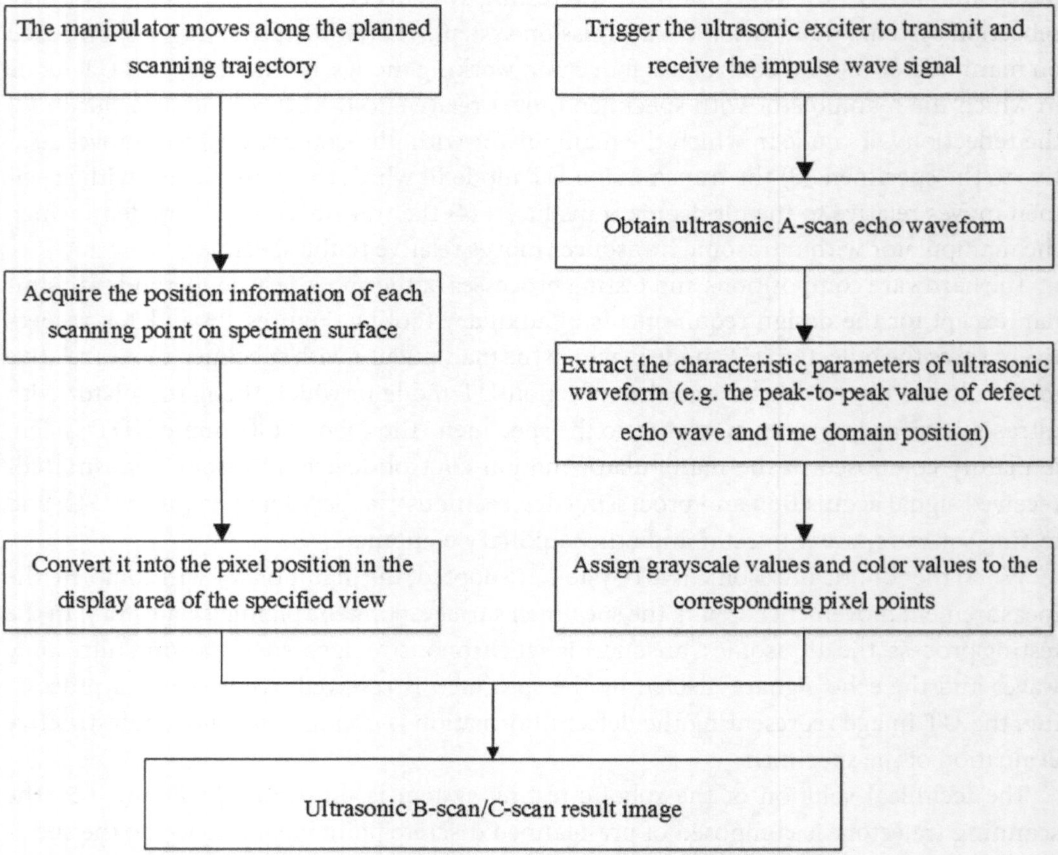

FIGURE 1.5 Workflow of a robotic testing system.

A-scan waveform corresponding to each scanning point stored in the ultrasonic acquisition card is displayed on the control interface of upper computer in real time. Meanwhile, the position information of each scanning point is converted into a pixel in the display area of the specified view. If each pixel is endowed with the corresponding ultrasonic characteristic information (e.g. the peak-to-peak value of defect echo wave, time domain position), the characteristic information of specimen surface/internal defect (e.g. location distribution and size parameter) can be continuously displayed in real time in the form of ultrasonic B-scan/C-scan image. The testing result of target specimen is composed of several sets of ultrasound B-scan/C-scan images. The scanning result view and test data of specimen section in the specified area can be provided to accurately evaluate the defects in the specimen.

In addition, low-frequency ultrasonic transducer is needed for the transmission detection of a fiber-reinforced composite component because of the large attenuation of ultrasonic sound in this medium. That is to say, two ultrasonic transducers are installed on both sides of the specimen, for transmitting ultrasonic signals on one side and for receiving the signals on the other side. Due to the use of two transducers, sometimes two manipulators, each holding an ultrasonic transducer, are needed to complete the scanning movement in the UT process. In this case, it is necessary to plan the synergic motion of the two manipulators and calibrate their reference/tool coordinate systems, so as to control their motion until they arrive at any trajectory point in the space.

1.3 MATHEMATICAL RELATIONSHIP BETWEEN THE COORDINATE SYSTEM AND EULER ANGLE

When the reference coordinate system of a manipulator is pre-defined, the spatial position and attitude of the manipulator effector can be expressed by its vector relative to the reference coordinate system. When the manipulator is holding the target object such as ultrasonic transducer or specimen, the motion of the effector satisfying the specified position and attitude requirement can be controlled according to the results of position and attitude conversion between coordinate systems. According to the requirements of manipulator kinematics, the final input to the controller should be the spatial position and attitude expression of the manipulator end-effector. According to its coordinate representation, the spatial position of the tool coordinate system on effector relative to the reference coordinate system and the Euler angle representing its attitude change can be given.

1.3.1 Definition of a Manipulator Coordinate System

In the description of a manipulator, several coordinate systems are important, including the joint coordinate system, world coordinate system, tool coordinate system and workpiece coordinate system. These coordinate systems constitute the basis of manipulator movement and control. The definition of each coordinate system is shown in Figure 1.6.

For a manipulator, the rotation angles of its six joints are under the most direct control and constitute its joint coordinate system. The joint coordinate system, as shown in Figure 1.6a, is to directly control the manipulator. However, it does not fit people's description habit. People are used to describe the position and attitude of an object in the Cartesian

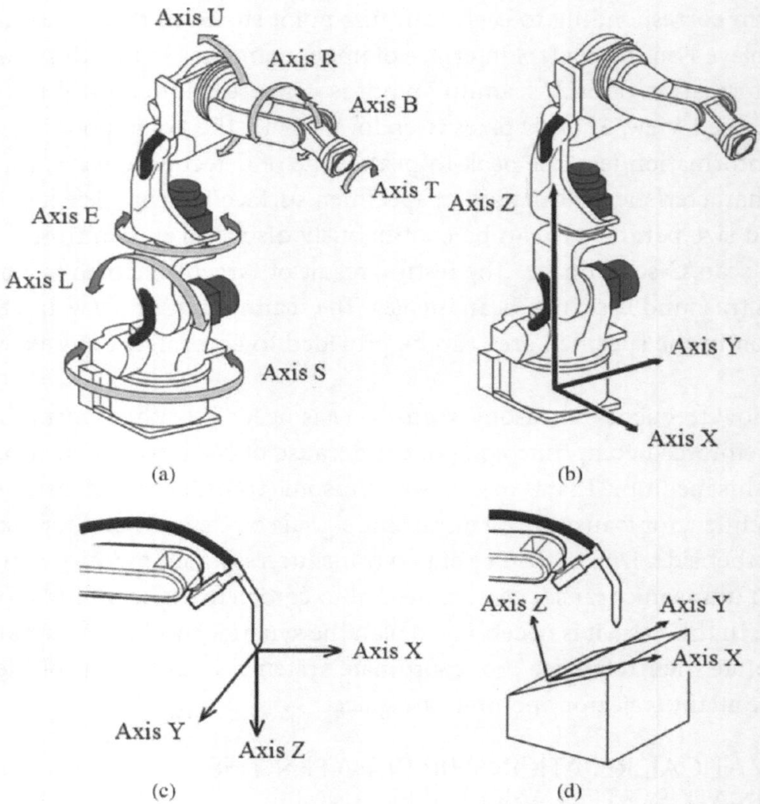

FIGURE 1.6 Distribution of the coordinate systems in a manipulator system. (a) Joint coordinate system, (b) Cartesian coordinate system, (c) tool coordinate system and (d) customized coordinate system.

coordinate system, so the world coordinate system has been introduced to the manipulator, as shown in Figure 1.6b. World coordinate system, also known as the base coordinate system, is generally fixed on the base of a manipulator. In the engineering application of the manipulator, the tool coordinate system is also defined, as shown in Figure 1.6c. The tool coordinate system is fixed on the end of manipulator tool and moves with the tool. In addition, the manipulator usually provides a customized coordinate system for the user's convenience, as shown in Figure 1.6d.

To ensure that the planned manipulator scanning trajectory meets the UT constraints, the end-effector should have correct position and attitude at each discrete point of the planned path on the surface of a complex curved component. For example, to keep the ultrasonic incidence angle, sound path length and other characteristic parameters constant in the testing process, the scanning trajectory of a manipulator should be reasonably planned. In this case, the distribution of each coordinate system in the manipulator system and the conversion of position and attitude should be relied on to obtain the kinematics control instructions. In the scanning mode in which the manipulator moves with the specimen held by it, the information on discrete path points based on the geometric model of specimen surface needs to be processed through coordinate transformation in order to

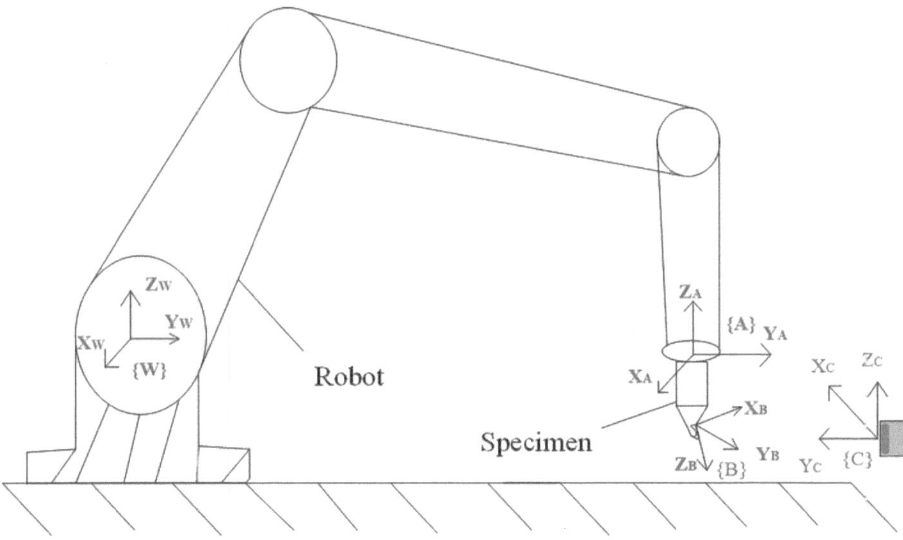

FIGURE 1.7 Distribution of the coordinate systems in a specimen-held testing system.

obtain the manipulator motion trajectory to be applied in UT. This transformation is realized through several coordinate systems defined in the system.

In the manipulator motion control, any coordinate system used for description will eventually be mapped to the world coordinate system, that is, it will be expressed in the base coordinate system. Then the position and attitude can be obtained by means of homogeneous coordinate transformation.

The distribution of each coordinate system in the testing system where the manipulator holds the specimen in scanning motion is shown in Figure 1.7. The clamping relation between the component under test and the manipulator end-effector is obtained by measurement. The tool coordinate system at the center of effector flange overlaps with the workpiece coordinate system through position and attitude compensation, as shown in the coordinate system $\{A\}$ below. The auxiliary coordinate system $\{B\}$ is constructed at each discrete point of the planned trajectory on the specimen, the user coordinate system $\{C\}$ is set at the position of ultrasonic transducer and the reference coordinate system $\{W\}$ is set for the manipulator. According to the robot kinematics theory, the position and attitude of the tool coordinate system $\{A\}$ relative to the reference coordinate system $\{W\}$ can be deduced from the D–H (Denavit–Hartenberg) parameter (if known) of the manipulator.

The distribution of the coordinate systems in the testing system where the manipulator holds a transducer is shown in Figure 1.8. The reference coordinate system of the manipulator is expressed by $\{W\}$. The tool coordinate system $\{M\}$ at the end of the manipulator is moved by translation according to the assembly position parameters (including clamping length and offset), until it coincides with the coordinate system of ultrasonic transducer (expressed by $\{C\}$). The workpiece coordinate system of the component under test is expressed as $\{A\}$. An auxiliary coordinate system $\{B\}$ is set up at the scanning trajectory point. The position and attitude of the auxiliary coordinate system $\{B\}$ in the coordinate

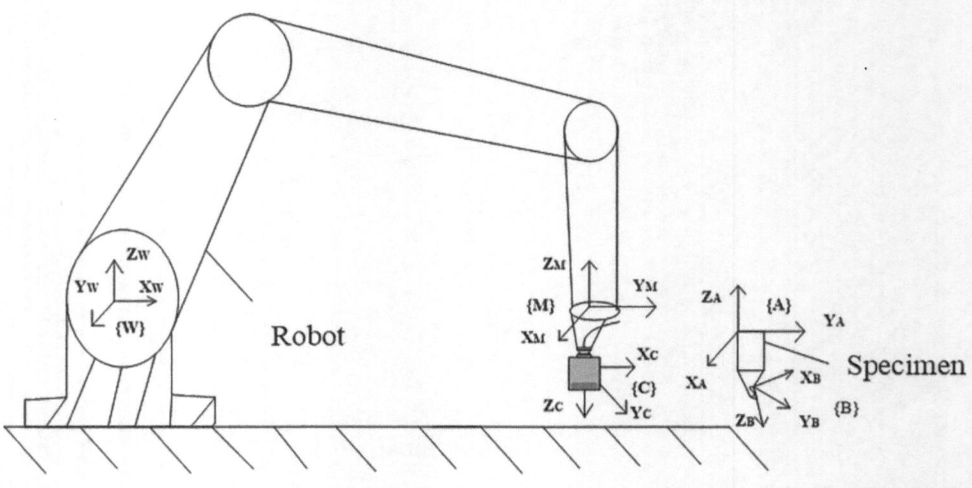

FIGURE 1.8 Distribution of the coordinate systems in a transducer-held testing system.

system $\{A\}$ are determined by specifying the origin of the coordinate system $\{B\}$ and the correspondence between the direction of each axis and the scanning trajectory point. Generally, the origin of the coordinate system $\{B\}$ is set at the scanning point $[x, y, z]$. Its Z-axis is along the normal direction of the scanning point. Its X-axis is along the tangent direction of the scanning trajectory (which can be approximated by the vector difference of two adjacent scanning points). Its Y-axis direction can be obtained by solving the vector product of the X-axis and the Z-axis.

After determining the coordinate system distribution in the robotic testing system, we can obtain the position and attitude transformation between different coordinate systems (including translation vector and rotation angle), the displacement vector relative to the previous position and attitude required by the manipulator to reach a certain position in space as well as the rotation angle around each axis.

1.3.2 Relationship between Position & Attitude and Coordinate System

1.3.2.1 Position Description

The position and attitude of a manipulator will be described through the world coordinate system. Once the world coordinate system is established, the position of any point can be described by a 3×1 position vector. In the coordinate system $\{A\}$, the position of any point P in three-dimensional space can be represented by the position vector ${}^{A}P$:

$$ {}^{A}P=\begin{bmatrix} p_x \\ p_y \\ p_z \end{bmatrix} \tag{1.1} $$

where p_x, p_y and p_z are the three coordinate components of the point P in the coordinate system $\{A\}$.

1.3.2.2 Attitude Description

When the manipulator is operated, the attitude of target object in the space, in addition to the position of a certain point in the space, needs to be described. To describe the orientation of the object *B*, the Cartesian coordinate system {*B*} fixed on the object *B* should be established. Denote the three unit vectors of the coordinate system {*B*} as *n*, *o* and *a*. Their expression in the reference coordinate system {*A*} will be:

$$ {}_B^A R = [\vec{n}\ \vec{o}\ \vec{a}] = \begin{bmatrix} n_x & o_x & a_x \\ n_y & o_y & a_y \\ n_z & o_z & a_z \end{bmatrix} \tag{1.2} $$

where $\|\vec{n}\| = \|\vec{o}\| = \|\vec{a}\| = 1, \vec{n} \times \vec{o} = \vec{a}$ and $\left({}_B^A R\right)^{-1} = \left({}_B^A R\right)^T$.

For the basic rotation around the coordinate axis, the resulting basic rotation matrix can be expressed as:

$$ R(x,\theta) = \begin{bmatrix} 1 & 0 & 0 \\ 0 & \cos\theta & -\sin\theta \\ 0 & \sin\theta & \cos\theta \end{bmatrix} \tag{1.3} $$

$$ R(y,\theta) = \begin{bmatrix} \cos\theta & 0 & \sin\theta \\ 0 & 1 & 0 \\ -\sin\theta & 0 & \cos\theta \end{bmatrix} \tag{1.4} $$

$$ R(z,\theta) = \begin{bmatrix} \cos\theta & -\sin\theta & 0 \\ \sin\theta & \cos\theta & 0 \\ 0 & 0 & 1 \end{bmatrix} \tag{1.5} $$

1.3.2.3 Spatial Homogeneous Coordinate Transformation

The spatial position and attitude of the object *B* can be fully described through the previous description of position and attitude. As shown in Figure 1.9, the position and attitude of the object *B* can be expressed as:

$$ \{B\} = \left[{}_B^A R \ \ {}_B^A P \right] \tag{1.6} $$

To facilitate the operation, the position and attitude of a manipulator are usually described by a homogeneous matrix in the manipulator kinematics. That is to say, one row is added to the original rotation transformation matrix and one column is added to the original translation transformation matrix to form a 4×4 matrix:

$$ {}_B^A T = \begin{bmatrix} {}_B^A R & {}_B^A P \\ 0 & 1 \end{bmatrix} \tag{1.7} $$

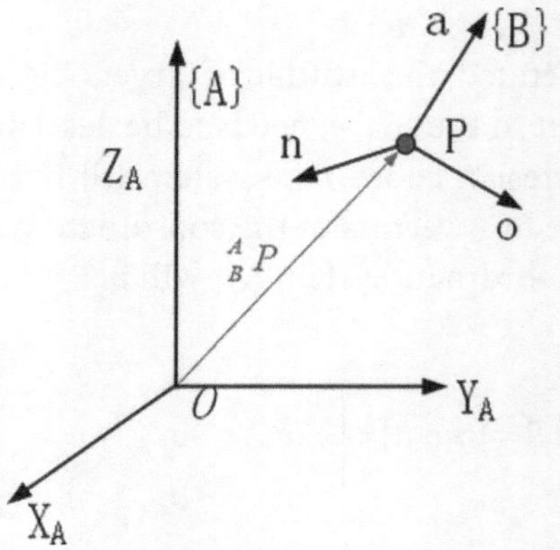

FIGURE 1.9 Description of the position and attitude of any object.

The transformation matrix represented by the matrix *T* can be decomposed into the product of several simple translation matrices and rotation matrices:

1. Translational homogeneous transformation:

$$\mathrm{Trans}(a,b,c)=\begin{bmatrix} 1 & 0 & 0 & a \\ 0 & 1 & 0 & b \\ 0 & 0 & 1 & c \\ 1 & 0 & 0 & 1 \end{bmatrix} \tag{1.8}$$

2. Rotational homogeneous transformation with the rotation angle θ around the *X*-axis:

$$\mathrm{Rot}(a,b,c)=\begin{bmatrix} 1 & 0 & 0 & 0 \\ 0 & \cos\theta & -\sin & 0 \\ 0 & \sin\theta & \cos\theta & 0 \\ 0 & 0 & 0 & 1 \end{bmatrix} \tag{1.9}$$

3. Rotational homogeneous transformation with the rotation angle θ around the *Y*-axis:

$$\mathrm{Rot}(y,\theta)=\begin{bmatrix} \cos\theta & 0 & \sin\theta & 0 \\ 0 & 1 & 0 & 0 \\ -\sin\theta & 0 & \cos\theta & 0 \\ 0 & 0 & 0 & 1 \end{bmatrix} \tag{1.10}$$

4. Rotational homogeneous transformation with the rotation angle θ around the Z-axis:

$$\mathrm{Rot}(z,\theta)=\begin{bmatrix} \cos\theta & -\sin\theta & 0 & 0 \\ \sin\theta & \cos\theta & 0 & 0 \\ 0 & 0 & 1 & 0 \\ 0 & 0 & 0 & 1 \end{bmatrix} \tag{1.11}$$

Through the combination of the above transformations, the position and attitude of a manipulator can be described. Usually, the position and attitude of the manipulator are described by the position and attitude matrix of the manipulator wrist end-effector coordinate system in the base coordinate system. For an industrial manipulator, its position is usually expressed by the position of the origin of the end-effector coordinate system in the base coordinate system, and its attitude is expressed by the attitude matrix (rotation matrix) of the end-effector coordinate system relative to the base coordinate system.

The coordinate transformation of a manipulator is usually the combination of the following four transformations:

1. Rotation angle θ around the Z-axis;
2. Transitional displacement d along the Z-axis;
3. Transitional displacement a along the X-axis;
4. Rotation angle a around the X-axis,

as expressed by:

$${}^{A}_{B}T=\mathrm{Rot}(z,\theta)\mathrm{Trans}(0,0,d)\mathrm{Trans}(a,0,0)\mathrm{Rot}(x,\theta) \tag{1.12}$$

1.3.3 Quaternion and Coordinate Transformation

The aforementioned position and attitude of discrete point of the manipulator scanning trajectory obtained by the Euler Angle coordinate transformation are often continuously first-order differentiable on the curved surface and will not cause the manipulator joint to appear in the singular position due to the existence of a first-order breakpoint. However, when the control vertices along a certain dimension direction are increased through discrete processing, a singular point or a motion space beyond the manipulator may appear at the boundary constraint. In this case, the position and attitude of any discrete point on this trajectory need to be recalculated with the dual quaternion method.

1. Definition of dual quaternion
   According to the concept of quaternion complex-number space proposed by William Hamiltonian, the rotation angle of a three-dimensional object can be calculated by using quaternions. Compared with a matrix, a quaternion can be regarded as the

combination of a scalar and a vector that is calculated more efficiently and is convenient for interpolation. Assuming that a quaternion is a hypercomplex number consisting of the real part *w* and the imaginary part (*i*, *j*, *k*), the normalized quaternion can be expressed as:

$$q = q_0 + q_1 i + q_2 j + q_3 k \tag{1.13}$$

Its norm is defined as:

$$|q|^2 = q_0^2 + q_1^2 + q_2^2 + q_3^2 \tag{1.14}$$

In a three-dimensional space, the position and attitude of a particular vector rotating around the specified axis are generally described by quaternions. For example, the rotation angle is expressed by α, and the direction cosine of the locating rotation axis is expressed by cos(βx), cos(βy) and cos(βz). Then, the elements in the above equation are:

$$q_0 = \cos(\alpha/2)$$

$$q_1 = \sin(\alpha/2)\cos(\beta_x)$$

$$q_2 = \sin(\alpha/2)\cos(\beta_y)$$

$$q_3 = \sin(\alpha/2)\cos(\beta_z)$$

Similarly, when the rotation is represented by unit quaternion q and the translation vector is labeled as d, the vector displacement in space can be expressed by the dual quaternion $\hat{q}$ as:

$$\begin{aligned} \hat{q} &= q + \varepsilon q^0 \\ q^0 &= (1/2)dq \end{aligned} \tag{1.15}$$

where ε is a dual unit, $\varepsilon^2 = 0$.

Usually, after the rotation of the quaternion $\boldsymbol{q}$ = [cos(α/2), sin(α/2) × $\vec{u}$] around the rotation axis $\vec{u}$, the 3D point p[0, (x, y, z)] can turn into the point P'[0, (x', y', z')] expressed by:

$$p' = qp^{-1}q \tag{1.16}$$

Similarly, when the vector $\boldsymbol{V}_1$ turns into $\boldsymbol{V}_2$ through rotation, it can be expressed by:

$$V_2 = V_1 \cos\alpha + (\vec{u} \times V_1)\sin\alpha + (\vec{u} \times V_1)(1 - \cos\alpha)\vec{u}$$

2. Manipulator position and attitude transformations based on dual quaternions

When the position and attitude transformations of the coordinate system are represented by dual quaternions, the Staubli TX90L manipulator follows the *X–Y–Z* Euler Angle rotation order. Suppose that the transformation composed of the rotation angle α around the *x*-axis, the transitional displacement *s* along the *x*-axis, the rotation angle β around the *y*-axis, the transitional displacement *c* along the *y*-axis, the rotation angle γ around the *z*-axis and the transitional displacement *h* along the *z*-axis can be defined as

$$q_{i-1}^{x}(\alpha_i, s_i)q_{i-1}^{y}(\beta_i, c_i)q_{i-1}^{z}(\gamma_i, h_i) \tag{1.17}$$

The transformation with the rotation angle α around the *x*-axis and the transitional displacement *s* along the *x*-axis can be defined as

$$\mathbf{q}_{i-1}^{x}(\alpha_{i-1}, s_{i-1}) = \mathbf{X}_{i-1} + \varepsilon \mathbf{X}_{i-1}^{0} = \mathbf{X}_{i-1} + \varepsilon \mathbf{S}_{i-1}\mathbf{X}_{i-1}/2$$

$$X_{i-1} = \cos(\alpha_i/2) + \sin(\alpha_i/2)(1\cdot i + 0\cdot j + 0\cdot k) = \cos(\alpha_i/2) + \sin(\alpha_i/2)\cdot i$$

$$\mathbf{S}_{i-1} = 0 + (s_i \cdot i + 0\cdot j + 0\cdot k) = s_i \cdot i$$

The transformation with the rotation angle β around the *y*-axis and the transitional displacement *c* along the *y*-axis can be defined as

$$\mathbf{q}_{i-1}^{y}(\beta_{i-1}, c_{i-1}) = \mathbf{Y}_{i-1} + \varepsilon \mathbf{Y}_{i-1}^{0} = \mathbf{Y}_{i-1} + \varepsilon \mathbf{C}_{\mathbf{i-1}}\mathbf{Y}_{i-1}/2$$

$$\mathbf{Y}_{i-1} = \cos(\beta_i/2) + \sin(\beta_i/2)(0\cdot i + 1\cdot j + 0\cdot k) = \cos(\beta_i/2) + \sin(\beta_i/2)\cdot j$$

$$\mathbf{C}_{i-1} = 0 + (0\cdot i + c_i \cdot j + 0\cdot k) = c_i \cdot j$$

The transformation with the rotation angle γ around the *z*-axis and the transitional displacement *h* along the *z*-axis can be defined as

$$\mathbf{q}_{i-1}^{z}(\gamma_{i-1}, h_{i-1}) = \mathbf{Z}_{i-1} + \varepsilon \mathbf{Z}_{i-1}^{0} = \mathbf{Z}_{i-1} + \varepsilon H_{i-1}\mathbf{Z}_{i-1}/2$$

$$\mathbf{Z}_{i-1} = \cos(\gamma_i/2) + \sin(\gamma_i/2)(0\cdot i + 0\cdot j + 1\cdot k) = \cos(\gamma_i/2) + \sin(\gamma_i/2)\cdot k$$

$$\mathbf{H}_{i-1} = 0 + (0\cdot i + 0\cdot j + h_i \cdot k) = h_i \cdot k$$

For the six-DOF joint manipulator, when the D–H parameters are known and the reference coordinate system at the base of the manipulator is converted into the tool coordinate system at the end-effector in the specified order, the transformation equation can be obtained with the dual quaternion method as

$$\mathbf{q}_{1}^{x}(\alpha_1, s_1)\mathbf{q}_{1}^{y}(\beta_1, c_1)\mathbf{q}_{1}^{z}(\gamma_1, h_1)\mathbf{q}_{2}^{x}(\alpha_2, s_2)\mathbf{q}_{2}^{y}(\beta_2, c_2)$$

$$\mathbf{q}_{2}^{z}(\gamma_2, h_2)\cdots\mathbf{q}_{6}^{x}(\alpha_6, s_6)\mathbf{q}_{6}^{y}(\beta_6, c_6)\mathbf{q}_{6}^{z}(\gamma_6, h_6)$$

where s_i, c_i and h_i are the structural parameters of the manipulator arm along the three-dimensional directions of the reference coordinate system; $\mathbf{q}_i^x(\alpha_i, s_i)$, $\mathbf{q}_i^y(\beta_i, c_i)$ and $\mathbf{q}_i^z(\gamma_i, h_i)$ are the dual quaternions indicating the rotation around and translation along the x-axis, y-axis and z-axis, respectively.

The D–H parameters of various joints of the Staubli manipulator are substituted into the above formula. Since only the joint angles in the initial attitude are unknown and other parameters are known, the dual-quaternion transformation matrix of the end-effector tool coordinate system relative to the reference coordinate system can be solved. After the elimination processing, the parameter values [s, c, h, α, β, γ] are obtained to characterize the position and attitude of the manipulator effector relative to the reference coordinate system.

To verify the singularity of discrete points of the planned manipulator trajectory, the Euler Angle rotation sequence under the same attitude transformation condition is taken as the reference and combined with the above equation to convert the position/attitude transformation equation of a six-joint manipulator into

$$\prod_{i=1}^{6}\left(\cos\frac{\alpha_i}{2}+\sin\frac{\alpha_i}{2}i\right)\left(\cos\frac{\beta_i}{2}+\sin\frac{\beta_i}{2}j\right)\left(\cos\frac{\gamma_i}{2}+\sin\frac{\gamma_i}{2}k\right)$$
$$\left(1+\frac{\varepsilon}{2}.s_i.i\right)\left(1+\frac{\varepsilon}{2}.c_i.j\right)\left(1+\frac{\varepsilon}{2}.h_i.k\right) \tag{1.18}$$

When the quaternion $q=q_0+q_1i+q_2j+q_3k$ is represented by a rotation matrix, its elements are as shown in the following equation. The modulus of this quaternion is $|q|=\sqrt{q_0^2+q_1^2+q_2^2+q_3^2}$.

$$\begin{bmatrix} 1-2\left(q_2^3+q_3^3\right) & 2\left(q_1q_2-q_0q_3\right) & 2\left(q_0q_2+q_1q_3\right) \\ 2\left(q_1q_2+q_0q_3\right) & 1-2\left(q_1^3+q_3^3\right) & 2\left(q_2q_3-q_0q_1\right) \\ 2\left(q_1q_3-q_0q_2\right) & 2\left(q_0q_1+q_2q_3\right) & 1-2\left(q_1^3+q_2^3\right) \end{bmatrix} \tag{1.19}$$

After substituting this equation into the above equation, the quaternion parameters [q, q_0, q_1, q_2] can be calculated through elimination because the initial position and attitude of the manipulator end-effector and their relationship with the base coordinate system can be solved by using the D–H parameters.

Then the parameters [q_1i, q_2j, q_3k] are assigned to [x, y, z]. Because the equation corresponds to the rotation matrix, the Euler Angle parameters required by the manipulator can be calculated:

$$\begin{bmatrix} \alpha \\ \beta \\ \gamma \end{bmatrix} = \begin{bmatrix} a\tan 2\left(2(wx+yz),1-2\left(x^2+y^2\right)\right) \\ \arcsin\left(2(wy-zx)\right) \\ a\tan 2\left(2(wz+xy),1-2\left(y^2+z^2\right)\right) \end{bmatrix}$$

By comparing the above equation with the Euler Angle calculated through coordinate transformation, we can analyze whether the manipulator's joints are in the accessible space, how stable the operation process is and what the occurrence probability of step point is. If necessary, the result obtained from the above equation can be used as the calibration coefficient to replace the Euler Angle data calculated by the coordinate transformation algorithm.

REFERENCES

1. Gharaibeh Y., Lee N., Ultrasonic guided waves propagation in complex structures[J], *Emerging Technologies in Non-destructive Testing*, 2012, 5:237–243.
2. Windisch T., Schubert F., Ultrasonic wave field determination of laser-acoustic sources with arbitrary shapes[J]. 2012 IEEE International Ultrasonics Symposium, 2012:2184–2186.
3. Leleux A., Micheau P., Castaings M., NDT process using LAMB waves generated/detected by ultrasonic phased array probes for the defect detection in metallic and composite plates[J], *Review of Progress in Quantitative Nondestructive Evaluation*, 2013, 1511(32):817–824.
4. Vijaya Lakshmi, M.R., Mondal, A.K., Jadhav, C.K., Ravi Dutta, B.V., Sreedhar, S., Overview of NDT methods applied on an aero engine turbine rotor blade[J], *Insight*, 2013, 55(9):482–486.
5. Frankenstein, B., Schulze, E., Weihnacht, B., Meyendorf, N., Boller, C., Structural health monitoring of major wind turbine components[J], *Structural Health Monitoring*, 2013, 1(2):2456–2462.
6. Tseng, P-Y., Chang, Y-f., Lin, C-M., Nien, W-J., Chang, C-H., Huang, C-C., A study of total focusing method for ultrasonic nondestructive testing[J], *Journal of Testing and Evaluation*, 2013, 41(4):557–563.
7. Yeun-Ho Y., Jin-Ho C., Jin-Hwe K., Dong-Hyun K., A study on the failure detection of composite materials using an acoustic emission[J], *Composite Structures*, 2006, 75:163–169.
8. Mahaut S., Leymarie N., Poidevin C., Fouquet T., Dupond O., Study of complex ultrasonic NDT cases using hybrid simulation method and experimental validations[J], *Insight*, 2011, 53(12):664–667.
9. Dworakowski Z., Ambrozinski L., Packo P., Dragan K., Stepinski T., Uhl, T., Application of artificial neural networks for damage indices classification with the use of lamb waves for the aerospace structures[J], *Smart Diagnostics*, 2014, 588:12–21.
10. Sun G., Zhou Z., Application of laser ultrasonic technique for non-contact detection of drilling- induced delamination in aeronautical composite components[J], *OPTIK*, 2014, 125(14):3608–3611.
11. Yang C. *Research on Automatic Ultrasonic Nondestructive Testing of Large Complex-shaped Aviation Composite Components [D]*. Hangzhou: Zhejiang University, 2005.
12. Khaled W., Reichling S., Ultrasonic strain imaging and reconstructive elastography for biological tissue [J], *Ultrasonics*, 2006, 44(1):199–202.
13. Wu R. *Research on Digital Ultrasonic Testing System and Key Technology [D]*. Hangzhou: Zhejiang University, 2004.
14. Jeong H., Time reversal-based beam focusing of an ultrasonic phased array transducer on a target in anisotropic and inhomogeneous welds[J], *Materials Evaluation*, 2014, 72(5):589–596.
15. Jiao J., Zhong X., He C., Wu B., Experiments on non-destructive testing of grounding grids using SH0 guided wave[J], *Insight*, 2012, 54(7):375–379.
16. Aljets D., Chong A., Wilcox S., Holford K., Acoustic emission source location on large plate-like structures using a local triangular sensor array[J], *Mechanical Systems and Signal Processing*, 2012, 30:91–102.

CHAPTER 2

Method of Acoustic Waveguide UT

ACOUSTIC WAVEGUIDE IS A method of acoustic guide-detection coupling that changes the propagation direction of acoustic wave through the tube bending by using the sound-conducting characteristic of the fluid or solid in the acoustic tube. This method is mainly used for ultrasonic import and export in the ultrasonic testing (UT) of narrow-space cavity-like components. The complex-shaped or special-shaped acoustic tube and the fluid or solid inside it is the core components of the acoustic waveguide detection system. The acoustic propagation path of ultrasound in an irregular acoustic tube is complex, where the ultrasound will encounter the acoustic reflection and transmission from the curved interface. When ultrasound is incident to the reflecting surface, several reflections, refractions, transmissions and wave mode conversions will occur and complex superimposed waves will be formed.

Based on the wave equation, the propagation characteristics of ultrasound in an irregular acoustic tube filled with fluid (usually water) are studied in this chapter. First, with regard to the reflection of ultrasound from liquid-solid interface during the propagation in an irregular acoustic tube, a half-space model is adopted to analyze the characteristics of ultrasonic reflection and transmission at the interface. Second, based on the COMSOL finite element software, the ultrasonic propagation in the tube is simulated, thus revealing the main law of ultrasonic wave propagation in the water waveguide tube. Finally, based on the sound-field measurement system of multi-degree of freedom (DOF) hydrophone, the acoustic pressure at the outlet of the waveguide tube is actually measured to obtain the sound field distribution of the acoustic waveguide.

2.1 WAVE EQUATION AND PLANE WAVE SOLUTION

2.1.1 Acoustic Wave Equation for an Ideal Fluid Medium

Acoustic wave equation is the mathematical representation of the relations between sound pressure and spatial position and between sound pressure and time, which is established based on the physical properties of sound propagation process. In the process of acoustic

DOI: 10.1201/9781003212232-2

disturbance, the changes of sound pressure ρ, particle displacement μ, particle velocity ν and density incremental ρ' are interrelated [1]. As a macroscopic physical phenomenon, sound vibration must satisfy three basic physical laws: Newton's second law, the law of conservation of mass and the equation of state that describes the relationship between state parameters such as pressure intensity, temperature and volume.

To simplify the problem, the following assumptions are made. However, these assumptions can be satisfied very well as a general rule:

1. The medium is an ideal fluid, that is, it has no viscosity. When traveling in this ideal medium, the sound wave has no energy loss.
2. In the absence of acoustic disturbance, the medium is static macroscopically, that is, its initial velocity is zero. In addition, the medium is homogeneous, so the static pressure intensity P_0 and static density ρ_0 in the medium are constants.
3. During the sound propagation, the dense and sparse processes in the medium are adiabatic, that is, the temperature difference due to sound propagation will not cause the heat exchange between the medium and the adjacent area.
4. What propagate in the medium are sound waves with small amplitude, whose acoustic variables are first-order micro-quantities. Their pressure ρ is much less than the static pressure intensity P_0 in the medium, that is, $\rho \ll P_0$. Their particle velocity ν is well below the sound speed c, that is, $\nu \ll c$. Their particle displacement ε is much shorter than the acoustic wave length λ, that is, $\varepsilon \ll \lambda$. Their media density increment ρ' is much less than the static density ρ_0, that is, $\rho' \ll \rho_0$ or their relative density increment $\Delta_\rho = \rho'/\rho_0$ is much smaller than 1, that is, $\Delta_\rho \ll 1$.

Let's start with the one-dimensional case, where the sound field is uniform in two directions in space so that only the motion in one direction (for example, x direction) needs to be considered.

Motion Equation

Take a volume element small enough from the sound field, as shown in Figure 2.1.

Its volume is Sdx (S is the section area of this volume element perpendicular to the x-axis). As the sound pressure varies with position, the force acting on the left side of the volume element is not equal to the force acting on the right side. The resultant force causes the particles in this volume element to move in the x direction. When a sound wave passes, the intensity of pressure on the left side is $P_0+\rho$. Therefore, the force acting on the left side of this volume element is $F=(P_0+\rho)S$. Because no tangential force exists in an ideal fluid medium and the internal pressure is always perpendicular to the surface taken, the direction of F_1 should be $+x$ direction. The intensity of pressure on the right side of this volume element is $P_0+\rho+d\rho$, where $dp=\dfrac{\partial p}{\partial x}dx$ is the change in sound pressure when the position is changed from x to $x+dx$. So, the force acting on the right side of the volume element is $F_2=(P_0+\rho+d\rho)S$ in the $-x$ direction. Considering that the static pressure intensity P_0

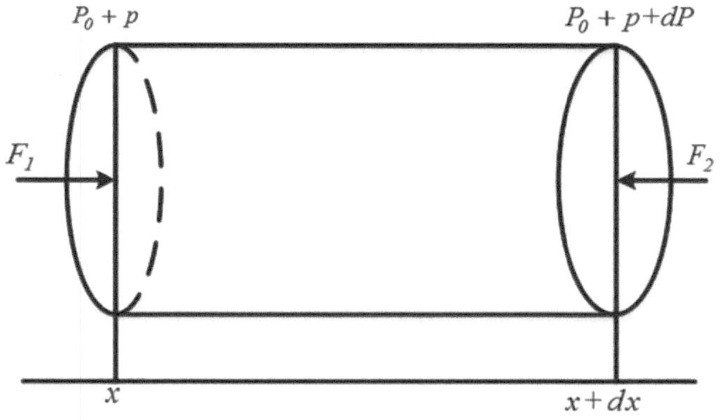

FIGURE 2.1 Force balance state of small volume element in sound field.

of the medium does not vary with x, the resultant force acting on the volume element in the x direction is $F = F_1 - F_2 = -S\frac{\partial p}{\partial x}dx$. The mass of the medium in this volume element is ρSdx, and its acceleration in the x direction under the force F is $\frac{dv}{dt}\left(\frac{d^2u}{dt^2}\right)$. So, according to Newton's second law, the following equation is obtained

$$pSdx\frac{dv}{dt} = -S\frac{\partial p}{\partial x}dx \tag{2.1}$$

$$pSdx\frac{d^2u}{dt^2} = -S\frac{\partial p}{\partial x}dx \tag{2.2}$$

After reorganization, it becomes

$$p\frac{dv}{dt} = -\frac{\partial p}{\partial x} \tag{2.3}$$

$$p\frac{d^2u}{dt^2} = -\frac{\partial p}{\partial x} \tag{2.4}$$

Here, $\rho = \rho_0 + \rho'$. ρ_0 is the static density of the medium without acoustic disturbance and is a constant, but the medium density increment ρ' is a variable, so ρ is still a variable. The acceleration $\frac{dv}{dt}\left(\frac{d^2u}{dt^2}\right)$ of a medium particle actually consists of two components: the acceleration obtained at a specified point in space due to the change of velocity at that position over time, i.e. the local acceleration $\frac{dv}{dt}\left(\frac{d^2u}{dt^2}\right)$; and the acceleration resulting from the velocity increment due to the change of velocity with position after the particle has migrated a certain distance in space, i.e. the migration acceleration $\frac{dv}{dx}\frac{dx}{dt} = v\frac{dv}{dx}, \frac{\frac{\partial u}{\partial t}}{dx}\frac{dx}{dt} = v\frac{\partial^2 u}{\partial t \partial x}$. So, Eq. (2.3) becomes

$$(\rho_0+\rho')\left(\frac{\partial v}{\partial t}+v\frac{\partial v}{\partial x}\right)=-\frac{\partial p}{\partial x} \tag{2.5}$$

After eliminating the micro-quantities of second or higher order, the equation can be simplified as

$$\rho_0\frac{\partial v}{\partial t}=\frac{\partial p}{\partial x} \tag{2.6}$$

Similarly, Eq. (2.4) becomes

$$(\rho_0+\rho')\left(\frac{\partial^2 u}{\partial t^2}+v\frac{\partial^2 u}{\partial t\partial x}\right)=-\frac{\partial p}{\partial x} \tag{2.7}$$

After eliminating the micro-quantities of third or higher order, the equation can be simplified as

$$\rho_0\frac{\partial^2 u}{\partial t^2}=-\frac{\partial p}{\partial x} \tag{2.8}$$

Eqs. (2.6) and (2.8) are the motion equations of the medium with acoustic disturbance. They describe the relationships between sound pressure ρ and particle velocity ν/particle displacement μ in the sound field.

Continuity Equation

The continuity equation describes the law of mass conservation of the medium in a small volume element during the propagation of a sound wave, which means the difference between the medium masses flowing into and out of the volume element in unit time shall be equal to the increase or decrease in the mass within the volume element. Take a volume element small enough from the sound field, as shown in Figure 2.2, whose volume is Sdx.

Suppose that, at the left side x of a volume element, the velocity of medium particle is $(\nu)_x$ and the density is $(\rho)_x$. Then the mass flowing through the left side into the volume element in unit time should be equal to the medium mass contained in a cylinder volume with the

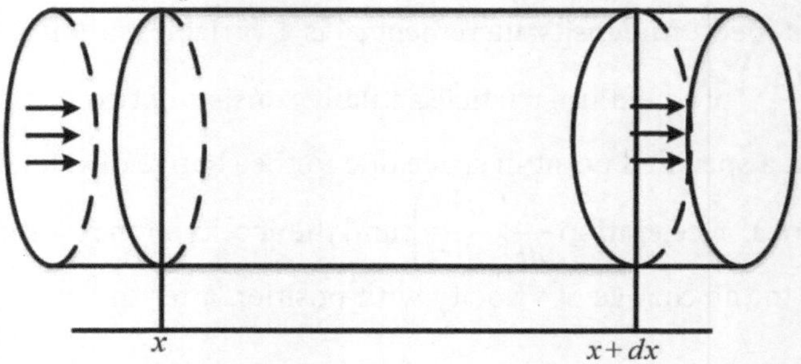

FIGURE 2.2 Conservation of mass of small volume element in sound field.

sectional area of S and the speed of $(\nu)_x$, namely $(\rho\nu)_x S$. The mass flowing out of the right side of the volume element in the same unit time is $-(\rho\nu)_{x+dx}S$, in which the "–" sign means flowing out. The first-order approximation of its Taylor expansion is $-\left[(\rho v)_x + \frac{\partial(\rho v)_x}{\partial x}dx\right]s$.

Therefore, the net mass flowing into the volume element in unit time is $-\frac{\partial(\rho v)}{\partial x}Sdx$ (where both ρ and ν are the functions of x and the subscript x is not added any more). On the other hand, the mass within the volume element increases, indicating an increase in its density. If its increment per unit time is $\frac{\partial \rho}{\partial t}$, then the increment of volume element mass per unit time will be $\frac{\partial \rho}{\partial t}Sdx$. Therefore, the mass increment of the volume element per unit time must be equal to the net mass flowing into the volume element, i.e.

$$-\frac{\partial(\rho v)}{\partial x}Sdx = \frac{\partial \rho}{\partial t}Sdx \tag{2.9}$$

After reorganization, it becomes

$$-\frac{\partial(\rho v)}{\partial x} = \frac{\partial \rho}{\partial t} \tag{2.10}$$

By substituting $\rho=\rho_0+\rho'$ into the above equation and eliminating the micro-quantities of second or higher order, the following equation can be obtained:

$$-\rho_0 \frac{\partial v}{\partial x} = \frac{\partial \rho'}{\partial t} \tag{2.11}$$

This is just the continuity equation of the medium in the sound field, which describes the relationship between particle velocity and medium density.

Equation of State

We still consider a volume element with a certain mass in the medium. In the absence of acoustic disturbance, its pressure intensity is P_0, its density is ρ_0, and its temperature is T_0. When the sound wave passes through this volume element, the pressure, density and temperature within the volume element will change. However, the changes of these three quantities are not independent but interrelated to each other. The change law of this medium state is described by the thermodynamic equation of state. The sound wave, even at lower frequency, still travels quite fast, and the period of volume compression and expansion is much shorter than that of heat conduction, so the specified medium has no time to exchange heat with the adjacent medium in the process of sound propagation. Therefore, the process of sound propagation can be considered as an adiabatic process. In this way, P is just a function of ρ, i.e.

$$P = P(\rho) \tag{2.12}$$

Therefore, the small increments of pressure and density due to acoustic disturbance satisfy

$$dP = \left(\frac{dP}{d\rho}\right)_s d\rho \tag{2.13}$$

where the subscript "s" indicates an adiabatic process. The pressure and density change in the same direction, that is, when the medium is compressed, both the pressure and density will increase ($dP>0$, $d\rho>0$); when the medium expands, both the pressure and density will decrease ($dP<0$, $d\rho<0$). So, the coefficient $\left(\frac{dP}{d\rho}\right)_s$ is always greater than 0. It is now represented by c^2, namely

$$dP = c^2 d\rho \tag{2.14}$$

This is just the equation of state of an ideal fluid medium with acoustic disturbance, which describes the relationship between the small change of pressure intensity (P) in the sound field and the small change of density (ρ).

Considering c^2 and the constant medium mass, so $\rho dV + V d\rho = 0$, i.e.

$$\left(\frac{d\rho}{d\rho}\right)_s = -\left(\frac{dV}{V}\right)_s \tag{2.15}$$

and

$$c^2 = \left(\frac{d\rho}{d\rho}\right)_s = \frac{dP}{\rho\left(\frac{d\rho}{d\rho}\right)_s} \tag{2.16}$$

After substituting Eq. (2.15) into Eq. (2.16), we obtain

$$c^2 = \frac{dP}{\rho\left(\frac{d\rho}{d\rho}\right)_s} = \frac{dP}{-\rho\left(\frac{dV}{V}\right)_s} = \frac{1}{\rho\beta_s} \tag{2.17}$$

where $\frac{dV}{V}$ is the relative increment of volume; $\frac{dV}{V}\beta_s = -\frac{\frac{dV}{V}}{dp}$ is adiabatic volume compression coefficient, representing the relative volume change caused by unit pressure-intensity change under adiabatic condition. The "–" sign indicates that the direction of pressure change is opposite to the direction of volume change.

For a sound wave with small amplitude, ρ' is small. Through the Taylor series expansion of $\rho\left(\frac{dP}{d\rho}\right)_s$ at the equilibrium state (P_0, ρ_0), we obtain:

$$\left(\frac{dP}{d\rho}\right)_s = \left(\frac{dP}{d\rho}\right)_{s,o} + \frac{1}{2}\left(\frac{d^2P}{d^2\rho}\right)_{s,o}(\rho-\rho_0)+\cdots \tag{2.18}$$

where the subscript "0" is the value at the equilibrium state. Since $\rho-\rho_0$ is very small, all the terms behind the second term in the above equation can be ignored to obtain the following simplified equation

$$\left(\frac{dP}{d\rho}\right)_s \approx \left(\frac{dP}{d\rho}\right)_{s,o} \tag{2.19}$$

After being combined with Eq. (2.16) and c^2 is represented by c_0^2, Eq. (2.19) will become

$$C^2_0 = \left(\frac{dP}{d\rho}\right)_{s,o} = \frac{1}{\rho_0\beta_s} \tag{2.20}$$

In addition, considering that, for a small-amplitude sound wave, the differential of pressure intensity in Eq. (2.14) is ρ and the differential of density is ρ', the equation of state of an ideal fluid medium with acoustic disturbance can be reduced to a linear equation, as shown below:

$$P = c_0^2\rho' \tag{2.21}$$

Wave Equation

For the derivation of wave equation, the partial derivative of Eq. (2.21) with respect to t is substituted into Eq. (2.11) to obtain:

$$\rho_0 c_0^2 \frac{\partial v}{\partial x} = \frac{\partial p}{\partial t} \tag{2.22}$$

The partial derivative of Eq. (2.22) with respect to t is

$$\rho_0 c_0^2 \frac{\partial^2 v}{\partial x \partial t} = \frac{\partial^2 p}{\partial t^2} \tag{2.23}$$

The partial derivative of Eq. (2.6) with respect to x is substituted into Eq. (2.23) to obtain:

$$\frac{\partial^2 p}{\partial x^2} = \frac{1}{c_0^2}\frac{\partial^2 p}{\partial t^2} \tag{2.24}$$

c_0 represents the longitudinal sound velocity in a fluid medium. This is the wave equation of small-amplitude sound wave in a uniform ideal fluid medium. It is also called linear acoustic equation because it is obtained after ignoring the micro-quantities of second or higher order.

Similarly, Eq. (2.22) is substituted into Eq. (2.11) to obtain

$$\rho_0 \frac{\partial v}{\partial x} = \frac{1}{c_0^2} \frac{\partial p}{\partial t} \tag{2.25}$$

The partial derivative of Eq. (2.25) with respect to x and the partial derivative of Eq. (2.6) with respect to t are, respectively,

$$\rho_0 \frac{\partial^2 v}{\partial x^2} = -\frac{1}{c_0^2} \frac{\partial^2 p}{\partial t \partial x} \tag{2.26}$$

$$\rho_0 \frac{\partial^2 v}{\partial x^2} = -\frac{\partial^2 p}{\partial t \partial x} \tag{2.27}$$

Eqs. (2.26) and (2.27) are turned into the following simultaneous equation:

$$\frac{\partial^2 v}{\partial x^2} = \frac{1}{c_0^2} \frac{\partial^2 v}{\partial t^2} \tag{2.28}$$

Then the partial derivative of Eq. (2.21) with respect to x is substituted into Eq. (2.6) to obtain:

$$\rho_0 \frac{\partial v}{\partial t} = -c_0^2 \frac{\partial \rho'}{\partial x} \tag{2.29}$$

The partial derivative of Eq. (2.29) with respect to x and the partial derivative of Eq. (2.11) with respect to t are, respectively,

$$\rho_0 \frac{\partial^2 v}{\partial t \partial x} = -c_0^2 \frac{\partial^2 \rho'}{\partial x^2} \tag{2.30}$$

$$\rho_0 \frac{\partial^2 v}{\partial t \partial x} = \frac{\partial^2 \rho'}{\partial t^2} \tag{2.31}$$

Eqs. (2.30) and (2.31) are turned into the following simultaneous equation:

$$\frac{\partial^2 \rho'}{\partial x^2} = \frac{1}{c_0^2} \frac{\partial^2 \rho'}{\partial t} \tag{2.32}$$

To obtain the relation between pressure and particle displacement, we define the coefficient of expansion as

$$\alpha_s \equiv \frac{\Delta V}{V} \tag{2.33}$$

Suppose the pressure on the left side of the volume element $dV = Sdx$ is ρ. When the pressure on the right side is $\rho + d\rho$, the volume element becomes $dV = s\left(1 + \partial u / \partial x\right)dx$. So

$$\alpha_s \equiv \frac{\Delta V}{V} = \frac{\partial u}{\partial x} \tag{2.34}$$

According to the definition of adiabatic volume compression coefficient $\beta_s = -\frac{dV}{V}\Big/dP$, this equation is combined with Eq. (2.20) into

$$p = -\frac{\alpha_s}{\beta_s} = -\frac{1}{\beta_s}\frac{\partial u}{\partial x} = -\rho_0 c_0^2 \frac{\partial u}{\partial x} \tag{2.35}$$

The partial derivative of Eq. (2.39) with respect to x is substituted into Eq. (2.8) to obtain:

$$\frac{\partial^2 u}{\partial x^2} = \frac{1}{c_0^2}\frac{\partial^2 u}{\partial t^2} \tag{2.36}$$

To sum up, Eqs. (2.24), (2.28), (2.32) and (2.36) are all the wave equations of small-amplitude sound waves in a uniform ideal fluid medium and are equivalent in linear approximation.

Therefore, according to the gradient theorem and divergence theorem in mathematics, the one-dimensional case can be generalized to the three-dimensional case very easily. For an ideal fluid medium with small amplitude, the three-dimensional wave equation represented by $\rho(x, y, z, t)$ can be written as [1,2]:

$$\nabla^2 p = \frac{1}{c_0^2}\frac{\partial^2 p}{\partial t^2} \tag{2.37}$$

where 2 represents the Laplace operator. In the rectangular coordinate system, it is expressed as:

$$\nabla^2 = \frac{\partial^2}{\partial x^2} + \frac{\partial^2}{\partial y^2} + \frac{\partial^2}{\partial z^2} \tag{2.38}$$

2.1.2 Plane Wave and Solutions of Wave Equations

It can be seen from the previous section that the acoustic wave equation is derived only by applying the basic physical properties of the medium, without involving the vibration condition of the specific sound source and the boundary condition. It reflects the general law of sound wave in an ideal medium. The specific acoustic propagation characteristics should be solved by considering the specific sound source and boundary conditions.

The ultrasonic transducers used in actual UT are usually plane-wave ultrasonic transducers. The ultrasonic wave excited by this ultrasonic transducer only propagates along the normal direction of the transducer chip, while all particles on the plane parallel to the transducer have the same amplitude and phase (that is, the wave fronts are the planes parallel to each other). This wave is just plane wave. The solution to the wave equation of a plane wave is also the sound field of the plane wave. It can be boiled down to solving the one-dimensional linear wave Eqs. (2.24), (2.28), (2.32) or (2.36).

Take the sound pressure solving as an example. According to the principle of the variables separation method for solving the second-order partial differential equations, the solution of Eq. (2.24) is assumed to be:

$$P = P(x)e^{jwt} \tag{2.39}$$

where w is the circular frequency of simple harmonic vibration of sound source, $w = 2\pi f = \frac{2\pi}{T}$.

By substituting Eq. (2.39) into Eq. (2.24), we obtain the ordinary differential equation with respect to $\rho(x)$

$$\frac{d^2 p(x)}{d^2 x} + k^2 p(x) = 0 \tag{2.40}$$

where k is called wave number, $k = \frac{w}{c_0}$.

Suppose the solution of Eq. (2.40) is

$$p(x) = Ae^{jkx} + Be^{-jkx} \tag{2.41}$$

where A and B are two arbitrary constants that are determined by boundary conditions.

Eq. (2.41) is substituted into Eq. (2.39) to obtain

$$p(t,x) = Ae^{jk(kx-wt)} + Be^{-j(kx+wt)} \tag{2.42}$$

In Eq. (2.42), the first term represents the sound pressure amplitude of the sound wave propagating along the $+x$ direction, while the second term represents the sound pressure amplitude of the sound wave propagating along the $-x$ direction. Since only the propagation of plane waves in the medium is studied here, the value of B will be $B=0$ if there are no reflectors on the propagation path (or no reflected waves appear). Therefore, Eq. (2.42) can be simplified as

$$p(t,x) = Ae^{jk(kx-wt)} \tag{2.43}$$

Suppose that the sound pressure $P_a e^{jwt}$ is generated in the adjacent medium when the sound source at $x=0$ vibrates, namely $A = \rho_a$. The sound pressure of any particle in the sound field is thus obtained:

$$p(t,x) = P_a e^{j(kx-wt)} \tag{2.44}$$

The velocity of any particle in the sound field can be derived from Eq. (2.6):

$$v_x = -\frac{1}{\rho_0}\int \frac{\partial p}{\partial x} dt \tag{2.45}$$

Eq. (2.44) is substituted into Eq. (2.45), whose integral is then calculated:

$$v(t,x) = -\frac{Pa}{\rho_0 c_0} e^{j(kx-wt)} \tag{2.46}$$

The solutions of Eqs. (2.44) and (2.46) are just the sound pressure and velocity of any particle in the plane-wave sound field of a uniform ideal medium.

From Eq. (2.46), the displacement of any particle in the sound field can be easily derived:

$$u(t,x) = \int \frac{P_a}{\rho_0 c_0} e^{j(kx-wt)} dt - \frac{Pa}{jw\rho_0 c_0} e^{j(kx-wt)} \tag{2.47}$$

It can be seen from Eq. (2.47) that, in the sound field, any particle can only oscillate back and forth at the equilibrium position.

2.2 ULTRASONIC REFLECTION AND TRANSMISSION AT THE INTERFACE

As known from the previous section, when the propagation direction of a plane wave is the normal line of its wave front (the plane formed by all the connected particles in the medium that have the same vibration phase at that time) along the axis, the plane wave can be expressed by Eq. (2.47). After careful consideration, we can find that the x in Eq. (2.47) is actually the projection of position vector **r** on the normal direction of wave front, which happens to be the x-axis, as shown in Figure 2.3a.

As shown in Figure 2.3b, the plane waves propagating in any directions may also have the same sound pressure or particle displacement when the position vectors at different positions of the wave front have the same projection on the normal direction of the wave front. Therefore, the x in Eq. (2.47) can be generally understood as the projection of the position vector **r** of a particle in the sound field onto the normal line of the wave front. It is equal to the scalar product of the unit vector $n = \cos\alpha i + \cos\beta j + \cos\gamma k$ and position vector $r = xi + yj + zk$ of wave front normal line, namely

$$x = n \cdot r \tag{2.48}$$

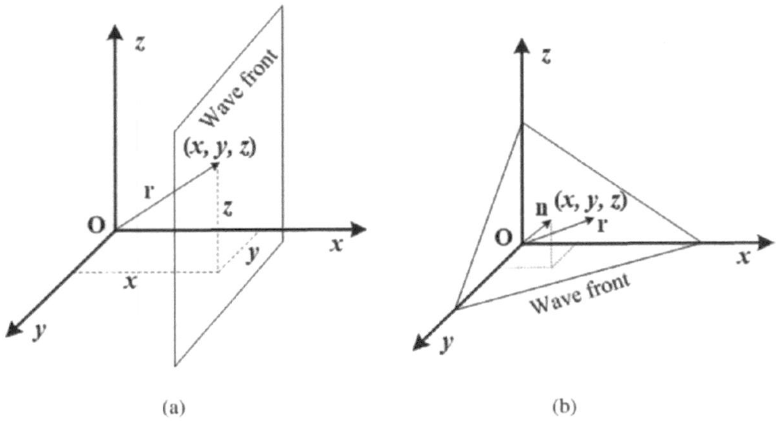

FIGURE 2.3 Schematic diagram of a wave front. (a) The wavefront normal coincides with the X-axis and (b) plane waves propagating in any direction.

Here, α, β and γ are the angles between wave front normal and x-axis/y-axis/z-axis, and cosα, cosβ and cos$r$ are the direction cosines of wave front normal. Considering the propagation modes of longitudinal and transverse waves in solid media, the displacement of a simple harmonic plane wave propagating along the direction n can be written as

$$u = Ade^{j(knr-wt)} = Ade^{jk(nr-ct)} \tag{2.49}$$

where A is the amplitude of sound wave, c is the propagation velocity of plane wave, d is the unit vector of particle motion direction, n is the unit vector of acoustic wave propagation and r is the position vector. If $d \cdot n = \pm 1$, that is, the vibration direction of a particle is parallel to the propagation direction of sound wave, Eq. (2.49) will represent the planar longitudinal wave. If $d \cdot n = 0$, that is, the vibration direction of a particle is perpendicular to the propagation direction of sound wave, Eq. (2.49) will represent the planar transverse wave.

For an ideal elastic solid with uniform isotropy, the relationships between strain ε and displacement μ/stress σ/strain ε in the rectangular coordinate system can be, respectively, expressed as:

$$\varepsilon_{ij} = \frac{1}{2}\left(u_{i,j} + u_{j,i}\right) \tag{2.50}$$

$$\sigma_{ij} = \lambda \varepsilon_{kk} \delta_{ij} + 2u\varepsilon_{ij} \tag{2.51}$$

where both i and j can be x, y, z; ε_{kk} follows the Einstein summation convention, namely $\varepsilon_{kk} = \varepsilon_{xx} + \varepsilon_{yy} + \varepsilon_{zz}$; λ and μ are the Lame constants of solid medium; δ_{ij} is Kronecker delta function,

$$\delta_{ij} = \begin{cases} 1, i = j; \\ 0, i \neq j; \end{cases} ; u_{i,j} = \frac{\partial^2 u_i}{\partial j}$$

According to Eqs. (2.50) and (2.51), the stress component corresponding to Eq. (2.49) can be expressed as the following displacement:

$$\sigma_{lm} = \left[\lambda \delta_{lm}\left(d_j n_j\right) + 2\mu\left(d_l n_m + d_m n_l\right)\right] kAe^{jk(nr-ct)} \tag{2.52}$$

where j, l and m can be x, y, z; $d_j n_j$ follows the Einstein summation convention.

As shown in Figure 2.4, when the plane wave is incident at a certain angle to the liquid-solid interface, the incident wave P_I and reflected wave P_R in the liquid medium 1 are both longitudinal waves because only longitudinal waves can propagate in the liquid. The particle vibration directions V_I and V_R are the same as the direction of acoustic wave propagation, and the sound velocity is C_{L1}. However, in the solid medium 2, both longitudinal wave and transverse wave can propagate, so the refracted longitudinal wave P_{TL} and the refracted transverse wave P_{TT} will exist simultaneously. When the refracted longitudinal wave propagates, the particle vibration direction V_{TL} is the same as the propagation direction of sound

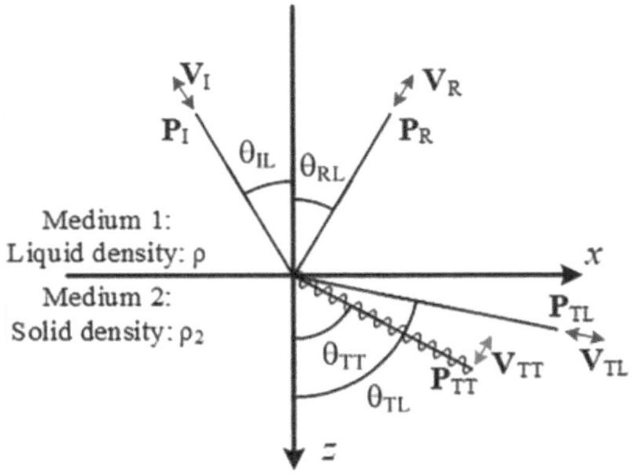

FIGURE 2.4 Reflection and transmission of an ultrasonic longitudinal wave at the liquid-solid interface.

wave. When the transverse wave propagates, the particle vibration direction V_{TT} is perpendicular to the propagation direction. In the two cases, the sound velocities are C_{L2} and C_{T2}, respectively. $\theta_{IL,}$ θ_{RL}, θ_{TL} and θ_{TT} represent the incident angle, the reflection angle, the refraction angle of longitudinal wave and the refraction angle of transverse wave, respectively. The densities of liquid medium 1 and solid medium 2 are ρ_1 and ρ_2, respectively.

As shown in Figure 2.4, in the rectangular coordinate system, the particle displacement vectors of incident wave P_I, reflected wave P_R, refracted longitudinal wave P_{TL} and refracted transverse wave P_{TT} can be expressed as:

$$u^I = A_I e^{ik_j(x\sin\theta_{IL} + z\cos\theta_{IL} - cL_{1t})} \begin{bmatrix} \sin\theta_{IL} \\ \cos\theta_{IL} \end{bmatrix}$$

$$u^R = A_R e^{ik_R(x\sin\theta_{RL} + z\cos\theta_{RL} - cR_2 t)} \begin{bmatrix} \sin\theta_{RL} \\ \cos\theta_{RL} \end{bmatrix}$$

$$u^{TL} = A_{TL} e^{ik_{TL}(x\sin\theta_{TL} + z\cos\theta_{TL} - cL_2 t)} \begin{bmatrix} \sin\theta_{TL} \\ \cos\theta_{TL} \end{bmatrix} \tag{2.53}$$

$$u^{TT} = A_{TL} e^{ik_{TT}(x\sin\theta_{TT} + z\cos\theta_{TT} - cT_2 t)} \begin{bmatrix} \sin\theta_{TT} \\ \cos\theta_{TT} \end{bmatrix}$$

where A_I, A_R, A_{TL}, A_{TT} and k_I, k_R, k_{TL}, k_{TT} represent the amplitudes and wave numbers of incident longitudinal wave, reflected longitudinal wave, refracted longitudinal wave and refracted transverse wave, respectively.

For particle displacement μ and stress σ, the continuity conditions of particle displacement and stress are satisfied at the interface $z=0$. According to the continuity conditions, we obtain

$$u_z : \left(u_z^l + u_z^R\right)\big|_{z=0} = \left(u_z^{TL} + u_z^{TT}\right)\big|_{z=0} \tag{2.54}$$

$$\sigma_{zz} : \left[-(P_1)\right]\big|_{z=0} = \left[\left(\rho_1 c_{L1}^2 \nabla \cdot \left(u^l + u^R\right)\right)\right]\big|_{z=0} = \left(\sigma_{zz}^{TL} + \sigma_{zz}^{TT}\right)\big|_{z=0} \tag{2.55}$$

$$\sigma_{zx} : \left(\sigma_{zx}^{TL} + \sigma_{zx}^{TT}\right)\big|_{z=0} = 0 \tag{2.56}$$

The left side $(P)_1$ of Eq. (2.55) is the sound pressure in the liquid medium 1. According to the constitutive equation of fluid medium, we can obtain $-(P)_1 = \rho_1 c_{L1}^2 \nabla \cdot \left(u^l + u^R\right)$. $\nabla \cdot u = \frac{\partial^2 u_x}{\partial x} + \frac{\partial^2 y}{\partial x} + \frac{\partial^2 u_z}{\partial z}$ is the divergence of the displacement vector $\boldsymbol{u}$. $\sigma_{zz}^{TL} + \sigma_{zz}^{TT}$ on the right side is the normal stress in the solid medium 2. Since the inward pressure is positive and the outward stress is positive, the left side of Eq. (2.55) is negative.

Eqs. (2.53)–(2.56) are true for all x and t values, which means that at $z=0$, the exponential terms of the four equations in Eq. (2.53) are equal, that is,

$$k_1 c_{L1} = k_R c_{R1} = k_{TL} c_{L2} = k_{TT} c_{T2} \tag{2.57}$$

$$k_1 \sin\theta_{IL} = k_R \sin\theta_{RL} = k_{TL} \sin\theta_{TL} = k_{TT} \sin\theta_{TT} \tag{2.58}$$

Because $k=\omega/c$, the result $\omega_{L1}=\omega_{R1}=\omega_{L2}=\omega_{T2}$ can be derived from Eq. (2.57). This indicates that when a simple harmonic plane wave is incident to the liquid-solid interface, the angular frequencies of reflected wave, refracted longitudinal wave and refracted transverse wave are the same as that of incident wave.

By substituting $k=\omega/c$ into Eq. (2.58), the relationships among sound velocity, incident angle, reflection angle and refraction angle can be obtained, namely

$$\frac{\sin\theta_{IL}}{c_{L1}} = \frac{\sin\theta_{RL}}{c_{R1}} = \frac{\sin\theta_{TL}}{c_{L2}} = \frac{\sin\theta_{TT}}{c_{T2}} \tag{2.59}$$

This equation is just the famous Snell theorem. Given the incident angle and the sound velocities in the materials on both sides of the interface, the reflected angle of incident side and the refraction angle of transmission side can be determined from this equation.

For reflection coefficient and transmission coefficient, their equation, as shown below, can be obtained by substituting the stress equation (2.52) and displacement equation (2.53) into continuity condition equations (2.54)–(2.56):

$$\begin{bmatrix} \cos\theta_{RL} & \cos\theta_{TL} & -\sin\theta_{TT} \\ -\rho_1 c_{L1}^2 k_R & k_{TL}(\lambda + 2u)\cos 2\theta_{TT} & k_{TT} u \sin 2\theta_{TT} \\ 0 & k_{TL}\sin 2\theta_{TL} & k_{TT}\cos 2\theta_{TT} \end{bmatrix} \begin{bmatrix} A_R \\ A_{TL} \\ A_{TT} \end{bmatrix} = A_l \begin{bmatrix} \cos\theta_{TL} \\ \rho_1 c_{L1}^2 k_l \\ 0 \end{bmatrix} \tag{2.60}$$

When the ultrasonic wave is incident to the liquid-solid interface, the reflection coefficient and transmission coefficient can be expressed by particle vibration displacement as: longitudinal-wave reflection coefficient $R_L(\theta)=A_R(\theta)/A_I(\theta)$, longitudinal-wave transmission coefficient $T_{TL}(\theta)=A_{TL}(\theta)/A_I(\theta)$ and transverse-wave transmission coefficient $T_{TT}(\theta)=A_{TT}(\theta)/A_I(\theta)$. Given the incident angle and the relevant parameters of the medium, the reflection coefficient and transmission coefficient can be calculated with Eqs. (2.59) and (2.60).

When using a special-shaped acoustic tube to test special-shaped components, the commonly used coupling liquid is water and the acoustic tube is stainless steel bend. The bend is usually 90°, so the ultrasonic wave will encounter the water-stainless steel interface at the acoustic tube bend, that is, the incident liquid-solid interface is just the water-stainless steel interface. The Table 2.1 lists the relevant characteristic parameters of water and stainless steel [3,4].

When the ultrasonic wave is incident from water to stainless steel surface, the curves of displacement-dependent reflection and transmission coefficients corresponding to different incident angles are as shown in Figure 2.5. As can be seen from the figure, when the ultrasonic wave is vertically incident, the reflection coefficient is 0.937, the transmission coefficient of longitudinal wave is 0.063 and the transmission coefficient of transverse wave is 0. This means that the vertical incidence of a longitudinal wave can only excite longitudinal wave in the stainless steel solid. When the incident angle gradually increases, the reflection coefficient and the longitudinal-wave transmission coefficient will basically remain unchanged, but the transverse-wave transmission coefficient will gradually increase. Meanwhile, a longitudinal wave and a transverse wave will be simultaneously excited in the stainless steel medium. When the incident angle increases to about 14.82°, the reflection coefficient will increase to 1. In this case, all the incident energy will be reflected. The longitudinal-wave transmission coefficient will be maximal, while the transverse-wave transmission coefficient will drop to 0. The refracted longitudinal wave will travel along the water-stainless steel interface (namely axis) at a certain angle. This angle is usually called the first critical angle, and the refracted longitudinal wave at this point is called critical refracted longitudinal wave. If the incident angle continues to increase, there will be $\sin\theta_{TL}>1$ according to Snell theorem, that is, the real angle θ_{TL} will not exist. This means that when the incident angle is greater than the first critical angle, the refracted longitudinal wave will no longer exist in the stainless steel medium, but the refracted transverse wave will be still present. When the incident angle increases to about 28.52°, the reflection coefficient will increase to 1 again, the transverse-wave refraction coefficient will increase to 0.2 and the energy will exist mainly in the form of critical refracted transverse wave. If the incident angle continues to increase, the reflection coefficient will be still 1. At this time, all the incident energy will be reflected into the water

TABLE 2.1 Characteristics of Water and Steel

Material	P (kg/m³)	C_L (m/s)	C_T (m/s)	λ (MPa)	μ (MPa)
Water	1.0×10^3	1.48×10^3	-	-	-
Stainless steel	7.91×10^3	5.79×10^3	3.1×10^3	1.13×10^5	7.57×10^4

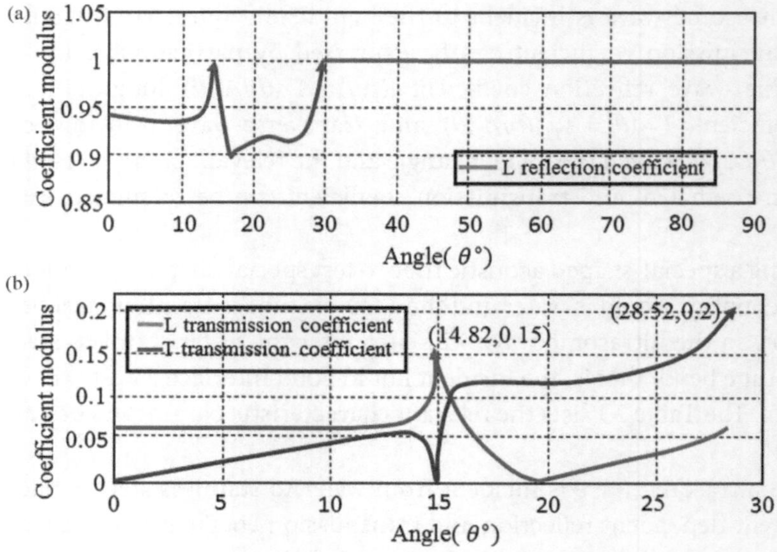

FIGURE 2.5 (a) Reflection and (b) transmission coefficients of the water-stainless steel interface.

and the refracted longitudinal/transverse waves will no longer be generated in the stainless steel medium. In addition, Figure 2.5 will no longer show the curves of the refracted longitudinal-wave and transverse-wave coefficient moduli calculated according to Eq. (2.60). It can also be seen from the longitudinal-wave reflection coefficient curve that, the reflection coefficient is always large, ranging from 0.9 to 1, due to the great difference in acoustic impedance between water and stainless steel.

According to the geometric-model simplification method of ultrasound propagation [5], a simplified ultrasonic propagation path has been drawn, as shown in Figure 2.6. It is easy to obtain the reflection angle at the bend of the acoustic waveguide tube. The reflection angles of all ultrasonic propagation paths have been drawn with MATLAB®, as shown in Figure 2.7. It can be seen that the minimum incident angle of ultrasonic wave at the bend of the acoustic waveguide tube is 39.52°, larger than the total reflection angle at the water-steel interface (28.52°). In other words, total reflection will occur in all positions of reflection surface inside the waveguide tube. Neither ultrasonic refracted longitudinal wave nor refracted transverse wave will be generated in the steel tube material. However, surface wave or Lamb wave will be generated in the tube wall at an appropriate angle. The latter is not the focus of this paper and thus will not be specially analyzed.

2.3 ANALYSIS OF SOUND FIELD IN AN ACOUSTIC WAVEGUIDE TUBE

Because of the complexity of ultrasonic propagation in tube, it is very difficult to establish a strict in-tube ultrasonic propagation model. Therefore, the numerical simulation of ultrasonic propagation has become an effective analysis method. The numerical simulation of basic acoustic wave equation can deepen the understanding of ultrasonic propagation

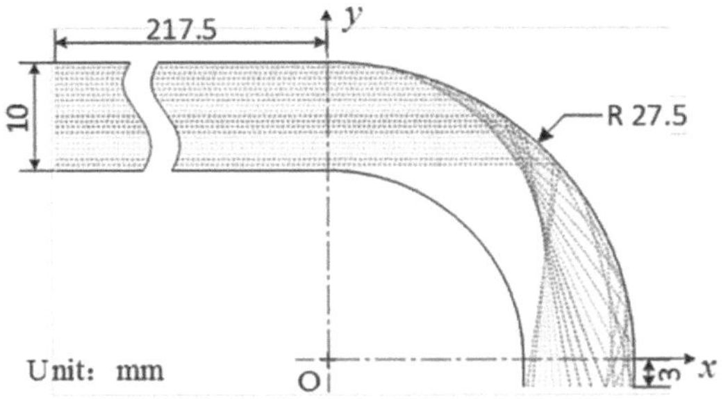

FIGURE 2.6 Simplified ultrasonic propagation path in an acoustic waveguide tube.

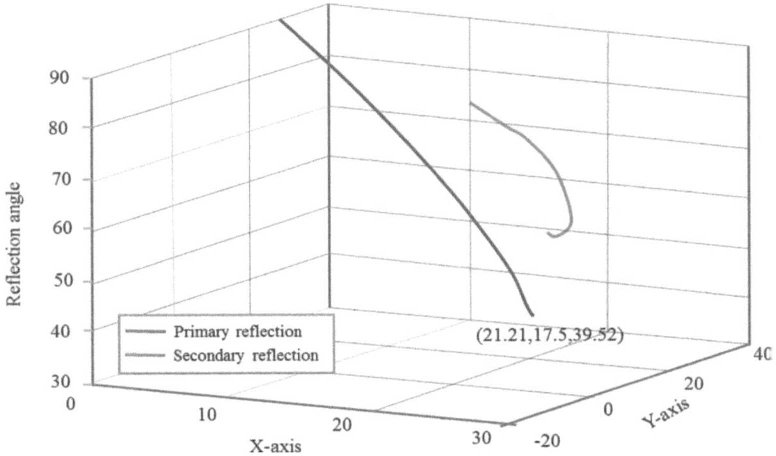

FIGURE 2.7 Changing curve of the reflection angle inside an acoustic waveguide tube.

mechanism theoretically and intuitively, simplify and visualize the propagation process of sound wave under complex background and thus reveal the main law of actual ultrasonic propagation in the tube. The numerical simulation of ultrasonic wave is based on acoustic wave equation. The numerical simulation results provide theoretical guidance for actual UT and reduce the blindness of testing [6,7].

As can be seen from Section 2.1, the sound pressure p, particle displacement μ, particle velocity ν and local density increment ρ' within an elastic medium are changing in the process of acoustic disturbance. The COMSOL Multiphysics software can use its Acoustics Module for convenient acoustical numerical simulation in nondestructive testing (NDT), ultrasonic transducer, ultrasonic flow meter and other aspects. In addition, it has a unique advantage in the multi-physics coupling simulation involving structural mechanics, fluid flow, heat conduction, electromagnetism, chemical engineering and other fields. Therefore, in order to visually characterize the ultrasonic propagation and sound pressure distribution inside and at the outlet of an acoustic waveguide tube, the propagation process of ultrasonic wave in the tube will be simulated by COMSOL Multiphysics finite element software and thus analyzed in this section. Before the numerical simulation, we assume that the coupling medium (water) in the acoustic waveguide tube is an ideal flowing medium. Although the water in the tube is flowing during testing, it is basically in the laminar flow state at a speed much smaller than the ultrasonic propagation speed in water, that is, the water speed is negligible. Therefore, such an assumption is reasonable.

The "pressure-acoustics, transient" interface in the COMSOL Multiphysics finite element software, which is suitable for the transient simulation of any transient field or source, can be used to simulate the fluid pressure change in a static background during the acoustic wave propagation. The standard control equation for "pressure-acoustics, transient" interface is an equation used to describe the propagation of sound in a fluid and derived from the control equations of fluid flow, namely, the equation of mass conservation represented by continuity equation; the equation of momentum conservation which is often referred to as Navier-Stokes equation; the energy conservation equation; the constitutive equation of model and the equation of state that describes the relationship between thermodynamic variables [8]. In addition, it tallies with the assumption for ideal flowing medium in Section 2.1.1. To derive the control equation for pressure-acoustics module, the complete Navier-Stokes equation is subject to linearization assumption (i.e. hypothesis of small acoustic disturbance) to obtain the linear Navier-Stokes equation at first, which, in turn, is subject to the assumption of no viscosity and heat conduction to obtain the linear Euler equation. Finally, the linear Euler equation is subject to the assumption of no flow (i.e. the fluid is assumed to be without viscosity, heat conduction and flow) to obtain the control equation for pressure-acoustics module. Of course, for the complex numerical simulation of sound field, we can also set up the fluid models for viscosity, heat conduction and other physical properties in COMSOL.

As the control equation of pressure-acoustics module is derived from the control equation of fluid flow and is not fully applicable to the simulation of ultrasonic propagation process inside the acoustic waveguide tube in this chapter, we have set up a liquid-solid coupled acoustic model based on the pressure acoustics module and solid mechanics module in COMSOL to solve the equation of sound wave in the coupling medium inside an acoustic waveguide tube in order to realize the numerical simulation and analysis of ultrasonic wave propagation in the tube.

The COMSOL software provides a set of consistent workflows that is simple and easy to use. Whether using the basic COMSOL Multiphysics modules (such as "acoustics module") alone or combining them with other products in the product portfolio to establish a multi-physics coupling model, you can follow a set of simple modeling and simulation workflows, which mainly includes the definition of global parameters, the establishment of geometric model, the setting of material parameters, the selection of an appropriate physical field interface, the definition of boundary and initial conditions, the grid generation, the setting and solution of solver parameters and the visualization of post-processing and solution results.

1. Definition of global parameters

   The setting of global parameters can define the variables in the whole model, make clear the parameters in the model and facilitate the later modification of model parameters. As the geometric simulation model of the acoustic waveguide tube is fixed, the section of global parameter definition only defines the following parameters of analogous ultrasonic transducer: excitation-source signal frequency $f_0=1\,\text{MHz}$, excitation-source signal cycle $t_0=5/f_0$ (that is, the cycles of excitation signal are five cycles).

2. Geometric modeling, multi-physics field and boundary conditions setting
According to the actual problems, the geometric modeling tool in simulation software or the interface function in COMSOL software is connected to the external software (such as CAD) to realize the synchronization of simulation model parameters. The reasonableness of simulation model establishment determines the success of simulation results. The establishment of geometric model should not only consider the actual situation but also analyze the actual problems. While meeting the requirements, the model is subject to some reasonable simplifications or other constraints, so as to obtain the simulation results closest to the actual situation with the minimum calculation burden.

In view of the computing capacity of the existing servers and by referring to the previous two-dimensional model established with appropriate boundary conditions to simulate the actual 3D model [9], the two-dimensional geometric model of axial section of a water-filled acoustic waveguide tube has been established in this chapter for the subsequent numerical simulation, as shown in Figure 2.8.

What is shown in Figure 2.8 is the geometric model of the established water-filled acoustic waveguide tube, which is composed of water area (pink area) and waveguide tube (blue area). The tube is made of steel, with an inner diameter of 10 mm, a thickness of 1.5 mm and an inner R angle of 17 mm. The material properties of water and acoustic waveguide tube are shown in the Table 2.1, Section 2.2. Its physical field is set as a multi-physics field with acoustic and structural interaction, whose water area, structural domain, excitation source, absorbing boundary and free boundary are set

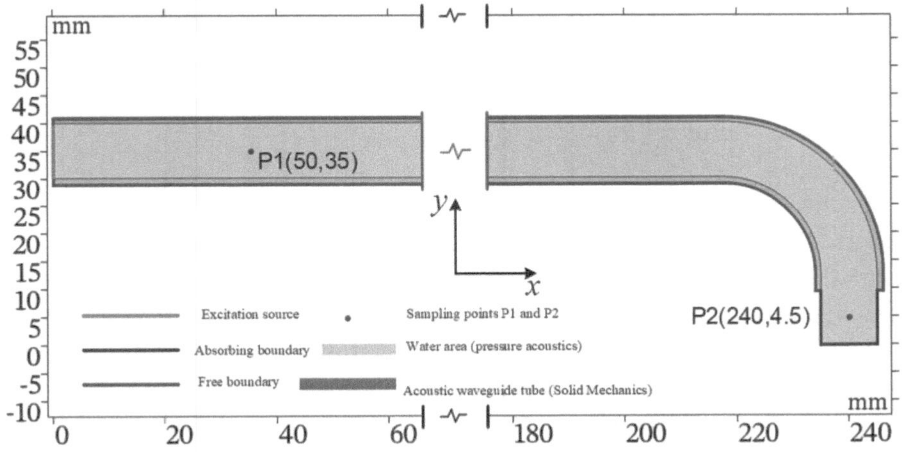

FIGURE 2.8 Geometric model of an acoustic waveguide tube.

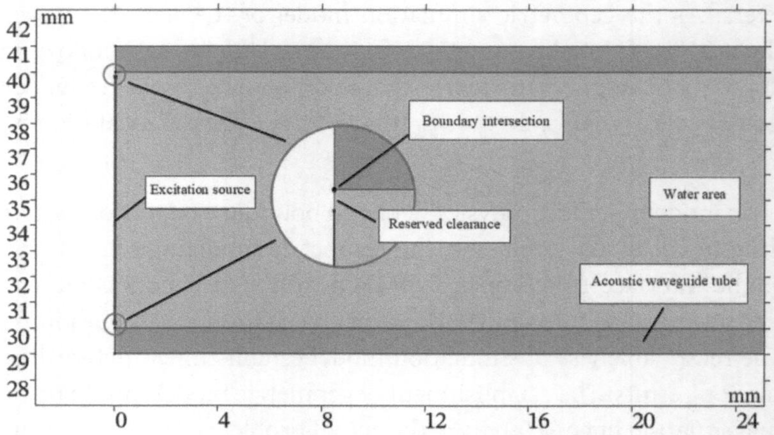

FIGURE 2.9 Clearance reserved between the excitation source and acoustic waveguide tube.

as shown in Figure 2.8. Besides, the sampling observation points *P*1 and *P*2 of A-scan signal are set at the point (50, 35) and the tube outlet point (240, 4.5), respectively, more than three times the near field distance.

Considering the small inner diameter of the acoustic waveguide tube, the above model is established in consideration of the calculation burden and the model accuracy and the water boundary at the outlet of the tube is set as absorbing boundary in order to prevent the reflected signals at the water boundary from affecting the signal analysis. In practice, the outer boundary of the acoustic waveguide tube is placed in the air. Due to the great difference in acoustic impedance between steel material and air, the outer boundary of the tube can be set as free boundary, where the steel tube is equivalent to a tube in the air [10,15]. The boundary where the water area and acoustic waveguide tube are connected is set as acoustic-structural boundary. In addition, it should be specially noted that, in order to prevent the excitation source from directly driving the boundary intersection between the left side of the acoustic waveguide tube and the water area to vibrate together and directly generating the sound wave in the tube, a 0.2 mm clearance has been reserved between the excitation source and the tube in the established geometric model and its boundary has been set as absorbing boundary, as shown in Figure 2.9.

3. Grid generation

The gridding plays a key role in finite element simulation. The accuracy of numerical simulation results is directly related to the grid resolution and quality of the finite element model. Free gridding is one of the most automated gridding techniques, with the advantages such as strong structural adaptability and easy node-density control. It can automatically generate triangular or quadrilateral grids on a plane or surface and tetrahedral grids on a body. Therefore, the technique of free triangular gridding is chosen for the acoustic waveguide tube and water area in this model. Through this technique, the number of grids can be controlled by the "size" and "distribution" in the settings of geometry and gridding. Specifically, the grid parameters such as maximum and minimum cell sizes, maximum cell growth rate, curvature factor and narrow area resolution can be set at the "size" node. The number of grids in the geometric model can be controlled by setting the grid distribution along the edge through the "distribution" function.

According to the COMSOL help file and the definition of unstructured grid, the grids generated by this technique are unstructured. Unstructured grid is a grid without regular topological relationship, which means that the points in the grid areas do not have the same adjacent cells. In other words, the number of grids connected with different points in the grid section is not exactly the same. Based on the Delaunay method or the method of advancing wave front, the triangle (tetrahedron) is used to subdivide the two-dimensional (three-dimensional) geometric model so as to eliminate the structural limitation of the nodes in the structural grids. In the unstructured grids, the nodes and cells can be well controlled, so the boundary can be dealt with very well and the model of a complex structure can be simulated. The generation of unstructured grid follows the relevant optimization judgment criteria. In this process, high-quality grids can be generated, and the grid size and node density can be simply controlled. In addition, a random data structure is adopted to the benefit of grid self-adaptation. As long as the distribution of grids on the boundary is specified during the grid generation, the grids can be generated automatically between the boundaries without the need for partitioning or user intervention and for the information transfer between subdomains. However, the unstructured grid also has disadvantages. The first disadvantage is that the viscosity problem cannot be handled well, because the boundary layer contains only triangular or tetrahedral grids, the number of which will be extremely huge. The second disadvantage is that the filling efficiency of grids is not high. So far, some methods of unstructured gridding also start to take advantage of the merits of structured gridding and add the processing function similar to structured gridding (such as the wall encryption function of ICEM (The Integrated Computer Engineering and Manufacturing) to the wall surface. Therefore, with the increasingly strengthening of computing power, the method of unstructured gridding has been highly valued and greatly developed in recent years.

To balance the requirements for simulation result correctness, calculation time, data amount and acoustic simulation calculation, the grid size needs to be controlled. The size of subdivided grid cell should not be too small, otherwise the requirements for computing power, computer memory and storage space will be stricter. The size of subdivided grid cell also should not be too large, otherwise the accuracy of simulation results will be seriously affected. Therefore, considering the computing capacity and reliable calculation accuracy of the existing servers, the size of subdivided grid units should ensure that the length of a wavelength in the propagation domain of sound wave (including the water area inside acoustic waveguide tube and the tube itself) contains 5–12 grids [9,11–15], namely,

$$L_e \leq \min\left(\frac{C_{water}}{nf}, \frac{C_{steel}}{nf}\right)$$

where L_S is the maximum size of the subdivided grid; C_{water} and C_{steel} are the longitudinal wave velocity of ultrasonic wave in water and steel medium; f is the frequency of ultrasonic excitation and n is a positive integer between 5 and 12, n=6 in this numerical simulation (i.e. L_e=0.25 mm). Figure 2.10 shows the finite element model of a water-filled acoustic waveguide tube divided by free triangles after controlling the maximum grid size. The subdivision generates 129,692 triangular grids, including 4279 boundary grids and 20 vertex grids.

Grid resolution and grid quality are two important issues to be considered when building an effective finite element model. Low grid resolution (often associated with solution and geometry changes) can lead to inaccurate simulation results. Poor grid quality (the irregularity of grid cell shape) may result in the grid inversion and a higher number of Jacobian matrix conditions, which, in turn, leads to the failure of solution convergence. Inverted grids are often found in the models with large grids and curved boundaries. The grid cells are wrapped from inside out or the grids have zero area (two-dimensional finite element model) or zero volume (three-dimensional finite element model). More precisely, for some coordinates, the determinant of the Jacobian matrix of the mapping from local coordinates to global coordinates is negative or zero. In most cases, the linear (straight-line) grids that you see in the grid diagram are not inverted grids, but the higher-order curved grid elements used for calculating the solution may be inverted grids. The inverted grid itself does not directly threaten the overall accuracy of the solution. However, if an iterative solver is used, the solution may not converge. During modeling, the presence of inverted meshes can be reduced by avoiding small curved boundaries (such as round corners)

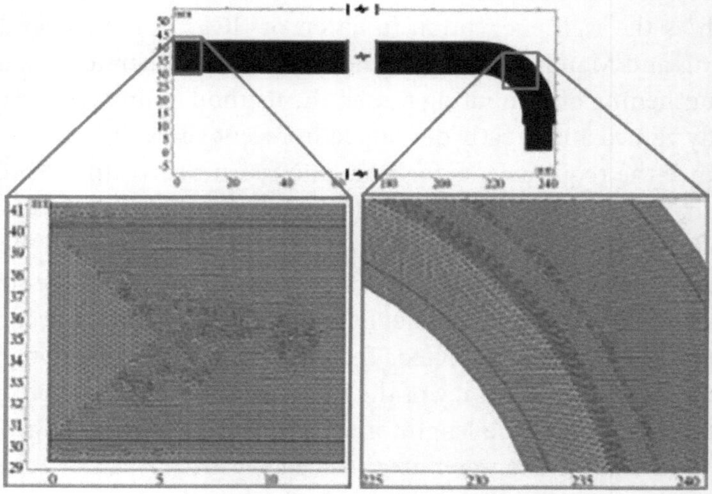

FIGURE 2.10 Finite element model of an acoustic waveguide tube.

and infinitesimal areas on domains and surfaces or avoided by using the software's "inverse-avoiding curved grids" feature (when using a "free tetrahedron" grid divider for grid subdivision).

The grid quality is a dimensionless quantity between 0 and 1. After selecting the rules on grid quality evaluation, 1 represents the grid completeness and 0 represents the grid degradation. The COMSOL Multiphysics includes the following statistical evaluation methods of grid quality:

- **Degree of skewness**: this is the default quality evaluation method for measuring the equiangular skewness, which is defined as the minimum value of the following quantity:

$$1-\max\left(\frac{\theta-\theta_e}{180-\theta_e},\frac{\theta_e-\theta_e}{\theta_e}\right) \tag{2.61}$$

  where θ is the angle at the grid vertex of a 2D finite element model or at the grid edge of a 3D finite element model and θ_e is the angle on the corresponding edge or vertex of an ideal element, and its minimum value is obtained at all the grid vertices of a 2D finite element model or at all the grid edges of a 3D finite element model.

- **Maximum angle**: the maximum angle in a grid cell. If no angle is greater than the maximum angle in the corresponding optimal grid cell, the quality will be 1; otherwise, the quality will be 0. Or, this evaluation will show how big the angle is. This grid-quality evaluation method is insensitive to the anisotropy of grid cells.
- **Volume versus circumcircle radius**: the quotient of grid cell volume and its circumcircle (or circumsphere) radius. This evaluation method is sensitive to large angle, small angle and anisotropy.
- **Volume versus length**: the quotient of grid cell edge length and grid cell volume. This evaluation method is mainly sensitive to anisotropy.
- **Number of conditions**: the number of conditions (Frobenius norm) of the matrix that converts a grid cell into a reference grid cell and is to divide the grid cell size.
- **Growth rate**: a measure of local (anisotropic) growth rate. If $S_a(E)$ is the size of the grid cell E in the spatial direction a and E_a is the grid cell adjacent to the grid cell E in the direction a, then the growth rate g can be expressed as:

$$g=\frac{\min\left(S_\alpha(E),S_\alpha(E_\alpha)\right)}{\max\left(S_\alpha(E),S_\alpha(E_\alpha)\right)} \tag{2.62}$$

where a is any direction.

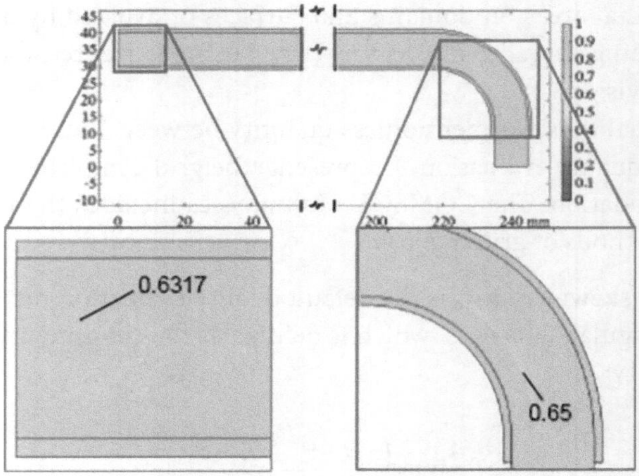

FIGURE 2.11 Cloud chart of grid quality.

The skewness evaluation method is used to statistically analyze the grid quality of finite element model of a water-filled acoustic waveguide tube. The minimum cell quality is 0.6137 and the average cell quality is 0.9545. The cloud chart of grid quality distribution is shown in Figure 2.11. The quality of most of the grid cells is close to 1 (shown in green), and only a few grid cells have poor quality near the excitation source and at the bend of the acoustic waveguide tube. The grids with the worst quality are near the excitation source, and the grid cells with poor quality at the bend of the tube have a minimum quality of about 0.65 (shown in light yellow). This is basically within a normal range and can be used for calculation. Next, the excitation signal and solver parameters will be set.

4. Setting of excitation signal
   To enable the numerical simulation to more realistically reflect the actual signal emitted by ultrasonic transducer, the single-frequency sinusoidal signal modulated by Hanning window is often used as the excitation signal at the excitation source in Figure 2.9 because this signal can effectively suppress the ultrasonic frequency dispersion and eliminate the high-frequency interference and energy spectrum leakage. Moreover, compared with sinusoidal wave signal, the modulated signal has a wider frequency band, narrower pulses and higher longitudinal resolution, thus ensuring the quality of excitation signal [15,16,20,25]. In addition, the period of excitation signal is also a very important parameter to be considered. When more excitation signal cycles are applied, the ultrasonic signals with stronger energy and narrower frequency band will be excited to facilitate the analysis of the reflected echo and improve the accuracy and reliability of numerical simulation results. However, if the excitation signal has too many cycles, the reflection echo signal and the excitation signal may overlap in the time domain to cause trouble to the signal analysis. If the excitation signal has too few cycles, the actual situation can't be truly reflected and the analysis of simulation results will be affected [17]. Therefore, in order to facilitate the analysis of simulation results, the sinusoidal sound pressure signal modulated by Hanning

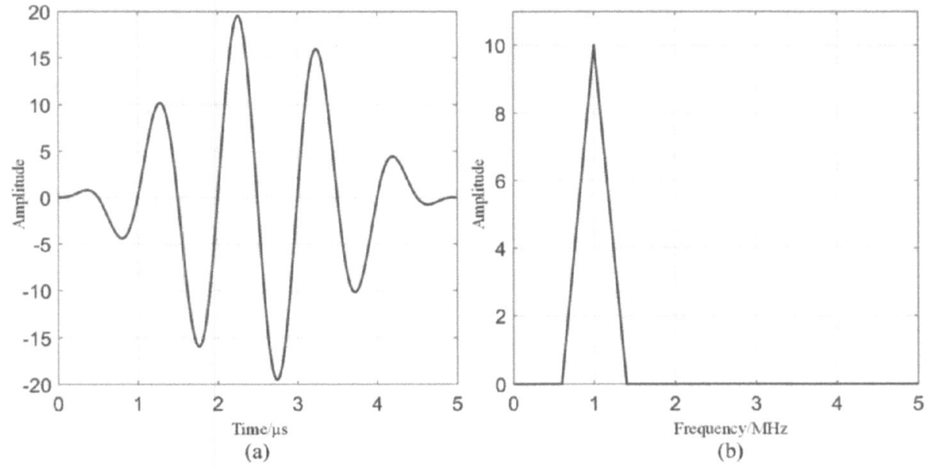

FIGURE 2.12 Waveform and spectrum distribution of the excitation signal. (a) Excitation signal waveform and (b) excitation signal spectrogram

window with an amplitude of 20 Pa, a frequency of $f_0 = 1$ MHz and five cycles is set as excitation source. The expression of excitation signal is as follows:

$$pw(t) = \begin{cases} 20\sin(2\pi f_0 t)\cdot\left\{\frac{1}{2}\left[1-\cos\left(\frac{2\pi f_0 t}{5}\right)\right]\right\}, 0 < t < 5/f_0 \\ 0, t > 5/f_0 \end{cases} \tag{2.63}$$

where $pw(t)$ is the required excitation signal, whose time domain waveform and spectrogram are shown in Figure 2.12a and b.

5. Setting of solver parameters
With respect to solver setting, the COMSOL will choose a direct solver or an iterative solver according to the physical field to be determined and the magnitude of the specific problem to be solved. The default iteration solver is selected based on the principle of maximum robustness and minimum memory without any user operation. The default solver in this model is MUMPS direct solver. The PARDISO and SPOOLES direct solvers can also be selected if needed. All the three solvers are based on LU (LU Factorization) decomposition.

For all well-conditioned finite element problems, these solvers can obtain the same solutions. This is their greatest advantage. They even support the solving of some ill-posed problems. The main difference between different direct solvers is relative velocity. The MUMPS, PARDISO and SPOOLES solvers can all take advantage of all the processor cores in a single computer. However, PARDISO is the fastest and SPOOLES is the slowest. Among direct solvers, the SPOOLES uses the least memory. Each direct solver requires a large RAM space, but MUMPS and PARDISO can store the solutions outside the core, that is, they can temporarily store some data on a hard disk. The MUMPS solver also supports the cluster computing so that the memory available to you is usually greater than the memory provided by any computer.

After the solver is selected, another key parameter to be set is solution time, including the total solution time and the time step. Theoretically, the smaller the time step is, the better the convergence effect of the model will be. However, the reduction of time step is at the sacrifice of the high consumption of server resources. Therefore, while ensuring the accuracy of simulation results, the time step Δt needs to satisfy [18] :

$$\Delta t \leq \frac{L_e}{c} \tag{2.64}$$

where c is the speed of sound in the medium. This model is more concerned with the ultrasonic propagation in an acoustic waveguide tube after the ultrasonic reflection from the tube bend. Considering the computing power of the existing server, we take $c = c_{\text{water}} = 1480\,\text{m/s}$, that is, $\Delta t = \frac{c_{\text{water}}}{nfc_{\text{water}}} = \frac{1}{nf} \approx 1.67\,\mu\text{s}$. Then according to the geometrical size of the acoustic waveguide tube and the ultrasonic velocity in water, we set the total solution time as 200 μs.

6. Analysis of numerical simulation results of ultrasonic propagation in the acoustic waveguide tube
   After the transient calculation of the finite element model established in the previous sections and the post-processing of calculation results of transient solver, we can observe the whole propagation process of ultrasonic wave inside the acoustic tube. Figure 2.13a–d show the sound pressure distributions in the tube and its internal water area at four time points (35.9, 163.66, 166 and 180.86 μs) after the activation of excitation source.

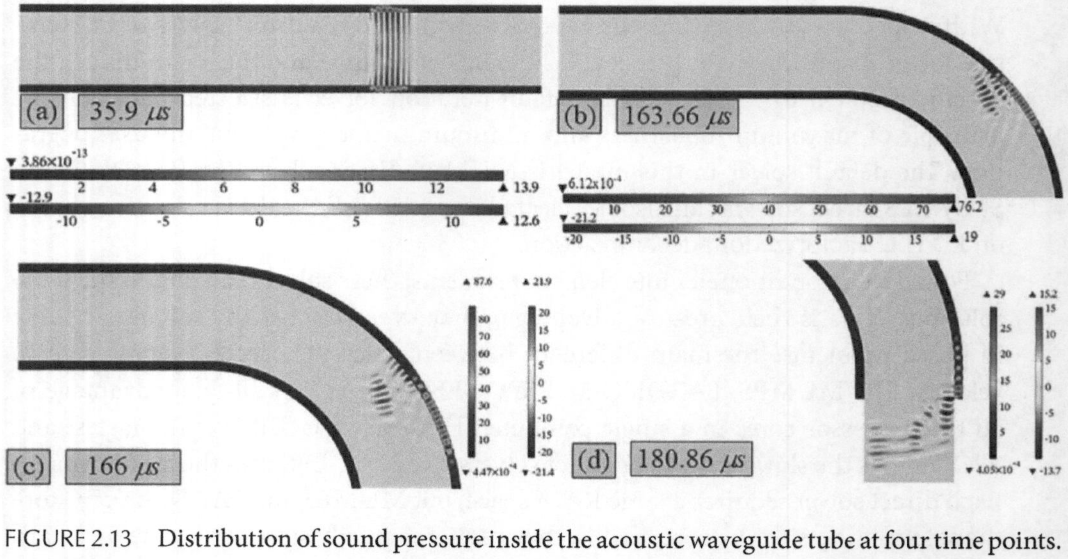

FIGURE 2.13 Distribution of sound pressure inside the acoustic waveguide tube at four time points.

It can be seen from Figure 2.13a that the ultrasonic wave advances smoothly forward in the straight section of the acoustic waveguide tube. At this time, the maximum positive peak of sound pressure in water is 12.6 *Pa*. Figure 2.13b shows the ultrasonic reflection occurring at the tube bend and the overlap of part of the reflected wave and the advancing wave. In addition, a surface wave or Lamb wave is generated in the tube wall at an appropriate incident angle during the propagation of sound wave in the tube bend. As the sound velocity in the tube wall is faster than that in water in the process of propagation, the area with a greater sound pressure in the tube wall is obviously farther away from the excitation source than that in the water, as shown in Figure 2.13b and more clearly in Figure 2.13c. In this case, the maximum positive peak of sound pressure in the water is 19 *Pa*. In Figure 2.13c, the area with a greater sound pressure in the water is smaller, which is also caused by the overlap of the reflected wave and the incident wave due to the fact that the time of ultrasonic reflection at different positions of the tube bend is different. In this case, the maximum positive peak of sound pressure in the water is 21.9 *Pa*. Figure 2.13d shows the sound pressure distribution when the ultrasonic wave propagates from the area with the maximum acoustic beam to the outer area of the acoustic waveguide tube after being reflected from the wall surface at the bend. It can be seen that the area with a greater sound pressure is basically changed back into the inner diameter of the tube. In this case, the maximum positive peak of sound pressure in the water is 15.2 *Pa* in a portion of the water area close to the right side at the tube outlet. This basically coincides with the ultrasonic propagation path in the acoustic waveguide tube simply shown in Figure 2.6, because the superposition of the reflected wave and the advancing wave is the most serious here and of course is also the most complex. The numerical simulation result is basically consistent with the simplified ultrasonic propagation path in the tube. They are complementary and mutually verified.

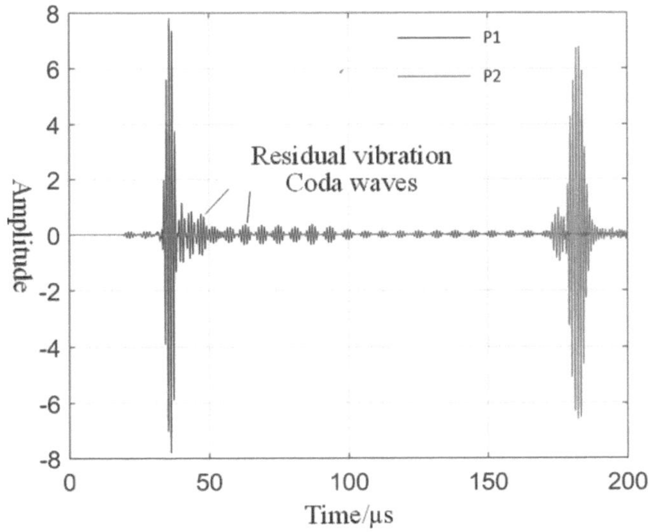

FIGURE 2.14 A-scan waveform signals at the sampling points *P*1 and *P*2.

To study the ultrasonic propagation at key positions more clearly, we extracted the A-scan signals from the sampling observation points *P*1 (50, 35) and *P*2 (240, 4.5) to observe their amplitude changes, as shown in Figure 2.14.

As can be seen from the figure, the sound pressure at the point *P*2 of tube outlet is basically the same as the sound pressure peak at the point *P*1 approximately three times the near field distance right ahead from the excitation source (ultrasonic transducer), without significant attenuation. Since the signal is continuously sampled for 200 μs, some residual vibration and wake waves are still affecting the point *P*1 after the main wave packet of the excitation signal passes the point *P*1, as seen from the A-scan signal waveform diagram. This is caused by the diffraction of the propagating ultrasonic wave generated by the excitation source. The diffracted ultrasonic wave continues to travel in the tube after being reflected from the inner tube wall, while some ultrasonic waves are reflected for many times in the tube to form a series of residual vibration and wake waves behind the main wave packet at the point *P*1, as shown in Figure 2.14.

To further observe whether the signal frequency component changed significantly after the ultrasonic signal was reflected from the curved surface at the bend of the acoustic waveguide tube, we performed spectral analysis on the A-scan signals extracted at the points *P*1 and *P*2. Meanwhile, in order to avoid the influence of residual vibration of other reflected signals, we only applied the Fourier transform to the main wave packets of two groups of A-scan signals and added the Hanning window function to them to reduce the energy leakage and improve the resolution of frequency range. The result of spectral analysis is shown in Figure 2.15.

As can be seen from the spectrogram, the main frequency of A-scan signal at the outlet of the acoustic waveguide tube is basically the same as that at the point *P*1 inside the tube. The former is 0.94 MHz, while the latter is 1 MHz. This minor change would be negligible for an ordinary longitudinal-wave ultrasonic transducer. For example, the bandwidth of

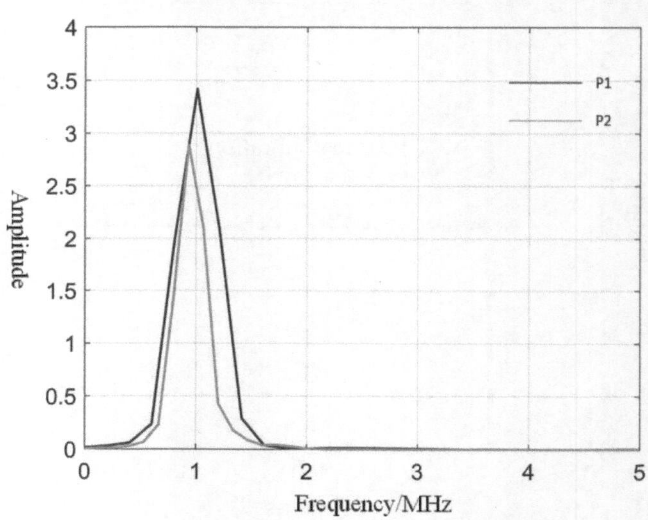

FIGURE 2.15 Spectrum analysis result of the A-scan signals sampled at *P*1 and *P*2.

OLYMPUS 1 MHz ultrasonic transducer used in this experiment is about 0.5 MHz at −6 dB, and its amplitude is only reduced by about 10%, so the testing result is basically not affected.

2.4 MEASUREMENT OF SOUND FIELD IN AN ACOUSTIC WAVEGUIDE TUBE

Understanding the distribution of acoustic field at the outlet of the acoustic waveguide tube is important to the improvement of testing accuracy and sensitivity and the optimization of testing process parameters. The existing Chinese standard GB/T 16540-1996 "Acoustics-Measurement and Characterization of Ultrasonic Fields in the Frequency Range 0.5–15 MHz - Hydrophone Method", which is equivalent to the international standard IEC1102:1991, also defines various acoustic parameters relating to the ultrasound field and its measurement in water and other liquids within the frequency range of 0.5–15 MHz, such as acoustic pulse waveform, positive peak sound pressure, scanning area, scanning plane, ultrasonic scan line and ultrasonic scan-line spacing. The standard also puts up with the requirements for the basic properties of the hydrophone measuring system, such as sensitivity and directivity response. In addition, it specifies the basic steps of sound field measurement. It can be seen that the sound field measurement is important to actual engineering applications.

The hydrophone method mentioned above is not the only method for sound field measurement. According to different measurement principles, the methods of sound field measurement can be divided into four types: the mechanical effect method, optical effect method, thermal effect method and chemical effect method. Among them, the mechanical effect method (such as hydrophone method [19–22], radiative force balance (RFB) method [23] or ball target method [22,24]) is the most commonly used in engineering practice. In particular, the hydrophone method is most widely used to measure the sound field. Hydrophone is a kind of sensor. It can convert the received sound pressure signal in a liquid medium (such as water) into voltage signal, and then amplify and input the voltage signal into the measuring system, so as to complete the measurement of the sound

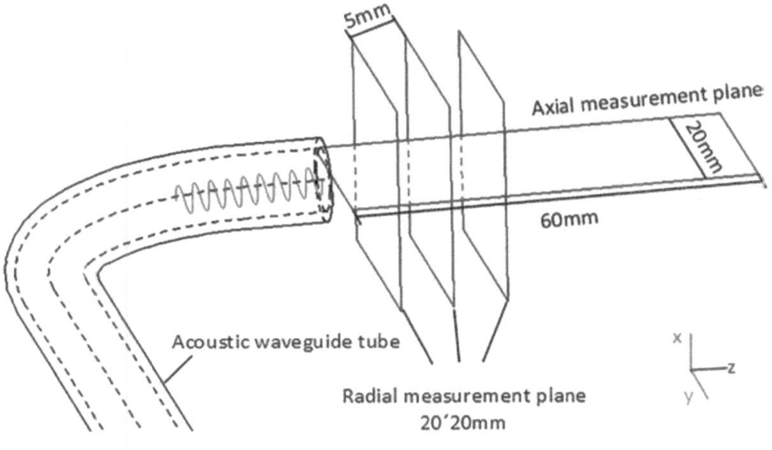

FIGURE 2.16 Sound pressure measurement plane at the outlet of the acoustic waveguide tube.

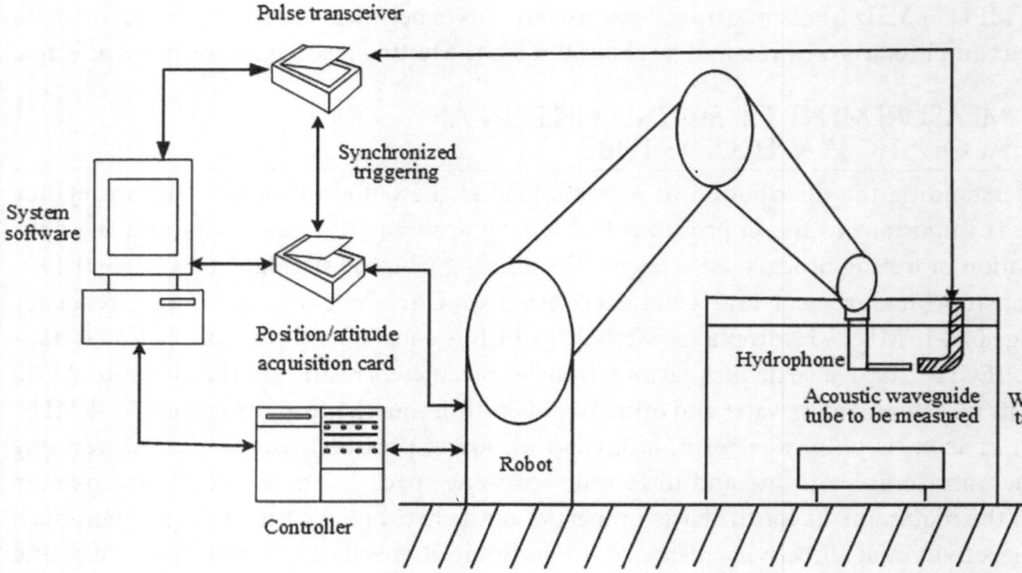

FIGURE 2.17 Hydrophone-based automatic sound field measurement system.

pressure value at a certain position in the sound field. As the sound field measurement based on the hydrophone method receives more and more research attention, the techniques of calibration and optimization of broadband hydrophone and high-frequency hydrophone and the low-cost and high-performance hydrophone fabrication technique have emerged at the right moment [25,26]. In this section, the hydrophone method will be used to measure the sound pressure distribution on the axial plane and three radial planes at the outlet of the acoustic waveguide tube (see Figure 2.16), among which the axial plane is 60×20 mm (length×width) and the radial planes are 20×20 mm (length×width) with the spacing of 5 mm.

The sound field measurement system based on manipulator is used to measure the sound field at the outlet of the acoustic waveguide tube. The sound field measurement system and its configuration are shown in Figure 2.17.

The main components of this system are as follows:

- One megahertz immersion-plane ultrasonic transducer, with the chip size of 0.5 in and the damping of 200 Ohm;
- Ultrasonic pulse transceiver, with the maximum excitation voltage of 400 V;
- Data acquisition card, with the maximum sampling frequency of 250 MHz;
- One millimeter pin-type hydrophone;
- Articulated manipulator, with six DOFs and
- Other auxiliary components, such as water jet, ultrasonic waveguide tube and water tank.

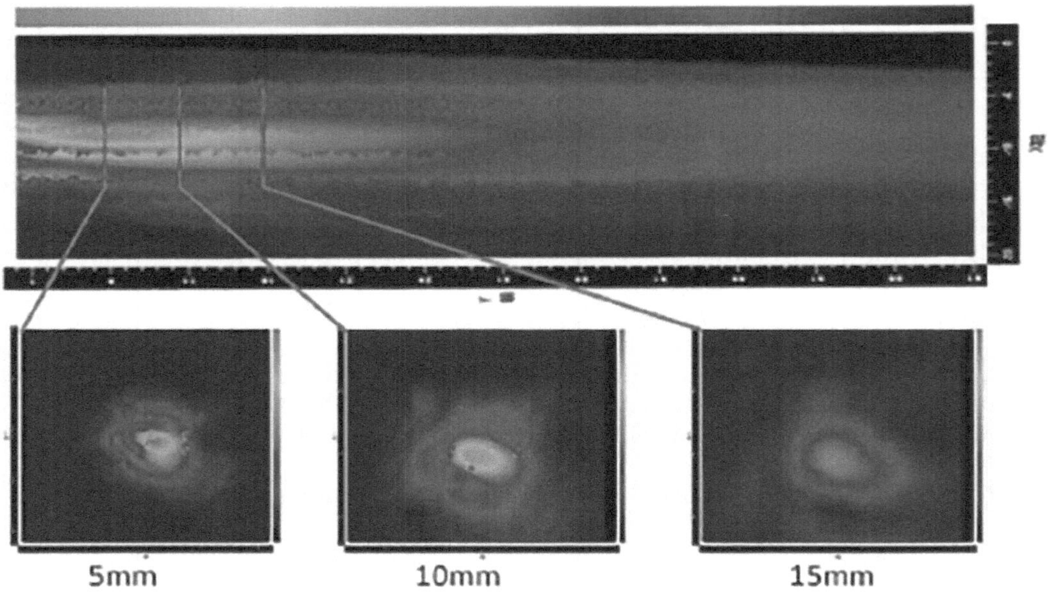

FIGURE 2.18 Measured results of sound fields at the outlet of the acoustic waveguide tube.

To show the gradient change of sound pressure on axial plane and radial planes more clearly, the excitation voltage of pulse transceiver was set as 300 V and the receiving gain was set as 12 dB. The measured results of sound fields on the axial plane and radial planes at the outlet of the acoustic waveguide tube are shown in Figure 2.18.

As shown in Figure 2.18, although the ultrasound propagates for a distance of 275 mm in the water jet and acoustic waveguide tube under the set parametric conditions, a strong sound pressure can still be detected at the tube outlet. The maximum sound pressure is located about 3–10 mm right ahead of the tube outlet. It is better to ensure that the distance from the tube outlet to the tested workpiece is within this distance range.

REFERENCES

1. Du G., Zhu Z., Gong X. *Fundamentals of Acoustics [M]*. 3rd edition. Nanjing: Nanjing University Press, 2012.
2. Yang C. *High-frequency Ultrasonic Testing Method for Multilayer Structures [D]*. Beijing:Beijing Institute of Technology, 2019.
3. Zheng H., Lin S. *Ultrasonic Testing [M]*. China Labor and Social Security Publishing House, 2008.
4. Feng R. *Ultrasonic Handbook [M]*. 1st edition. Nanjing: Nanjing University Press, 1999.
5. Zhang W., Gang T., Development of UT analysis, calculation and simulation software based on VB [J], *Nondestructive Testing*, 2003(01):8–11.
6. Declercq N.F., Degrieck J., Leroy O., Ultrasonic polar scans: numerical simulation on generally anisotropic media [J], *Ultrasonics*, 2006, 45(1):32–39.
7. Chassignole B., Guerjouma R.E., Ploix M.A., Ultrasonic and structural characterization of anisotropic austenitic stainless steel welds: towards a higher reliability in ultrasonic nondestructive testing [J], *NDT&E International*, 2010, 43(4):273–282.

8. COMSOL. Multi-physics Simulation of the Interaction between Sound Field and Flow Field [EB/OL] (2020). http://cn.comsol.com/video/modeling-flow-acoustic-interaction-in-comsol-multiphysics-webinar-cn.
9. Liu J., Xu G., Ren L., et al., Simulation analysis of ultrasonic detection for resistance spot welding based on COMSOL Multiphysics [J], *International Journal of Advanced Manufacturing Technology*, 2017, 93(5–8):2089–2096.
10. Zhao K. *Research on COMSOL-based In-rail Ultrasonic Guided Wave Propagation Model and Defect Detection Method [D]*. Xi'an University of Technology, 2016.
11. Zhou G. *Research on Ultrasonic Quantitative Testing and Quality Evaluation of Stainless-steel Lapped Laser Welding Joints [D]*. Jilin University, 2018.
12. Wang S. *Research on Lamb Wave Mode Control and Imaging Testing Method of Array Transducer [D]*. Beijing Institute of Technology, 2018.
13. Liu L., Chen Y., Wang Z., Quantitative evaluation of ultrasonic debonding testing based on COMSOL [J], *Journal of Test and Measurement Technology*, 2016, 30(06):467–470.
14. Fu X., COMSOL simulation for ultrasonic TOFD testing of pressure vessel welds [J], *Nondestructive Testing*, 2018, 40(07):9–14.
15. Gao J. *Cylinder Sound Field Simulation and Flaw Echo Characteristic Analysis Based on Finite Element [D]*. North University of China, 2016.
16. Ma X. *Ultrasonic Detection Mode Recognition of Material Defects Based on COMSOL Simulation [D]*. Nanchang Hangkong University, 2019.
17. Ma S. *Numerical Simulation and Experimental Research on Propagation of Ultrasonic Lamb Wave in Welds [D]*. Harbin Institute of Technology, 2017.
18. Ghose B., Balasubramaniam K., Krishnamurthy C.V. et al., Two Dimensional FEM Simulation of Ultrasonic Wave Propagation in Isotropic Solid Media Using COMSOL [C]. The COMSOL Conference, 2010, India.
19. Peng, G. *Robotic Measurement Technology of Sound Field Induced by Ultrasonic Transducer [D]*. Beijing Institute of Technology, 2017.
20. Zhai F. *Research on Measurement and Modeling Methods of Complex Ultrasonic Field [D]*. Lanzhou University, 2015.
21. Du J. *Research on Ultrasound Field Distribution Measurement Based on Hydrophone [D]*. China Jiliang University, 2012.
22. Han M., Yang P., Comparative study on sound field parameters of ultrasonic probe measured by pulse echo method and hydrophone method [J], *Acta Metrologica Sinica*, 2015, 36(02):166–170.
23. Shou W., 30 years of development of ultrasonic measurement technology and standardization in China [J], *Technical Acoustics*, 2011, 30(1):21–26+45.
24. Sha Z., Gang T., Zhao X., Research on simulation and visualization of sound field induced by ultrasonic transducer [J], *Nondestructive Testing*, 2011, 33(05):2–6.
25. Umchid S., Gopinath R., Srinivasan K. et al., Development of calibration techniques for ultrasonic hydrophone probes in the frequency range from 1 to 100 MHz [J], *Ultrasonics*, 2009, 49(7):306–311.
26. Lee J.W., Ohm W.S., Kim Y.T., Development of disposable membrane hydrophones for a frequency range from 1 MHz to 10 MHz [J/OL], *Ultrasonics*, 2017, 81:50–58.

CHAPTER 3

Planning Method of Scanning Trajectory on Free-Form Surface

The advantage of robotic nondestructive testing (NDT) technique is the automatic testing of complex curved components. The premise of realizing the automatic programming of NDT scanning trajectory is to unify the manipulator coordinate system, workpiece coordinate system and clamping-tool coordinate system into a testing coordinate system. Therefore, the conversion between different coordinate systems is the basis of automatic programming.

3.1 MAPPING RELATIONS BETWEEN MULTIPLE COORDINATE SYSTEMS

3.1.1 Translation, Rotation and Transformation Operators

The general mathematical expression for the point transformation between different coordinate systems is called operator. The main operators of points in robotics include translation operator, vector rotation operator and translation and rotation operator [1]. In this section, the three operators will be mathematically described.

3.1.1.1 Translation Operator

Translation is a certain distance of a point in space moving in the direction of a known vector. The actual translation of a point in space is only described by one coordinate system. The translation of a point in space has the same mathematical description as the mapping of this point to another coordinate system. Therefore, it is very important to figure out the mathematical meaning of mapping. This differentiation process is very simple. The forward movement of a vector relative to a coordinate system can be considered as either the forward motion of the vector or the backward motion of the coordinate system, both of which have the same mathematical expression but different observation places. Figure 3.1

DOI: 10.1201/9781003212232-3

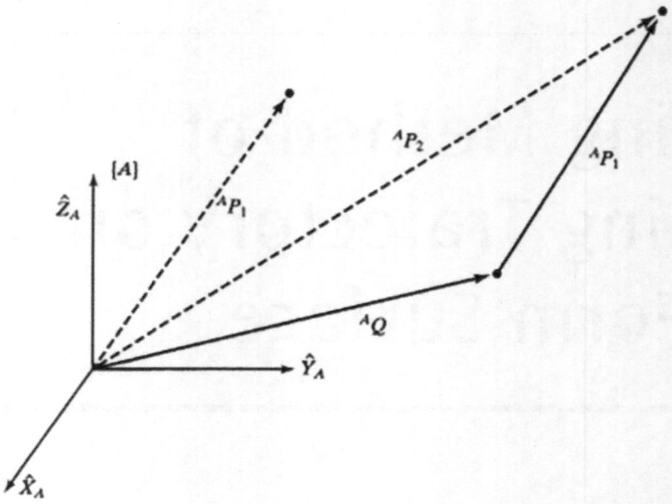

FIGURE 3.1 Translation operator.

shows the translation of the vector $^AP_1$ along the vector ^AQ, which gives the translation information.

After the translation of the vector $^AP_1$, the new vector $^AP_2$ will be obtained and can be calculated with the following formula:

$$^AP_2 = {}^AP_1 + {}^AQ \tag{3.1}$$

Then Eq. (3.1) is converted into a homogeneous matrix, which is expressed by

$$^AP_2 = D_Q(q) \cdot {}^AP_1 \tag{3.2}$$

where q is a vector translational along the vector Q, and the operator D_Q is a special homogeneous transformation.

$$D_Q(q) = \begin{bmatrix} 1 & 0 & 0 & q_x \\ 0 & 1 & 0 & q_y \\ 0 & 0 & 1 & q_z \\ 0 & 0 & 0 & 1 \end{bmatrix} \tag{3.3}$$

where q_x, q_y and q_z are the components of the translation vector Q on the coordinate axes, $q = \sqrt{q_x^2 + q_y^2 + q_z^2}$. Now the operator notation D_Q is introduced and can be used to describe the translation mapping of points.

3.1.1.2 Rotation Operator

Rotation matrices can also be described by rotation transformation operator. Through the rotation R, the vector $^AP_1$ can be transformed into the new vector $^AP_2$. In general, when a rotation matrix acts as an operator, its subscript or superscript doesn't have to be written, because the rotation does not involve two coordinate systems. Therefore, ${}^{A}P_2$ can be written into

$$ {}^{A}P_2 = R \cdot {}^{A}P_1 \tag{3.4} $$

For this reason, we need to obtain the rotation matrix acting as operator. The rotation matrix of a vector obtained through the rotation R is the same as that of a coordinate system obtained through the rotation R relative to the reference coordinate system. Although it is a simple way to treat the rotation matrix as an operator, we will define the rotation operator with another notation to specify which axis is the rotation axis:

$$ {}^{A}P_2 = R_K(\theta) \cdot {}^{A}P_1 \tag{3.5} $$

where $R_K(\theta)$ is a rotation operator, indicating the rotation around the axis K by the angle θ. This operator can be written into a homogeneous transformation matrix, in which the position vector component is 0. For example, the operator representing the rotation around the axis Z by the angle θ is

$$ R_Z(\theta) = \begin{bmatrix} \cos\theta & -\sin\theta & 0 & 0 \\ \sin\theta & \cos\theta & 0 & 0 \\ 0 & 0 & 1 & 0 \\ 0 & 0 & 0 & 1 \end{bmatrix} \tag{3.6} $$

3.1.1.3 Transformation Operator

Like vector translation matrix and rotation matrix, a coordinate system can also be defined by transformation operator. In this definition, only one coordinate system is involved, so the operator T has no superscript or subscript. The vector ${}^{A}P_1$ is translational and rotated by the operator T and then turns into a new vector

$$ {}^{A}P_2 = T \cdot {}^{A}P_1 \tag{3.7} $$

where the operator in the form of homogeneous matrix represents the mapping from a vector to another vector through rotation and translation, so we know how to obtain the homogeneous transformation matrix acting as an operator.

The rotation matrix of a vector obtained through the rotation R and the translation Q is the same as that of a coordinate system obtained through the rotation R and the translation Q relative to the reference coordinate system. A transformation operator is generally considered as a homogeneous transformation matrix composed of a generalized rotation matrix and translation vector components.

3.1.2 Equivalent Rotation and Quaternion Equation

In the above case, attitude is represented by a 3×3 rotation matrix, which is a special unit matrix with orthogonal columns. Therefore, the determinant of any rotation matrix is always +1 and is called standard orthogonal matrix. Since rotation matrix can be considered

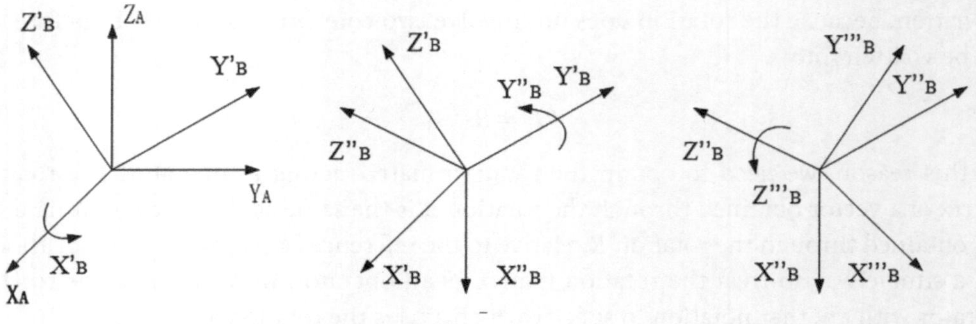

FIGURE 3.2 Rotation sequences of an *X–Y–Z* fixed angular coordinate system.

as both an operator and an attitude descriptor, undoubtedly it has different representations in different applications. Rotation matrix can be used as an operator. When multiplied by a vector, it plays the role of rotation operator. However, it is still a little complicated to determine the attitude with a rotation matrix containing nine elements, especially when determining the position and attitude changes of a large number of data points. Next, the coordinate-system attitude rotation method using three numbers to represent the position and attitude changes of a manipulator will be introduced:

3.1.2.1 Representation in an Angular Coordinate System

As there are as many as 24 sequence combinations of axial rotations around the axes of the reference coordinate system, we will take the *X–Y–Z* fixed angular coordinate system as an example to present the sequences of each coordinate system rotating around the axes of the fixed reference coordinate system $\{A\}$, as shown in Figure 3.2.

At first, the coordinate system $\{B\}$ overlaps with the known reference coordinate system $\{A\}$. $\{B\}$ is rotated around $\hat{X}_A$ by the angle γ, then around $\hat{Y}_A$ by the angle β and finally around $\hat{Z}_A$ by the angle α. Here, γ, β and α are called angle of revolution, angle of pitch and angle of deflection, respectively. The equivalent rotation matrix ${}^A_B R_{XYZ}(\gamma,\beta,\alpha)$ can be derived directly, because all the rotations are around the axes of the reference coordinate system, namely

$$R_{X-Y-Z} = R_Z(\alpha) R_Y(\beta) R_X(\gamma) = \begin{bmatrix} \cos\gamma & -\sin\gamma & 0 \\ \sin\gamma & \cos\gamma & 0 \\ 0 & 0 & 1 \end{bmatrix} \times \begin{bmatrix} \cos\beta & 0 & \sin\beta \\ 0 & 1 & 0 \\ -\sin\beta & 0 & \cos\beta \end{bmatrix} \begin{bmatrix} 1 & 0 & 0 \\ 0 & \cos\alpha & -\sin\alpha \\ 0 & \sin\alpha & \cos\alpha \end{bmatrix} \tag{3.8}$$

The rotation matrix is considered as an operator to rotate successively (from right) in the sequence of "$R_x(\gamma) \to R_Y(\beta) \to R_Z(\alpha)$.

3.1.2.2 Representation in an Equivalent Axial Angular Coordinate System

The above representation with three rotations around the axes in a certain order is the representation in an angular coordinate system. However, when the equivalent axial angular coordinate system is used for representation, any orientation can be obtained by selecting the appropriate axis and angle. The coordinate system $\{B\}$ is represented by overlapping the coordinate system $\{B\}$ and the known reference coordinate system $\{A\}$ and then rotating $\{B\}$ around the vector ${}^{A}\hat{K}$ (called the equivalent axis for finite rotation) by the angle θ according to the right-hand rule. The general attitude of $\{B\}$ relative to $\{A\}$ can be expressed by ${}^{A}_{B}R\left(\hat{K},\theta\right)$ or $R_K(\theta)$.

Only two parameters are needed to define the vector ${}^{A}\hat{K}$, whose length is always 1. The third parameter is determined by angle. Attitude, denoted by K, is often described by a 3×1 vector multiplying the rotation amount θ by the unit direction vector $\hat{K}$. If the rotation axis is a general axis, then the equivalent rotation matrix will be

$$R_k(\theta)=\begin{bmatrix} k_x k_x v\theta + c\theta & k_x k_y v\theta - k_z s\theta & k_x k_z v\theta + k_y s\theta \\ k_x k_y v\theta + k_z s\theta & k_y k_y v\theta + c\theta & k_y k_z v\theta - k_x s\theta \\ k_x k_z v\theta - k_y s\theta & k_y k_z v\theta + k_x s\theta & k_z k_z v\theta + c\theta \end{bmatrix} \tag{3.9}$$

where $c\theta=\cos\theta$, $s\theta=\sin\theta$, $v\theta=1-\cos\theta$ and $\vec{k}=\left[k_x,k_y,k_z\right]^T$. The sign of θ is determined by right-hand rule, that is, our thumb is pointing to the positive direction of ${}^{A}\hat{K}$. Eq. (3.9) transforms the representation in the equivalent axial angular coordinate system into that in rotation matrix. Therefore, for any rotation axis and any angle, the equivalent rotation matrix can be easily constructed.

3.2 SURFACE SPLIT AND RECONSTRUCTION BASED ON NUBRS

In the ultrasonic testing (UT) application, the discrete points on manipulator scanning path obtained with the above coordinate transformation method and CAD/CAM technique may be fewer than those to be matched during high-frequency ultrasonic acquisition (the trigger frequency for tiny tested parts can be higher than 1 KHz). Therefore, it is necessary to process, through discrete interpolation, the trajectory formed by the above discrete point data in order to obtain a smaller step.

3.2.1 Parametric Spline Curve and Surface Split Method

When a parametric spline curve is used to describe the geometrical elements of the target object, it can be discretized based on parametric curve constraints and split into several groups of curves and surfaces so that the interpolation processing of the target geometrical elements can be implemented. Currently, the commonly used trajectory discretization methods include the method of equal chord length, the method of equal interval and the method of equal error, as shown in Figure 3.3.

Among them, the method of equal interval is to equalize the distances (Δx) between discrete points projected on the specified parametric axis. This algorithm can be easily

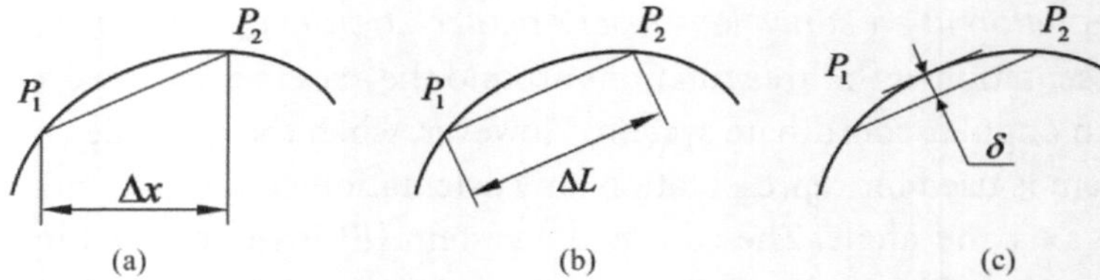

FIGURE 3.3 Trajectory discretization methods. (a) Method of equal interval, (b) method of equal chord length and (c) method of equal error.

realized by programming calculation. However, in this algorithm, the curved sections with a larger curvature also have a larger fitting error, and the chords approximating the trajectory have a different length. The method of equal chord length is to ensure that the linear distances between points are equal and that the chord lengths between discrete points are relatively uniform. The method of equal error is to ensure that the errors (δ) between two adjacent discrete points and the curved sections are almost the same. In this method, the highest accuracy can be achieved when approximating the curve, but the distances between the generated trajectory points may not be the same. Moreover, this algorithm is relatively complex. Considering the equal displacements of various discrete points on the planned tool trajectory and the characteristics of the above discretization methods, the method of equal chord length is chosen to discretize the planned manipulator trajectory.

3.2.2 Scanning of Non-uniform Rational B-Splines (NURBS)

Non-uniform rational B-spline (NURBS) surfaces and B-splines are usually used to describe and analyze free-form curves and surfaces. They are also widely used in surface split techniques such as curve segmentation, surface subdivision construction and local modification [1–3]. Therefore, they can also be discretized under the constraint of equal chord length. The functional expression of rational polynomial of NURBS is [4–6]

$$P(u)=\frac{\sum_{i=0}^{n} B_{i,k}(u)W_iV_i}{\sum_{i=0}^{n} B_{i,k}(u)W_i} \tag{3.10}$$

where V_i represents the control point; W_i is the weight factor of each point; $B_{i,k}(u)$ is the kth-order B-spline basis function and is derived from the following recursive formula:

$$\begin{cases} B_{i,0}(u)=\begin{cases} 1, & u_i \le u \le u_{i+1} \\ 0, & \text{in other cases} \end{cases} \\ B_{i,k}(u)=\dfrac{u-u_i}{u_{i+k}-u_i}B_{i,k-1}(u)+\dfrac{u_{i+k+1}-u}{u_{i+k+1}-u_{i+1}}B_{i+1,k-1}(u), k\ge 1 \\ \dfrac{0}{0}=0 \end{cases} \tag{3.11}$$

Similarly, the $k \times l$-order rational polynomial function of NURBS surface in the space is defined as

$$\mathbf{S}(u,v)=\frac{\sum_{i=0}^{n}\sum_{j=0}^{m}B_{i,k}(u)B_{j,l}(v)W_{i,j}V_{i,j}}{\sum_{i=0}^{n}\sum_{j=0}^{m}B_{i,k}(u)B_{j,l}(v)W_{i,j}} \tag{3.12}$$

where u and v are, respectively, the parametric variables in two dimensions of NURBS surface; $B_{i,k}(u)$ is the kth-order non-rational B-spline basis function in the direction u; $B_{j,l}(v)$ is the lth-order non-rational B-spline basis function in the direction v; $V_{i,j}$ represent the control points in the two-dimensional directions, which form quadrilateral grids, and $W_{i,j}$ represent the weight factors of all points.

When the trajectory of free-form surface is planned, the control vertex and its weight factor are set according to the convergence speed of the directional derivative of each dimension of NURBS surface, and the target surface to be measured is divided into several small areas. Spatial quadrilateral grids can be obtained with the masked method of constructing the section curve of each area and unifying the powers and node vectors of all section curves. The path points on the planned surface are generated by approximating each surface area through iterative calculation.

Similar to the discrete points on NURBS surfaces, the discrete points on the scanning trajectory planned by CAD/CAM simulation technique can be considered as control vertices. In the discretization processing, these control vertices are required to meet certain constraints, such as collineation, coplane and in proportion (determined by the requirements of parametric curves and surfaces). Considering the necessity for control vertices to satisfy the boundary constraints when a curved surface is formed by the scanning parametric curve – especially considering the great influence of the control vertices at the surface boundary on surface smoothness and continuity – boundary constraints should be applied on those control vertices to ensure the continuous differentiability and convergence of the surface boundary. The following section will focus on the analysis of (partial) derivative vector constraint, normal vector constraint, parametric curve/surface patch constraint and other constraints. Figure 3.4 shows that the surface formed under the above curve constraints highly approximates the target surface.

3.2.3 Surface Construction Based on Differential Equation and Interpolation Algorithm

When the scanning is based on parametric curves and surfaces, boundary constraints need to be limited and analyzed. In consideration of the characteristics of NUBRS curves, the constraints on guide curve and non-uniform B-spline curve will be analyzed below.

1. Partial lerivative vector constraint

   In the design of curves and surfaces, the position and attitude of (partial) derivative vector at each discrete point need to be controlled. If the tangent vector of the

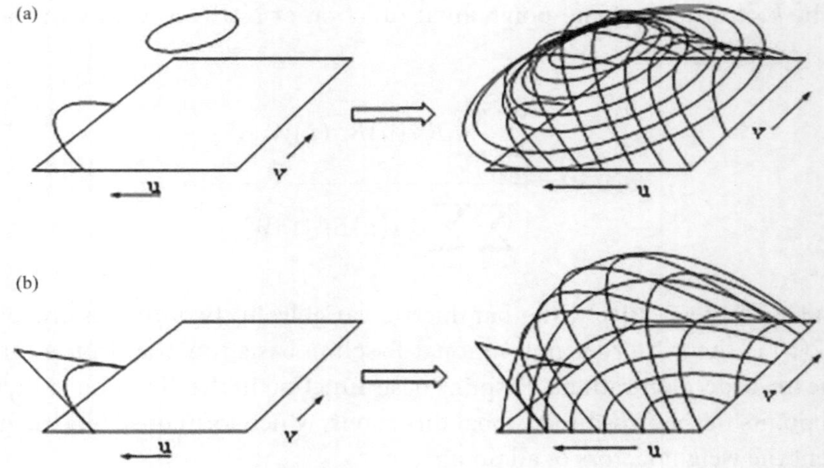

FIGURE 3.4 Surface generation strategy based on spline curve with discrete trajectory. (a) Surface generation strategy based on elliptic curve and (b) surface generation strategy based on B-spline curve.

surface $\mathbf{w}(u, v)$ at the parameter (u^*, v^*) in the direction u is T, this constraint can be defined as $\mathbf{w}_u(u^*, v^*) = T$. By combining it with the derivation functions of first-order derivative vector $\mathbf{w}_u(u)$ and second-order derivative vector $\mathbf{w}_{uu}(u)$ of $\mathbf{w}(u)$, $\mathbf{w}_u(u)$ and $\mathbf{w}_{uu}(u)$ can be solved:

$$\begin{cases} \mathbf{w}_u(u) = \sum\limits_{i=0,m} V_i \mathrm{B}'_{i,s}(u) \\ \mathbf{w}_{uu}(u) = \sum\limits_{i=0,m} V_i \mathrm{B}''_{i,s}(u) \end{cases} \Rightarrow \sum_{\substack{i=0,mu \\ j=0,mv}} V_{i,j} \mathrm{B}'_{i,su}(u^*) \mathrm{B}_{j,sv}(v^*) = T \tag{3.13}$$

where V is the control vertex of B-spline surface $\mathbf{w}(u, v)$; $B_{i,s}(u)$ is B-spline basis function and is determined by the power s and the node vector $\mathbf{U} = [u_0, u_1, \ldots, u_{m+s+1}]$.

Similarly, the second-order partial derivative vector constraint of the surface in the direction u can be derived:

$$\sum_{\substack{i=0,mu \\ j=0,mv}} V_{i,j} B''_{i,su}(u^*) B_{j,sv}(v^*) = T \tag{3.14}$$

The mixed partial derivative vector constraint of the surface can be defined as:

$$\sum_{\substack{i=0,mu \\ j=0,mv}} V_{i,j} B'_{i,su}(u^*) B'_{j,sv}(v^*) = T \tag{3.15}$$

All of the above constraints are linear equality constraints and linear functions using the position information at each discrete point as parametric variable. In the interpolation calculation, we only need to determine whether the discrete points are continuously differentiable.

2. Normal vector constraint
   Similarly, if the normal vector direction of the surface $\mathbf{w}(u, v)$ at the parameter (u^*, v^*) is N, the constraint will be

$$\mathbf{w}_u\left(u^*,v^*\right)\times\mathbf{w}_v\left(u^*,v^*\right)=kN \tag{3.16}$$

   By combining it with the expression of B-spline curve

$$\mathbf{w}(u,v)=\sum_{\substack{i=0,mu\\ j=0,mv}} V_{i,j}B_{i,su}(u)B_{j,sv}(v) \tag{3.17}$$

   the constraints in the normal vector direction can be solved:

$$\left\{\begin{aligned}
\mathbf{w}_u(u,v)&=\sum_{\substack{i=0,mu\\ j=0,mv}} V_{i,j}B'_{i,su}(u)B_{j,sv}(v)\\
\mathbf{w}_v(u,v)&=\sum_{\substack{i=0,mu\\ j=0,mv}} V_{i,j}B_{i,su}(u)B'_{j,sv}(v)\\
\mathbf{w}_{uu}(u,v)&=\sum_{\substack{i=0,mu\\ j=0,mv}} V_{i,j}B''_{i,su}(u)B_{j,sv}(v)\\
\mathbf{w}_{vv}(u,v)&=\sum_{\substack{i=0,mu\\ j=0,mv}} V_{i,j}B_{i,su}(u)B''_{j,sv}(v)\\
\mathbf{w}_{uv}(u,v)&=\sum_{\substack{i=0,mu\\ j=0,mv}} V_{i,j}B'_{i,su}(u)B'_{j,sv}(v)
\end{aligned}\right. \tag{3.18}$$

$$\Rightarrow\left\{\begin{aligned}
&\sum_{\substack{i=0,mu\\ j=0,mv}}\left(N_x\cdot X_{i,j}+N_y\cdot Y_{i,j}+N_z\cdot Z_{i,j}\right)B'_{i,su}\left(u^*\right)B_{j,su}\left(v^*\right)=0\\
&\sum_{\substack{i=0,mu\\ j=0,mv}}\left(N_x\cdot X_{i,j}+N_y\cdot Y_{i,j}+N_z\cdot Z_{i,j}\right)B_{i,su}\left(u^*\right)B'_{j,su}\left(v^*\right)=0
\end{aligned}\right.$$

   The above equation only contains the unknown position quantity (X, Y or Z) of the first-order term. It can be judged that the normal vector at each discrete point also meets the linear equality constraint.

3. Parametric curve constraint
   To accurately track the geometric features of the measured surface and control the approach of NURBS curve or surface to the target profile, the NURBS curve

is sometimes required to pass the specified spatial parametric curve and its control vertex. Therefore, it is necessary to analyze the isoparametric curve constraint represented by B-spline curve and the arbitrary parametric curve constraint represented by non-B-spline curve.

Suppose the isoparametric line of the surface $\mathbf{w}(u, v)$ at v^* in the direction u is

$$c(u)=\sum_{i=0,mu} P_i B_{i,su}(u) \tag{3.19}$$

Then the isoparametric curve constraint can be expressed by $\mathbf{w}(u, v^*)=c(u)$, which is substituted into Eq. (3.5) to obtain

$$\sum_{\substack{i=0,mu\\j=0,mv}} V_{i,j} B_{i,su}(u) B_{j,sv}(v^*)=\sum_{i=0,mu} P_i B_{i,su}(u) \tag{3.20}$$

In this case, the constraints on the parametric curve can be transformed into the linear equality constraints on $mu+1$ discrete points (control vertices), that is, the constraint on each discrete point only contains the first-order term of V.

Similarly, when the constraint on B-Spline parametric curve is $c(t)=\mathbf{w}(u(t), v(t))$, the energy parameter value of constraint curve can be based on to convert any parametric curve into

$$\int \left[c(t)-\mathbf{w}(u(t),v(t))\right]^2 dt=0 \tag{3.21}$$

After substituting Eq. (3.21) into Eqs. (3.18) and (3.19), we can obtain

$$\begin{aligned}&\sum_{\substack{i=0,m\\j=0,m}} P_i P_j B_{i,s}(t) B_{j,s}(t) dt - 2\sum_{\substack{i=0,m\\j=0,mu\\k=0,mv}} P_i V_{j,k}\int B_{i,s}(t) B_{j,su}(u(t)) B_{k,sv}(v(t)) dt+\\&\sum_{\substack{i=0,m\\j=0,mv\\k=0,mu\\l=0,mv}} V_{i,j} V_{k,l}\int B_{i,su}(u(t)) B_{j,sv}(v(t)) B_{k,su}(u(t)) B_{l,sv}(v(t)) dt=0\end{aligned} \tag{3.22}$$

In this case, the constraint on the curve is a quadratic function expressed by the control vertex V and is nonlinear equality constraint.

4. Parametric surface patch constraint
During the surface joining, transition and prolongation, the constraint on a surface patch can be Eq. (3.23) for B-spline curve or Eq. (3.24) for any surface patch. Then parametric surface patch constraint is used as an essential condition for surface generation:

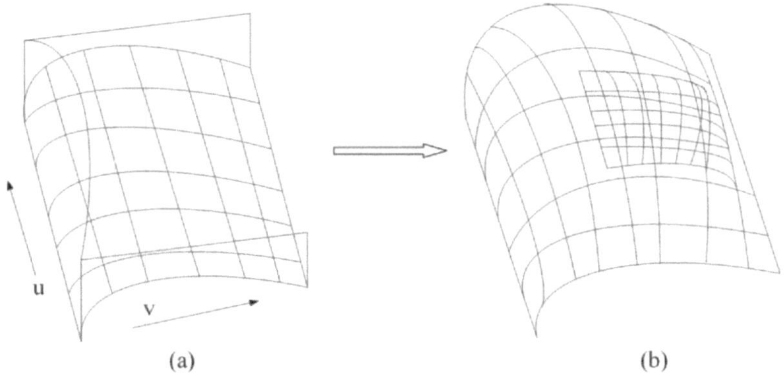

FIGURE 3.5 Generation strategy of a B-spline surface patch used for constraint. (a) B-spline surface and (b) discrete processing.

$$\int\left[\sum_{\substack{i=0,mu1\\j=0,mv1}} P_{i,j}B_{i,su1}(s)B_{j,sv1}(t)-\sum_{\substack{i=0,mu2\\j=0,mv2}} V_{i,j}B_{i,su2}(u(s))B_{j,sv2}(v(t))\right]^2 ds\,dt=0 \tag{3.23}$$

$$\int\left[\mathbf{w}_1(s,t)-\sum V_{i,j}B_{i,su2}(u(s))B_{j,sv2}(v(t))\right]^2 dsdt=0 \tag{3.24}$$

Figure 3.5 shows the surface model generated under the parametric surface patch constraint. Since the discrete points on the planned manipulator trajectory can be generated through CAD/CAM post-processing, local surface areas at the discrete points (considered as control vertices) need to be regenerated and the data of control vertices needs to be extracted when choosing B-spline surface patch as constraint. Then through the discretization processing under the parametric curve constraint, the discrete points on the trajectory are increased and the high-frequency ultrasonic trigger signal is matched, so as to achieve the interpolation discretization of the planned manipulator scanning trajectory.

The curve discretization and B-spline surface reconstruction results of the data point cloud generated by manipulator trajectory planning are shown in Figure 3.6b. Compared with the point cloud data obtained in CAD/CAM simulation software as shown in Figure 3.6a, a denser fitting surface composed of discrete point data with smoother transition can be obtained and control vertices can be identified and extracted more easily. Therefore, the discretization method is suitable for processing the data of manipulator scanning trajectory.

3.3 SURFACE SCANNING TRAJECTORY ALGORITHM BASED ON CAD/CAM

Usually, the tested components are curved structures in different shapes. Especially for an asymmetric complex component (such as blade), an accurate geometric trajectory is hard to be obtained from its surface by scanning or approximating a fixed spline curve. Since most

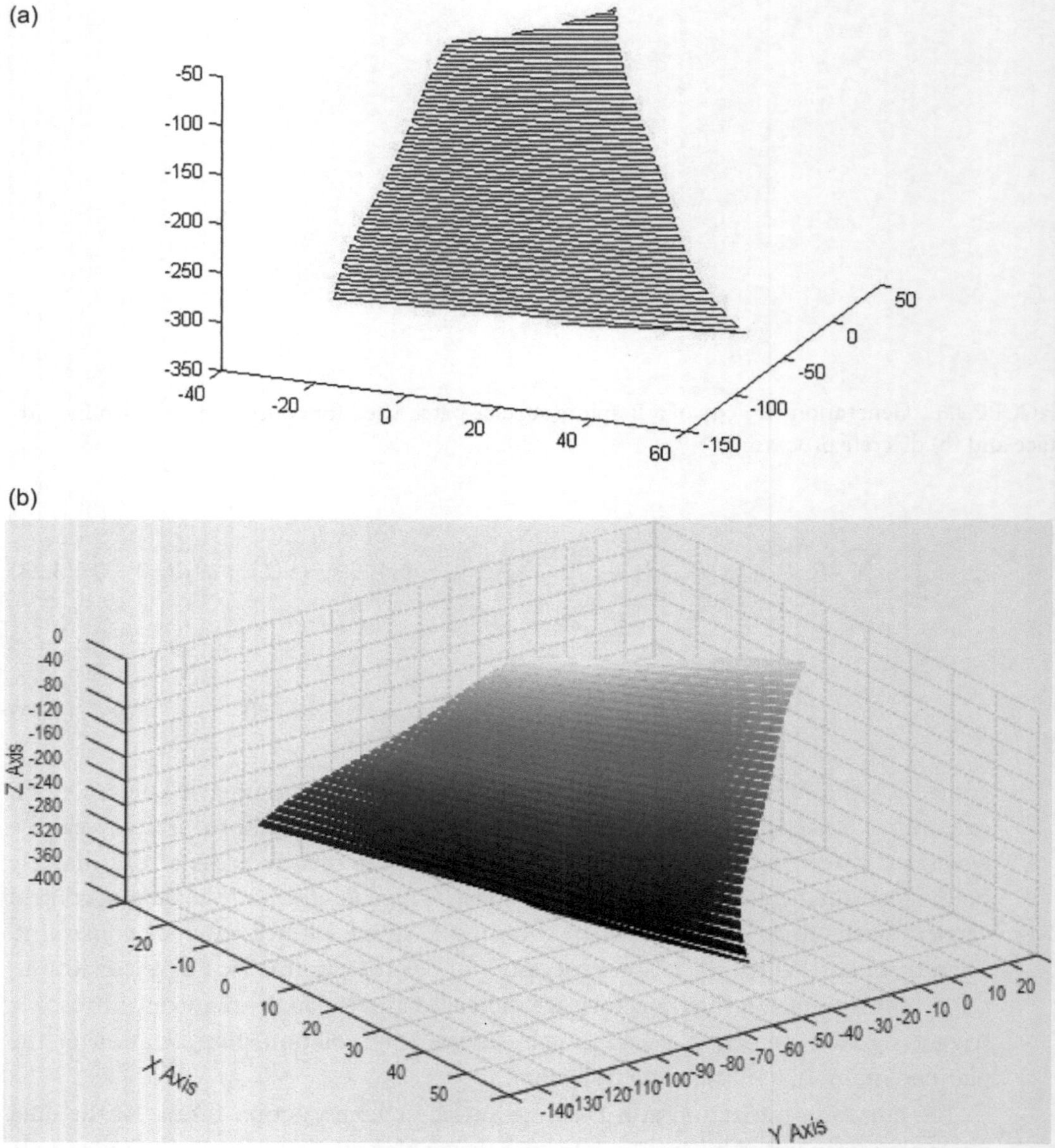

FIGURE 3.6 Before and after the blade trajectory discretization based on B-spline surface GeneratioStrategy. (a) Manipulator trajectory planned by CAM software before discretization and (b) trajectory after the discretization of B-spline surface data.

of the geometric information of component surface can be obtained from design drawing and mathematical model or the CAD model of the tested component can be established through profiling measurement and surface reconstruction, the data of discrete points on cutting tool path obtained by CAD/CAM is transformed here by coordinate transformation algorithm into the scanning trajectory of the manipulator tool coordinate system (TCP) during testing.

FIGURE 3.7 Tool path in blade machining simulation.

3.3.1 Generation of Discrete Point Data of Free-Form Surface

Take aero-engine turbine blade (see Figure 3.7) as an example. We can use the CAD/CAM simulation technique to obtain the tool path on the machined blade surface, read the spatial position and normal vector of each discrete point relative to the workpiece coordinate system from the post-processing file and build the position and attitude matrix characterizing the normal vector of each discrete point to guide the generation of spatial scanning trajectory of manipulator end-effector.

In MATLAB®, the coordinates (x, y, z) and normal vector $[\psi_x, \psi_y, \psi_z]$ of each discrete point are established. In addition, an auxiliary coordinate system is built to characterize the changes of position and attitude of normal vector relative to the origin of the workpiece coordinate system. Figure 3.8 shows the cloud data of discrete trajectory points obtained from the CAM post-processing file.

To build the position and attitude matrix of discrete points with known normal vector and coordinates, we can first set the displacement vector $[\xi_x, \xi_y, \xi_z]$ of two adjacent discrete points as the tangential direction (the X-axis of the auxiliary coordinate system $\{B\}$) of the

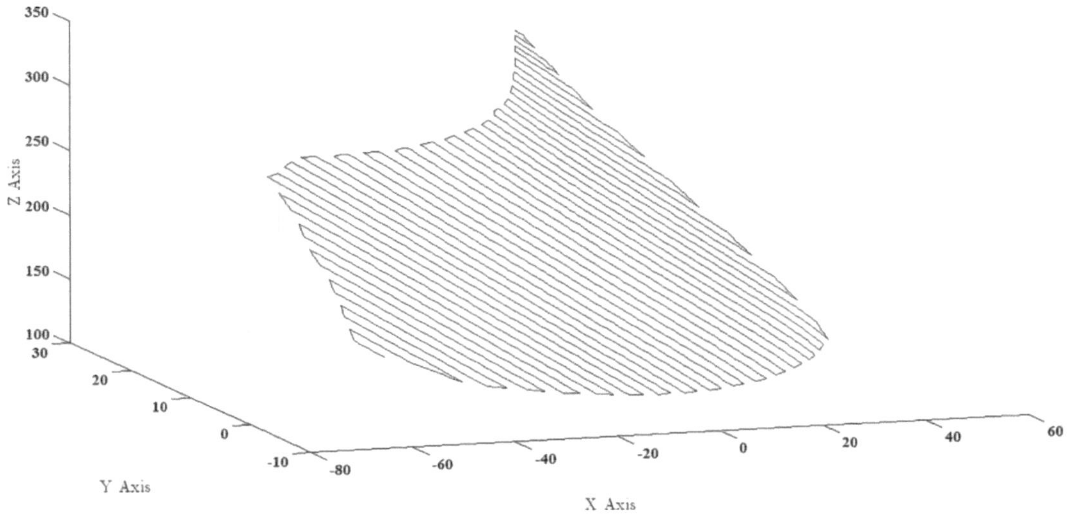

FIGURE 3.8 Discrete point cloud of tool trajectory on the blade surface.

target discrete point, and then calculate the vector $[\varphi_x, \varphi_y, \varphi_z]$ of another axis (the *Y*-axis of the auxiliary coordinate system {*B*}), namely the cross product of normal and tangential vectors.

The discrete point data from the CAD/CAM processing simulation postprocessor is the position/attitude information based on the workpiece coordinate system. The normal vector of the discrete point with respect to the coordinate system {*A*} is expressed by ${}^{A}P$. Similarly, the normal vectors with respect to the coordinate systems {*B*} and {*C*} are expressed by ${}^{B}P$ and ${}^{C}P$, respectively. To meet the requirement of incident angle in the ultrasonic longitudinal-wave reflection method (taking vertical incidence as an example), the normal vector ${}^{B}P$ of discrete point should be consistent with the acoustic beam axis ${}^{C}P$ of the ultrasonic transducer, that is, the position and attitude of the tool coordinate system {*A*} when the coordinate system {*B*} rotates to the position and attitude of the coordinate system {*C*} should be determined. When the auxiliary coordinate system rotates from {*B*} to {*C*}, the following conditions are met:

$$ {}^{A}P = {}^{A}_{B}T\,{}^{B}_{C}T\,{}^{C}P = {}^{A}_{C}T\,{}^{C}P \tag{3.25}$$

$$ {}^{C}_{A}T = \begin{bmatrix} {}^{C}_{A}R & {}^{C}P_{\mathrm{AORG}} \\ 0 & 1 \end{bmatrix} = \left[\begin{array}{c|c} {}^{C}_{B}R^{T}\,{}^{B}_{A}R^{T} & {}^{C}_{B}R\,{}^{B}P_{\mathrm{AORG}} + {}^{C}P_{\mathrm{BORG}} \\ \hline 0\quad 0\quad 0 & 1 \end{array}\right] $$
$$ = \left[\begin{array}{c|c} {}^{C}_{B}R^{T}\,{}^{B}_{A}R^{T} & -{}^{C}_{B}R\,{}^{A}_{B}R^{T}\,{}^{A}P_{\mathrm{BORG}} + {}^{C}P_{\mathrm{BORG}} \\ \hline 0\quad 0\quad 0 & 1 \end{array}\right] \tag{3.26}$$

The position and attitude transformation matrix ${}^{A}_{C}T$ between the coordinate system {*A*} and the coordinate system {*C*} can be decomposed into the solutions to the rotation matrix ${}^{A}_{C}R$ and to the position translation vector ${}^{C}P_{\mathrm{AORG}}$, as shown in Eq. (3.26).

When the data of discrete points on the machined component surface is extracted from CAD/CAM post-processor, the position and attitude of the coordinate system {*B*} relative to the tool coordinate system {*A*} are known (including the position and normal-vector direction cosine of each discrete point relative to the workpiece coordinate system) and are denoted as the transformation matrix ${}^{A}_{B}R$:

$$ {}^{A}_{B}R^{T} = {}^{A}P\,{}^{B}P^{-1} = \begin{bmatrix} \xi_X & \varphi_X & \psi_X \\ \xi_Y & \varphi_Y & \psi_Y \\ \xi_Z & \varphi_Z & \psi_Z \end{bmatrix} \tag{3.27}$$

where

$\xi = [\xi_X, \xi_Y, \xi_Z]^T$ is the description of *X*-axis unit direction vector of the coordinate system {*B*} in the coordinate system {*A*} (consistent with the tangential vector of this point);

$\varphi = [\varphi_X, \varphi_Y, \varphi_Z]^T$ is the description of Y-axis unit direction vector of the coordinate system $\{B\}$ in the coordinate system $\{A\}$ (obtained from the cross product of normal and tangential vectors);

$\psi = [\psi_X, \psi_Y, \psi_Z]^T$ is the description of Z-axis unit direction vector of the coordinate system $\{B\}$ in the coordinate system $\{A\}$ (consistent with the normal vector of this scanning point).

When the acoustic beam axis of the ultrasonic transducer coincides with the normal direction of the planned discrete point on the tested component, that is, the Y-axis of the coordinate system $\{B\}$ coincides completely with the Y-axis of the coordinate system $\{C\}$ (the rotation matrix ${}^B_C R$ is a unit matrix) or they are collinear in reverse along the Y-axis vector direction (the ultrasonic beam axis is also perpendicular to the tested surface), the rotation matrix ${}^B_C R$ can be solved:

$$ {}^B_C R \, {}^C P = {}^B P \tag{3.28} $$

When the coordinate system $\{B\}$ coincides with the coordinate system $\{C\}$, ${}^B_C R$ can be defined as a unit matrix:

$$ {}^B_C R = \begin{bmatrix} 1 & 0 & 0 \\ 0 & 1 & 0 \\ 0 & 0 & 1 \end{bmatrix} \tag{3.29} $$

In this case, ${}^C_A T$ can be calculated:

$$ {}^C_A T = \left[\begin{array}{ccc|c} & {}^B_A R^T & & -{}^B_A R \, {}^A P_{\text{BORG}} + {}^C P_{\text{BORG}} \\ \hline 0 & 0 & 0 & 1 \end{array} \right] \tag{3.30} $$

The motion of each joint of a six-degree of freedom (DOF) manipulator follows the rotation order of X–Y–Z Euler angle, as shown in Figure 3.9. That is to say, the coordinate system $\{X_A, Y_A, Z_A\}$ rotates around the X-axis, Y-axis and Z-axis in the last rotation attitude by the angles α, β and γ, respectively, to obtain the final coordinate system $\{X_B''', Y_B''', Z_B'''\}$ attitude. Therefore, the attitude transformation information of each trajectory point can be obtained only by solving the Euler angle between trajectory points. In the rotation process of the coordinate system, the transformation matrix required for rotating the position and attitude of the coordinate system $\{A\}$ to those of the coordinate system $\{C\}$ is shown in Eq. (3.31), where cosα and sinα are abbreviated to $c\alpha$ and $s\alpha$ to help observe and calculate the matrix parameters.

$$R_{X-Y-Z} = R_X(\alpha)R_Y(\beta)R_Z(\gamma) = \begin{bmatrix} 1 & 0 & 0 \\ 0 & \cos\alpha & -\sin\alpha \\ 0 & \sin\alpha & \cos\alpha \end{bmatrix}$$

$$\times \begin{bmatrix} \cos\beta & 0 & \sin\beta \\ 0 & 1 & 0 \\ -\sin\beta & 0 & \cos\beta \end{bmatrix} \begin{bmatrix} \cos\gamma & -\sin\gamma & 0 \\ \sin\gamma & \cos\gamma & 0 \\ 0 & 0 & 1 \end{bmatrix}$$

$$= \begin{bmatrix} c\beta c\gamma & -c\beta s\gamma & s\beta \\ s\alpha s\beta c\gamma + c\alpha s\gamma & -s\alpha s\beta s\gamma + c\alpha c\gamma & -s\alpha c\beta \\ -c\alpha s\beta c\gamma + s\alpha s\gamma & c\alpha s\beta s\gamma + s\alpha c\gamma & c\alpha c\beta \end{bmatrix} \tag{3.31}$$

By substituting the rotation matrix ${}_A^B R^T$ in Eq. (3.30) into Eq. (3.31), the Euler angle parameters $[\alpha, \beta, \gamma]$ meeting the testing requirements and the demand of the manipulator tool coordinate system $\{A\}$ in rotation can be solved. Considering that Euler angle parameters and position information of each scanning point must be input to control the manipulator end-effector to arrive at the specified position and attitude, the required input parameters are solved below:

$$\beta = a\tan 2\left(-\psi_x, \sqrt{\xi_x^2 + \varphi_x^2}\right)$$

$$\alpha = a\tan 2\left(-\varphi_x/\cos\beta, \xi_x/\cos\beta\right)$$

$$\gamma = a\tan 2\left(-\psi_y/\cos\beta, \psi_z/\cos\beta\right)$$

$${}^C P_{\text{AORG}} = -{}_A^B R \, {}^A P_{\text{BORG}} + {}^C P_{\text{BORG}}$$

where ${}^C P_{\text{AORG}}$ is displacement translation vector, namely the displacement required to move the coordinate system $\{A\}$ to the ultrasonic transducer $\{C\}$ through translation. In the actual testing process, an ultrasonic water-coupled longitudinal-wave focused transducer, rather than a contact probe, is usually adopted in consideration of manipulator trajectory calibration, error and other factors. Therefore, the constant distance d (sound propagation length) should be kept from any discrete point planned on the component surface under test to the center of piezoelectric wafer in the ultrasonic transducer (i.e. the origin of the user coordinate system). In this case, the motion parameters input to manipulator controller should be $[{}^C P_{\text{AORGX}} - d_x, {}^C P_{\text{AORGY}} - d_y, {}^C P_{\text{AORGZ}} - d_z, \alpha, \beta, \gamma]$.

3.3.2 Coordinate Transformation under the Constraint of Ultrasonic Testing (UT) Principle

In the course of manipulator trajectory planning, the position and attitude of discrete points should be defined according to the UT constraints (including ultrasonic incident angle and sound path length), and the relevant factors influencing the manipulator position and attitude transformation matrix should be considered in the calculation of trajectory generation. The deflection angle of the ultrasonic transducer beam axis relative to the normal vector of the tested surface is defined as θ^*, which is similar to the front rake affecting the tool path in the process of CAM machining simulation. When the ultrasonic longitudinal-wave reflection method is used during testing, θ^* should be selected according to the requirement for ultrasonic incident angle, the defect type and the detection requirements.

When the defect to be detected is located inside the tested component, the ultrasonic beam axis is required to be perpendicular to the tested surface, that is, $\theta^*=0$, in order to obtain the maximum reflection echo energy. As required by the position and attitude constraints of the coordinate system transformation matrix, when the ultrasonic wave is vertically incident, the normal vectors of the discrete-point coordinate system $\{B\}$ and user coordinate system $\{C\}$ should be parallel or opposite, as shown in Eq. (3.29) for a unit matrix or Eq. (3.32) for a positive definite matrix rotating 180° around the Z-axis.

$$ {}_{C}^{B}R=\begin{bmatrix} 1 & 0 & 0 \\ 0 & -1 & 0 \\ 0 & 0 & -1 \end{bmatrix} \quad \text{or} \quad {}_{C}^{B}R=\begin{bmatrix} -1 & 0 & 0 \\ 0 & 1 & 0 \\ 0 & 0 & -1 \end{bmatrix} \tag{3.32} $$

In all these cases, the position and attitude constraints for ultrasonic beam axis can be satisfied. However, considering the limitation on the initial position and attitude and working space of the manipulator-end tool coordinate system, the rotation matrix with the minimum transformation amplitude relative to the last position and attitude of the manipulator tool coordinate system is usually selected in accordance with the trajectory optimization principle of "shortest travel" and the spatial position of the user coordinate system.

In the actual testing process, considering the possibility of existence of some near-surface defects, the method of ultrasonic longitudinal-wave oblique incidence is sometimes required to acquire the transverse-wave echo signals purely refracted by defects, so as to improve the waveform signal-to-noise ratio and identify the defect information. Suppose the deflections of ultrasonic incident angle along the X-axis, Y-axis and Z-axis of the manipulator tool coordinate system are ξ, φ and ψ, respectively. As shown in Figure 3.9, the vector $\mathbf{R}$ is obtained after R_0 is deflected by the angles ξ, φ and ψ successively and can be used to find the analytical solution of each deflection angle after orthonormalization. According to the definition of the coordinate system transformation matrix, the rotation matrix of the coordinate system $\{B\}$ relative to $\{C\}$ is:

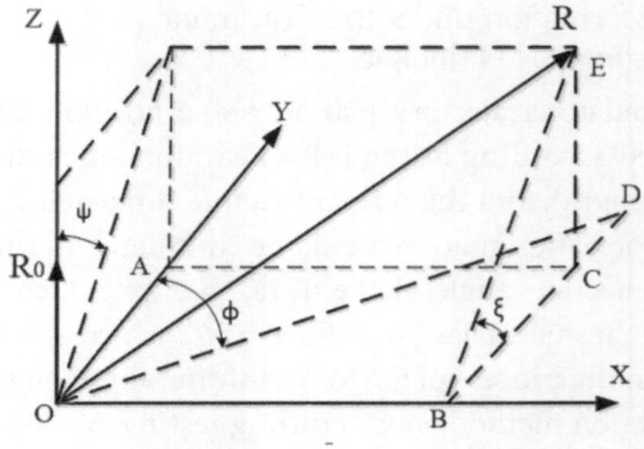

FIGURE 3.9 Deflection of an ultrasonic incident angle.

$$
{}^{B}_{C}R=\begin{bmatrix}1 & 0 & 0\\ 0 & \cos\xi & \sin\xi\\ 0 & -\sin\xi & \cos\xi\end{bmatrix}\begin{bmatrix}\cos\varphi & 0 & -\sin\varphi\\ 0 & 1 & 0\\ \sin\varphi & 0 & \cos\varphi\end{bmatrix}\begin{bmatrix}\cos\psi & \sin\psi & 0\\ -\sin\psi & \cos\psi & 0\\ 0 & 0 & 1\end{bmatrix}
$$

$$
=\begin{bmatrix}c\varphi c\psi & c\varphi s\psi & -s\varphi\\ c\xi s\varphi c\psi - c\xi s\psi & c\xi s\varphi s\psi + c\xi c\psi & s\xi c\varphi\\ c\xi s\varphi c\psi + s\xi s\psi & c\xi s\varphi s\psi - s\xi c\psi & c\xi c\varphi\end{bmatrix} \tag{3.33}
$$

Then it is substituted into Eq. (3.33) to solve the ultrasonic constraints satisfying the above deflection relation. As the ultrasonic oblique incidence is usually required to deflect from the acoustic beam axis by a fixed angle, the deflection angle ξ, φ or ψ should be equal to the angle of incidence depending on the normal vector datum in the actual planning of manipulator trajectory.

Similarly, when the ultrasonic transducers of different frequencies are used for NDT, their sound path length needs to be adjusted according to the focal length and transducer center frequency. If the sound path length of the ultrasonic transducer is d, there should be a fixed distance between the target point B at which the manipulator with a specimen should arrive and the coordinate system $\{C\}$ where the ultrasonic transducer is located during trajectory planning, as shown in Figure 3.10. Then the ultrasound path length

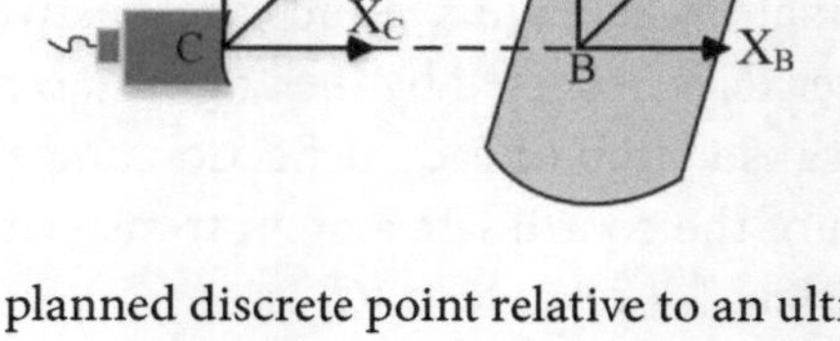

FIGURE 3.10 Position of the planned discrete point relative to an ultrasonic transducer.

$d=[d_x, d_y, d_z]$ with respect to the coordinate system $\{C\}$ is substituted into Eq. (3.34) to obtain the position translation vector under oblique incidence:

$$^{C}P_{\text{AORG}}=-{}^{C}_{B}R\,{}^{B}_{A}R\,{}^{A}P_{\text{BORG}}+{}^{C}P_{\text{BORG}} \tag{3.34}$$

That is, the position information input to the manipulator controller is $\left[{}^{C}P_{\text{AORGx}}+d_x, {}^{C}P_{\text{AORGy}}+d_y, {}^{C}P_{\text{AORGz}}+d_z\right]$, so as to obtain the offset of the coordinate system $\{B\}$ and ensure a fixed ultrasonic path distance between a discrete point on the planned manipulator trajectory and the ultrasonic transducer.

3.4 SCANNING TRAJECTORY SMOOTHNESS JUDGMENT AND DATA DISCRETIZATION PROCESSING

Most of the curved surfaces encountered in surface modeling are quadrangular, but sometimes non-quadrangular surfaces (such as single-sided surface, two-sided surface and N-sided surface) may be seen. In this case, when the curves converge or diverge to the control vertices to form a surface under the control of the weights of data points (control vertices) derived from the above parametric spline curve constraints, they may wrinkle or become sharp. In this paper, wavelet transform and grid energy method are used to smooth the reconstructed surface after trajectory discretization.

3.4.1 Wavelet Processing Method of Surface Data

As the most important curve/surface interpolation method in CAD/CAM domain, wavelet analysis has been widely used and verified in the computer graphics applications, including curve and surface editing, image editing and so on [7–10]. For example, the method of wavelet transform (including wavelet decomposition and wavelet reconstruction) can be used to store a small number of important coefficients of curves and surfaces. While keeping the overall shape of curves and surfaces unchanged, this method can compress the data without losing the detailed features of curves.

In general, approximation error can be compared with a specified smoothing tolerance, and then the resulting percentage can be evaluated to determine whether the wavelet decomposition should continue. For example, suppose $\gamma(t)$ $(0\leq t\leq 1)$ is a B-spline curve defined by m control vertices and node vectors. The quasi-uniform B-spline curve $\gamma_n(t)$ is used to approximate the curve $\gamma(t)$. The set of all the node vectors of $\gamma(t)$ is expressed by U. Then the approximation error ε_n can be defined as:

$$\varepsilon_n=\max\left|\gamma(t)-\gamma_n(t)\right| \tag{3.35}$$

For a given smoothing tolerance $\varepsilon(\varepsilon>\varepsilon_n)$, we assume that $\gamma_j(t)(j\leq n)$ is the approximation curve of $\gamma_n(t)$ and satisfies the following constraints of the approximation error and smoothing tolerance:

$$\begin{cases} \varepsilon_j=\max\limits_{t\in U}\left|\gamma(t)-\gamma_j(t)\right|\leq\varepsilon \\ \varepsilon_{j-1}=\max\limits_{t\in U}\left|\gamma(t)-\gamma_{j-1}(t)\right|\geq\varepsilon \end{cases} \tag{3.36}$$

Then $\gamma_j(t)$ can be decomposed by wavelet into $\gamma_{j-1}(t)$ and the detail curve $\beta_{j-1}(t)$, whose scaling factor is

$$s_j = 1 - \frac{\varepsilon - \varepsilon_j}{\max_{t \in U} \left| \beta_{j-1}(t) \right|} \tag{3.37}$$

As a result, an appropriate scaling factor can be set. By using the low-resolution discrete points obtained from manipulator trajectory planning and the detail curves satisfying the above geometric constraints, the target curves and surfaces can be precisely approximated [11–13]. While keeping the curve details, the wavelet decomposition algorithm reduces the data to be saved. It is suitable for the storage and post-processing of a large amount of testing data.

Similarly, if there is a bicubic interpolation B-spline surface $\mathbf{S}(u, v)$ $(0 \le u, v \le 1)$ with m control vertices in the direction u and n control vertices in the direction v, then the two positive integers $n1$ and $n2$ will satisfy $2^{n1} + 3 \ge m$ and $2^{n2} + 3 \ge n$. The quasi-uniform bicubic B-spline surface $\mathbf{S}_{n1,\, n2}(u, v)$ with $(2^{n1} + 3) \times (2^{n2} + 3)$ control vertices is used to approximate the target surface S(u, v). The set of parameter points in the definition domain $S(u, v)$ is represented by V. Then the approximation error $\varepsilon_{n1,\, n2}$ is

$$\varepsilon_{n_1, n_2} = \max_{(u,v) \in V} \left| \mathbf{S}(u,v) - \mathbf{S}_{n_1, n_2}(u,v) \right| \tag{3.38}$$

According to the wavelet decomposition theory, $\mathbf{S}_{j1,\, j2}(u, v)$ can be decomposed into the low-resolution approximation surface $\mathbf{S}_{j1-1,\, j2-1}(u, v)$ and the detail surface $T_{j1-1,\, j2-1}(u, v)$. When the approximation error and the smoothing tolerance satisfy Eq. (2.27), the scaling factor $\mathbf{S}_{j1,\, j2}$ of the detailed surface can be solved:

$$\begin{cases} \varepsilon_{j_1, j_2} = \max_{(u,v) \in V} \left| S(u,v) - S_{j_1, j_2}(u,v) \right| \le \varepsilon \\ \varepsilon_{j_1 - 1, j_2 - 1} = \max_{(u,v) \in V} \left| S(u,v) - S_{j_1 - 1, j_2 - 1}(u,v) \right| \ge \varepsilon \end{cases} \tag{3.39}$$

$$s_{j_1, j_2} = 1 - \frac{\varepsilon - \varepsilon_{j_1, j_2}}{\varepsilon_{j_1 - 1, j_2 - 1} - \varepsilon_{j_1, j_2}} = \frac{\varepsilon_{j_1 - 1, j_2 - 1} - \varepsilon}{\varepsilon_{j_1 - 1, j_2 - 1} - \varepsilon_{j_1, j_2}} \tag{3.40}$$

When wavelet is used to decompose smooth curves and surfaces, only the low-frequency part and the scaling factor of detail surface are kept to reduce the discrete points and compress the stored data without changing the original surface shape. In addition, this algorithm is easy to implement. It can efficiently smooth the curves and surfaces composed of a large number of control vertices and quickly complete the offline data compression and discretization of the planned manipulator trajectory to improve the data retrieval efficiency.

3.4.2 Handling and Judgment of Surface Smoothness

The surface reconstructed after being split in several parts and rejoined may not be as smooth as the original one at the transitions, so the smoothness of the reconstructed surface needs to be judged.

1. Skin surface smoothing based on the grid energy method
   Some complex-shaped specimens may be the combined surfaces composed of multiple B-spline surfaces. In this case, the wavelet decomposition algorithm cannot satisfy the boundary constraints of individual surfaces at the same time [14–17]. In this paper, the grid energy method is used to smooth the surface, that is, to shift from the judgment of surface boundary smoothness to the research on the optimal solution of curve grids (to ensure that the overall surface energy is the minimum under the constraints of curves and surfaces) and the optimization of several clusters of iso-parametric grid lines on the surface in order to ensure the surface smoothness.
   The boundary constraint on each control vertex of the planned trajectory can be expressed as the following optimal objective function:

$$\begin{cases} \min E(V) \\ D(V) = \sum_{i=1}^{n}\sum_{j=1}^{m}\left(V_{ij} - V_{ij}^{0}\right)^2 \le \varepsilon \end{cases} \tag{3.41}$$

where V is the planned discrete point on the target surface, $E(V)$ is the objective function (energy function), V_{ij}^0 is the control vertex located on the original surface, V_{ij} is the control vertex to be solved, $D(V)$ is the square of the coordinate deviation of each discrete point on the smoothed surface relative to the original planned surface and ε is the preset tolerance.

If a local surface is simplified as an elastic sheet, its strain energy can be solved:

$$E = c\iint\left\{\left[\frac{\partial^2 F}{\partial x^2} + \frac{\partial^2 F}{\partial y^2}\right]^2 - 2(1-v)\left[\frac{\partial^2 F}{\partial x^2}\frac{\partial^2 F}{\partial y^2} - \left(\frac{\partial^2 F}{\partial x \partial y}\right)^2\right]\right\}dxdy \tag{3.42}$$

Considering that the curvature of a surface changes in different dimensions (only two dimensions are considered below), Eq. (3.42) is simplified as

$$E = \iint\left[\left(\frac{\partial k_1}{\partial s_1}\right)^2 + \left(\frac{\partial k_2}{\partial s_2}\right)^2\right]dS \tag{3.43}$$

where $\partial k_1/\partial s_1$ represents the differential of arc length in the direction of principal curvature. Similarly, the rate of change in the other curvature direction can be defined as $\partial k_2/\partial s_2$. According to the above principle, the grid energy method is needed when the target surface is applied in multiple curvature directions, that is, the energy sum of all grid line is used to approximate the surface energy in order to determine the minimum surface energy that can be achieved under the specified constraints.

Similarly, suppose that each control vertex on the smoothed surface **S**(u, v) is defined as V_{ij} and that the tested component has the given bicubic B-spline surface:

$$\mathbf{S}_0(u,v)=\sum_{i=0}^{n+2}\sum_{j=0}^{m+2}V_{ij}^0 B_{ip}(u)B_{jq}(v) \tag{3.44}$$

where p and q are the powers of the surface in the directions u and v, respectively, $B_{ip}(u)$ and $B_{jq}(v)$ are the B-spline basis functions defined by the two node vectors, respectively, and $V_{ij}{}^0(i=0, 1, \ldots, n+2; j=0, 1, \ldots, m+2)$ are the control vertices of the surface before smoothing. Then the curve energy of **S**(u, v) in the two-dimensional directions is defined as

$$\begin{cases} E_s^u=\displaystyle\sum_{i=3}^{n+3}\int k\big(\mathbf{S}(u_i,v)\big)^2\left\|\mathbf{S}_v(u_i,v)\right\|dv \\ E_s^v=\displaystyle\sum_{j=3}^{m+3}\int k\big(\mathbf{S}(u,v_j)\big)^2\left\|\mathbf{S}_u(u,v_j)\right\|du \end{cases} \tag{3.45}$$

where $k\big(S(u_i,v)\big)$ and $k\big(S(u,v_j)\big)$ represent the curvatures of the isoparametric line **S**(u_i, v) and **S**(u, v_j), respectively, and $\|\cdot\|$ is the modulus length of the vector. According to the definition of grid line energy, Eq. (3.45) can be simplified as

$$\begin{cases} E_s^u=\displaystyle\sum_{i=3}^{n+3}\int_0^1\mathbf{S}_{vv}(u_i,v)^2\,dv \\ E_s^v=\displaystyle\sum_{j=3}^{m+3}\int_0^1\mathbf{S}_{uu}(u,v_j)^2\,du \end{cases} \tag{3.46}$$

Then the total energy of grid lines is

$$E_s=E_s^u+E_s^v=\sum_{i=3}^{n+3}\int_0^1\mathbf{S}_{vv}(u_i,v)^2\,dv+\sum_{j=3}^{m+3}\int_0^1\mathbf{S}_{uu}(u,v_j)^2\,du \tag{3.47}$$

To minimize the surface grid energy E_s under the constraint (Eq. 3.37), the weight factor ω is introduced to simplify it into an optimization problem considering the constraint. The corresponding objective function $F(V)$ consists of two parts: (1) grid smoothing term, namely energy function $E(V)$; (2) approximation term: the product of the variance $D(V)$ of the position deviation of each discrete point on the smoothed surface relative to the original surface and the introduced weight factor ω.

$$D=\sum_{i=0}^{n+2}\sum_{j=0}^{m+2}\left(V_{ij}-V_{ij}^{0}\right)^{2}\leq\varepsilon \tag{3.48}$$

$$\begin{aligned}F(V)&=E(V)+\omega D(V)\\&=\sum_{i=3}^{n+3}\int_{0}^{1}\mathbf{S}_{vv}(u_i,v)^{2}dv+\sum_{j=3}^{m+3}\int_{0}^{1}\mathbf{S}_{uu}(u,v_j)^{2}du+\sum_{i=0}^{n+2}\sum_{j=0}^{m+2}\left(V_{ij}-V_{ij}^{0}\right)^{2}\end{aligned} \tag{3.49}$$

By solving the second-order partial derivatives of the parameters u and v in Eq. (3.44), we can obtain

$$\begin{cases}\mathbf{S}_{uu}=\sum_{i=0}^{n}\sum_{j=0}^{m}B_{ip}''(u)B_{jq}(v)V_{ij}\\ \mathbf{S}_{uv}=\sum_{i=0}^{n}\sum_{j=0}^{m}B_{ip}'(u)B_{jq}'(v)V_{ij}\\ \mathbf{S}_{vv}=\sum_{i=0}^{n}\sum_{j=0}^{m}B_{ip}(u)B_{jq}''(v)V_{ij}\end{cases} \tag{3.50}$$

Then $D(V)$ is expanded into a binomial

$$\begin{aligned}D&=\sum_{i=0}^{n+2}\sum_{j=0}^{m+2}\left(V_{ij}-V_{ij}^{0}\right)^{2}\\&=\sum_{i=0}^{n}\sum_{j=0}^{m}\left(V_{ij}^{2}-2V_{ij}V_{ij}^{0}+\left(V_{ij}^{0}\right)^{2}\right)\\&=\sum_{i=0}^{n}\sum_{j=0}^{m}V_{ij}^{2}-2\sum_{i=0}^{n}\sum_{j=0}^{m}V_{ij}V_{ij}^{0}+\sum_{i=0}^{n}\sum_{j=0}^{m}\left(V_{ij}^{0}\right)^{2}\end{aligned} \tag{3.51}$$

By substituting Eq. (3.50) into Eq. (3.49), the energy function $E(V)$ can be obtained:

$$E(V)=\sum_{i=0}^{n}\sum_{j=0}^{n}\sum_{k=0}^{m}\sum_{l=0}^{m}\left(V_{ij},V_{kl}\right)C(i,j,k,l) \tag{3.52}$$

where

$$\begin{aligned}C(i,j,k,l)=&\int B_{ip}(u)B_{kp}(u)du\int B_{jq}''(v)B_{lq}''(v)dv\\&+\int B_{ip}''(u)B_{kp}''(u)du\int B_{jq}(v)B_{lq}(v)dv\end{aligned}$$

If $V_{(m+1)i+j} = V_{ij}$, then

$$E(V)=\sum_{i=0}^{n}\sum_{j=0}^{n}\sum_{k=0}^{m}\sum_{l=0}^{m}\left(V_{(m+1)i+j}, V_{(m+1)k+l}\right)C(i,j,k,l) \tag{3.53}$$

If $V=(V_0, \ldots, V_{(m+1)n+m})^T$, then the above equation can be expressed as

$$E(V)=V^TAV \tag{3.54}$$

where the element of the matrix $A=\left(a_{ij}\right)$ is $a_{(m+1)i+j,(m+1)k+l}=C(i,j,k,l)$. Therefore, A is a symmetric matrix.

If $c=\sum_{i=0}^{n}\sum_{j=0}^{m}\left(V_{ij}^0\right)^2$, Eq. (3.47) can be simplified as:

$$D(V)=V^TV-2V^TV^0+c \tag{3.55}$$

According to Eqs. (3.53)–(3.55), we can obtain

$$\mathrm{F}(V)=E(V)+\omega D(V)=V^T(A+\omega I)V-2\omega V^TV^0+\omega c \tag{3.56}$$

Let $B=A+\omega I$ and $C=\omega V^0$. Because I is the unit matrix, the above equation can be simplified as:

$$F(V)=V^TBV-2V^TC+\omega c \tag{3.57}$$

Given that B is a positive definite symmetric matrix, it can be seen from the above equation that the minimum value of the objective function $F(V)$ can be obtained if and only if $BV=C$. In this case, the objective function of the optimization problem to be solved by the grid energy method can be obtained. Based on its analytic solution, the control vertices of the smoothed surface $S(u, v)$ can be determined.

The smoothing result of discretized trajectory surface (see Figure 3.5b) obtained with the grid energy method shows that the curvature changes uniformly at the transition points of the surface, and that the curves are connected stably and smoothly, as shown in Figure 3.11. After solving the objective function $F(V)$, we find that each control vertex has no step change because the total energy of surface grids is the minimum. This ensures the second-order geometric continuity of surface tangent direction and curvature vector as well as lower strain energy after smoothing. As a result, the discretized trajectory point cloud surface approaches to the original model surface.

2. Surface smoothness check
   For the surface treated by the grid energy method, smoothness check and evaluation is still needed. The smoothness evaluation is mainly based on four methods: the curvature method, contour line method, lighting model method and surface

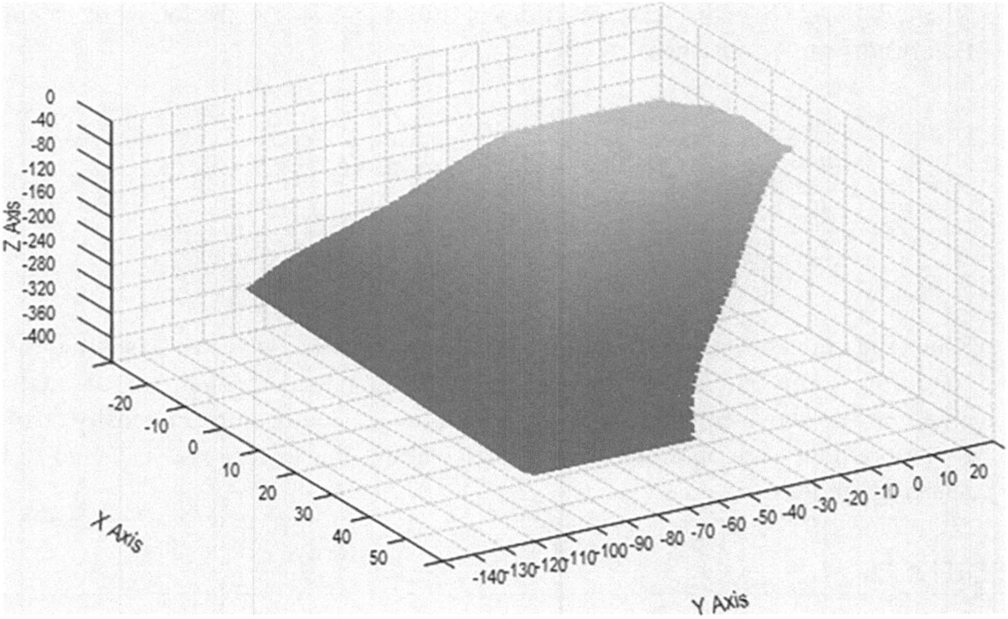

FIGURE 3.11 Smoothing effect of the blade surface.

transformation method [18]. Among them, the curvature method needs to display a large amount of data, including the focusing surface and the parametric curve representing the curvature change of a section line. In addition, the calculation of small change of surface shape is more complicated. In the lighting model method, a simulative light source is needed, so the offline processing of the manipulator trajectory planned in this paper is hard to implement. The method of auxiliary surface is usually efficient in detecting the areas with zero Gaussian curvature but is inefficient in judging other information. The contour line method can display the information on extreme points, which is helpful to judge the fluctuation of control vertices (data points) on the surface [19,20]. Therefore, this method is selected below to evaluate the surface smoothness.

Suppose the contour line equation is $\mathbf{p}(s)=\mathbf{S}(u(s),v(s))$ (s is arc length, $s>0$) and the unit normal vector of its section is $n(u,v)$. If $\varphi(u,v)=(n(u,v),S(u,v))$, then the contour line satisfies the equation $\varphi(u,v)=c$, where c is a constant. Through its differential calculation, we obtain

$$\frac{\partial\varphi}{\partial u}du+\frac{\partial\varphi}{\partial v}dv=0 \tag{3.58}$$

According to the first basic form of differential geometry

$$ds^2=Edu^2+2Fdudv+Gdv^2 \tag{3.59}$$

where E, F and G are the basic quantities of the surface $S(u, v)$, the following calculation results can be obtained:

$$\begin{aligned} \frac{du}{ds} &= \frac{du}{\sqrt{Edu^2 + 2Fdudv + Gdv^2}} = \pm \frac{\varphi_v}{\sqrt{E\varphi_v^2 - 2F\varphi_u\varphi_v + G\varphi_u^2}} \\ \frac{dv}{ds} &= \frac{dv}{\sqrt{Edu^2 + 2Fdudv + Gdv^2}} = \pm \frac{\varphi_u}{\sqrt{E\varphi_v^2 - 2F\varphi_u\varphi_v + G\varphi_u^2}} \end{aligned} \tag{3.60}$$

When the initial point $S(u(0), v(0)) = S(u_0, v_0)$ on contour lines is used as initial value condition, a contour line of the surface can be obtained through the calculation based on the above equation. If several contour lines with uniform density change at the boundary are obtained through calculation, the surface is considered to be relatively smooth.

REFERENCES

1. Craig J.J., Yun Chao (translator). *Introduction to Robotics [M]*. Beijing: China Machine Press, 2006.
2. Ding X. *Calibration Algorithm and Experimental Research of Staubli Industrial Robot [D]*. Hangzhou: Zhejiang Sci-Tech University, 2014.
3. Montanini R., Freni F., Rossi G.L., Quantitative evaluation of hidden defects in cast iron components using ultrasound activated lock-in vibrothermography [J], *Review of Scientific instruments*, 2012, 83(9):1–5.
4. Ren B., Liang Z., Kong M., Implementation of planning algorithm of manipulator position and attitude on arc trajectory in space [J], *Journal of Harbin Institute of Technology*, 2012, 44(7):26–31.
5. Li X., Xu Z., Hu H., Zhou X., Research on automatic probe alignment in ultrasonic measurement of curved workpiece [J], *China Mechanical Engineering*, 2008, 19(11):1289–1292.
6. Schmerr L. *Ultrasonic Nondestructive Evaluation Systems Models and Measurements [M]*. Berlin: Springer, 2007.
7. Asmatulu R., Movva V., Evaluation of fiber reinforced aircraft composites via nondestructive testing techniques [J], *International Mechanical Engineering Congress and Exposition*, 2013, 1:257–265.
8. Yang R., He Y., Zhang H., progress and trends in nondestructive testing and evaluation for wind turbine composite blade [J], *Renewable & Sustainable Energy Reviews*, 2016, 60:1225–1250.
9. Khan M.I., Evaluation of non-destructive testing of high strength concrete incorporating supplementary cementitious composites [J], *Resources Conservation and Recycling*, 2012, 61:125–129.
10. Jiang W., Bai F., Research on phased-array ultrasonic testing of aero-engine turbine blades [J], *Journal of Changchun University of Science and Technology*, 2011, 34(4):66–70.
11. Ji J. *Research on Key Technologies of NC Machining of Complex Integral Impellers [D]*. Nanjing: Nanjing University of Aeronautics and Astronautics, 2009.
12. Li Z. *Research and Application of Several Key Technologies in Five-axis NC Machining of Blades [D]*. Wuhan: Huazhong University of Science and Technology, 2007.

13. Shan C. *Research on Cutting Point Trajectory Planning for Spiral Milling of Blades [D]*. Xi'an: Northwestern Polytechnical University, 2004.
14. Mao X. *Research on Spatial Motion Trajectory Control of Pneumatic Manipulator [D]*. Harbin: Harbin Institute of Technology, 2009.
15. Kang Y., Zhang H., Measurement system based on laser line scanning sensor for large free-form surfaces [J], *Transducer and Microsystem Technologies*, 2012, 31(12):79–82.
16. Pelivanov I., Buma T., Xia J., Wei C., Shtokolov A., O'Donnell M., Non destructive evaluation of fiber-reinforced composites with a fast 2D fiber-optic laser-ultrasound scanner [J], *41ST Annual Review of Progress in Quantitative Nondestructive Evaluation*, 2015, 1650(34):43–50.
17. Santospirito S.P., Lopatka R., Cerniglia D., Slyk K., Luo B., Panggabean D., Rudlin J., Defect detection in laser power deposition components by laser thermography and laser ultrasonic inspections [J], *Frontiers in Ultrafast Optics: Biomedical, Scientific, and Industrial Applications XII*, 2013, 8611:3–5.
18. Pelivanov I., Buma T., Xia J., Wei C., O'Donnell M., A new fiber-optic non-contact compact laser-ultrasound scanner for fast non-destructive testing and evaluation of aircraft composites [J], *Journal of Applied Physics*, 2014, 115(11):105–113.
19. Miguel L., Alberto L., Agustin C., Iban A., Non-destructive testing of composite materials by means of active thermography-based tools [J], *Infrared Physics & Technology*, 2015, 71:113–120.
20. Li X., Yang Y., Hu H., Ni P., Yang C., Path calibration method in ultrasonic testing of complex surface components [J], *China Mechanical Engineering*, 2010, 21(15):1775–1779.

CHAPTER 4

Single-Manipulator Testing Technique

THE REFLECTION FROM THE DEFECT INTERFACE during ultrasonic propagation is used for the ultrasonic reflection nondestructive testing (NDT) on one side of a component. Usually, the defects in metal components are interface defects such as cracks, shrinkage cavities, shrinkage porosities and sand holes. There is always a certain inclination angle between the reflection surface of such a defect and the acoustic beam axis of the ultrasonic transducer, which results in the decrease of effective reflection area [1]. To evaluate the size and position of defects in the components more effectively, the motion device of the detection system must have high three-dimensional positioning accuracy and spatial attitude accuracy. Manipulator has been extensively applied in industrial manufacturing, assembly, painting, welding and other industrial fields, and its machining error, assembly error, zero error, mechanical wear and other aspects are the focus of attention [2]. However, the manipulator control techniques used for NDT and measurement (such as position accuracy control, probe attitude control and coordinate origin zeroing) are becoming very important and critical to the overall detection accuracy, efficiency and effect of the NDT system.

Ultrasonic wave is a kind of mechanical wave, usually in the form of the mechanical vibration traveling in medium, which is generated by the piezoelectric wafer of an ultrasonic probe under the excitation of electric pulse. Ultrasonic testing (UT) is a method to detect the defects inside the tested material by using the acoustic properties shown by defects (such as attenuation and directional change) when the ultrasonic wave propagates in the tested material. When there is a layer of air between the probe and the workpiece, the reflectivity of ultrasonic wave is 100%. Even a very thin layer of air can prevent ultrasonic wave from entering the workpiece, so the ultrasonic transducer and the workpiece under test must be coupled with water or other liquid medium. The water-coupling modes mainly include water immersion coupling and water spray coupling. Generally, water-spray UT is used for larger workpieces, and water-immersion UT is used for smaller workpieces. Depending on how the ultrasonic transducer receives ultrasonic signals, the modes of UT

DOI: 10.1201/9781003212232-4

can be divided into ultrasonic transmission testing and ultrasonic reflection testing. By collecting and processing the reflected or transmitted signals, the state of defect in the workpiece can be determined.

4.1 COMPOSITION OF A SINGLE-MANIPULATOR TESTING SYSTEM

To meet the requirement of automatic testing of curved components (especially curved rotary components), a testing system composed of a manipulator with a transducer is proposed. The component under test is placed on a turntable and fastened with a three-jaw chuck. Then the manipulator holding an ultrasonic transducer tracks and scans the curved contour of the component. For a curved rotary component, when the manipulator moves along the component generatrix, the turntable rotates simultaneously to realize spiral scanning so as to further improve the detection efficiency. This robotic UT system with an additional turntable has further expanded the scope of testing objects, which includes not only complex curved components such as blades and boxes but also rotary components such as turbine disks, vehicle hubs and shafts.

4.1.1 Workflow of a Testing System

The workflow of a robotic testing system is shown in Figure 4.1. The scanning trajectory is composed of the pre-planned discrete point clouds covering the surface of the tested component. The excitation waveform of the ultrasonic transducer is triggered synchronously based on the manipulator position information or in the mode of equal time interval, and the ultrasonic A-scan waveform corresponding to each scanning point stored in the ultrasonic acquisition card is displayed on the control interface of upper computer in real time. At the same time, the position information of each scanning point on the trajectory is converted into a pixel in the display area of the specified view. Each pixel is assigned with the corresponding ultrasonic feature information (such as the peak-to-peak value of defect echo wave and the position of time domain) so that the feature information (such as location distribution and dimension parameters) of superficial/internal defects of the tested component can be continuously displayed in real time in the form of ultrasonic B-scan/C-scan image. The testing result of the target component is composed of several groups of ultrasound B-scan/C-scan images, which can provide the cross-section scanning view and testing data of the specified component area so as to accurately evaluate the defects inside the component.

In addition, for a component made of fiber-reinforced composite material, a low-frequency ultrasonic transducer is needed for its transmission testing because of the large ultrasonic attenuation in this medium [3]. In other words, two ultrasonic transducers, one for ultrasonic signal transmission and the other for signal reception, should be installed on both sides of the tested component. Because of the use of double transducers, sometimes two manipulators, each holding an ultrasonic transducer, are needed to complete the UT scanning. In this case, it is necessary to plan the synergic movement of the two manipulators and calibrate their reference/tool center frames (TCPs) [4,5]. In this chapter, the theories and key technologies of "manipulator + transducer" robotic testing are proposed based on the single-manipulator testing system. They also hold true in the industrial applications of double manipulators.

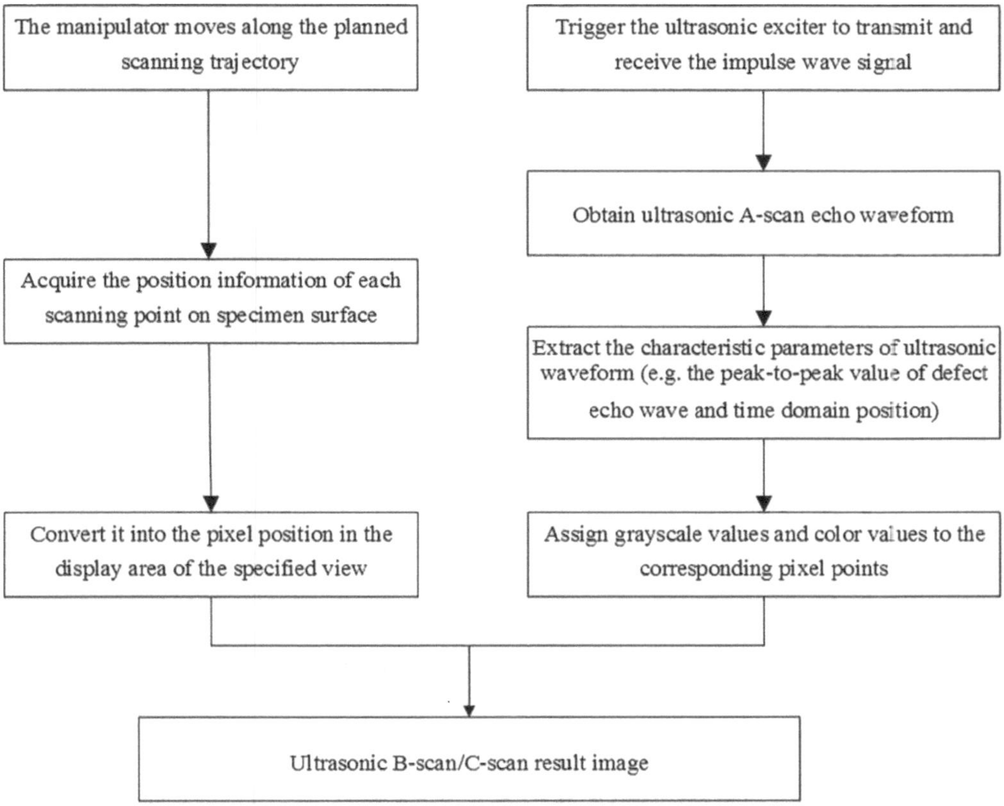

FIGURE 4.1 Workflow of a robotic testing system.

4.1.2 Principle of Equipment Composition

The hardware of the robotic UT system mainly includes a motion control unit (including six-degree of freedom [DOF]-articulated manipulator and its controller), ultrasonic transceiving devices (including pulse transceiver or signal generator, high-speed A/D data acquisition card and ultrasonic transducer), an industrial personal computer (IPC), an ultrasonic transducer/a clamping device and other mechanical support structures, a storage tank as well as other auxiliary equipment. Among them, the motion control unit is the core of the whole robotic testing system and plays a vital role in the realization of complex-component scanning. It can track, cover and scan the profile of the tested curved component along the programmed trajectory to complete the testing task. The ultrasonic transceiving devices are mainly responsible for the excitation, acquisition, transmission and storage of ultrasonic signals. Ultrasonic echo signals can represent the material integrity in the scanned component area. The IPC can realize the control of manipulator motion, the data processing and visualized operation of the collected signals, as well as the storage of the detection result images. The auxiliary tooling is necessary for ensuring the positional precision of the scanning trajectory. The alignment of the position/attitude relationship between the TCP on the manipulator and the user coordinate system where the transducer is located also needs the information on tooling position.

1. Manipulator and its controller

The manipulator in workspace is characterized by flexible attitude, fast movement speed, high response sensitivity and high positioning accuracy. Compared with traditional Cartesian five-axis scanning frame mechanism (or parallel mechanism), the manipulator has a more compact structure and can reach any point position in the workspace. Therefore, it is more suitable for automatic testing of curved components. Especially, the end-effector of six-DOF manipulator has high positional accuracy and its load capacity can meet the testing demand of most of blades, so this manipulator is suitable for holding a blade to complete the spatial scanning motion in NDT. The model of this manipulator and the distribution of its six joints are shown in Figure 4.2.

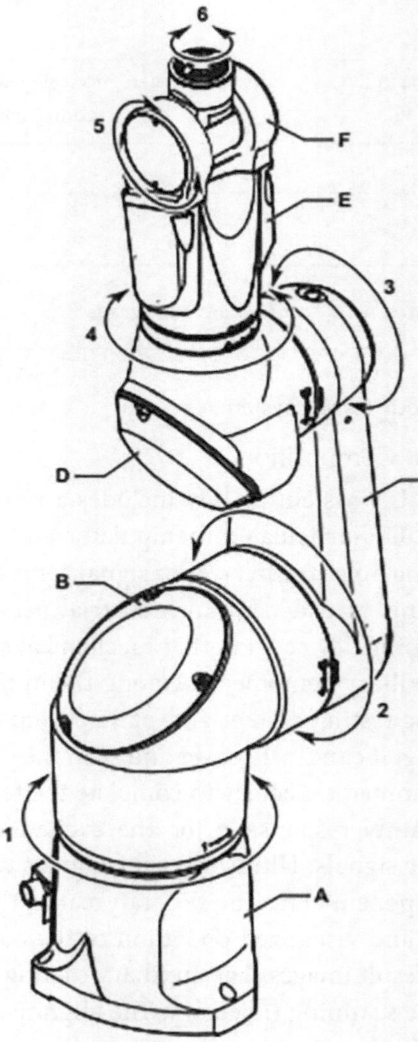

FIGURE 4.2 Manipulator model.

The six joints of manipulator connected in series are all rotary joints, each of which is driven by a servo motor to rotate. The servo motors, reducers and bearings of the manipulator are integrated. This design can save space, simplify the wiring arrangement and improve the waterproofing grade by ensuring that all the joint motors are wrapped in the manipulator without being exposed. In this paper, water is used as coupling agent in the testing process. The waterproofing grade of the manipulator body is IP65 and that of the wrists is IP67, high enough to meet the waterproof requirements of ultrasonic NDT.

2. Synchronous acquisition device for manipulator position/attitude data and ultrasonic signals

   The ultrasonic nondestructive evaluation of complex components (such as blades) is based on ultrasonic A-scan waveform, C-scan image and other testing results. Whether the defect information can be displayed accurately depends on whether the ultrasonic echo signal from the corresponding scanning position can be collected accurately. Especially when the positional accuracy of the scanning trajectory is high enough, the accurate matching between the position/attitude information of the scanning point and the corresponding ultrasonic echo signal is a necessary condition to ensure the reliability of UT results. Therefore, the position information of any trajectory scanning point measured by the robotic testing system and the ultrasonic echo signal (read and stored by the data acquisition card) from the corresponding point should be synchronously collected to correctly describe the defect information in ultrasonic C-scan imaging.

   In the traditional multi-axis-linked scanning device, the upper computer sends a control signal to the servo motor, which drives the ultrasonic transducer on guide rail to the designated position in the Cartesian coordinate system. At this point, the spatial position of the ultrasonic transducer relative to the tested component is known. As the resolution of motor encoder is generally high (131072P/R), the transducer can arrive at the specified position accurately. Moreover, the uniform motor rotation can ensure the same time interval between two adjacent scanning points. Thus, the corresponding ultrasonic signals excited and received at the same repetition frequency can be accurately matched and synchronously collected. This synchronous data acquisition based on time-triggered ultrasonic pulse excitation signal is widely used in the existing multi-axis-linked scanning devices and some robotic testing systems. However, when it is applied to the NDT of small defects in complex components, especially when the scanning trajectory planned for manipulator changes with the curvature of the target surface to cause the non-uniform spatial distribution of discrete points, the consistency of the time intervals to arrive at various points is hard to achieve. Therefore, the mode of synchronous trigger and acquisition with equal time interval is not suitable for the detection of small defects in the curved components under test.

   In this paper, absolute rotary photoelectric encoder is used for each joint of the manipulator. Through Data Systems Inquiry (DSI), the numerical values are transmitted from each encoder to the controller to realize the servo control. When the method of synchronous acquisition based on time-triggered ultrasonic pulse excitation signal is adopted, the update frequency of the real-time manipulator position provided by the programming function (such as VAL3) is too low (usually no less than 4 ms/cycle) to meet the requirement of high-speed high-efficiency NDT. The method to improve the position update frequency is usually to interpolate the collected position feedback information so as to meet the requirement of high-frequency trigger and sampling of ultrasonic acquisition card [6]. However, the interpolation calculation by upper computer programming is delayed relative to the sampling period of ultrasonic acquisition card. Without post-processing, the synchronous acquisition of position information and high-frequency ultrasonic signals will be hard to guarantee. To realize the high-frequency acquisition of manipulator position information, this system adopts an external position acquisition card based on the digital control of field-programmable gate array (FPGA) to read the angle data of each encoder in real time, so as to sample the manipulator position information at the frequency of >10 kHz. Therefore, the system can adapt to the need of high-frequency A/D acquisition card better.

During the manipulator operation, a FPGA decoding card is used to decode the binary data of six joint encoders in the manipulator and calculate the real-time position information of the end-effector through positive kinematics solution. Meanwhile, it judges whether the end-effector displacement meets the condition of triggering an ultrasonic excitation signal in the testing process. If so, the FPGA will send a trigger pulse to the pulse transceiver and upload the real-time manipulator position and attitude via PCI bus to the IPC for storage. Because the CLCOK frequency of an encoder is as high as 1 MHz and basically meets the needs of high-frequency ultrasonic repetitive excitation, the synchronous acquisition and data matching between manipulator position information and ultrasonic signals can be guaranteed in high-speed detection.

The flow chart of synchronous trigger and excitation of ultrasonic signals based on manipulator position information is shown in Figure 4.3. The IPC sends instructions to control the manipulator motion. Meanwhile, the FPGA position acquisition card reads the manipulator position information in real time and completes the pulse excitation of the corresponding ultrasonic signals according to the judgment of end-effector displacement increment, thus realizing the synchronous acquisition of scanning point information and ultrasonic signals. This method can effectively improve the update frequency of scanning-point position information and ultrasonic triggering repetition frequency during testing. Therefore, it can detect the cracks/defects smaller than those detected in the time-triggered mode and reduce the omission ratio of complex curved components. It is especially suitable for the nondestructive testing and evaluation of tiny components.

3. Ultrasonic transceiving devices

The ultrasonic transceiving devices are used to excite and receive the signal back from the tested component and feed the collected ultrasonic A-scan waveform information back to IPC for data storage and visualization. The signal acquisition process is shown in Figure 4.4. In the testing process, the manipulator moves along

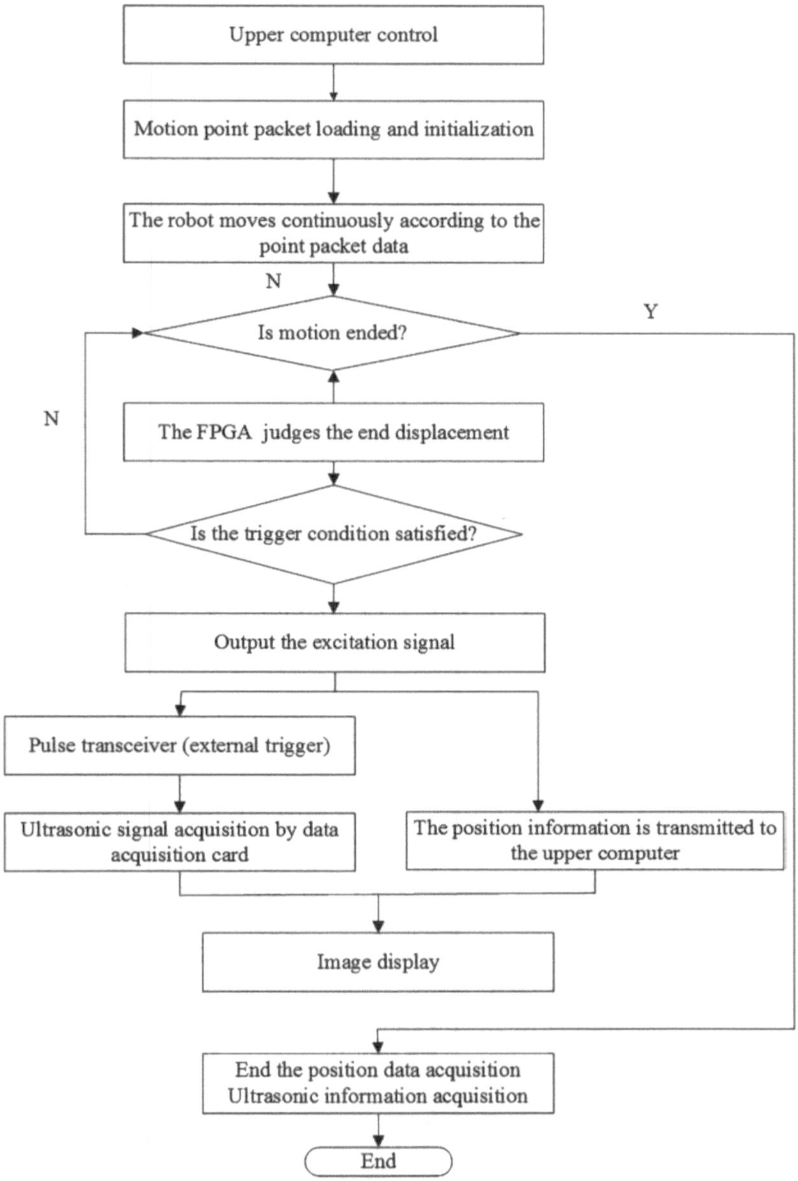

FIGURE 4.3 Flow chart of synchronous data acquisition.

the planned scanning path. The pulse transceiver excites the ultrasonic transducer according to the trigger frequency given by the external FPGA position card, amplifies the ultrasonic echo signal that is fed back and then provides the signal to the data acquisition card. The upper computer software reads the waveform information saved in the card and then stores it in the specified memory address. During ultrasonic B-scan and C-scan imaging, the upper computer reads the corresponding characteristic parameters of time-domain waveform, thus identifying the defect information in the component.

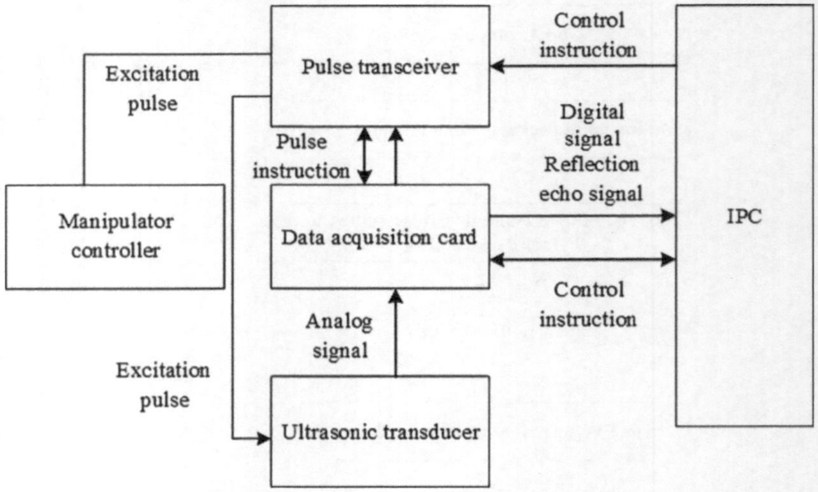

FIGURE 4.4 Signal acquisition and processing of a robotic UT system.

The main ultrasonic transceiver devices used in the system include:

1. Data acquisition card

During testing, the data acquisition card takes the AD samples from the received analog signals and converts them into the digital signals that can be recognized by IPC. To ensure the integrity of ultrasonic signals, the sampling frequency of the acquisition card is generally more than five times the center frequency of the ultrasonic transducer. Therefore, a higher sampling frequency should be set to improve the detection resolution of tiny defects in complex curved components. The center frequency of the ultrasonic transducer chosen for the robotic UT system is usually between 5 and 35 MHz.

The high-speed data acquisition card should meet the sampling demand of a high-frequency ultrasonic transducer and support three modes: software self-trigger, external trigger and encoder trigger. If the mode of external trigger is adopted, the triggering of an ultrasonic signal should depend on whether the displacement of the manipulator end-effector reaches the preset threshold, in order to ensure the synchronous acquisition of scanning point position data and the corresponding ultrasonic information. The acquisition card supports not only the hardware data gate but also the hardware tracking gate. During testing, the tracking gate can track the ultrasonic peak signals within the specified time domain and effectively capture the ultrasonic echo characteristic signals that have a position offset in the time domain due to the manipulator trajectory error. As a result, the detection accuracy is improved. In addition, the acquisition card can support high-speed acquisition. After the trigger waveform signals are saved in the internal storage, the IPC extracts the signals according to the "first-in first-out" stack principle to reuse the internal storage, so as to improve the signal acquisition efficiency without losing ultrasonic waveform signals.

2. Pulse transceiver

In this testing system, the ultrasonic transceiver with the function of single square-wave pulse excitation is chosen. With the voltage gain range of −59 dB ~ 59 dB, this device is suitable for testing the materials with large thickness or high attenuation. The transceiver has a pulse repetition frequency range of 100 Hz–10 kHz. It supports two modes of external trigger, including the single-transducer mode (self-transmission and self-reception) and double-transducer mode (one transducer for ultrasonic excitation or transmission and the other for reception).

During testing, the ultrasonic transducer should be selected according to the thickness and material properties of the workpiece to be tested. For example, to detect the crack/defect with a width of 0.15 mm inside a nickel-based alloy blade, the ultrasonic wave length λ propagating in the blade is required to be not greater than the crack width (usually 1/4–1/2 of the crack width). In addition, the propagation velocity (V) of ultrasonic wave in nickel-based alloy is known to be about 5000 m/s. According to Eq. (2.1), the center frequency of the ultrasonic transducer required to detect a 0.15 mm defect can be calculated:

$$f = \frac{V}{\lambda} = \frac{5000 \times 1000\,\text{m/s}}{0.15\,\text{mm}} = 35\,\text{MHz} \tag{4.1}$$

Accordingly, when the diameter D of the ultrasonic focusing transducer is 16 mm, the underwater acoustic path (within the focal range) can be calculated according to the focal beam diameter d of ultrasonic longitudinal wave (which should not be larger than the equivalent defect diameter):

$$F = \frac{2Rd}{\lambda} = \frac{2 \times 8 \times 0.15}{0.15} = 16\,\text{mm} \tag{4.2}$$

This system adopts the robotic ultrasonic longitudinal-wave reflection/oblique incidence method and selects and uses the liquid-immersion ultrasonic focusing transducer with the center frequency range of 10–35 MHz and the focal length of 0.5–4.5 inches, which can meet the testing needs.

4. Auxiliary tooling

Although the manipulator end-effector has a high repeated positioning accuracy, position calibration is still needed before accurately locating it to a specific discrete point in the specified position and attitude in space. Especially in the NDT of small complex curved components, the position and attitude of the manipulator reference coordinate system relative to the user coordinate system in the target position need to be accurately aligned. Thus, we can obtain the ultrasonic A-scan waveform feedback signal with a stable time-domain waveform and a high signal-to-noise ratio and satisfying the constraints such as ultrasonic incident angle and acoustic path length.

According to the scanning characteristics of the manipulator holding the tested component, ultrasonic A-scan waveform is used in this system to observe in real time the changes of position and attitude of several feature points on the component, in order to determine the relative position and attitude of various coordinate systems and their distribution in the manipulator detection system. In the manual mode of the manipulator, the accurate position and attitude at target point location in the reference coordinate system can be found by adjusting the position and attitude of TCP on the end-effector. Before testing, ultrasonic transceiving devices are connected. By observing the distance between the solid-liquid interface waveform (ultrasonic head wave) on ultrasonic time-domain axis and the excitation source, the accuracy of the position of manipulator trajectory relative to the transducer can be evaluated and the correct position of the target feature point on the tested component can be found. Meanwhile, the amplitude of ultrasonic head wave can characterize whether the attitude of the target point meets the requirement of ultrasonic incident angle. When the ultrasonic wave is vertically incident to the component surface, the acoustic beam of its reflection echo has the strongest energy and the largest amplitude. Therefore, through the real-time observation and adjustment of ultrasonic signals at several feature points before testing, the correct position and attitude of the target points can be found, and the position and attitude of the component coordinate system can be determined. In addition, the coordinate transformation algorithm is used to deduce the position and attitude error of the component coordinate system relative to the planned trajectory coordinate system. After the error compensation, the accuracy of manipulator scanning trajectory can be improved. When the position and attitude of the coordinate system are determined, the position and attitude of each planned point can meet the testing requirements.

To adjust the position and attitude of the component held by the manipulator end-effector, a three-DOF angular displacement table is designed to connect the component and the flange at the end of manipulator. When the target position (transducer coordinate system) is aligned under the manual manipulator mode, the change of the amplitude of ultrasonic head wave crest that characterizes the solid-liquid interface of ultrasonic A-scan waveform can be observed in real time to adjust the angular displacement table and change the position and attitude of the manipulator. Thus, the manipulator holding the component can arrive at the specified trajectory point in the specified position on the ultrasonic beam axis. At this time, the amplitude of ultrasonic echo (including ultrasonic head wave and bottom wave) is the maximum and the sound attenuation is relatively small. The amplitude of ultrasonic sound pressure at this position is the maximum and the signal-to-noise ratio of the echo is relatively strong so that the characteristic information such as defect echo peak and time-domain position can be extracted conveniently.

To meet the needs of automatic testing, automatic clamping devices (such as those for the ultrasonic transducer and specimen) can also be added into the testing system. For the complex components such as aero-engine blades, a set of tools applying to all the tested components is hard to design due to the difference in component curvature. Therefore, this system provides a device connected with both the manipulator end flange and the angular displacement table. In addition, a special fixture can be designed according to the blade rabbet structure to realize the connection and alignment with manipulator. In addition to mechanical joint elements, auxiliary equipment such as water tank, water circulation system and manipulator load-bearing platform should also be built according to the requirements of different detection methods in the experiment.

4.2 PLANNING OF SCANNING TRAJECTORY

Different from the NDT method for traditional regular reflectors, the automatic testing of complex components needs to consider the scattering and absorption characteristics of ultrasonic waves caused by the surface profile with varying thickness and the tiny defects of different types. In addition, it has higher requirements for the center frequency, focal length and acoustic path length (underwater acoustic path), sampling rate, automatic scanning trajectory accuracy and ultrasonic incident angle of the ultrasonic transducer. To detect the superficial and internal defects of complex components, we need to study in depth the ultrasonic propagation path in the blades and the characteristic parameters of sound field, in order to find the NDT method and parameter setting suitable for the high-efficiency high-precision automatic detection and evaluation of small defects/cracks. Thus, the service life and product reliability of manufactured goods can be improved.

4.2.1 Ultrasonic/Electromagnetic Testing Parameters

1. Longitudinal-wave vertical incidence detection of internal defects

   During the blade detection, the analog signal collected by the ultrasonic transducer is converted by the acquisition card into a digital signal, which is then displayed as an A-scan echo signal with time-domain waveform. The principle of characterizing the defects when the ultrasonic wave propagates in the tested component is illustrated in Figure 4.5.

   It can be seen from Figure 4.5 that the traveling ultrasonic wave will be scattered when encountering the interface and that the energy loss of ultrasonic wave depends on the material of the tested component and the size and type of the target

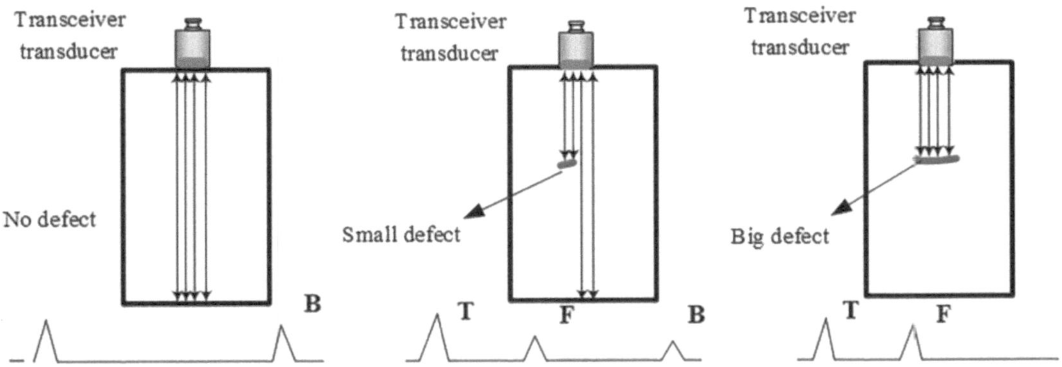

FIGURE 4.5 Ultrasonic A-scan time domain waveforms and defect characterization methods.

defect [7,8]. Since the amplitude of defect echo is significantly lower than that of excitation wave, the defect location in the component (the depth of a defect is determined by its location in the time domain) and the defect size can be determined by tracking the amplitude of defect echo and capturing the signal. In order to obtain a strong echo signal and reduce the energy attenuation caused by scattering, the defects inside a blade should be detected by an ultrasonic reflected longitudinal wave with vertical incidence. The wave equation of this ultrasonic wave can be established based on Eq. (4.3):

$$\begin{cases} \dfrac{\partial^2 \mathbf{u}_x}{\partial x^2} = \dfrac{1}{\mathbf{C}_\mathrm{L}^2} \dfrac{\partial^2 \mathbf{u}_x}{\partial t^2} \\ \dfrac{\partial^2 \mathbf{u}_y}{\partial x^2} = \dfrac{1}{\mathbf{C}_\mathrm{T}^2} \dfrac{\partial^2 \mathbf{u}_y}{\partial t^2} \end{cases} \tag{4.3}$$

The compressed harmonic solution can be obtained:

$$\mathbf{u}_x = A_1 e^{i(kx-\omega t)} + A_2 e^{-i(kx+\omega t)} \tag{4.4}$$

where k is the wave number, and $\mathbf{C}_\mathrm{L}$ and $\mathbf{C}_\mathrm{T}$ are the propagation velocities of ultrasonic wave in liquid and solid, respectively;

$$k = \frac{\omega}{\mathbf{C}_\mathrm{L}} \mathbf{C}_\mathrm{T}^2 = \frac{\mu}{\rho} \mathbf{C}_\mathrm{L}^2 = \frac{\lambda + 2\mu}{\rho} \tag{4.5}$$

Suppose the incident wave is

$$\mathbf{u}_x^{(L)} = I e^{i(k_1 x - \omega t)} \tag{4.6}$$

Then the displacements of the reflection field and the transmission field in the second medium (which is the solid sample here) are

$$\begin{cases} \mathbf{u}_x^{(R)} = A_R e^{-i(k_1 x + \omega t)} \\ \mathbf{u}_x^{(T)} = A_T e^{-i(k_2 x - \omega t)} \end{cases} \tag{4.7}$$

where A_R and A_T are, respectively, the reflection and transmission coefficients to be solved.

According to the boundary constraints of liquid-solid interface, we obtain

$$(\lambda_1 + 2\mu_1) \cdot k_1 [I - A_k] = (\lambda_2 + 2\mu_2) k_2 A_T \tag{4.8}$$

where the wave number satisfies

$$\begin{cases} \lambda_n + 2\mu_n = \rho_n \cdot \left[c_L^{(n)}\right]^2 \\ I + A_R = A_T \\ k_n = \dfrac{\omega}{c_L^{(n)}} (n=1,2) \end{cases} \tag{4.9}$$

The coefficients of reflection field and transmission field of ultrasonic waves can be deduced from the simultaneous Eqs. (4.8) and (4.9). According to the values of the two coefficients, we can determine the position/attitude deflection angle that may be allowed by ultrasonic incident angle when the manipulator moves while holding the component. It is generally believed that the higher the reflection field coefficient is, the more likely the vertical incidence of ultrasonic wave will be.

$$\begin{cases} A_R = \dfrac{\rho_1 \mathbf{c}_L^{(1)} - \rho_2 \mathbf{c}_L^{(2)}}{\rho_1 \mathbf{c}_L^{(1)} + \rho_2 \mathbf{c}_L^{(2)}} . I \\ A_T = \dfrac{2\rho_2 \mathbf{c}_L^{(1)}}{\rho_1 \mathbf{c}_L^{(1)} + \rho_2 \mathbf{c}_L^{(2)}} . I \end{cases} \tag{4.10}$$

where I is the intensity of ultrasonic wave. Therefore, the reflection coefficient is only related to the propagation velocity of ultrasonic wave in the medium and the acoustic impedance of the medium. When the propagation velocities of ultrasonic wave in different liquid and solid media are known, the change of reflection echo amplitude relative to excitation signal amplitude can be determined by solving the above equation, so that the detection sensitivity of the pulse reflection method to the defects inside the tested component can be determined. When the reflection field coefficient is much larger than the transmission field coefficient, the ultrasonic incident angle can be considered meeting the requirement of vertical longitudinal-wave incidence.

2. Oblique-incidence pure-refraction transverse wave detection of surface defects

When an ultrasonic wave is obliquely incident to the interface between two materials, one part of the wave will be refracted according to the Snell's law and experience the wave mode conversion (generating the refracted longitudinal and transverse waves at the interface), and the other part of the wave will be reflected. To detect the defects in the near-surface layer, considering that a stable echo signal is hard to be received in the use of the vertical incidence method due to the scattering phenomenon, the oblique incidence method should be introduced.

According to the principle of wave pattern conversion, when the ultrasonic wave enters the liquid-solid interface at an incident angle greater than the first critical angle, only the pure refracted transverse wave will propagate in the solid, and the longitudinal wave will be converted into interfacial wave under the influence of beam diffusion. Therefore, to detect the blade surface defects, the inclination angle ${}^{A}P_2 = R \cdot {}^{A}P_1$ of incident longitudinal wave should be greater than the first critical angleα_i.

$$\alpha_L \geq \alpha_I = \sin^{-1}\left(\frac{c_L}{c_T}\right) \tag{4.11}$$

As shown in Figure 4.6, α_L and α_{LL} are, respectively, defined as the incident angle and reflection angle of longitudinal wave. β_{LT} and β_{LL} are, respectively, defined as the refraction angles of transmitted transverse wave and transmitted longitudinal wave. C_{LI}, C_{LT} and C_{LL} are, respectively, defined as the propagation velocity in the liquid and the propagation velocities of transmitted transverse wave and transmitted longitudinal wave in the blade.

Then, according to Snell's law, we obtain

$$\frac{\sin\alpha_L}{\sin\beta_{LT}} = \frac{\mathbf{C_{LI}}}{\mathbf{C_{LL}}} \tag{4.12}$$

That is, the refraction angle of transverse wave should satisfy the following constraint:

$$\beta_{LT} \geq \sin^{-1}\left(\frac{\mathbf{c_{LL}}}{\mathbf{c_{LT}}}\right) \tag{4.13}$$

Generally, for the ultrasonic NDT of complex curved components, the echo energy of acoustic line defects needs to be estimated based on the energy reflection/refraction coefficients, which are defined as

$$\text{RLL} = \frac{B_{rl}}{A_{il}} \quad \text{DLL} = \frac{B_{tl}}{A_{il}} \quad \text{DLT} = \frac{B_{ts}}{A_{il}} \tag{4.14}$$

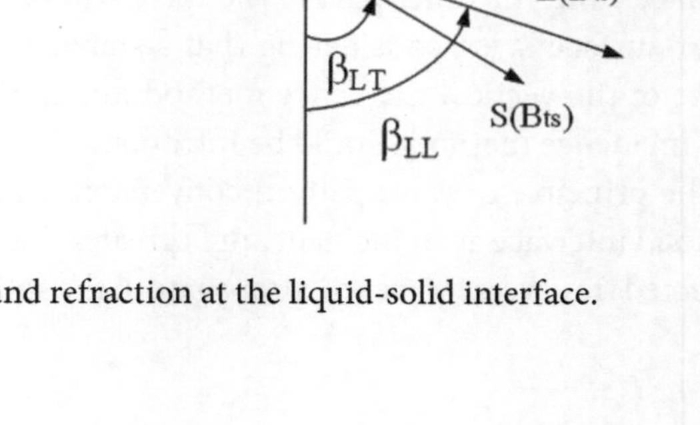

FIGURE 4.6 Reflection and refraction at the liquid-solid interface.

The left-hand side of each equation consists of three letters. The first letter represents the reflected wave *R* or the transmitted wave *D*. The second letter represents the incident transverse wave *T* or the incident longitudinal wave *L*. The third letter represents the transverse wave *T* or longitudinal wave *L* reflected (transmitted) from the solid-liquid interface. On the right-hand side of the equations, A_{iL} is the amplitude of incident longitudinal wave, and B_{mn} is the reflected or transmitted (i.e. $m = r$ or $m = t$) wave (for longitudinal wave, $n = l$; for transverse wave, $n = s$). Then the equation of reflection and refraction coefficients under the liquid-solid boundary condition is

$$\begin{bmatrix} -\cos\alpha_{LL} & -\cos\beta_{LL} & \sin\beta_{LT} \\ 0 & k_{L2}(\lambda_2+2\mu_2)\cos 2\beta_{LT} & -k_{T2}\mu_2\sin 2\beta_{LT} \\ 0 & -k_{L2}\mu_2\sin 2\beta_{LL} & -k_{T2}\mu_2\cos 2\beta_{LT} \end{bmatrix} \begin{bmatrix} \text{RLL} \\ \text{DLL} \\ \text{DLT} \end{bmatrix} = \begin{bmatrix} -\cos\alpha_L \\ k_{L1}\lambda_1\cos 2\alpha_L \\ 0 \end{bmatrix} \quad (4.15)$$

where k_{L1}, k_{L2} and k_{T2} are defined as the wave numbers of the incident longitudinal wave, reflected longitudinal wave and transmitted transverse wave, respectively. By combining this equation with Eq. (4.12), the corresponding relationship among the incident angle of longitudinal wave α_L, the refraction angle of transmitted transverse wave β_{LT} and the refraction angle of transmitted longitudinal wave β_{LL} can be obtained, so as to determine the appropriate incident angle of longitudinal wave according to the defect location.

Taking the conversion of an ultrasonic longitudinal wave incident to blade into a pure refracted transverse wave (see Figure 4.7) as an example, we calculate the ultrasonic incident angle α_L required to obtain the returning pure transverse wave signal after the refraction from a defect under two constraints, namely the given underwater

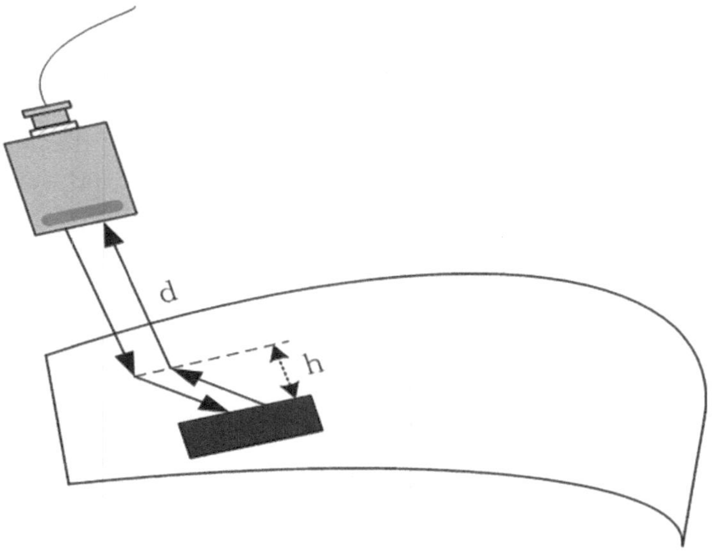

FIGURE 4.7 Calculation of refraction angle of longitudinal wave at liquid-solid interface.

acoustic path length d (determined by the focal length of the transducer) and the depth h from the defect to be detected to the surface.

As can be known from Figure 4.6, there are four boundary constraints on both sides of the liquid-solid interface, including normal displacement, tangential displacement, normal stress and shear stress. According to the strain-displacement correspondence of isotropic media on two-dimensional plane and the Snell's law, the wave mode relation satisfying all boundary constraints can be solved:

$$\begin{cases} A_{il}\cos\alpha_L - B_{rl}\cos\alpha_{LL} + B_{ts}\cos\beta_{LT} + B_{tl}\cos\beta_{LL} = 0 \\ A_{il}\sin\alpha_L - B_{rl}\sin\alpha_{LL} + B_{ts}\sin\beta_{LT} + B_{tl}\sin\beta_{LL} = 0 \\ A_{il}C_{LI}\cos 2\alpha_{LL} - B_{rl}C_{LI}\sin 2\alpha_{LL} - B_{tl}C_{LL}\left(\dfrac{\rho_1}{\rho_2}\right)\cos 2\beta_{LT} - B_{ts}C_{LL}\left(\dfrac{\rho_1}{\rho_2}\right)\sin 2\beta_{LT} = 0 \\ \rho_1 C_{LI}\left[(A_{il} - B_{rl})\sin 2\alpha_{LL}\right] - \rho_2 C_{LT}^2\left[B_{ts}\left(\dfrac{C_{LI}}{C_{LT}}\right)\sin 2\beta_{LT} - B_{tl}\left(\dfrac{C_{LI}}{C_{LL}}\right)\cos 2\beta_{LL}\right] = 0 \end{cases} \tag{4.16}$$

where the normal stress constraints on the two media satisfy equal $\sum\left((\lambda+2\mu)\dfrac{\partial u}{\partial x}+\lambda\dfrac{\partial v}{\partial y}\right)$, namely the following elastic constant identity holds true:

$$\lambda_1 + 2\mu_1\cos^2\alpha_L = (\lambda_1 + 2\mu_1)\cos(2\alpha_{LL}) \tag{4.17}$$

The constraint condition of equal shear stress is that the shear stresses of the two media, namely $\sum\left(\mu\left(\dfrac{\partial v}{\partial x}+\dfrac{\partial u}{\partial y}\right)\right)$ and $\sum\left(\mu\left(\dfrac{\partial w}{\partial x}+\dfrac{\partial u}{\partial z}\right)\right)$, are equal, respectively, where ***u***, ***v*** *and* ***w*** are the displacements of simple harmonic particles in the three-dimensional directions, respectively. The underwater acoustic path d and defect depth h are substituted into the trigonometric functions of the longitudinal-wave incident angle α_L and refraction angle β_{LL}. An appropriate ultrasonic oblique incidence angle α_L under specified constraints can be determined by solving Eq. (4.16).

3. Study on signal attenuation and gain compensation

In the actual detection process, the amplitudes of defect echo and bottom waves are significantly lower than that of head wave due to the influence of component material properties, measurement error and random error on the amplitude of ultrasonic echo. Especially for the components with complex curvature (such as blades), the attenuation of detection signal energy may cause the feature information to be easily overwhelmed by noise signals. To reduce the interference of power-frequency noise signals and improve the signal-to-noise ratio of ultrasonic waveform, it is necessary to compensate the gain of echo amplitude and extract useful signals so as to accurately evaluate the defect features of the tested component.

To reduce the signal energy attenuation, the power-frequency interference signals need to be reduced to improve the amplitude of ultrasonic characteristic signal and the signal-to-noise ratios of defect echo and bottom waves, in addition to ensuring the accuracy of position and attitude of the ultrasonic transducer in the scanning process. The power frequency interference can't be completely avoided, although in most cases the measures such as grounding and shielded wire are taken by the testing system to prevent electromagnetic interference and reduce the ultrasonic noise signals brought by environmental interference. In this case, the echo signal gain should be compensated by adding the algorithm of signal denoising to the software. As mentioned above, the cross-correlation calculation can not only determine the delay time between ultrasonic head wave and bottom wave in the system but also identify periodic power-frequency interference signals and extract the signals mixed in the noise.

Suppose the sampling signal $S(t)$ contains the weak ultrasonic echo signal $x(t)$ to be detected and the power-frequency interference signal $x_i(t)$, which is a periodic function:

$$X_i(t)=C\sin(\omega_i t+\phi) \tag{4.18}$$

According to the trigonometric function formula, the above equation can be expanded into

$$X_i(t)=C\sin\phi\cos\omega_i t+C\cos\phi\sin\omega_i t \tag{4.19}$$

If the sampling time interval is T_0 and the recorded sampling time length $t_p = NT_0$ is an integer multiple of the power frequency period, the sampling time series $S(nT_0)$ can be discretized into

$$\mathbf{S}(nT_0)=x(nT_0)+\hat{A}\cos\hat{\omega}_i nT_0+\hat{B}\sin\hat{\omega}_i nT_0 \tag{4.20}$$

where $\hat{A}$, $\hat{B}$ and $\hat{\omega}_i$ are the estimated values of A, B and ω_i respectively, $A=C\sin\phi$, $B=C\cos\phi$, $n = 0, 1, 2, \ldots, N-1$.

The cross-relation operation between the sampling sequence values $S(n)$ and $\cos\hat{\omega}_i n$ is performed, and the cross-correlation function value $\hat{R}_A(0)$ at $\tau = 0$ (τ is the delay time) is determined:

$$\mathbf{R}_A(\tau)=\lim_{t_p\to\infty}\frac{1}{t_p}\int_0^{t_p}\mathrm{S}(t)\cos\omega_i(t+\tau)dt \tag{4.21}$$

In the collected ultrasonic signal, the discrete time sequence values are $S(t)$ and $\cos\omega_i t$, that is, Eq. (4.21) can be converted into a summation operation:

$$\hat{R}_A(kT_0)=\frac{1}{N}\sum_{n=0}^{N-1}S(nT_0)\cos\hat{\omega}_i(n+k)T_0 \tag{4.22}$$

Eq. (4.19) is substituted into Eq. (4.22) to obtain the value of cross-correlation function $\hat{R}_A(0)$ at $\tau = kT_0 = 0$:

$$\hat{\mathbf{R}}_A(0) = \frac{1}{N}\sum_{n=0}^{N-1}\left[\mathbf{x}(n) + \hat{A}\cos\hat{\omega}_i n + \hat{B}\sin\hat{\omega}_i n\right]\cos\hat{\omega}_i n \tag{4.23}$$

Since $x(n)$ is independent of $\cos\hat{\omega}_i n$ and the trigonometric functions $\sin\hat{\omega}_i n$ and $\cos\hat{\omega}_i n$ are orthogonal, the above equation can be simplified as

$$\hat{\mathbf{R}}_A(0) = \frac{1}{N}\sum_{n=0}^{N-1}\left[\hat{A}\cos^2\hat{\omega}_i n\right] = \frac{\hat{A}}{2N}\sum_{n=0}^{N-1}\left[1 + \cos 2\hat{\omega}_i n\right] \approx \frac{\hat{A}}{2} \tag{4.24}$$

Similarly, the cross-relation operation between the sampling sequence values $\mathbf{S}(n)$ and $\sin\hat{\omega}_i n$ is performed to calculate the cross-correlation function value $\hat{\mathbf{R}}_B(0)$ at $\tau = 0$ and obtain its discrete time expression:

$$\hat{\mathbf{R}}_B(0) = \frac{1}{N}\sum_{n=0}^{N-1} S(n)\sin\hat{\omega}_i n \tag{4.25}$$

so

$$\begin{cases} \hat{A} = 2\hat{R}_A(0) = \dfrac{2}{N}\displaystyle\sum_{n=0}^{N-1}\mathbf{S}(n)\cos\hat{\omega}_i n \\ \hat{B} = 2\hat{R}_B(0) = \dfrac{2}{N}\displaystyle\sum_{n=0}^{N-1}\mathbf{S}(n)\sin\hat{\omega}_i n \end{cases} \tag{4.26}$$

From the above equation, it can be seen that the real-time angular frequency $\hat{\omega}_i$ of power-frequency interference signal is a unique variable. Then, $\hat{\omega}_i$ is searched and iterated. When $\hat{A}$ and $\hat{B}$ are determined, the actual angular frequency of power frequency interference can be determined according to the peak value of the correlation between $\hat{C} = \sqrt{\hat{A}^2 + \hat{B}^2}$ ($\hat{C}$ is the modulus of $\hat{A}$ and $\hat{B}$) and $\hat{\omega}_i$. In this case, the power-frequency interference signal is

$$\mathbf{x}_i(t) = C\sin\phi\cos\omega_i t + C\cos\phi\sin\omega_i t \tag{4.27}$$

Since the discrete time series $\mathbf{x}(n)$ of the useful signal $\mathbf{x}(t)$ to be detected is determined by the difference between the sampling sequence value $\mathbf{S}(n)$ and the sequence value of power-frequency interference signal $\mathbf{x}_i(n)$, the power-frequency interference signal can be removed and the desired ultrasonic signal can be separated and extracted.

$$\mathbf{x}(n) = \mathbf{S}(n) - \mathbf{x}_i(n) \quad n = 0,1,2,\ldots N \tag{4.28}$$

The ultrasonic A-scan waveform signal used in the blade testing is shown in Figure 4.8a. After the model of the ultrasonic transducer and the characteristic parameters of sound field are determined, the noise signals are hard to be eliminated by only adjusting the gain range and input impedance. However, the cross-correlation calculation of the stored full-wave signals (as shown in Figure 4.9b) can reduce the noise signals caused by power frequency interference and facilitate the extraction and analysis of defect echo. In addition, the post-processing of ultrasonic A-scan full-wave waveform signal can increase the peak value of defect echo, facilitate the capture of defect information by use of time-domain gate and improve the resolution of the ultrasonic NDT system in the detection of small defects.

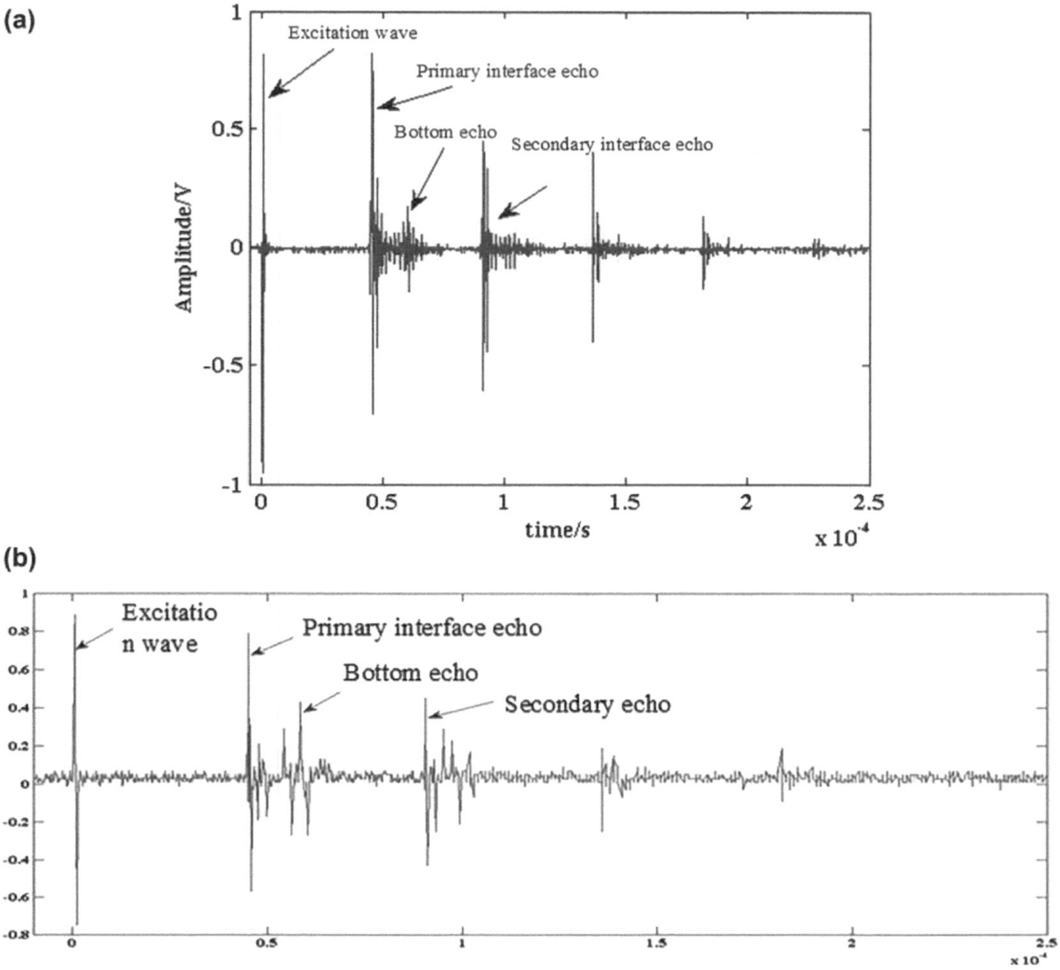

FIGURE 4.8 Denoising of ultrasonic A-scan waveform. (a) Initial ultrasound A-scan waveform and (b) full-wave waveform after denoising.

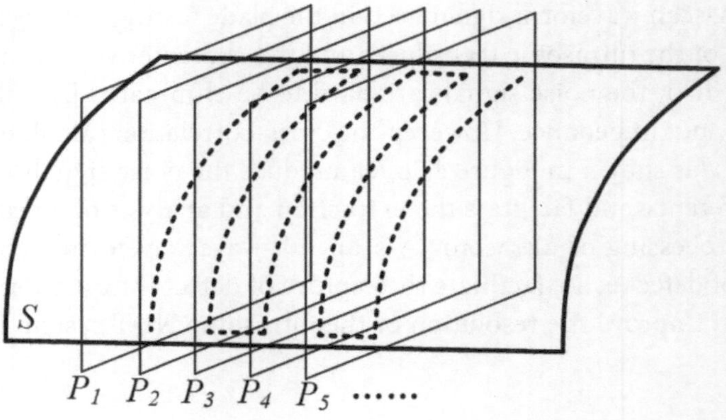

FIGURE 4.9 Isoplanar path planning. The free-form surface S, the parallel planes (P1, P2, P3…).

4.2.2 Trajectory Planning Parameters

Before UT, the surface under test should be discretized into a series of paths. This process is called detection path generation. This process is generally realized through CNC machining simulation software by using the methods such as equal parameter (isoparametric method), equal plane (isoplanar method), constant residual height, spatial filling curve and continuous curve generation. UT is not to remove materials but to ensure that the command points are distributed on the tested surface as evenly as possible, so that the scanning image can reflect actual defect morphology without distortion. The isoplanar method is more in line with this requirement. As shown in Figure 4.9, the process of dividing the free-form surface *S* by a series of parallel planes (*P*1, *P*2, *P*3…) to obtain the intersection curve (i.e. isoplanar path) is just isoplanar path planning. The parameters involved in path planning are mainly step interval and scanning interval:

1. Step interval

   During the CNC machining, the path is planned according to cutting speed, cutting depth, feed rate and other parameters to ensure the ideal machining accuracy. During UT, the path is planned based on the focal length F, focal beam diameter Φ, material sound velocity C_M and underwater sound velocity C_W, in order to obtain more ideal detection sensitivity and resolution. Take the path planning of surface detection based on the focusing probe (shown in Figure 4.10) as an example. The distance between the transducer surface and the incident surface is just the water path h, which can be calculated by

$$h = F - t\left(C_M / C_W\right) \tag{4.29}$$

   where t is the depth of focus. The water path can be calculated with Eq. (4.29) according to the requirement of detection depth, in order to locate the focus at the detection depth. For the path planning of surface detection shown in Figure 4.11, the TCP $\{T\}$

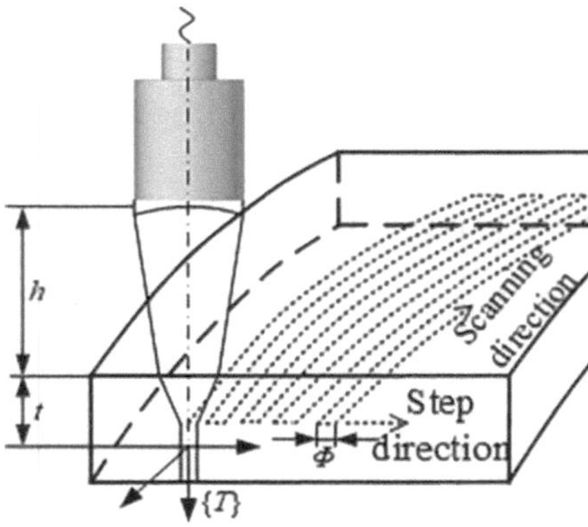

FIGURE 4.10 Surface scanning path of an ultrasonic transducer.

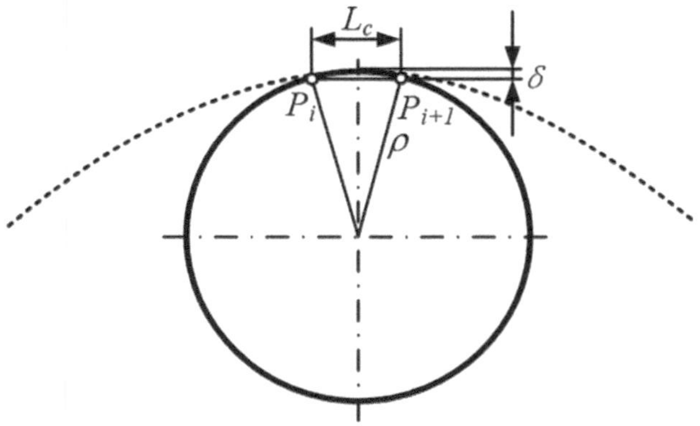

FIGURE 4.11 Linearized approximation of the scanning curve.

is established at the focus, the water path is h, the step interval is approximate to or slightly larger than the focal beam diameter Φ and the incident angle is 0°, in order to better cover the whole detection area.

2. Scanning interval

In Figure 4.11, the dashed line is part of the scanning curve, and the spatial distance between the two adjacent command points P_i and P_{i+1} is the scanning interval. In the detection process, the tracking of the surface profile is realized by using linear interpolation to approximate the scanning curve, so the fitting error of linear interpolation will affect the tracking accuracy of the surface profile. In path planning, the scanning interval must be set reasonably according to the detection requirement of the smallest defect, so as to ensure the detection efficiency and give full play to the detection performance of the system. If the radius of curvature within a scanning interval is ρ and the scanning interval is L_c, a triangular relationship will exist:

$$\rho^2 = \left(\frac{L_c}{2}\right)^2 + (\rho - \delta)^2 \tag{4.30}$$

Then the error of linearized approximation is $\delta = \rho - \sqrt{(L_c/2)^2 - \rho^2}$. The smaller the error is, the higher the tracking accuracy of the surface profile will be.

4.3 CALIBRATION AND ALIGNMENT OF ASSEMBLY ERROR

The trajectory accuracy of the manipulator end-effector is affected by many factors, including the forward and inverse solution calculation errors caused by the measurement errors in the structural parameters and motion variable parameters of each joint, the measurement and alignment errors of the user coordinate system, the positioning errors of the tooling/fixture clamping the component under test, the position/attitude errors of discrete points on the scanning path caused by the geometry information inconsistency between model and real object and the uncertainties brought by measurement equipment (measurement disturbance). This paper mainly studies the influence of the measurement and alignment error of the user coordinate system on the trajectory accuracy. After the compensation and calibration of the user coordinate system, ultrasonic signals can be collected accurately.

4.3.1 Method of Coordinate System Alignment

Before testing, the manipulator should identify the user-defined coordinate system placed in the workspace (this coordinate system is located at the ultrasonic transducer when the manipulator holds the tested component for scanning), so as to accurately identify and reach the position and attitude of the target point on the planned scanning trajectory. However, when the component under test is held by the tooling or fixture in connection with the manipulator end-effector, the complete overlap between the component coordinate system and manipulator TCP (or a relatively clear position/attitude relationship between them) will be hard to achieve in the planned position and attitude due to the existence of clamping error. This is especially true when the surface profile of the tested component is complex and its feature points are difficult to mark. It is common in robotic detection systems that the accuracy of manipulator trajectory is reduced by the alignment deviation of the coordinate system. According to the suggestions of this paper, the characteristic parameters of ultrasonic echo (head wave amplitude, time-domain position, etc.) shall be used to evaluate the alignment accuracy of the manipulator coordinate system. The position/attitude information of the target point shall be read and its difference from the planned value shall be used as the calibration factor. The settings of manipulator TCP shall be corrected by referring to the position/attitude information of the planned points. In this way, the correct spatial position of the user coordinate system can be found, and the position and attitude of feature points can be identified to ensure the scanning trajectory accuracy of the manipulator.

The process of aligning the TCP of the tested component is shown in Figure 4.12. Six feature points are selected according to the CAD mathematical model of the component. The position information of each point $P_0(x_i, y_i, z_i)$ $(i = 1, \ldots, 6)$ as well as the corresponding position/attitude information of TCP at manipulator end $P_1(x_i', y_i', z_i', \alpha_i, \beta_i, \gamma_i)(i = 1, \ldots, 6)$ obtained after matrix transformation are recorded. During testing, the specified position/

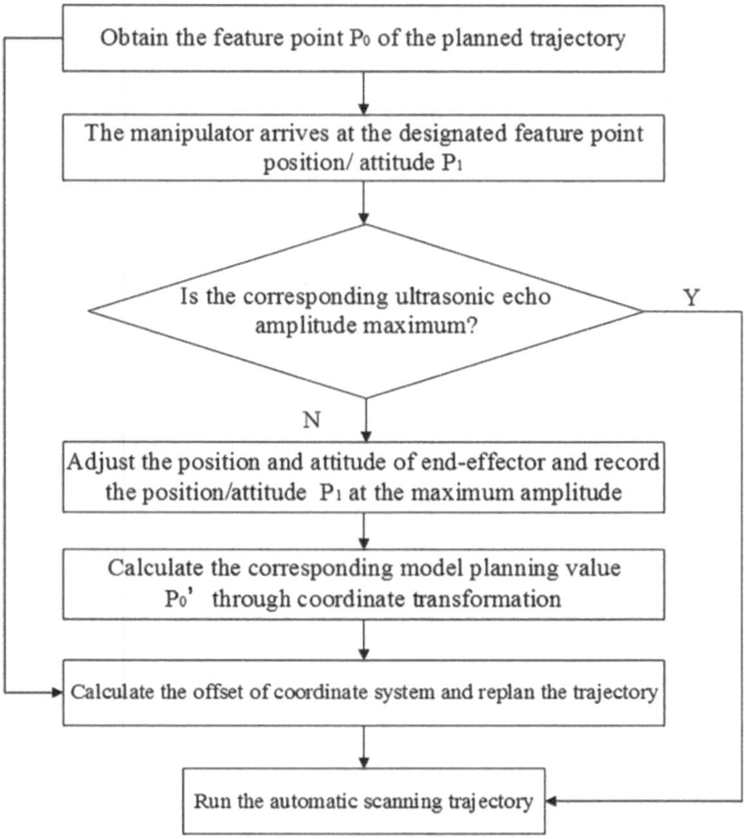

FIGURE 4.12 Alignment process of a manipulator coordinate system.

attitude information P_1 is input with the manipulator in manual mode. When the manipulator arrives at the predetermined position according to the designated position and attitude of the target point, the inspector should observe the amplitude of ultrasonic pulse echo signal and the time-domain position of the first interface echo to judge whether the ultrasonic incident angle and sound field length meet the planning requirements, as shown in Figure 4.13.

According to the principle of UT, when the method of ultrasonic pulse reflection is used for detection, the sound attenuation caused by the scattering of an incident ultrasonic longitudinal wave perpendicular to the tested surface will be minimized and thus the returned ultrasonic energy will be maximized. The position and attitude of the manipulator can be moved or adjusted locally to achieve the relatively largest amplitude of the first interface echo. At this time, the reflected echo energy is the maximum. The current position and attitude information $P_1^{'}(x_i'', y_i'', z_i'', \alpha_i', \beta_i', \gamma'_i)$ $(i=1,...,6)$ of the TCP at the end of the manipulator should be recorded.

From the p_1' information of feature points, the position information of the corresponding discrete points on the tested component is obtained, and the rotation matrix ${}_A^BR^T$ is derived. Then the position/attitude transformation matrix ${}_C^AT$ for changing the user

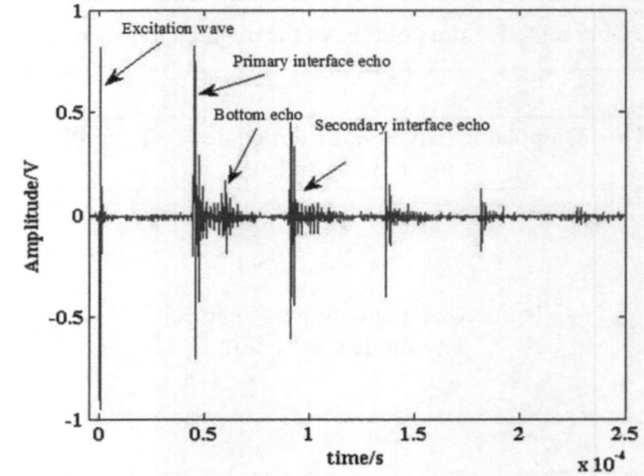

FIGURE 4.13 Waveform of ultrasonic pulse echo in time domain.

coordinate system {C} into TCP {A} should be solved. From the coordinate transformation algorithm, the following equation can be derived:

$$ {}^{A}_{C}T = \left[\begin{array}{ccc|c} \multicolumn{3}{c|}{{}^{A}_{B}R\,{}^{B}_{C}R} & {}^{A}_{B}R\,{}^{B}R_{\mathrm{CORG}} + {}^{A}R_{\mathrm{BORG}} \\ \hline 0 & 0 & 0 & 1 \end{array}\right] \tag{4.31} $$

By substituting the simultaneous Eqs. (4.32) and (4.34) into the above equation, ${}^{A}_{C}T$ can be determined. When the coordinate system {B} coincides with the coordinate system {C}, ${}^{B}P_{\mathrm{CORG}}$ should be a unit matrix.

By substituting Eq. (4.31) into the following equation, we can determine the position/attitude of the discrete point with the maximum pulse interface echo amplitude read by the manipulator in manual mode relative to the workpiece coordinate system {A}, which is denoted as $P_0'\left(\hat{x}_i, \hat{y}_i, \hat{z}_i\right)$.

$$ {}^{C}P = {}^{C}_{A}T\,{}^{A}P \Rightarrow {}^{A}P\,{}^{C}_{A}T^{1}\,{}^{C}P = {}^{A}_{C}T\,{}^{C}P \tag{4.32} $$

By substituting the discrete point position/attitude $p_1'\left(x_i'', y_i'', z_i'', \alpha_i', \beta_i', \gamma_i'\right)$ $(i = 1, 2 \ldots, 6)$ read by the manipulator into the ${}^{C}P$ in Eq. (4.32), the position information ${}^{A}P$ (i.e. $P_0'\left(\hat{x}_i, \hat{y}_i, \hat{z}_i\right)$) of the corresponding discrete point on component surface in the current position and attitude relative to TCP can be obtained, as shown in Eq. (4.33).

$$ \begin{bmatrix} \hat{x}_i \\ \hat{y}_i \\ \hat{z}_i \end{bmatrix} = \left[\begin{array}{ccc|c} \multicolumn{3}{c|}{{}^{A}_{B}R\,{}^{B}_{C}R} & {}^{A}_{B}R\,{}^{B}R_{\mathrm{CORG}} + {}^{A}R_{\mathrm{BORG}} \\ \hline 0 & 0 & 0 & 1 \end{array}\right] \begin{bmatrix} x_i'' \\ y_i'' \\ z_i'' \end{bmatrix} \tag{4.33} $$

By comparing the P_0' information with the $P_0\left(x_i, y_i, z_i\right)$ information, the mean offsets of six discrete points in three-dimensional directions are determined, as shown in Eq. (4.34).

When enough discrete points are collected, the position error of the user coordinate system relative to the manipulator tool/reference coordinate system can be determined accurately.

$$\left(\sum_{i=0}^{j}\left(\frac{\hat{x}_i - x_i}{j}\right), \sum_{i=0}^{j}\left(\frac{\hat{y}_i - y_i}{j}\right), \sum_{i=0}^{j}\left(\frac{\hat{z}_i - z_i}{j}\right)\right)(j=3,\ldots 6) \tag{4.34}$$

4.3.2 Alignment Method Based on Ultrasonic A-Scan Signal

The positions and attitudes of the manipulator TCP and user coordinate system are determined through the magnitude of ultrasonic reflection echo. By comparing the position information obtained during the alignment of the manipulator coordinate system with its planned value (as shown in a section of manipulator scanning trajectory in Figure 4.15a), the accurate position and attitude of each scanning point can be determined. Suppose the offsets to be solved in the three-dimensional directions of the coordinate system are $[\Delta x, \Delta y, \Delta z]$, because the measured values $[x_i, y_i, z_i]$ $(i=1,2\ldots,k)$ of each discrete point are known from the positive kinematics solution of rotational angle of each joint and the planned manipulator trajectory data $[x_i', y_i', z_i']$ can be exported by the CAD/CAM software postprocessor (but not corresponding to the measured values), the planned and measured offsets between the two coordinate systems can be calculated according to the principle of the least square method of data fitting. The approximate function $\varphi(x_i', y_i', z_i')$ shown in Eq. (4.35) is the polynomial fitting objective function reckoning in the offset of the TCP coordinate system at an end-effector caused by the coordinate transformation of the planned trajectory values.

$$F(x,y,z) = \min_{\varphi \in H} \sum_{i=1}^{k}\left[\left(x_i - \varphi(x_i')\right)^2 + \left(y_i - \varphi(y_i')\right)^2 + \left(z_i - \varphi(z_i')\right)^2\right] \tag{4.35}$$

Polynomial fitting is carried out for the discrete points originating from actual measurement in order to obtain the theoretical approximate position that is determined through the calculation of planned discrete points to minimize the objective function and that should be reached by the manipulator:

$$\begin{cases} \varphi(x') = a_0 + a_1 x' + \cdots + a_m x'^m \\ \varphi(y') = a_0 + a_1 y' + \cdots + a_m y'^m \\ \varphi(z') = a_0 + a_1 z' + \cdots + a_m z'^m \end{cases} \tag{4.36}$$

When the distance between the position that should be reached by the theoretically planned trajectory through coordinate transformation and the measured value is the shortest, the polynomial coefficient can be determined by iterative calculation. Then the origins of the measured coordinate system and the planned coordinate system can be identified. After the displacement compensation, the manipulator TCP in almost the correct position and attitude can be fixed so as to eliminate the trajectory error caused by the misalignment of the coordinate system.

In order to simplify unnecessary iterative calculation when using the least square method for alignment, we take a section of the arc trajectory planned for manipulator as an example to calculate the alignment error of TCP. The position distribution of the measured trajectory points after the displacement compensation for the coordinate system offset is shown in Figure 4.14b. Compared with 4.14a, the manipulator trajectory after displacement compensation is closer to the planned value, so the normal vector of the tested component can satisfy the constraints on ultrasonic incident angle in the detection process.

Although the polynomial coefficient in Eq. (4.36) can be obtained through iterative calculation, the solution accuracy and the number of iterations are affected by the power of the polynomial. Therefore, quite a lot of discrete points need to be collected to accurately

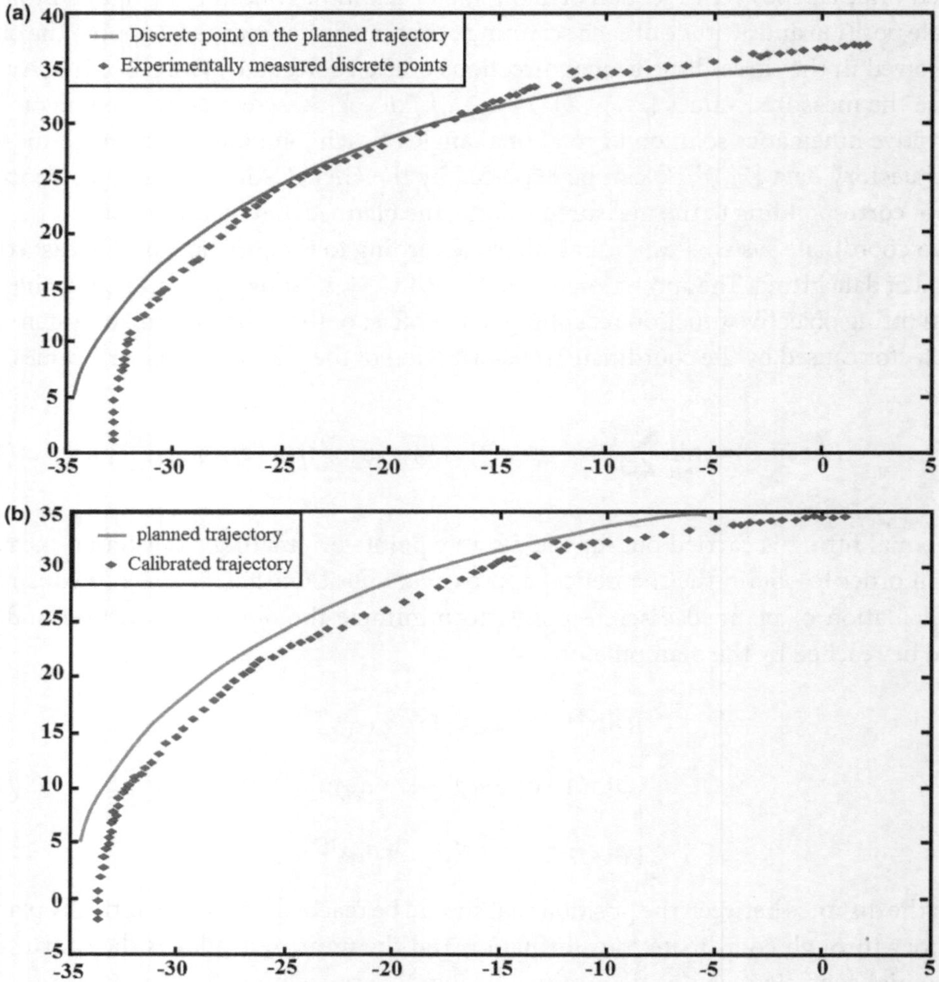

FIGURE 4.14 Distribution of discrete points before and after the error compensation of a manipulator trajectory coordinate system. (a) Measured trajectory points and planned path before the alignment of a coordinate system and (b) measured trajectory points and planned path after the alignment of a coordinate system.

calculate the displacements in three-dimensional directions of the coordinate system. For the coordinate system of a regular sample, its alignment operation is not too complicated. The trajectory distribution of a surface sample after the alignment of its coordinate system is shown in Figure 4.15. The discrete point cloud on the surface highly fits the planned value, thus ensuring that the scanning trajectory and the position/attitude meet the testing requirements. On the other hand, in the practical application to complex components, the optimal value of manipulator TCP obtained through iteration sometimes also needs the fine-tuning of attitude. Under the limitation of ultrasonic incident angle and underwater acoustic path length, the position and attitude of the component coordinate system are adjusted in real time according to the amplitude and time-domain position of ultrasonic pulse echo, so as to obtain a more accurate scanning trajectory. In this case, a movable/rotatable multi-DOF fixture needs to be designed to meet the need of adaptive detection. In this system, a three-DOF angular displacement platform that can rotate along the *rx*, *ry* and *rz* directions is used as the position/attitude adjustment device connecting the component fixture to the manipulator end-actuator. Through its rotation and the translational motion of manipulator end flange (whose position relative to the user coordinate system can be changed by the setting of the coordinate system during trajectory planning), the attitudes in three-dimensional directions can be adjusted.

The alignment of manipulator TCP is an important factor affecting the positioning accuracy of the measured trajectory. The alignment of the user coordinate system relative to manipulator TCP should be completed before the start of formal flaw detection, and the setting of the user coordinate system should be modified during the planning of component trajectory.

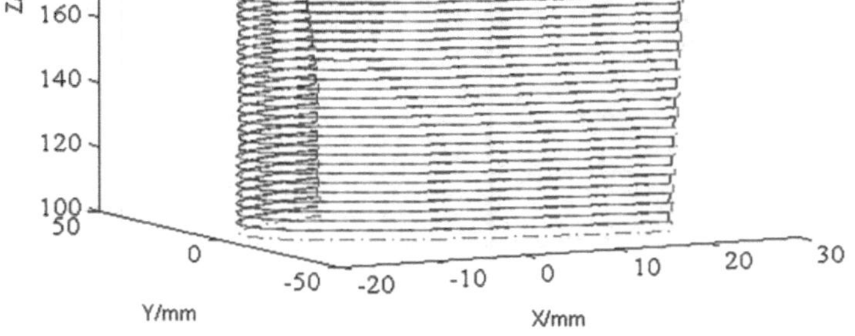

FIGURE 4.15 Manipulator trajectory distribution after the error compensation of tool center frame of 3D sample.

4.3.3 Error Compensation Strategy and Gauss-Seidel Iteration

As shown in Figure 4.16, because the tested component is connected to the flange at the manipulator end-effector through fixture, its locating datum may not coincide completely with the TCP on the flange. In this case, the coordinate system has a positioning error. Suppose the actual coordinate system of a reference workpiece is $\{M\}$. In the detection process, the displacement deviation between the TCP $\{A\}$ obtained by the coordinate system alignment and the reference coordinate system $\{M\}$ can be determined so as to calculate and compensate for the positioning error brought by the fixture.

If the user coordinate system $\{C\}$ where the ultrasonic transducer is located is known relative to the manipulator reference coordinate system $\{W\}$, then:

$$ {}^{W}P = {}^{W}_{C}T\,{}^{C}P = {}^{W}_{A}T\,{}^{A}_{M}T\,{}^{M}_{B}T\,{}^{B}_{C}T\,{}^{C}P = {}^{W}_{A}T\,{}^{A}_{C}T\,{}^{C}P = {}^{W}_{M}T\,{}^{M}_{C}T\,{}^{C}P \tag{4.37} $$

where ${}^{W}_{A}T$ can be derived from the positive kinematic solution of the manipulator, and then ${}^{A}_{C}T$ can be determined. According to the transformation matrix algorithm of the coordinate system, we can obtain:

$$ {}^{W}_{M}T = \begin{pmatrix} {}^{W}_{M}R & {}^{W}P_{\mathrm{MORG}} \\ 0 & 1 \end{pmatrix} = \left[\begin{array}{ccc|c} \multicolumn{3}{c|}{{}^{W}_{C}R\,{}^{B}_{C}R\,{}^{B}_{M}R} & {}^{W}_{C}R\,{}^{C}_{B}R\,{}^{B}P_{\mathrm{MORG}} + {}^{W}P_{\mathrm{BORG}} \\ \hline 0 & 0 & 0 & 1 \end{array}\right] $$

$$ \left[\begin{array}{ccc|c} \multicolumn{3}{c|}{{}^{W}_{C}R\,{}^{C}_{B}R\,{}^{B}_{M}R} & {}^{W}_{C}R\,{}^{C}_{B}R\,{}^{B}P_{\mathrm{MORG}} + {}^{W}_{C}R\,{}^{C}P_{\mathrm{BORG}} + {}^{W}P_{\mathrm{CORG}} \\ \hline 0 & 0 & 0 & 1 \end{array}\right] \tag{4.38} $$

The transformation matrix ${}^{M}_{C}T$ can be obtained according to the feature point information read during the coordinate system alignment. By combing this equation with Eq. (4.37),

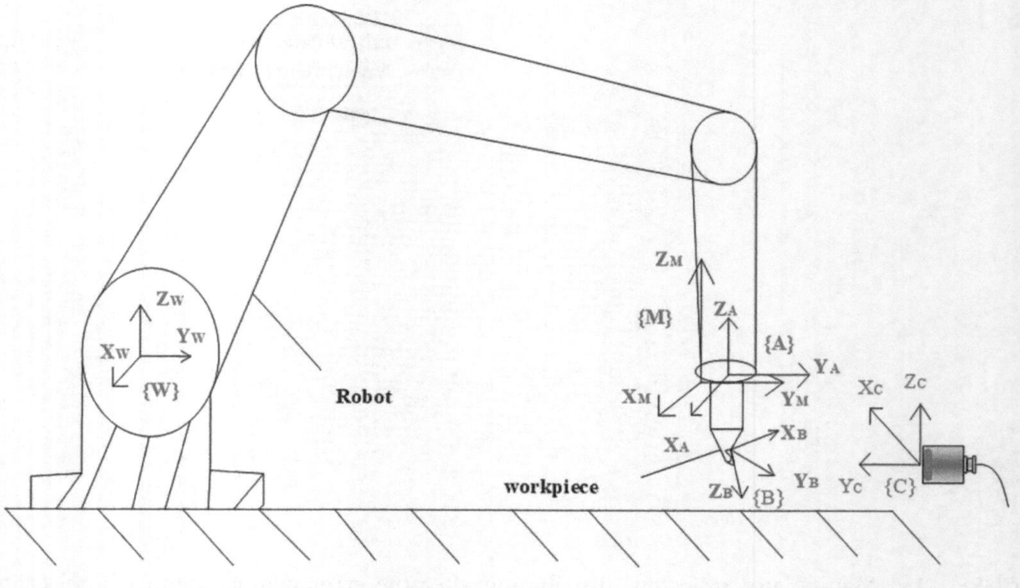

FIGURE 4.16 Distribution of a manipulator coordinate system with clamping error.

the position/attitude transformation matrix ${}^{A}_{M}T$ between the coordinate system of clamped component $\{M\}$ and the TCP $\{A\}$ can be solved:

$$ {}^{A}_{M}T = {}^{A}_{C}T\,{}^{M}_{B}T^{-1}\,{}^{B}_{C}T^{-1} = {}^{W}_{A}T^{-1}\,{}^{W}_{M}T\,{}^{M}_{C}T\,{}^{M}_{B}T^{-1}\,{}^{B}_{C}T^{-1} \tag{4.39} $$

where ${}^{M}_{B}T$ is the rotation matrix of the discrete point trajectory planned for manipulator and is equivalent to ${}^{A}_{B}R$. By combing Eq. (4.39) with Eq. (4.38), the Euler angle $[\alpha^{*}, \beta^{*}, \gamma^{*}]$ and displacement vector $[\Delta x, \Delta y, \Delta z]$ necessary for the TCP $\{A\}$ at the manipulator end-effector to rotate from a certain position/attitude to the actual component coordinate system $\{M\}$ can be calculated:

Considering that the measurement error usually found in the feature point information read under the manual mode of manipulator (especially the 1°–2° error in the deflection angle characterizing the position and attitude) can't be determined by observing the amplitude, time-domain location and other information of ultrasonic reflection echo, the position of the workpiece coordinate system which is close to the real value and takes the clamping and positioning error into account should be obtained through numerical iteration. Suppose the discrete point $P_0'(\hat{x}_i, \hat{y}_i, \hat{z}_i)$ on specimen surface corresponding to the maximum echo amplitude is the initial point and the search step along the direction of $[\Delta x, \Delta y, \Delta z]$ displacement vector is $\tau = [\Delta x/k, \Delta y/k, \Delta z/k](k=100)$. According to the Gauss-Seidel iteration algorithm, we can obtain:

$$ \left(x^{(k+1)}, y^{(k+1)}, z^{(k+1)}\right) = M\left(x^{(k)}, y^{(k)}, z^{(k)}\right) + g(k=0,1,2,\ldots) \tag{4.40} $$

where M is an iteration matrix, $M = (D-L)^{-1}U$ and g is the additional term $(D-L)^{-1}b$.

$D = \mathrm{diag}(a_{11}, a_{22}, \ldots, a_{nn})$. U and L are upper and lower triangular matrices, respectively. Then

$$ (x_1, y_1, z_1) = \left(\left(b_1 - \sum_{j=2}^{n} a_{1j} x_j^{(0)}\right)\Big/ a_{11}, \left(b_1 - \sum_{j=2}^{n} a_{1j} y_j^{(0)}\right)\Big/ a_{11}, \left(b_1 - \sum_{j=2}^{n} a_{ij} z_j^{(0)}\right)\Big/ a_{11}\right) $$

$$ (x_1, y_1, z_1) = \left(\left(b_i - \sum_{j=1}^{i-1} a_{1j} x_j - \sum_{j=i+1}^{n} a_{ij} x_j^{(0)}\right)\Big/ a_{ii}, \left(b_i - \sum_{j=1}^{i-1} a_{ij} y_j - \sum_{j=i+1}^{n} a_{ij} y_j^{(0)}\right)\Big/ a_{ii},\right. $$

$$ \left.\left(b_i - \sum_{j=1}^{i-1} a_{ij} z_j - \sum_{j=i+1}^{n} a_{ij} z_j^{(0)}\right)\Big/ a_{ii}\right) $$

$$ (x_n, y_n, z_n) = \left(\left(b_n - \sum_{j=1}^{n-1} a_{ij} x_j\right)\Big/ a_{nn}, \left(b_n - \sum_{j=1}^{n-1} a_{nj} y_j\right)\Big/ a_{nn}, \left(b_n - \sum_{j=1}^{n-1} a_{nj} z_j\right)\Big/ a_{nn}\right) $$

After the numerical iteration points obtained according to the above equation are successively converted into the positions and attitudes of the manipulator end-effector, the ultrasonic path length and interface echo peak amplitude corresponding to each discrete point on the workpiece surface are observed under the manual mode of the manipulator. The discrete point closest to the planned acoustic path length and with the maximum echo amplitude is selected as the optimal solution, and the corresponding coordinate system is the actual workpiece coordinate system $\{M\}$ in the clamped state. Then the positioning error caused by the fixture can be compensated for through the position and displacement vector $[\Delta x, \Delta y, \Delta z]$ of this coordinate system.

4.3.4 Positioning Error Compensation

When the alignment error of the manipulator coordinate system is excluded, the inconsistency between the installation position/attitude of the tested component and the planned position/attitude due to the existence of clamping error may lead to the position and attitude error of a discrete point on the measured manipulator scanning trajectory relative to the planned value. In particular, attitude deviation can lead to the change of ultrasonic incident angle at each sampling point and then affect the energy and amplitude of the received ultrasonic reflection echo. In severe cases, the stable pulse echo may not be obtained, which directly affects the reliability and resolution of the detection result.

To ensure that the waveform characteristic parameters obtained from each sampling point during ultrasonic C-scan imaging accurately correspond to each discrete point on the workpiece under test, the angular value of each joint read from the manipulator encoder and the spatial position and attitude of TCP obtained from the positive kinematic solution should be converted into those of the workpiece coordinate system that have taken the clamping error into account. If the angular displacement platform connecting the tested component and the manipulator end flange rotates by $[\alpha, \beta, \gamma]$ along the rx, ry and rz directions, the position/attitude of TCP, denoted as $E(V) = \sum_{i=0}^{n}\sum_{j=0}^{n}\sum_{k=0}^{m}\sum_{l=0}^{m}\left(V_{(m+1)i+j}, V_{(m+1)k+l}\right)C(i,j,k,l)$, can be converted into that of the workpiece coordinate system (${}^{M}P$) offset by the clamping displacement $[\Delta x, \Delta y, \Delta z]$ that can be measured:

$$
{}^{M}P = {}^{M}_{T}R^{T}P = \begin{pmatrix} c\beta c\gamma & -c\beta s\gamma & s\beta \\ s\alpha s\beta c\gamma + c\alpha s\gamma & -s\alpha s\beta s\gamma + c\alpha c\gamma & -s\alpha c\beta \\ -c\alpha s\beta c\gamma + s\alpha s\gamma & c\alpha s\beta s\gamma + s\alpha c\gamma & c\alpha c\beta \end{pmatrix} {}^{T}P \tag{4.41}
$$

The position and attitude of TCP can be derived from the discrete point information of the planned scanning path and from the positive solution of manipulator kinematics and then converted into the position and attitude of trajectory points in the workpiece coordinate system with clamping offset (displacement) through the above coordinate transformation. The obtained discrete point data can be compared with the planned value more easily to verify the offset between the clamped workpiece and the manipulator end-effector.

Figure 4.17 shows a certain trajectory in the scanning process of manipulator TCP and its position in the workpiece coordinate system when it is rotated to the clamping position through coordinate transformation (it is rotated by –90° along the z-axis direction and moved by 110 mm in translation during actual clamping). After determining this positioning error, the position and attitude of discrete points in the manipulator TCP can be converted into those of scanning points characterized by the workpiece coordinate system, so as to make necessary preparation for the definition of scanning display area during

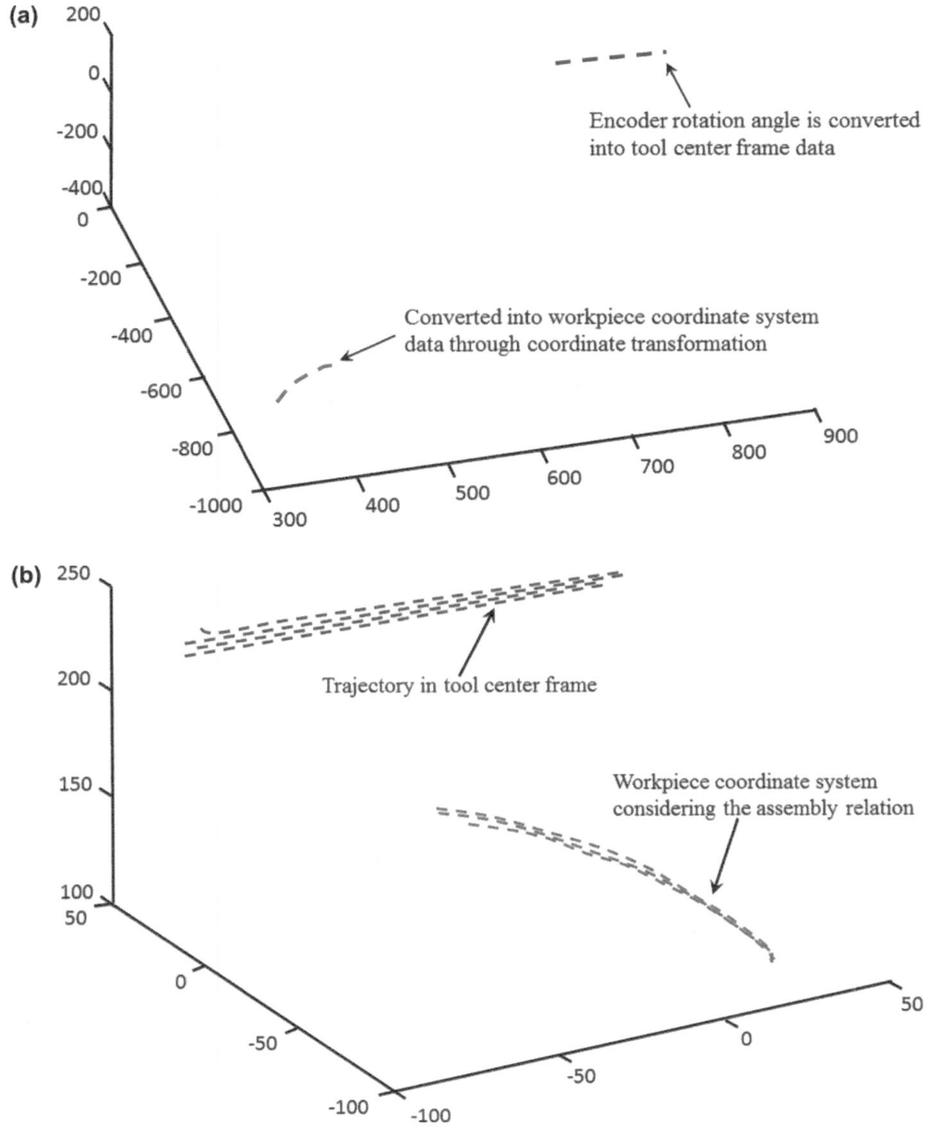

FIGURE 4.17 Trajectory in the manipulator tool center frame and the workpiece coordinate system considering the clamping relation. (a) Position distribution of the tool center frame and initial workpiece coordinate system and (b) position distribution of the tool center frame and the transformed workpiece coordinate system.

ultrasonic C-scan imaging. If and only if the position/attitude data updated by the manipulator in real time is converted into the spatial position measured based on the workpiece coordinate system, the inspector can correctly input the relevant display parameters of the resulting detection images and change the observation angle more intuitively.

4.4 MANIPULATOR POSITION/ATTITUDE CONTROL AND COMPENSATION

In addition to the above alignment error and clamping error of the manipulator coordinate system, the change of manipulator structure parameters will also affect the position and attitude of manipulator end joint. For example, to expand the range of motion at the end of the manipulator, a joint extension shaft connected to the end flange is sometimes added to meet the needs of industrial applications. In this case, the transformation matrix parameters of positive and inverse solutions of manipulator end position and attitude will be changed, so the errors in structural parameters may be brought to reduce the trajectory positioning accuracy. In addition, in order to meet the needs of high-frequency UT, the encoder in this system collects the rotation angle value of each joint to determine the position and attitude of TCP at the manipulator end. The collected numerical value information needs to be converted into joint rotation angles through the binary system. The numerical truncation error introduced in this process will also affect the solution accuracy of positive solution of end position and attitude. In the following section, we will analyze the position/attitude transformation matrix of the end-effector through positive kinematics solution under the constraint conditions of UT (taking the method of ultrasonic signal acquisition based on position trigger as an example) to explore the way of controlling and calibrating the position and attitude of manipulator trajectory points, so as to ensure that the trajectory accuracy meets the needs of ultrasonic NDT.

4.4.1 Kinematics Analysis

The problem of manipulator kinematics is divided into positive kinematics solution and inverse kinematics solution. Positive kinematics solution is to determine the position and attitude of the end-effector TCP relative to the reference coordinate system when the angular displacement of each joint and the D–H geometric parameters of each connecting rod are known. On the contrary, inverse kinematics solution is to determine the angular displacement of each joint when the position and attitude of the end-effector TCP are known. Whereas a problem of positive solution has one and only one solution, a problem of inverse solution may have multiple solutions (or no solution). Therefore, the optimized analysis of trajectory is required.

1. Model of positive kinematics solution

   In the manipulator, all the connecting rods are connected in series. Starting from the shoulder joint fixed onto manipulator base, the six-DOF joints such as big arm, elbow, forearm and wrist are defined as the i-th connecting rod in succession. Figure 4.19 shows the distribution of the coordinate systems $\{i\}$ ($i = 1, \ldots, 6$) fixed onto them and the connecting-rod parameters (referred to as D–H (Denavit-Hartenberg) parameters, including connecting-rod offset d_i, joint angle θ_i, connecting-rod length a_i and rotation angle α_i) for describing the kinematic relationships between joint mechanisms.

From the D–H parameters of articulated manipulator and the angular displacement of each joint read from the encoder, the transformation matrix of the end-effector TCP relative to the reference coordinate system (including rotation matrix and displacement matrix) can be derived. At the same time, based on this transformation matrix, the positive kinematics solution can be solved, and the transformation relationship between the position/attitude of the manipulator end-effector and the rotation angle of each joint can be derived.

Table 4.1 lists the D–H parameters of manipulator in this system. Figure 4.18 shows the distribution and transformation of coordinate systems at the joints. According to D–H parameters, the homogeneous transformation matrix between the connecting rods i and $B = A + \omega I$ can be established:

$$T_{i-1}^{i} = \text{Rot}(Z_{i-1}, \theta_i)\text{Trans}(0,0,d_i)\text{Trans}(a_i,0,0)\text{Rot}(X_i,\alpha_i)$$

$$= \begin{pmatrix} c\theta_i & -s\theta_i c\alpha_i & s\theta_i s\alpha_i & a_i c\theta_i \\ s\theta_i & c\theta_i c\alpha_i & -c\theta_i s\alpha & a_i s\theta_i \\ 0 & s\alpha_i & c\alpha_i & d_i \\ 0 & 0 & 0 & 1 \end{pmatrix} \tag{4.42}$$

where cos is abbreviated as "c" and sin is abbreviated as "s" to facilitate the observation of each element in the matrix. When the connecting-rod offset d_i, connecting-rod length a_i and rotation angle α_i are all known from the manipulator D–H parameters, the joint angle θ_i can be determined successively through the homogeneous transformation matrix between the manipulator base coordinate system and TCP. By defining the coordinate system of base, shoulder, big arm, elbow, forearm, wrist and flange as $\text{F}(V) = V^T BV - 2V^T C + \omega c$, respectively, the transformation matrix between joints can be derived.

TABLE 4.1 D–H Parameters of Manipulator

Arm Level	a_i (°)	a_i (mm)	d_i (mm)	θ_i (°)	Joint Limit
1	−90	50	0	θ_1	−180°~180°
2	0	550	0	θ_2	−130°~148°
3	90	0	50	θ_3	−145°~145°
4	−90	0	500	θ_4	−270°~270°
5	90	0	0	θ_5	−115°~140°
6	0	0	100	θ_6	−270°~270°

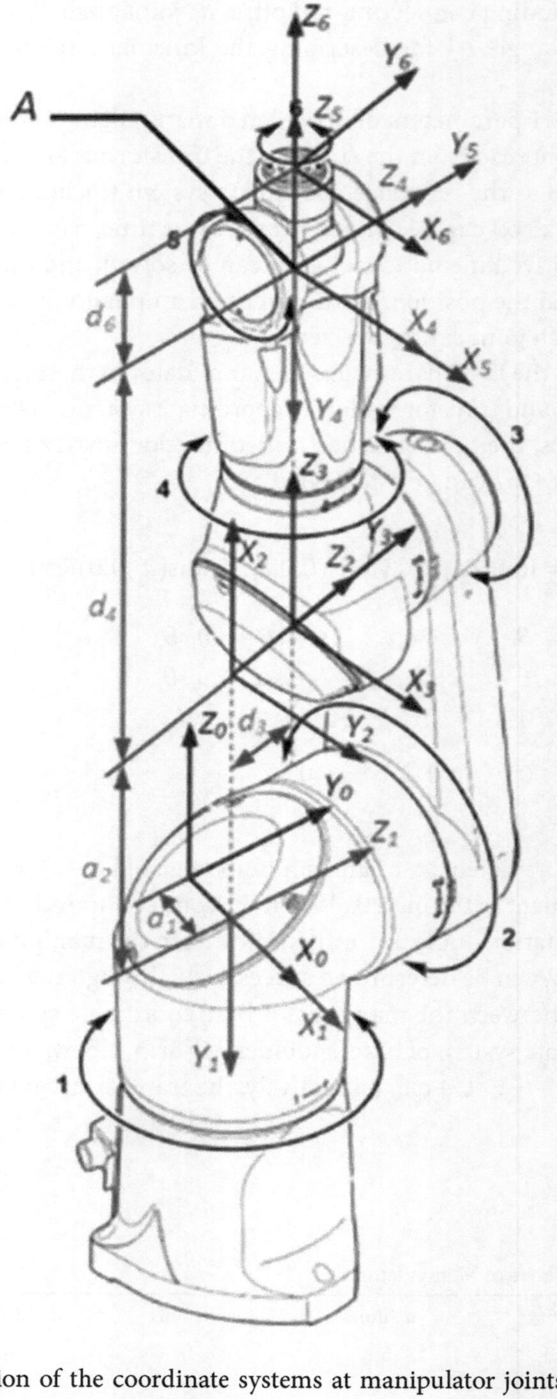

FIGURE 4.18 Distribution of the coordinate systems at manipulator joints, and transformation relations between them.

$$T_0^i = \begin{pmatrix} c\theta_i & 0 & s\theta_1 & a_i c\theta_i \\ s\theta_i & 0 & c\theta_1 & a_i s\theta_i \\ 0 & -1 & 0 & 0 \\ 0 & 0 & 0 & 1 \end{pmatrix} \quad T_1^2 = \begin{pmatrix} c\theta_2 & -s\theta_2 & 0 & a_2 c\theta_2 \\ s\theta_2 & c\theta_2 & 0 & a_2 s\theta_2 \\ 0 & 0 & 1 & 0 \\ 0 & 0 & 0 & 1 \end{pmatrix}$$

$$T_2^3 = \begin{pmatrix} c\theta_3 & 0 & s\theta_3 & 0 \\ s\theta_3 & 0 & -c\theta_3 & 0 \\ 0 & 1 & 0 & d_3 \\ 0 & 0 & 0 & 1 \end{pmatrix} \quad T_3^4 = \begin{pmatrix} c\theta_4 & 0 & -s\theta_4 & 0 \\ s\theta_4 & 0 & c\theta_4 & 0 \\ 0 & -1 & 0 & d_4 \\ 0 & 0 & 0 & 1 \end{pmatrix}$$

$$T_4^5 = \begin{pmatrix} c\theta_5 & 0 & s\theta_5 & 0 \\ s\theta_5 & 0 & -c\theta_5 & 0 \\ 0 & 1 & 0 & 0 \\ 0 & 0 & 0 & 1 \end{pmatrix} \quad T_5^6 = \begin{pmatrix} c\theta_6 & -s\theta_6 & 0 & 0 \\ s\theta_6 & c\theta_6 & 0 & 0 \\ 0 & 0 & 1 & d_6 \\ 0 & 0 & 0 & 1 \end{pmatrix}$$

The transformation matrix characterizing the position/attitude relation between the end-effector TCP and base coordinate system can be derived:

$$ {}_6^0T = {}_1^0T\,{}_2^1T\,{}_3^2T\,{}_4^3T\,{}_5^4T\,{}_6^5T = \begin{pmatrix} n_x & o_x & a_x & p_x \\ n_y & o_y & a_y & p_y \\ n_z & o_z & a_z & p_z \\ 0 & 0 & 0 & 1 \end{pmatrix} \tag{4.43}$$

where

$$n_x = c_1\left[c_{23}\left(c_4c_5c_6 - s_4s_6\right) - s_{23}s_5c_6\right] - s_1\left[s_4c_5c_6 + c_4s_6\right]$$

$$n_y = s_1\left[c_{23}\left(c_4c_5c_6 - s_4s_6\right) - s_{23}s_5c_6\right] - c_1\left[s_4c_5c_6 + c_4s_6\right]$$

$$n_z = s_{23}\left(s_4s_6 - c_4c_5c_6\right) - c_{23}s_5c_6$$

$$o_x = c_1\left[-c_{23}\left(c_4c_5c_6 + s_4c_6\right) + s_{23}s_5s_6\right] + s_1\left(s_4c_5s_6 - c_4c_6\right)$$

$$o_y = s_1\left[-c_{23}\left(c_4c_5c_6 + s_4c_6\right) + s_{23}s_5s_6\right] + c_1\left(s_4c_5s_6 - c_4c_6\right)$$

$$O_z = s_{23}(s_4c_6 + c_4c_5c_6) + c_{23}s_5s_6$$

$$a_x = c_1(s_{23}c_5 + c_{23}c_4s_5) - s_1s_4s_5$$

$$a_y = s_1(s_{23}c_5 + c_{23}c_4s_5) + s_1s_4s_5$$

$$a_z = c_{23}c_5 - s_{23}c_4s_5$$

$$P_x = c_1\left[a_1 + a_2c_2 + d_4s_{23} + d_6(s_{23}c_5 + c_{23}c_4s_5)\right] - s_1\left[d_3 + d_6s_4s_5\right]$$

$$P_y = s_1\left[a_1 + a_2c_2 + d_4s_{23} + d_6(s_{23}c_5 + c_{23}c_4s_5)\right] + c_1\left[d_3 + d_6s_4s_5\right]$$

$$P_z = d_4c_{23} + d_6(c_{23}c_5 - s_{23}c_4s_5) - a_2s_2$$

where c_i is $\cos\theta, s_i$ is $\sin\theta, c_{ij}$ is $\cos(\theta_i + \theta_j), s_{ij}$ is $\sin(\theta_i + \theta_j)$

When the angular displacement of each encoder is known, the position and attitude information of TCP relative to the manipulator base coordinate system (i.e. the positive kinematics solution of the manipulator) can be obtained from the above transformation matrix. Thus, the manipulator can receive the point packet information based on the data information of the Cartesian coordinate system except for joint angle parameters, which can control the end-effector to arrive at the specified position and attitude.

2. Inverse kinematics model

The problem of inverse kinematics will affect the motion control and position/attitude change of a manipulator. In particular, when a discrete point on the planned trajectory has multiple inverse solutions, or when it has no solutions due to the potential of motion arm to reach the workspace limit or the existence of singularity, the optimal inverse analytical solution should be determined through the relevant algorithm based on the stroke optimization and D–H parameters of motion arm.

According to Pieper's solving process, when the last three connecting rods of a manipulator intersect at the point A, the end-effector workspace is a sphere with the point A as the center and the connecting-rod offset d_6 at the end joint as the radius, as shown in Figure 4.19. Therefore, the position coordinates of the point A are only affected by the first three joint angles rather than the joint angles θ_4, θ_5 and θ_6, that is, arm-wrist separation.

The calculation results of the above manipulator trajectory generation algorithm are the position (P_x, P_y, P_z) and attitude (α, β, γ) of the end-effector TCP. When the manipulator holds a workpiece and moves with it, a relatively defined position/attitude relationship will exist between the TCP and discrete-point coordinate system (they should coincide with each other if the existence of underwater acoustic path is not considered). Next, we will use the method of Euler angle coordinate transformation

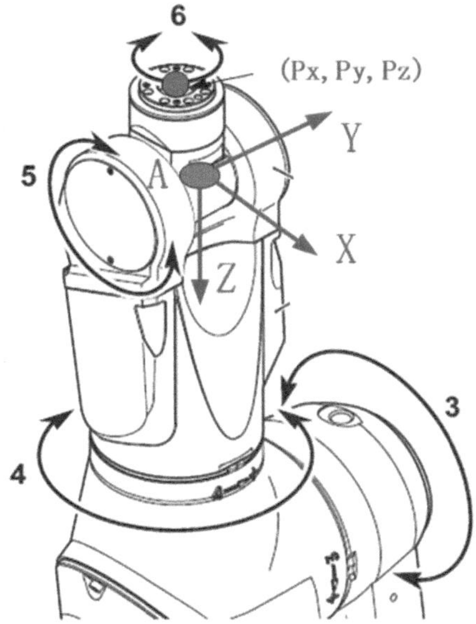

FIGURE 4.19 Manipulator wrist structure and arm-wrist separation method.

to calculate the angular displacement of each joint under the constraint of this position (attitude).

When the position and attitude of TCP are known, its transformation matrix relative to the manipulator reference coordinate system is as shown in Eq. (4.44). In this case, the position of point A (P_{ax}, P_{ay}, P_{az}) and the position of TCP (P_x, P_y, P_z) have the following relation:

$$\begin{cases} P_{ax} = p_x - d_6 a_x \\ p_{ay} = p_y - d_6 a_y \\ p_{az} = p_z - d_6 a_z \end{cases} \tag{4.44}$$

3. Calculate the first joint angle θ_1

Because the position and attitude of the point A are only related to the D–H parameters of the first three joints, its homogeneous transformation matrix is

$$T_0^{\ 4} = T_0^{\ 1} T_1^{\ 2} T_2^{\ 3} T_3^{\ 4} = \begin{bmatrix} n_{ax} & o_{ax} & a_{ax} & p_{ax} \\ n_{ay} & o_{ay} & a_{ay} & p_{ay} \\ n_{az} & o_{az} & a_{az} & p_{az} \\ 0 & 0 & 0 & 1 \end{bmatrix}$$

$$= \begin{bmatrix} c_1c_{23}c_4 - s_1s_4 & -c_1s_{23} & -c_1c_{23}s_4 - s_1c_4 & d_4c_1s_{23} - d_3s_1 + a_1c_1 + a_2c_1c_2 \\ s_1c_{23}c_4 + c_1s_4 & -s_1s_{23} & -s_1c_{23}s_4 + c_1c_4 & d_4s_1s_{23} + d_3c_1 + a_1s_1 + a_2c_2s_1 \\ -s_{23}c_4 & -c_{23} & s_{23}s_4 & d_4c_{23} - a_2s_2 \\ 0 & 0 & 0 & 1 \end{bmatrix} \tag{4.45}$$

By eliminating the unknown elements for the position A (P_{ax}, P_{ay}, P_{az}) from the above equation,

$$p_{ay}c_1 - p_{ax}s_1 = d_3 \tag{4.46}$$

Then the first joint angle θ_1 can be determined:

$$\theta_1 = A\tan 2\left(p_{ay}, p_{ax}\right) - A\tan 2\left(d_3, \pm\sqrt{p_{ax}^2 + p_{ay}^2 - d_3^2}\right) \tag{4.47}$$

1. Calculate the second joint angle θ

   From the fourth column of elements in Eq. (4.45), we can deduce

$$p_{ax}c_1 + p_{ay}s_1 = d_4s_{23} + a_1 + a_2c_2 \tag{4.48}$$

   Considering the corresponding elements of P_{az} in the equation,

$$\begin{cases} s_{23} = \dfrac{\left(p_{ax}c_1 + p_{ay}s_1\right) - a_1 - a_2c_2}{d_4} \\ c_{23} = \dfrac{p_{az} + a_2s_2}{d_4} \end{cases} \tag{4.49}$$

   According to the trigonometric function theorem,

$$\left(p_{ax}c_1 + p_{ay}s_1 - a_1\right)c_2 - p_{az}s_2 = \frac{\left(p_{ax}c_1 + p_{ay}s_1 - a_1\right) + a_2^2 + p_{az}^2 - d_4^2}{2a_2} \tag{4.50}$$

   Then the second joint angle θ_2 can be determined:

$$\theta_2 = A\tan 2\left(p_{ax}c_1 + p_{ay}s_1 - a_1, p_{az}\right) - A\tan 2\left(t_1, \pm\sqrt{\left(p_{ax}c_1 + p_{ay}s_1 - a_1\right)^2 + p_{az}^2 - t_1^2}\right) \tag{4.51}$$

where $t_1 = \dfrac{\left(p_{ax}c_1 + p_{ay}s_1 - a_1\right)^2 + a_2^2 + p_{ax}^2 - d_4^2}{2a_2}$

2. Calculate the third joint angle θ_3

   According to Eq. (4.45),

$$\tan(\theta_2+\theta_3)=\frac{p_{ax}c_1+p_{ay}s_1-a_1-a_2c_2}{p_{az}+a_2s_2} \tag{4.52}$$

Then the third joint angle θ_3 can be determined:

$$\theta_3 = A\tan 2\left(p_{ax}c_1+p_{ay}s_1-a_1-a_2c_2, p_{az}+a_2s_2\right)-\theta_2 \tag{4.53}$$

3. Calculate the fifth joint angle θ_5

   When the position of the point A is deduced from the first three joint angles, the position and attitude of TCP can be calculated from, $R_0^6 = R_0^3 R_3^6$, that is, only the position/attitude transformation matrix from wrist to TCP, denoted as, $R_3^6=\left(R_0^3\right)^{-1}R_0^6$ needs to be calculated. Then considering the D–H parameters of manipulator, we can obtain

$$R_3^6=\left(R_0^3\right)^{-1}R_0^6=\begin{bmatrix} n_{wx} & o_{wx} & a_{wx} \\ n_{wy} & o_{wy} & a_{wy} \\ n_{wz} & o_{wz} & a_{wz} \end{bmatrix} \tag{4.54}$$

$$R_3^6=R_3^4R_4^5R_5^6=\begin{bmatrix} c_4c_4c_6-s_4s_6 & -c_4c_5s_6-s_4c_6 & c_4s_5 \\ s_4c_5c_6+c_4s_6 & -s_4c_5s_6+c_4c_6 & s_4s_5 \\ -s_5c_6 & s_5s_6 & c_5 \end{bmatrix} \tag{4.55}$$

Considering the corresponding elements in Eqs. (4.54) and (4.55),

$$\begin{cases} a_{wx}^2+a_{wy}^2=s_5^2 \\ a_{wz}=c_5 \end{cases} \tag{4.56}$$

Then the fifth joint angle is

$$\theta_5 = A\tan 2\left(\pm\sqrt{a_{wx}^2+a_{wy}^2}, a_{wz}\right) \tag{4.57}$$

4. Calculate the fourth joint angle θ_4

   According to Eq. (4.56), if $\theta_5 \neq 0$, then

$$\theta_4 = A\tan 2\left(a_{wy}/s_5, a_{wx}/s_5\right) \tag{4.58}$$

When $\theta_5 = 0$, joints 4 and 6 are on the same line and the manipulator is in a singular position. Thus, there is a unique solution to the attitude of the connecting rod at the end of the manipulator. According to Eq. (4.58), all the solutions that meet the requirements in this case satisfy the sum or difference of

θ_4 and θ_6. To avoid the occurrence of singularities on the manipulator trajectory, here we usually need to select the initial value of the fourth joint or the expected value under the current attitude (determined by the attitude of the previous motion point).

5. Calculate the sixth joint angle θ_6

   According to Eq. (4.56), if $\theta_5 \neq 0$, then

$$\theta_6 = A\tan 2\left(o_{wz}/s_5, -n_{wz}/s_5\right) \tag{4.59}$$

When $\theta_5 = 0$, the joint is in a singular position. Similarly, the initial value or the current value should be selected in this case.

It can be seen from the above solving process that, θ_1, θ_3 and θ_5 each have two solutions, so the manipulator joints corresponding to the same position and attitude have eight inverse solutions. Among these solutions, some are real and some are imaginary. Appropriate solutions need to be determined according to the motion space of each joint in actual operation (as shown in Table 4.2) and the previous motion position and attitude. For example, when the principle of "shortest stroke" is used to control the joint motion, the evaluation function Q_n can be introduced to evaluate multiple analytic solutions. Taking the minimum sum of mean square errors of angular displacements of all the joints as the optimal solution, the stability and advantages and disadvantages of trajectories in the eight inverse solutions can be judged.

$$Q_n = \sum_{i=1}^{6} p_i\left(\theta_{i,n} - \theta_{i,n-1}\right)^2 \tag{4.60}$$

where p_i: stroke energy weight coefficient of the i-th joint;

$\theta_{i,n}$: rotation angle of the joint i at the position and attitude of the n-th trajectory point;

$\theta_{i,n-1}$: rotation angle of the joint i at the position and attitude of the n−1-th trajectory point.

4.4.2 End-Effector Position Error and Compensation Strategy

According to the D–H model proposed by Hayati et al., the homogeneous transformation matrix of the adjacent joints can be obtained:

$$T_i^{i-1} = \text{Rot}(Z,\theta_i)\text{Trans}(Z,d_i)\text{Trans}(X,a_{i-1})\text{Rot}(X,\alpha_{i-1})\text{Rot}(y,\beta_i) \tag{4.61}$$

That is to say, the parameter β rotating around the y-axis is added to the typical manipulator D–H model to evaluate the parallelism error between adjacent joints. According to the principle of homogeneous coordinate transformation,

$$T_i^{i-1} = \begin{bmatrix} \cos\theta_i & -\sin\theta_i & 0 & 0 \\ \sin\theta_i & \cos\theta_i & 0 & 0 \\ 0 & 0 & 1 & 0 \\ 0 & 0 & 0 & 1 \end{bmatrix} \begin{bmatrix} 1 & 0 & 0 & 0 \\ 0 & 1 & 0 & 0 \\ 0 & 0 & 1 & d_i \\ 0 & 0 & 0 & 1 \end{bmatrix} \begin{bmatrix} 1 & 0 & 0 & a_i \\ 0 & 1 & 0 & 0 \\ 0 & 0 & 1 & 0 \\ 0 & 0 & 0 & 1 \end{bmatrix}$$

$$\begin{bmatrix} 1 & 0 & 0 & 0 \\ 0 & \cos\alpha_i & -\sin\alpha_i & 0 \\ 0 & \sin\alpha_i & \cos\alpha_i & 0 \\ 0 & 0 & 0 & 1 \end{bmatrix} \begin{bmatrix} \cos\beta_i & 0 & -\sin\beta_i & 0 \\ 0 & 1 & 0 & 0 \\ \sin\beta_i & 0 & \cos\beta_i & 0 \\ 0 & 0 & 0 & 1 \end{bmatrix}$$

$$= \begin{bmatrix} c\theta_i c\beta_i + s\theta_i s\alpha_i s\beta_i & -s\theta_i c\alpha_i & -c\theta_i s\beta_i + s\theta_i s\alpha_i c\beta & a_i c\theta_i \\ s\theta_i c\beta_i - c\theta_i s\alpha_i s\beta_i & c\theta_i c\alpha_i & -s\theta_i s\beta_i - c\theta_i s\alpha_i c\beta_i & a_i s\theta_i \\ c\alpha_i s\beta_i & s\alpha_i & c\alpha_i c\beta_i & d_i \\ 0 & 0 & 0 & 1 \end{bmatrix} \quad (4.62)$$

It can be seen that when a joint is added or the manipulator D–H parameters are changed, the position and attitude of the end-effector will change with the parameters of rotation matrix. If the D–H parameters have the errors $\Delta\theta_i, \Delta d_i, \Delta a_i, \Delta\alpha_i$ and $\Delta\beta_i$, the manipulator cannot arrive at the specified position and attitude accurately according to the previously planned scanning point data. When the position of TCP at the end of the manipulator is deduced from the angle information read by the joint encoders, its error should be analyzed, compensated for and calibrated.

According to the characteristics of structural parameters of articulated manipulator, the position/attitude compensation algorithms can be divided into the following two categories:

1. Error model compensation algorithm based on manipulator parameter identification technique. The structural parameter error of each joint needs to be calculated, which is essentially to compensate for the position/attitude error of the end-effector TCP by solving the variables of each joint. Correct position and attitude are derived through the optimization calculation of inverse kinematics solution and the parameter feedback. It is a common practice to add a correction factor to the manipulator position/attitude matrix, or to compensate for the joint variables. The correct position and attitude need to be approached through multiple iterations;
2. lgorithm of spatial error fitting (interpolation). Several nodes are measured in the manipulator's working space. The errors in the adjacent areas of each point are fitted through interpolation operation.

As the structural parameters of each joint are known, the algorithm of structural parameters identification will be introduced below to compensate for the position/attitude error of the end-effector TCP. The joint motion variables will be used as feedback values to guide the manipulator to adjust the position and attitude of the TCP and compensate for the attitude error caused by the change of structural parameters. The compensation process is shown in Figure 4.20.

For the position and attitude of the end-effector TCP, the strategy and steps of error compensation can be described as below:

1. At first, when the position and attitude at target point, denoted as $P_d\left(x_i', y_i', z_i', \alpha_i, \beta_i, \gamma_i\right)$, are known, the corresponding joint angle θ_i can be determined through the inverse kinematic solution algorithm;
2. The real-time manipulator position update system reads the actual position and attitude at target point $P_c\left(x_i'', y_i'', z_i'', \alpha_i', \beta_i', \gamma_i'\right)$, namely the position and attitude of TCP obtained from the nominal joint angle under the action of structural parameter error, as well as the partial derivative $\frac{\partial P_i}{\partial \theta_i}(i=1,\ldots,6)$;
3. From $p_d - p_c = \sum_{i-1}^{6} \frac{\partial P_i}{\partial \theta_i} \Delta\theta_i$, the joint angle change $\Delta\theta_i$ when P_c is changed into P_d through compensation can be derived;

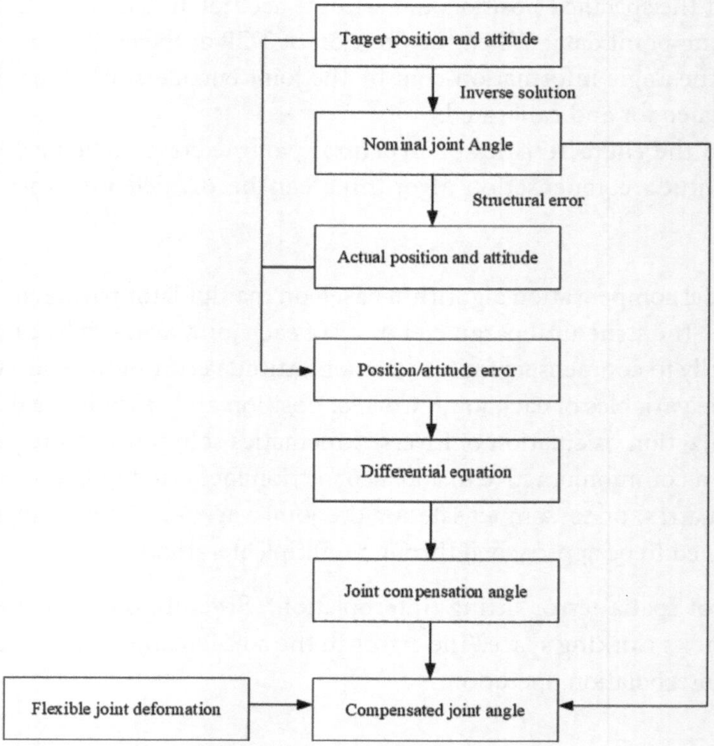

FIGURE 4.20 Error compensation flowchart for an end-effector TCP.

4. The joint angle after compensation is $\theta_c = \theta_d + \Delta\theta_i$, and the joint angle considering flexible deformation is $\theta_i^d = \theta_i^c \cdot \delta\theta_i \cdot \delta\theta_i'$, where $\delta\theta_i'$ is the flexible deformation of the joint;
5. The compensated joint angle is moved to the target position and attitude through the manipulator's positive solution (Figure 4.20).

Because the manufactured and delivered manipulator has high positional accuracy with a small structural parameter error for each joint, it can be approximately replaced by a differential kinematics model. After the complete-differential calculation of Eq. (4.61), the position error of the end connecting rod (end-effector) can be approximately expressed as

$$\Delta_p = \sum_{i=1}^{6} \frac{\partial P}{\partial \theta_i} \Delta\theta_i + \sum_{i=0}^{5} \frac{\partial P}{\partial \alpha} \Delta\alpha_i + \sum_{i=0}^{5} \frac{\partial P}{\partial a_i} \Delta a_i + \sum_{i=1}^{6} \frac{\partial P}{\partial d_i} \Delta d_i + \sum_{i=1}^{6} \frac{\partial P}{\partial \beta_i} \tag{4.63}$$

The corresponding attitude error can be approximately expressed as

$$\Delta R = \sum_{i=1}^{6} \frac{\partial R}{\partial \theta_i} \Delta\theta_i + \sum_{i=0}^{5} \frac{\partial R}{\partial \alpha} \Delta\alpha_i + \sum_{i=0}^{5} \frac{\partial R}{\partial a_i} \Delta a_i + \sum_{i=1}^{6} \frac{\partial R}{\partial d_i} \Delta d_i + \sum_{i=1}^{6} \frac{\partial R}{\partial \beta_i} \tag{4.64}$$

Eq. (4.63) is expressed in matrix form as:

$$\begin{cases} \Delta p = J_\delta \Delta\delta \\ \Delta p = p_c - p_d \end{cases} \tag{4.65}$$

where $\Delta\delta$ is a 25×1 geometric error parameter vector, that is,

$$\Delta\delta = \left(\Delta a_0 \ldots \Delta a_5, \Delta d_1 \ldots \Delta d_6, \Delta\alpha_0 \ldots \Delta\alpha_5, \Delta\theta_1 \ldots \Delta\theta_6, \Delta\beta\right) \tag{4.66}$$

J_δ is a 3×25 error coefficient matrix, namely identification Jacobian matrix:

$$J_\delta = \begin{bmatrix} \frac{\partial p_x}{\partial a_0} & \cdots & \frac{\partial p_x}{\partial a_6} \frac{\partial p_x}{\partial d_0} & \cdots & \frac{\partial p_x}{\partial d_6} \frac{\partial p_x}{\partial \alpha_0} & \cdots & \frac{\partial p_x}{\partial \alpha_6} \frac{\partial p_x}{\partial \theta_0} & \cdots & \frac{\partial p_x}{\partial \theta_6} \frac{\partial p_x}{\partial \beta_0} \\ \frac{\partial p_y}{\partial a_0} & \cdots & \frac{\partial p_y}{\partial a_6} \frac{\partial p_y}{\partial d_0} & \cdots & \frac{\partial p_y}{\partial d_6} \frac{\partial p_y}{\partial \alpha_0} & \cdots & \frac{\partial p_y}{\partial \alpha_6} \frac{\partial p_y}{\partial \theta_0} & \cdots & \frac{\partial p_y}{\partial \theta_6} \frac{\partial p_y}{\partial \beta_0} \\ \frac{\partial p_z}{\partial a_0} & \cdots & \frac{\partial p_z}{\partial a_6} \frac{\partial p_z}{\partial d_0} & \cdots & \frac{\partial p_z}{\partial d_6} \frac{\partial p_z}{\partial \alpha_0} & \cdots & \frac{\partial p_z}{\partial \alpha_6} \frac{\partial p_z}{\partial \theta_0} & \cdots & \frac{\partial p_z}{\partial \theta_6} \frac{\partial p_z}{\partial \beta_0} \end{bmatrix} \tag{4.67}$$

Linear correlations may exist between the columns so that some parameters may have multiple solutions and cannot be identified. Khalil [9] proposed the QR decomposition of J_δ to obtain the identifiable parameters and the use of the customized numerical tolerance

τ to define the unidentifiable parameters. In general, multiple iterations are needed to find a solution that satisfies the full-rank condition. It can be seen from the homogeneous transformation matrix Eq. (4.62) that in the articulated manipulator used in this system, only the joint angle θ_i is variable, while the other three parameters d_i, a_i and α are known. Therefore, Eq. (4.33) has only one unknown variable and thus a unique solution.

4.4.3 Method of Joint Position/Attitude Feedback

In the robotic ultrasonic detection system, the signals are synchronously collected by triggering ultrasonic pulse excitation according to the end-effector position information, that is, the position information of the end-effector TCP is obtained through the positive kinematics solution calculation of the manipulator joint angle θ_i that is read from each joint encoder in real time at high frequency. When the displacement is judged meeting the triggering condition (usually determined by the defect size to be detected, the manipulator speed and the planned scanning-point spacing), an excitation signal will be sent to the ultrasonic signal transceiver to meet the need for ultrasonic high-frequency triggering.

Since the manipulator reference coordinate system $\{W\}$ is defined at the center of the mounting base and no reference point can be accurately captured, the compensation for the alignment and clamping errors of the coordinate system before actual testing is completely based on TCP. Therefore, the accuracy of position and attitude of the end-effector and its TCP has great influence on the reliability and accuracy of detection results. Next, the position and attitude at the target point P_c will be calculated through the positive kinematics solution based on the real-time rotation angle obtained by the joint encoders during testing.

The partial-differential calculation result of Eq. (4.63) is

$$\begin{cases} \dfrac{\partial p_i}{\partial \theta} = p_i \cdot Q_\theta \\ \dfrac{\partial p_i}{\partial a} = p_i \cdot Q_a \\ \dfrac{\partial p_i}{\partial \alpha} = p_i \cdot Q_\alpha \\ \dfrac{\partial p_i}{\partial \theta} = p_i \cdot Q_\theta \end{cases} \tag{4.68}$$

where

$$Q_\theta = \begin{bmatrix} s\theta c\beta + c\theta s\alpha s\beta & -c\theta c\alpha & -s\theta s\beta + c\theta d\alpha c\beta & -a_i s\theta \\ c\theta c\beta - s\theta s\alpha s\beta & -s\theta c\alpha & -c\theta s\beta + s\theta s\alpha c\beta & a_i c\theta \\ 0 & 0 & 0 & 0 \\ 0 & 0 & 0 & 0 \end{bmatrix} \quad Q_d = \begin{bmatrix} 0 & 0 & 0 & 0 \\ 0 & 0 & 0 & 0 \\ 0 & 0 & 0 & 1 \\ 0 & 0 & 0 & 0 \end{bmatrix}$$

$$Q_a = \begin{bmatrix} 0 & 0 & 0 & c\theta \\ 0 & 0 & 0 & s\theta \\ 0 & 0 & 0 & 0 \\ 0 & 0 & 0 & 0 \end{bmatrix} \quad Q_\alpha = \begin{bmatrix} s\theta s\beta c\alpha & s\theta s\alpha & s\theta c\alpha c\beta & 0 \\ c\theta c\alpha s\beta & -c\theta s\alpha & -c\theta c\alpha c\beta & 0 \\ -s\alpha s\beta & c\alpha & -s\alpha c\beta & 0 \\ 0 & 0 & 0 & 0 \end{bmatrix} \tag{4.69}$$

Then Eq. (4.63) can be converted into

$$dp_i = p_i\left(Q_a \cdot a_i + Q_\alpha \cdot \alpha_i + Q_d \cdot d_i + Q_\theta \cdot \theta\right) \tag{4.70}$$

The transformation matrix between the TCP planned by the manipulator and the TCP calculated by the joint encoder is, which is expressed as

$${}_v^cT = \begin{bmatrix} 1 & -\delta z & \delta y & dx \\ \delta z & 1 & \delta x & dy \\ -\delta y & \delta x & 1 & dz \\ 0 & 0 & 0 & 1 \end{bmatrix} \tag{4.71}$$

where $dx, dy, dz, \delta x, \delta y$ and δz, respectively, represent the small translations and rotations around three coordinate axes between the TCP planned by the manipulator and the end-effector TCP read and deduced by the encoders. Then the position/attitude at planned target point P_d, the position/attitude of the end-effector TCP P_c calculated by the encoder and the position/attitude at actual arrival point P_v have the following relations:

$$\begin{cases} \Delta p = p_c - p_d \\ p_v = {}_v^c T p_c \end{cases} \tag{4.72}$$

so

$$\begin{aligned} p_v - p_d &= {}_v^c T p_c - p_d \\ &= {}_v^c T\left(p_d + \Delta p\right) - p_d \\ &= \left({}_v^cT - E\right)p_d + \Delta p + \left({}_v^cT - E\right)\Delta p \end{aligned} \tag{4.73}$$

If the higher-order term $\left({}_v^cT - E\right)\Delta p$ in the above equation is ignored, Eq. (4.43) can be simplified as

$$\begin{aligned} p_v - p_d &= \left({}_v^cT - E\right)p_d + \Delta p \\ &= \left({}_v^cT - E\right)p_d + \begin{bmatrix} M_\theta & M_d & M_a & M_\alpha \end{bmatrix} \begin{bmatrix} \Delta\theta \\ \Delta d \\ \Delta a \\ \Delta\alpha \end{bmatrix} \end{aligned} \tag{4.74}$$

where M_θ, M_d, M_a and M_α are 3 × 6 matrices, where the elements are the functions composed of the geometric parameters of each joint, as shown in Eq. (4.45):

$$\begin{bmatrix} \xi_{i+1}^{v} \times d_x & \xi_{i+1}^{v} \times d_y & \xi_{i+1}^{v} \times d_z & \xi_{i+1}^{v} \cdot \delta_x & \xi_{i+1}^{v} \cdot \delta_y & \xi_{i+1}^{v} \cdot \delta_z \\ \varphi_{i+1}^{v} \times d_x & \varphi_{i+1}^{v} \times d_y & \varphi_{i+1}^{v} \times d_z & \varphi_{i+1}^{v} \cdot \delta_x & \varphi_{i+1}^{v} \cdot \delta_y & \varphi_{i+1}^{v} \cdot \delta_z \\ \psi_{i+1}^{v} \times d_x & \psi_{i+1}^{v} \times d_y & \psi_{i+1}^{v} \times d_z & \psi_{i+1}^{v} \cdot \delta_x & \psi_{i+1}^{v} \cdot \delta_y & \psi_{i+1}^{v} \cdot \delta_z \end{bmatrix} \tag{4.75}$$

In the above equation, $[\xi, \iota, \psi]$ are the vector components of the target point along the three axes of TCP. Their cross products with the translation vector in the coordinate system are the three-dimensional vectors of the rotation matrix ${}_{v}^{c}T$, and their dot products with the rotation vector are the rotations along the corresponding axes of TCP.

Take several structural feature points of the workpiece under test as examples. Table 4.2 lists the position/attitude information at the planned target trajectory point, the nominal arrival point calculated from the joint rotation angle read by encoder and the actual arrival point after error compensation. Figure 4.21 shows the error information of nominal arrival point, target point and actual post-compensation point that has been obtained in the experiment. Since the traditional measurement of actual manipulator position/attitude is mostly achieved by laser tracker or three-coordinate measuring machine, this system uses an ultrasonic feedback waveform signal to judge the accuracy of manipulator motion trajectory. The manipulator trajectory with position/attitude compensation should produce an ultrasonic pulse echo signal with more stable time-domain waveform position and higher-interface wave amplitude, so as to ensure the accurate extraction of ultrasonic characteristic information and the reliability of detection results.

4.5 METHOD OF SYNCHRONIZATION BETWEEN POSITION AND ULTRASONIC SIGNAL

The purpose of synchronous trigger is to accurately excite and receive ultrasonic signals according to the triggering conditions set by the system (typically based on a certain time interval or displacement interval), then collect via a circuit the ultrasonic signals and

TABLE 4.2 Coordinates of Target Point, Nominal Arrival Point and Compensation Point

	Planned Target Point			Nominal Arrival Point			Compensation Point		
No.	X (mm)	Y (mm)	Z (mm)	X (mm)	Y (mm)	Z (mm)	X (mm)	Y (mm)	Z (mm)
1	4.715	36.865	83.877	4.923	36.998	83.771	4.820	36.932	83.820
2	3.222	36.932	83.910	3.142	36.407	84.035	3.182	36.669	83.973
3	2.620	36.437	84.023	2.424	36.802	84.014	2.522	36.620	84.019
4	1.710	36.938	83.950	1.890	37.334	83.852	1.800	37.136	83.901
5	0.888	37.175	83.888	0.918	36.885	83.985	0.903	37.030	83.937
6	−0.918	36.885	83.985	−0.913	36.765	83.966	−0.915	36.825	83.976
7	−2.423	36.801	84.014	−2.620	36.437	84.023	−2.522	36.620	84.019
8	−3.530	36.637	84.055	−3.142	36.407	84.036	−3.336	36.520	84.045

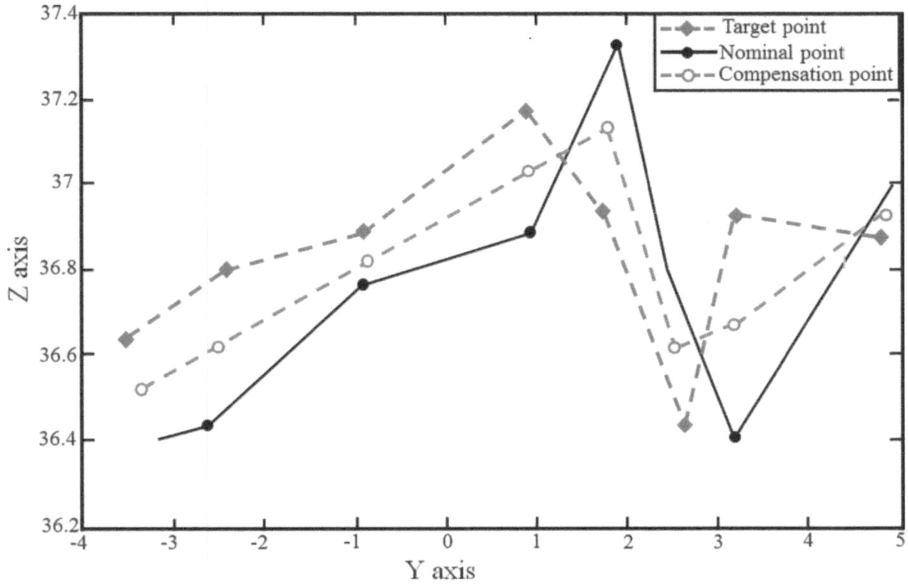

FIGURE 4.21 Difference in the positions and attitudes of a manipulator at the initial target point, nominal point and compensation point.

convert them into digital signals and finally acquire accurate waveform information for subsequent data processing. If the processes of signal excitation, receiving, acquisition and conversion are not synchronized accurately, the defect position determination after signal conversion will go awry. This is just the origin of synchronous triggering function. The robotic UT system designed in this study can realize synchronous trigger in three ways, including time trigger, position trigger and turntable trigger, each of which has its own applicable conditions.

1. Time trigger

   Square-wave pulses are sent to the excitation circuit at a certain clock frequency (repetition frequency). The trigger of each square-wave pulse includes the excitation, reception, sampling, A/D conversion and data processing of ultrasonic signal. Whenever the excitation circuit identifies a square-wave pulse, it will emit a narrow pulse signal to excite the ultrasonic transducer. Take pulse reflection detection as an example. An ultrasonic wave needs to go through a period of delay before arriving at the tested component via the coupling agent and then is reflected by each surface of the component to form the reflection echoes, which are received by the ultrasonic transducer finally. There is no valuable waveform information in the echo signal in the period from the generation of a narrow-pulse signal to the reception of the echo wave reflected by the surface of the component under test. Sampling can be initiated prior to the reception of the first reflection echo (i.e. upper-surface reflection echo) from the component under test. The period prior to initial sampling is the delay time. The total number of samples collected by the A/D conversion circuit after the start of sampling is just the sampling depth. When the sampling frequency is known, the sampling depth can also be converted into the sampling time. By extracting the gate characteristic value and judging whether the alarm information is generated, the sampled digital signal is processed in the data processing circuit. Finally, this detection cycle is completed, and the processor is expected to send the next square-wave pulse. The working process of time trigger is illustrated in Figure 4.22 in the form of sequence chart. This triggering mode is generally applicable to manual detection.

2. Position trigger

For the UT system using manipulator as the scanning device, more attention should be paid to the accurate locating of defect location, in addition to whether the component under test contains defects. The manipulator end-effector holding an ultrasonic transducer does not always move at a constant speed along the component surface to track the component contour. Therefore, the time-based calculation of motion displacement will seriously affect the positioning accuracy, thus introducing the position trigger. This triggering method is widely used in the scanning and testing of curved components.

To realize the contour tracking along the surface of the component under test, the path of the ultrasonic transducer should be planned before detection. For example, for the curved surface area shown in Figure 4.23, the path planning of UT is similar to that of curved-surface machining with five-axis numerical control (NC) machine tool. The whole surface is covered by the zig-zag motion of the ultrasonic transducer in two directions, namely scanning direction and step direction. The spatial distance between two adjacent command points in the scanning direction is defined as scanning interval, that is, L_c. Similarly, the spatial distance between two adjacent command points in the step direction is defined as step interval, that is, L_t. Position trigger interval L_p is a spatial distance covered by the end-effector before the ultrasonic flaw detector receives a Transistor-Transistor Logic (TTL) trigger signal from the position acquisition card. It can be seen that the scanning interval and step interval affect the surface tracking accuracy, while the position trigger interval affects the detection

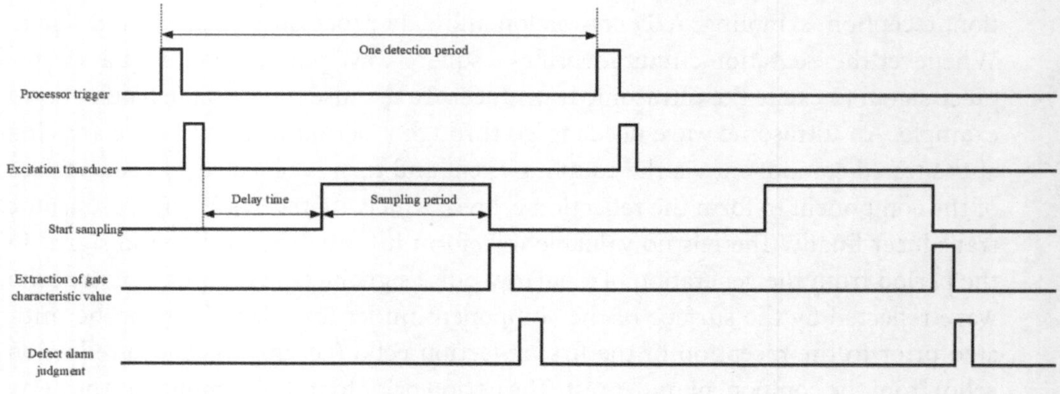

FIGURE 4.22 Time-triggered working sequence.

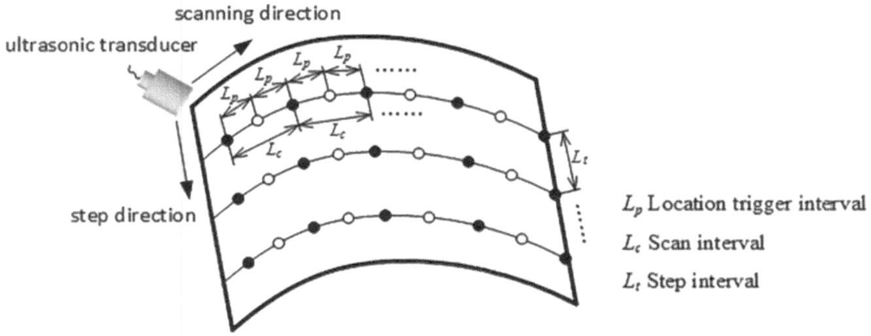

FIGURE 4.23 Relationship between surface scanning and position triggering.

resolution. The smaller the scanning interval and stepping interval are, the better the fitting between motion path and surface contour will be. The smaller the position trigger interval is, the higher the detection resolution will be. Scanning interval and step interval are generally set as an equal value. The preset position trigger interval is generally smaller than or equal to scanning interval and step interval in an integral-multiple relationship.

3. Turntable trigger

   Turntable has been used frequently in the frame-type moving devices but seldom in the UT systems with manipulator as moving device. In this paper, the turntable trigger method based on articulated manipulator and extension shaft is proposed. It is especially suitable for the testing of curved rotary components and can significantly improve the detection efficiency. In Figure 4.24, the end-effector holds an ultrasonic transducer and moves along the generatrix of a rotary component, while the component is driven by the turntable to rotate. The two motions are combined into spiral scanning. The generatrix of the rotary component is the step direction, and the spatial distance between two command points along the step direction is the step interval. Turntable trigger interval is a rotation angle covered by the turntable before the ultrasonic flaw detector receives a TTL trigger signal. The step interval affects the motion path accuracy, namely the degree of fitting to the generatrix. The resolution of detection system includes lateral resolution (in the rotation direction) and longitudinal resolution (in the generatrix direction).

As shown in Figure 4.24, the lateral resolution is determined by turntable trigger interval. The smaller the interval is, the higher the lateral resolution will be. The longitudinal resolution is determined by helix pitch. The smaller the pitch is, the higher the longitudinal resolution will be. The pitch is affected by both the turntable speed and manipulator speed. When the manipulator speed is fixed, a faster turntable speed will lead to a smaller pitch. When the turntable speed is fixed, a faster manipulator speed will lead to a bigger pitch.

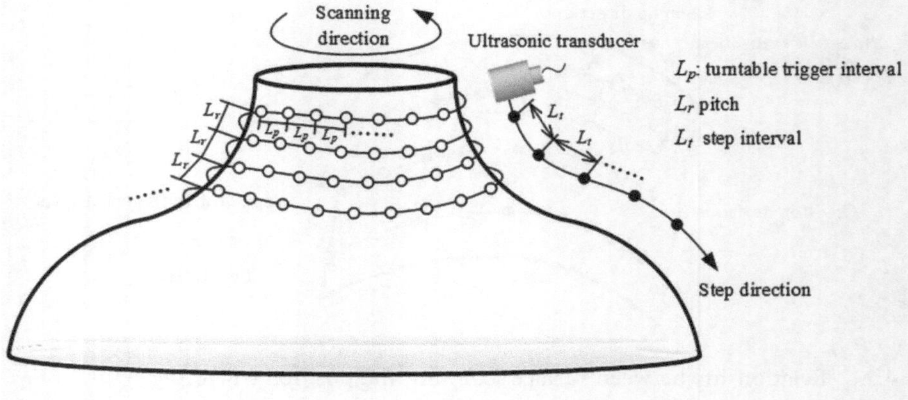

FIGURE 4.24 Relationship between spiral scanning and turntable triggering.

REFERENCES

1. Juan F., Florez Ospina, H.D., Benitez-Restrepo, toward automatic evaluation of defect detectability in infrared images of composites and honeycomb structures [J], *Infrared Physics & Technology*, 2015, 71:99–112.
2. Alvarenga A.V., Silva C.E.R., Costa-Felix R.P.B., Estimation of ultrasonic beam parameters uncertainty from NDT immersion probes using Monte Carlo method [J], *International Congress on Ultrasonics*, 2012, 1433:656–659.
3. Shin S.W., Popovics J.S., Oh T., Cost effective air-coupled impact-echo sensing for rapid detection of delamination damage in concrete structures [J], *Advances in Structural Engineering*, 2012, 15(6):887–895.
4. Hayati S, Mirmirani M. Improving the absolute positioning accuracy of robot manipulators [J], *Journal of Robotic Systems*, 1985, 2(4):397–413.
5. Wu Z, Tan D. An effective trajectory planning method for spatial arc motion of manipulator [J], *Robot*, 1999, 21(1):8–12.
6. Kadlec M, Ruzek R. A comparison of laser snearography and C-scan for assessing a glass/epoxy laminate impact damage [J], *Applied Composite Materials*. 2012, 19(3–4):393–407.
7. Boukani H.H., Chentouf S.M., Viens M., Tahan A., Gagnon M., The effect of martensitic stainless steel microstructure on the ultrasonic inspection of turbine runner joints [J], *41ST Annual Review of Progress in Quantitative Nondestructive Evaluation*, 2015, 1650:909–916.
8. Zhao X., Wang Z., Liu B., Research on phase deviation correction for ultrasonic phased array detection echo signal [J], *Journal of Measurement Science and Instrumentation*, 2015, 6(1):47–52.
9. Khalil W, Dombre E. *Modeling, Identification and Control of Robots [M]*. Oxford: Butterworth-Heinemann, 2004.

CHAPTER 5

Dual-Manipulator Testing Technique

THE TRANSMISSION NONDESTRUCTIVE TESTING (NDT) method is applicable to the NDT conditions where the components are large in size; the sound attenuation of material is great or the signal reflection cannot be performed, such as X-ray detection, ultrasonic transmission detection of thick plates or composite materials with great sound attenuation; thermal-wave transmission infrared detection and other NDT conditions. The defects in flat plates or simple mechanical components can be detected by using the automatic scanning devices with a traditional or Cartesian coordinate system. For the automatic detection of complex-shaped structures, such as the wave-transparent fairing and complex composite components at the front end of an aircraft, dual-manipulator detection technique is the best solution.

For complex-shaped components in ultrasonic transmission detection or X-ray transmission detection, it is important that the receiving transducer can accurately receive the penetrating ultrasonic or X-ray signal energy and judge the existence and size of defects in the components according to the magnitude of received transmission energy and the location of transmission scanning. Therefore, the position accuracy and space attitude of the transceiving transducer and the accuracy of trajectory planning are very important in transmission detection. The spatial position accuracy of the dual manipulator implementing nondestructive transmission detection or measurement, the attitude of dual transceiving probe, the zeroing of origin of coordinates in the detection system and other control techniques are particularly important and critical. They determine the overall detection accuracy, efficiency and effect of this NDT system.

DOI: 10.1201/9781003212232-4

5.1 BASIC PRINCIPLE OF ULTRASONIC TRANSMISSION DETECTION

5.1.1 Basic Principles of Ultrasonic Reflection and Ultrasonic Transmission

5.1.1.1 Basic Principle of Ultrasonic Reflection Detection

Ultrasonic reflection detection is to detect the workpiece defect by observing the wave reflected from internal defect or workpiece bottom after the ultrasonic probe transmits a pulse wave into the workpiece. If the workpiece is intact, only the initial wave and the bottom echo will appear. If the workpiece has a defect smaller than sound beam section, a defect wave will appear between the initial wave and the bottom wave. If the workpiece has a defect larger than sound beam section, the whole ultrasonic beam will be reflected by the defect, that is, only the initial wave and the defect wave will appear, while the bottom wave will disappear. The position of a defect in the workpiece can be determined according to its position in the time axis. The height of defect echo can reflect the size of a defect to some extent. When there is a defect, the height of bottom wave will decrease. The pulse reflection signals obtained when the workpiece has no defect, a small defect or a big defect, are shown in Figure 5.1.

5.1.1.2 Basic Principle of Ultrasonic Transmission Detection

In the ultrasonic transmission detection, two probes are usually used and are placed on both sides of the workpiece. The probe on one side transmits a pulse wave into the workpiece, while the probe on the other side receives the transmitted wave. From the change in the pulse wave energy received by the probe at the receiving end, the condition of a defect in the workpiece can be judged. When the workpiece has no defect, the transmitted ultrasonic signal will propagate in the workpiece generally at a small attenuation rate. The probe at the other end of the workpiece can receive the transmitted signal, whose magnitude depends on the material and thickness of the workpiece. When the workpiece has pores, cracks and other defects, the ultrasonic wave will be greatly attenuated. As a result, a weak signal or even no transmission signal is received at the other end of the workpiece. By comparing this signal with the signal transmitted in a workpiece without defect, the information on the detected defect can be obtained Table 5.1. The ultrasonic propagations in the workpieces with or without defect are illustrated in Figure 5.2.

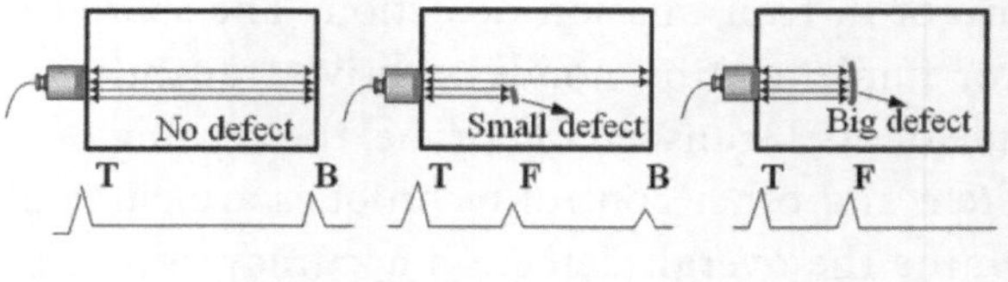

FIGURE 5.1 Pulse reflection signals detected by ultrasound.

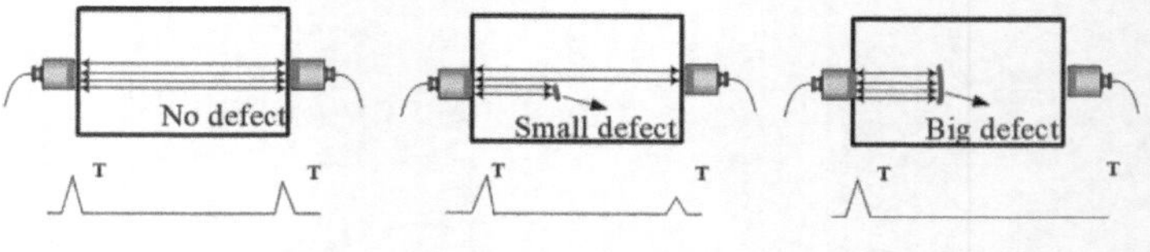

FIGURE 5.2 Pulse transmission signals detected by ultrasound.

5.1.1.3 Comparison between the Elements of an Ultrasonic Reflection Method and Those of an Ultrasonic Transmission Method

TABLE 5.1 Comparison between the Elements of the Ultrasonic Transmission Method and Those of the Ultrasonic Reflection Method

Comparison Item	Ultrasonic Transmission Method	Ultrasonic Reflection Method
Probe type	Because the attenuation rate of a signal in the composite material is very large and increases with the frequency, only the low-frequency probe can be used in the transmission method.	The reflection method is to analyze the signal reflected from the workpiece bottom, so the attenuation rate is small and the probe frequency is limited to a small extent.
Defect locating	The information on defect depth is not available.	The defect can be precisely located through the accurate calibration of scanning time baseline based on the linear relation between propagation time and distance, or according to the position of pulse wave on the time baseline and the known sound velocity.
Motion accuracy	The two probes are always aligned with each other during the motion, which requires a high motion precision for the mechanical system.	Only one probe is used, so the mechanical system doesn't need high motion accuracy.
Adaptability	Only open or semi-open specimens can be tested.	Only one surface of the specimen is tested, so the specimen structure is not heavily limited.
Scope of application	This method applies to the defects on and near surface. It can detect any defect in the way of sound. It provides a small sound attenuation rate, which is beneficial to the materials with high attenuation.	This method has high sensitivity, easy operation and extensive applications. It can detect small continuous defects and accurately locate them.

5.1.2 Ultrasonic Transmission Testing of Curved Workpieces

5.1.2.1 Principle of Reflection and Transmission of Ultrasonic Wave Incident on Curved Workpieces

Similar to light, ultrasonic wave will be not only reflected and refracted but also focused and defocused when it is incident on a curved workpiece. The degree of focusing and defocusing depends on the surface curvature and the sound velocities in the two media.

When a plane wave is incident on a surface with the radius of r (see Figure 5.3) along the acoustic axis, it will be reflected along the original path. When the wave is not incident along the acoustic axis, it will be reflected along the path similar to that of light reflection. Since the incident plane is curved, the reflection rays will converge on a focal line. At this time, the focal length F is:

$$F = \frac{r}{2} \tag{5.1}$$

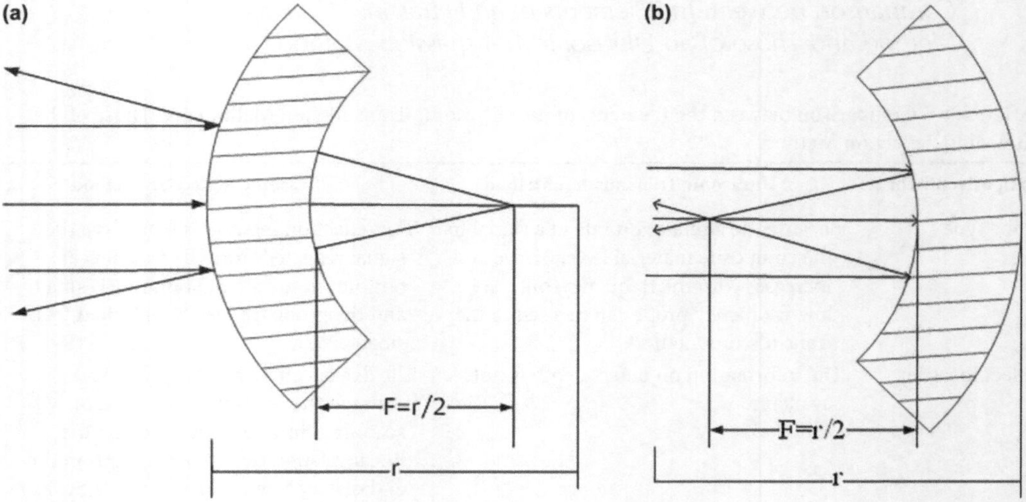

FIGURE 5.3 Reflection of plane waves incident on a curved surface. (a) Focusing and (b) defocusing.

When the incident surface is concave, the reflected waves will be focused on a real focus. When the incident surface is convex, the reflected waves will be defocused from a virtual focus, namely the convergence point of incident wave extensions.

5.1.2.2 Principle of Refraction of Ultrasonic Wave Incident on a Curved Surface

When a plane wave is incident on a surface with the radius of r, its refracted wave will also be focused or defocused, as shown in Figure 5.4. Whether the refracted wave will be focused or defocused depends on not only the concaveness and convexness of the surface but also the velocities of sound in the media on both sides of the surface. The focal length of refracted wave, denoted as F, is:

$$F = \frac{r}{1 - \frac{C_2}{C_1}} \tag{5.2}$$

where

C_1: velocity of ultrasonic wave in the curved workpiece;
C_2: velocity of ultrasonic wave in the emergent medium

5.2 COMPOSITION OF A DUAL-MANIPULATOR TESTING SYSTEM

Curved composite workpieces are widely used in aerospace industry. The inaccurate and incomprehensive NDT of a composite material may cause the product non-conformance and even major safety hazards. At present, the NDT of curved composite workpieces mainly relies on manual testing, which is inefficient and requires the inspector's rich testing experience. Due to the disadvantages of manual testing such as non-uniformity, easy detection missing, and difficult preservation of detection results, the replacement of traditional manual testing by a robotic ultrasonic NDT system is imperative.

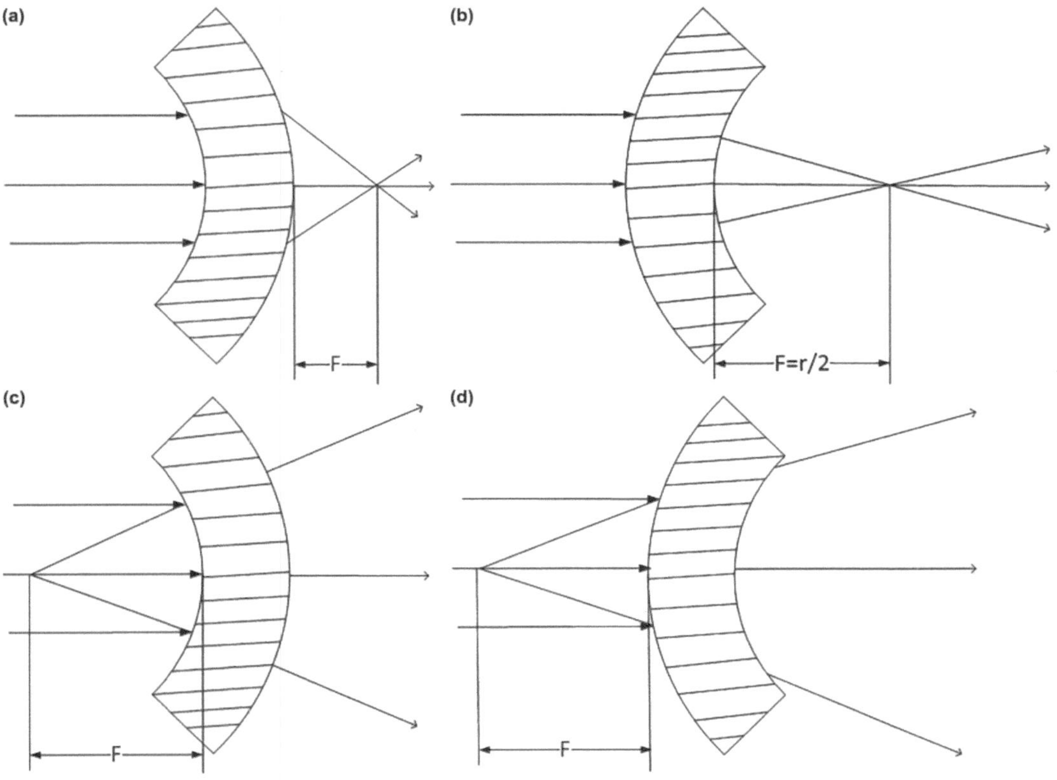

FIGURE 5.4 Refraction of plane waves incident on a curved surface. (a) Focusing, $c_1 > c_2$, (b) focusing, $c_1 < c_2$, (c) defocusing, $c_1 < c_2$ and (d) defocusing, $c_1 > c_2$.

In view of the characteristics of ultrasonic testing (UT) for curved composite materials, this chapter presents an automatic ultrasonic NDT system based on dual-manipulator coordination technique. In the whole system, the position and attitude of end-effectors of two six-degree of freedom (DOF) manipulators are adjusted in the working space so that the ultrasonic transducer can scan the curved surface. In order to overcome the high sound attenuation and impenetrability of composite materials, two manipulators in coordinated motion are used to realize the transmission scanning of large curved composite materials. The whole testing system mainly includes a hardware system and a software system. The software system is divided into upper computer software and lower computer software. The organic combination of the three parts is the foundation of automatic testing of curved workpieces and is also the core of dual-manipulator testing system.

5.2.1 Hardware Structures in a Dual-Manipulator Testing System

The structural diagram of the dual-manipulator robotic ultrasonic NDT system is shown in Figure 5.5. The hardware structures in this system mainly include two six-DOF articulated manipulators, a manipulator controller, an industrial personal computer (IPC), a high-speed acquisition card, an ultrasonic pulse transceiver, a water-circulation transducer

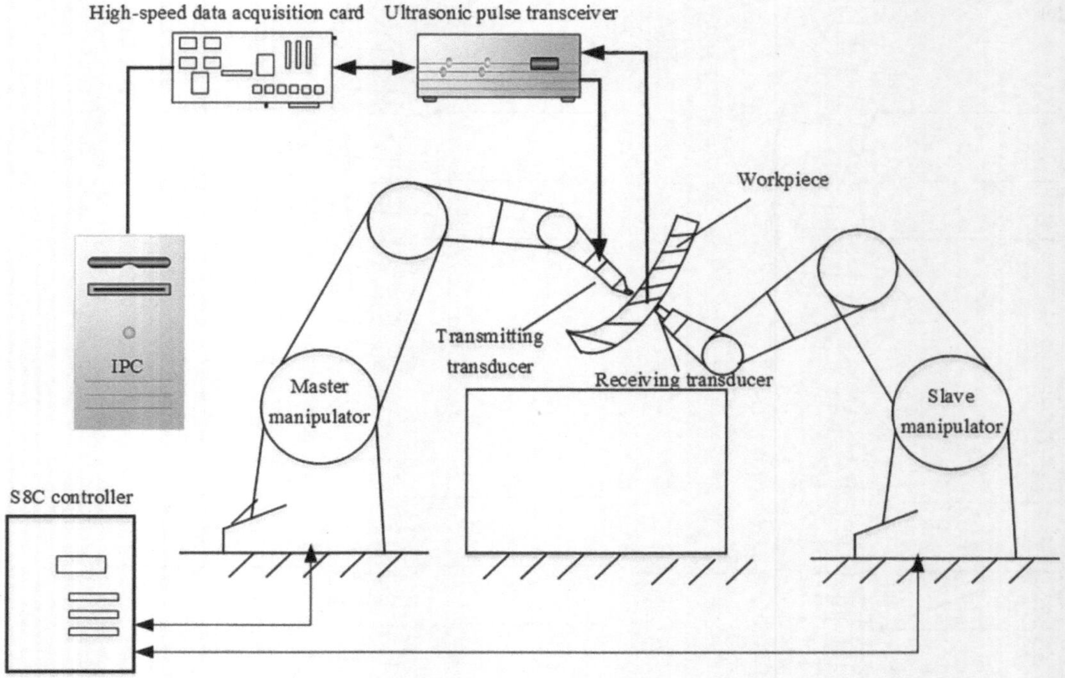

FIGURE 5.5 Structural diagram of a dual-manipulator testing system.

coupling system and other mechanical support structures. As the core unit of the whole testing system, the control unit of manipulator motion plays a vital role in the testing of curved workpieces. The motion unit consists of two high-precision six-DOF manipulators, whose flexibility is based on to track the trajectory of a curved workpiece. The two manipulators include one master manipulator and one slave manipulator. The one holding the transmitting transducer is set as the master manipulator, while the one holding the receiving transducer is set as the slave manipulator. Data acquisition card, pulse transceiver and ultrasonic transducers are mainly used for transceiving, acquiring and processing ultrasonic signals.

After the hardware structures are connected, the supporting software system is also needed to control them. The software system is also an important part of the testing system. At the beginning of a detection task, the software system initializes the detection parameters, such as scanning mode, scanning area, sampling parameters and display mode. Then it completes the process of automatic scanning and finally obtains the C-scan defect image as the detection result.

5.2.1.1 Six-DOF Articulated Manipulator

The six-DOF tandem-type articulated manipulator has been widely used in many manufacturing fields, such as vehicle assembly line, welding, painting, unloading and handling. The detection system studied in this chapter mainly makes use of the flexibility of manipulators to achieve the automatic detection of a curved workpiece, that is, ultrasonic transducers are clamped by flexible manipulators to move along the preset scanning trajectory so as to complete the detection task. The manipulator in workspace is characterized by flexible motion attitude, fast motion speed, high sensitivity and high motion accuracy, so the motion control unit with two manipulators is the core unit of the testing system. The structure of a six-DOF tandem-type articulated manipulator can be divided into six parts that are connected with each other through joints. Each part can rotate around the joint. By specifying the angular displacements of the six joints, the center point of manipulator end flange can move to the specified point in space, and the flange will point to the specified direction. Compared with other five-axis scanning frame structures or parallel structures used for automatic ultrasonic NDT, the six-DOF manipulator has the features such as compact structure, fast motion speed and high motion accuracy and the ability to realize the complex position and attitude of the ultrasonic transducer. Therefore, it fits the automatic detection of curved workpieces better.

The real manipulator used in the testing system is shown in Figure 5.6. Its structure mainly includes the following parts:

A. **Base**: The base is located at the bottom of the manipulator mainly to anchor the manipulator. Generally, the base of a manipulator is mounted on the ground. However, depending on the overall requirements for platform construction, the base can be mounted on the side wall or the ceiling, or even on a movable platform to form a mobile robotic working platform.

B. **Shoulder**: Shoulder is the first joint of the manipulator mainly to realize its horizontal angular displacement. This joint has a strong load capacity and high rigidity. In addition, its internal space is large to facilitate the wiring of joint motors behind the manipulator. Due to the influence of the load, this joint has a slow angular velocity.

C. **Upper arm**: Upper arm is the second joint of a manipulator structure. It is a slender structure whose length directly determines the range of working space. The manipulators of the same type and of different sizes are mainly distinguished by the signals on upper arms.

D. **Elbow**: Elbow is the third joint of a manipulator mechanism. It is the most flexible joint in the whole manipulator structure. Its structure is mainly to provide a carrier for the manipulator motion on two vertical planes. Almost all the manipulator motion depends on the displacement of this joint.

E. **Forearm**: Forearm is the fourth joint of a manipulator structure. Its main role is to orient the flange at the manipulator end. The velocity of this joint directly determines the normal angular velocity of manipulator end and plays a vital role in determining its position and attitude.

F. **Wrist**: Small as it is, the wrist joint includes the manipulator's two joints. In particular, end flange is an important end-effector of the manipulator. Each workpiece is directly installed in the end flange. The flange only changes the rotation direction of tool center frame. Since a transducer is an axisymmetric rotating body, the rotation angle of this joint will not affect the testing result but will affect the overall speed of the manipulator.

The six components of a manipulator are connected by joint motors. All of the six joints are rotary joints, as shown in the figure. Each joint motor also contains an accelerator, a reducer, an encoder and a brake. Such a design can save space and simplify the wiring

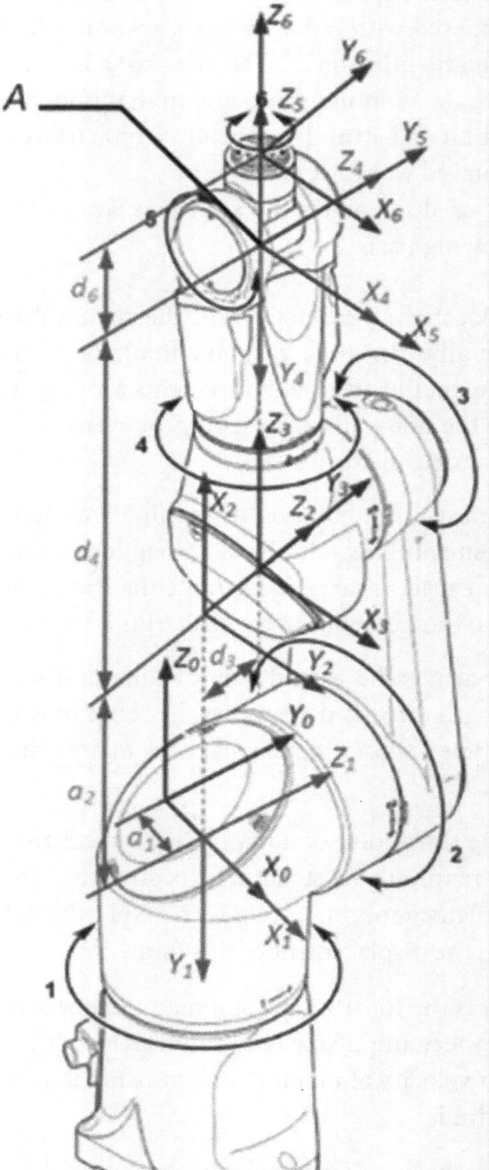

FIGURE 5.6 Structural diagram of a manipulator.

of joint motors, so the joint motors are all wrapped in the components without being exposed. In this way, the waterproof grade of a manipulator can be improved. Water is the coupling agent in the detection process. The waterproof grade of a manipulator can reach IP65, which meets the waterproof requirement of ultrasonic NDT. The encoder in each joint motor provides the information on the angular displacement of each joint to the manipulator controller, which, in turn, designs the acceleration or deceleration process according to the positions and attitudes of the current position point and the target point.

5.2.1.2 Manipulator Controller

The manipulator controller in this testing system contains a processor, field buses, a motion control card, as well as input and output interfaces and a network interface. In addition, it is equipped with a manual control pendent (MCP) as the human-machine interaction window.

The manipulator controller mainly consists of a digital power amplifier, an RPS (Redundant Power System) power supply, an ARPS (Advanced Regional Prediction System) power supply, an RSI (Robot Sensor Interface) board, a CPU, a main power switch and a PSM (Peripheral Switching Module) . It is a programmable structure connected to IPC through the serial communication protocol between Ethernet Modbus server and RS232/422. The controller has a variety of field buses, such as Profibus, CANopen, DeviceNET, Modbus and ProfiNET, which can increase the communication efficiency between the controller and the IPC. The controller also has a 16/16 digital input-output board, which can transmit the trigger pulses or receive external pulse signals. It allows the pre-compiled control program to be imported into the manipulator controller, while providing a 64 MB storage space and a 128 MB dynamic storage space for the program operation.

Each controller is equipped with an MCP. MCP is mainly used to power up and down the manipulator and control its motion. On the MCP interface, the operating mode of the manipulator can be chosen from local mode, manual mode, remote network mode and full-speed automatic mode. When the manipulator is powered on and works in manual mode, the end-effector can be instructed to choose the desired motion mode from "joint motion", "motion of base coordinate system", "motion of tool center frame" and "motion of point coordinates". According to different motion paths, different motion modes are selected to achieve the purpose of teaching. In addition, the MCP has a display screen, which consists of three areas: status bar, work interface and menu bar. The man-machine interaction on the display screen can not only facilitate the browsing of joint information but also execute the compiled manipulator program. The online programming of the manipulator can also be done through MCP, but the programming speed is limited by keystrokes. Therefore, the MCP is generally used to adjust the offline programs.

Each of the arm joints has an absolute encoder system. Through DSI, the numerical values in each encoder are transmitted to the controller as the position feedback of arm motion.

5.2.1.3 Data Acquisition Card

Data acquisition card is the core of the ultrasonic signal acquisition system. In the testing system, its main function is to sample the ultrasonic analog signal returned by pulse transceiver and then convert it into the digital signal that can be used by IPC. In order to ensure the integrity of ultrasonic signal without distortion, the sampling frequency of data acquisition card is generally more than five times the center frequency of ultrasonic signal. The transducer frequency selected for this testing system is between 0.5 and 5 MHz, while the selected data acquisition card is high-speed type, which can meet the requirements of transducer signal sampling in this system.

The high-speed acquisition card supports three trigger modes: software trigger, external trigger and encoder trigger. Software trigger is generally used for debugging ultrasonic signals, while external trigger and encoder trigger are used to meet high requirements for position trigger. In the operation process of this testing system, the trigger signal is provided by the manipulator controller. The acquisition card supports not only the hardware gate but also the hardware tracking gate. The tracking gate can eliminate the signal excursion caused by difference distances between workpiece surface and transducers during testing.

Another important feature of high-speed data acquisition card is that it supports high-speed signal acquisition. That is to say, the trigger signals are stored in the internal storage, where the signals are extracted by IPC in the stack principle of first-in first-out and can be reused to ensure no loss of trigger signals in the detection process.

5.2.1.4 Ultrasonic Signal Transceiver System

The ultrasonic signal transceiver unit mainly consists of the ultrasonic pulse transceiver and ultrasonic transducer, which, together with high-speed data acquisition card and IPC, constitute the ultrasonic signal acquisition system, as shown in Figure 5.7. It can be seen from the figure that the main function of the ultrasonic pulse transceiver is to excite the ultrasonic transducer and amplify the received echo signal or transmitted wave signal in hardware. To improve the detection accuracy, the position and attitude of the manipulator can be triggered in real time. The position and attitude of manipulator arm in motion are monitored by manipulator controller in real time. When the arm moves to the specified trajectory point, the controller will send out a trigger pulse, and the ultrasonic pulse transceiver and the acquisition card will work simultaneously to collect an ultrasonic signal.

The ultrasonic pulse transceiver has a single square pulse for excitation. Its voltage gain is between −59 and 59 dB, which is more favorable for the testing of composite materials with larger thickness or higher attenuation. With the pulse repetition frequency of 100 K–5 KHz, the transceiver supports the mode of external trigger and two modes of ultrasonic excitation and reception, including the self-transmission and self-receiving of a single transducer and the separate transmission and receiving of two transducers.

Because the objects to be tested are the composite workpieces with large thickness, the center frequency of the ultrasonic transducer should not be high. The testing system needs to select the ultrasonic transducer according to the thickness and material properties of the tested workpiece. In this system, the immersion-type ultrasonic transducer with the central frequency of 0.5–5 MHz is mainly used.

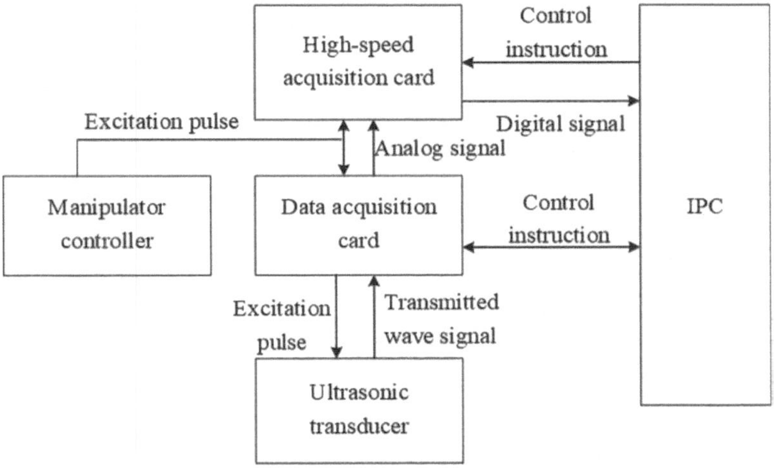

FIGURE 5.7 Ultrasonic signal transceiver system.

5.2.1.5 Water-Coupled Circulation System

Since ultrasonic attenuation in space is large, water or other media should be used as a coupling agent between the transducer and tested workpiece. Limited by the workpiece size and manipulator waterproofing, the full immersion method is not fit for the workpiece. Thus, spray coupling is used for the transducer.

The structure of a water-coupled circulation system is shown in Figure 5.8. The spray coupling of the ultrasonic transducer has a high requirement for water quality, which, if inferior, will weaken the received ultrasonic signal. The role of filter screen is just to filter out the particles and impurities in the water. Bubbles in water are also an important factor affecting the intensity of received signals. In order to reduce the generation of bubbles, the water in the catch basin is first channeled to the water storage tank through a submersible pump. The water flow at the bottom of the tank is stable without obvious bubbles. At the bottom of the tank, the coupling agent meeting the requirements is provided by centrifugal pump to the ultrasonic transducer. In the testing process, the flow rate of water spray nozzle should be adjusted and controlled through two flow valves. When the water flows too fast, the rapid flow will splash at the contact point between water column and workpiece surface so that no stable coupling effect will be yielded and the incident energy of acoustic beam will be affected. When the water flows too slowly, the water column will bend under the gravity action and can't contact with the workpiece surface.

The nozzle of the transducer-holding device is illustrated in Figure 5.9. The design parameters of this nozzle will be given in detail in the following chapters. The nozzle is mainly divided into two parts: fluid storage chamber and mounting bracket. A fluid pool is formed in the storage chamber to ensure that all transducers can be locally immersed in the fluid pool. The effect of local liquid immersion is beneficial to the emission, transmission and reception of ultrasonic signals. The mounting bracket mainly plays a supporting role to keep the position and attitude relationship between the transducer and manipulator end-effector unchanged, while providing an installation space for the transducer.

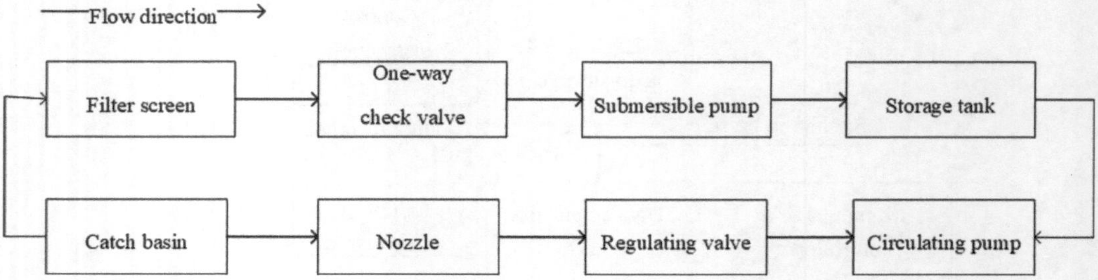

FIGURE 5.8 Water-coupled circulation system.

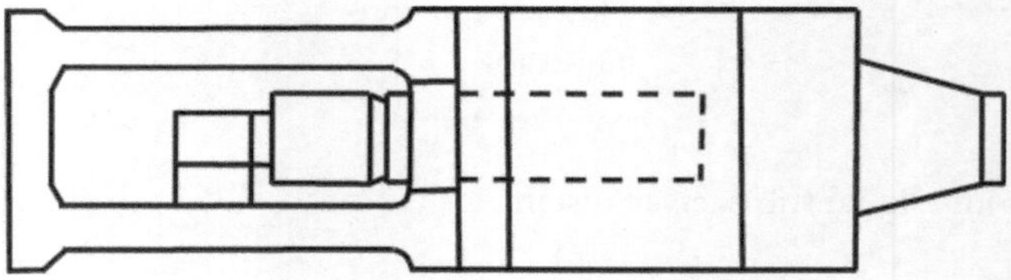

FIGURE 5.9 Schematic diagram of a water nozzle.

5.2.2 Upper Computer Software of a Dual-Manipulator Testing System

If the whole dual-manipulator ultrasonic NDT system is likened to a person, then the hardware system will be the human body, while the software system will be the human soul. The software system and the hardware system complement each other and are indispensable in the detection task. The whole dual-manipulator software system includes upper computer software and lower computer software. IPC, also called upper computer, is the main control computer for human-computer interaction in the testing system. CS8C manipulator controller, known as lower computer, is used for some real-time simple control operations. The upper and lower computers communicate and exchange information with each other through the Ethernet port to realize feedback and control.

5.2.2.1 Overall Design of Upper Computer Software

When selecting the programming language and development tool of the upper computer software system, we should first consider whether each unit in the hardware system supports the API (Application Programming Interface) of the corresponding language. The manipulator controller in this testing system supports the API interfaces of several languages, but the AL12250 high-speed data acquisition card currently only provides the development package of .NET Framework. Therefore, .NET Framework 4.0 is selected as the development tool of the software system. The integrated compilation platform Visual Studio 2010 is selected as the compilation tool, and C# language is selected as the programming language through comprehensive consideration. The testing of a curved workpiece is a multi-tasking process, involving motion control, ultrasonic signal acquisition, data processing and imaging algorithm. Therefore, it is necessary to consider whether the development tool and programming language support the multi-tasking process. However, the C#

language in the .NET Framework is an object-oriented programming language that supports multithreaded processing. In addition, the program is packaged completely and is not crashed easily. Finally, the graphical user interface (GUI) of .NET Framework program is intuitive and friendly. Owing to the use of rich graphical interfaces, the color display of detection results becomes more realistic. This programming language is easy to use and the developed GUI is intuitive and friendly.

The upper computer of the dual-manipulator ultrasonic NDT system is based on the idea of modular programming and is mainly composed of four modules: data acquisition, manipulator motion control, automatic scanning and data processing. The overall software structure of this system is shown in Figure 5.10.

5.2.2.2 Data Acquisition

Before the A-signal data is collected by data acquisition card, the A-signal parameters such as trigger channel, trigger frequency, time delay and sampling length should be set. Then the A-signal waveform data is obtained by the called API method and is displayed in the A-signal window through waveform imaging algorithm. The whole data parameter setting and A-signal waveform display window are shown in Figure 5.11. In the illustrated waveform display area, the yellow gate is a tracking gate, and the red gate is a data gate. The time difference between the initial sampling point of data gate and the first peak value of tracking gate is a constant value.

5.2.2.3 Synchronous Control of Dual Manipulator

The synchronous motion control module of dual manipulator is mainly developed by calling the dynamic link library (DLL) of manipulator controller. This module is an important part of the testing system. Its main functions and parameter settings include manipulator login, trajectory point initialization, master/slave manipulators zeroing, synchronous motion start-stop and manipulator status query. The software control interface is shown in Figure 5.12.

By adding the DLL to the reference list of the whole project, the software can call the API interface function of manipulator controller conveniently. In whatever development environment for manipulator control design, the communication connection, command sending, status query feedback and other processes of manipulator controller always follow the same model. The software is mainly divided into the following parts:

1. **Login:** Before being controlled, the manipulator controller should be connected to the upper computer to ensure that its IP is the same as that in the IP textbox of the interface.
2. **Power-up and power-down:** The manipulator must be powered on to ensure that it can move. At the end of the program, the manipulator should be powered down to ensure its safety.
3. **Zeroing:** Prior to the scanning motion, the manipulator should move to the zero point specified by the program to set and optimize the acquisition parameters. This

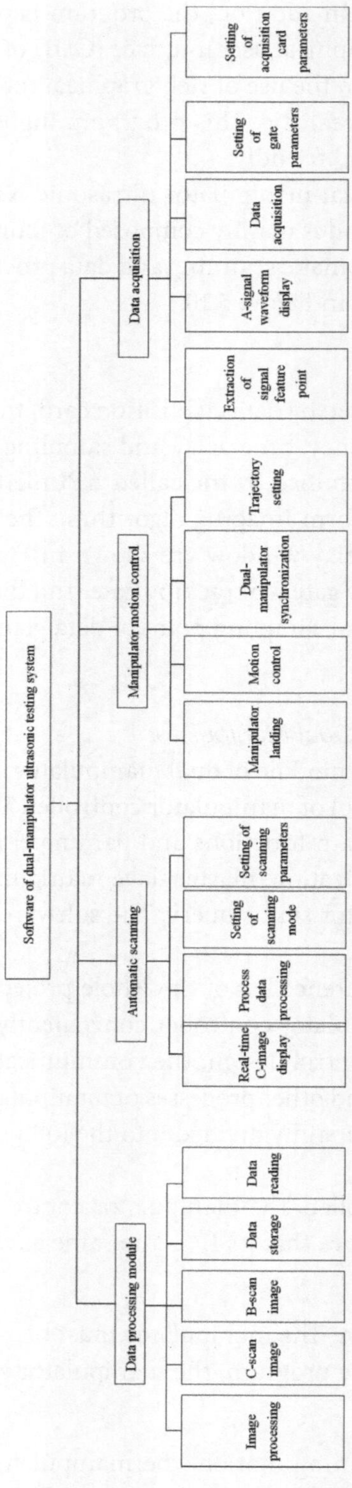

FIGURE 5.10 Software structure of a dual-manipulator ultrasonic NDT system.

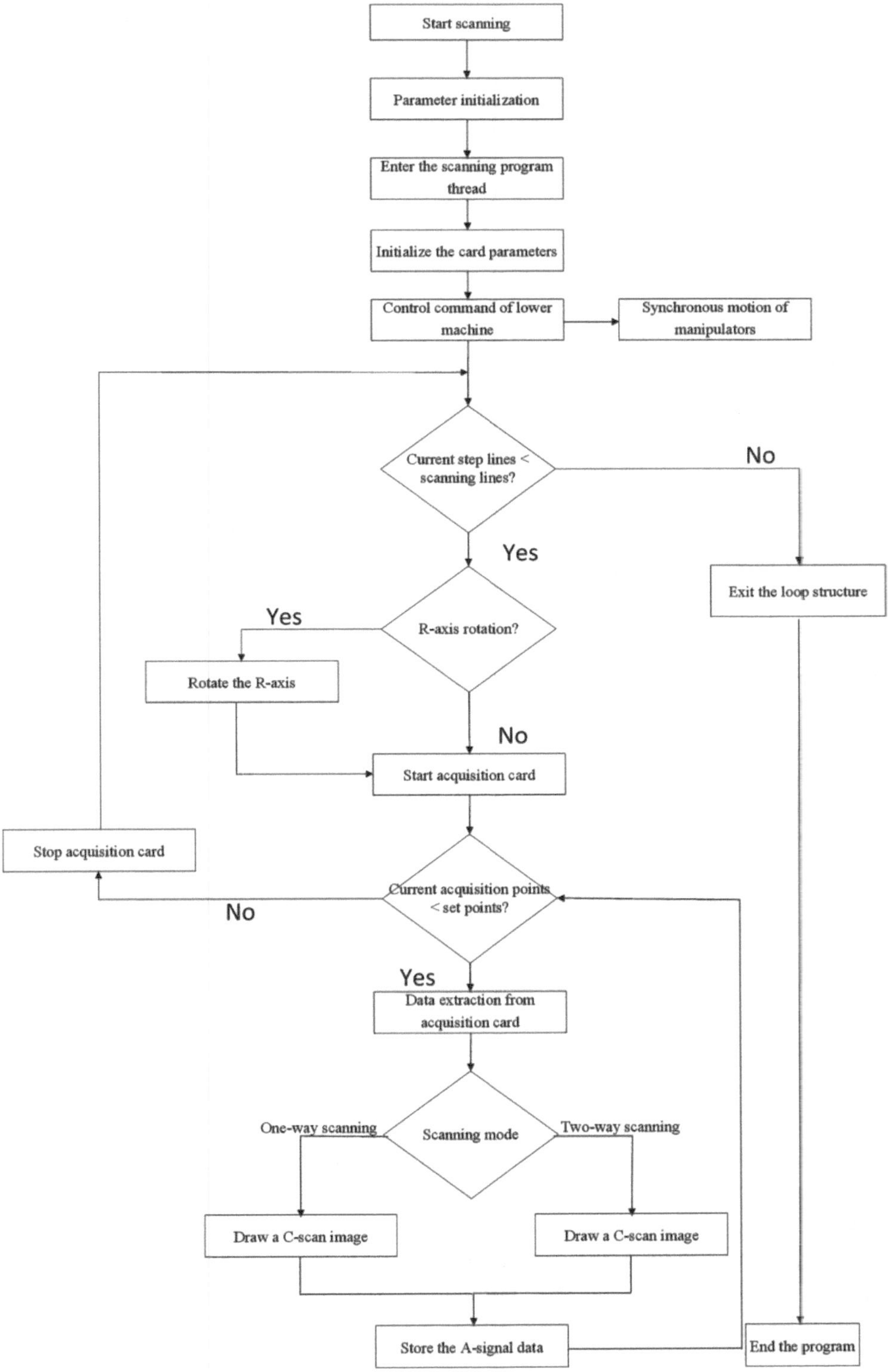

FIGURE 5.11 Data acquisition loop module.

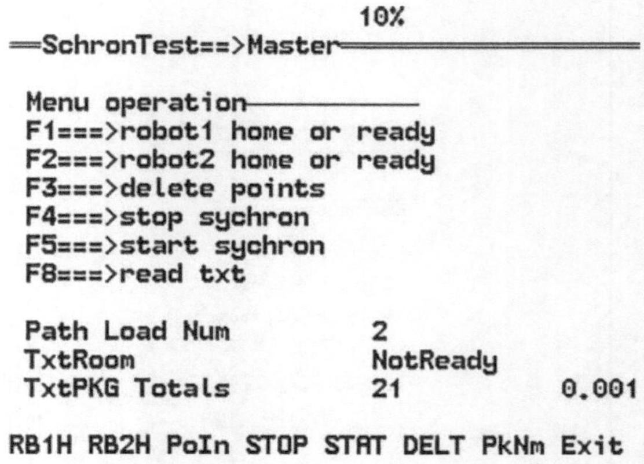

FIGURE 5.12 Human-computer interaction interface of a lower computer program.

software controls two manipulators, thus providing two functions, namely master manipulator zeroing and slave manipulator zeroing.

4. **Trajectory point initialization:** During testing, the manipulator should move according to the preset trajectory points that are initialized before testing.

5. **Synchronous motion:** The synchronous motion of dual manipulator is started under control. The start-stop of synchronous motion is to verify whether the motion states of the two manipulators are synchronized on the given trajectory.

6. **Point library clearing:** The point packets reserved in the manipulator memory need to be cleared at the end of manipulator motion to ensure that the next scan will not be affected.

5.2.2.4 Automatic Scanning Imaging Module

The evaluation of detection result is based on the C-scan imaging result of a defect. In order to intuitively evaluate the detection result, the detection data of each scanning point in the scanning area should be displayed in the form of images. Before the start of automatic scanning, the scanning mode, imaging algorithm, size of scanning area, scanning accuracy and other parameters should be set for each scanning area. The automatic scanning imaging module will initialize the size of C-image according to these parameters, calculate the corresponding scale to fit the display area and create the corresponding ruler. The automatic scanning C-imaging window is shown in Figure 5.13.

In the process of testing, the software reads the A-signal full-time-domain waveform data of each scanning point through the loop module, extracts the characteristic information of A-signal waveform inside the data gate, converts it into C-image color information through the imaging algorithm and adds the color information of trajectory points to C-image through the coloring algorithm.

In the process of automatic scanning, the loop module should be used to read the data in the acquisition card in sequence. The sequences of A-signal data in the acquisition card

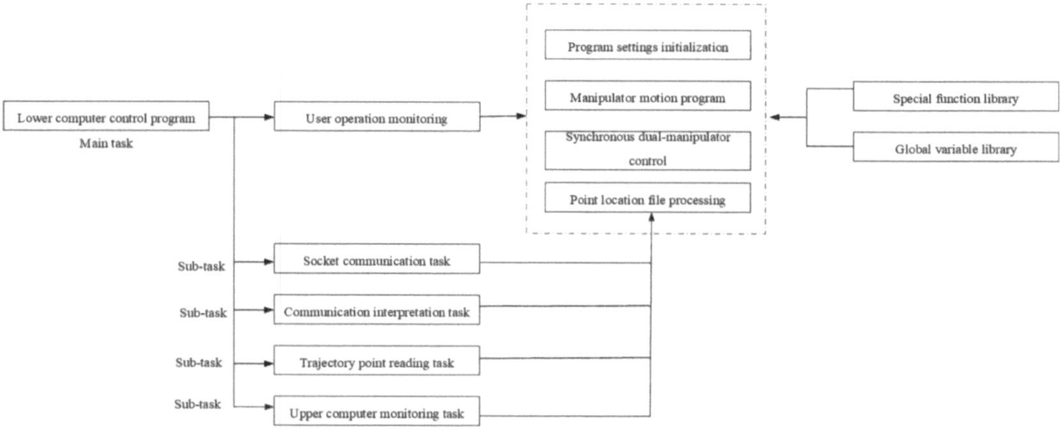

FIGURE 5.13 Structure of a lower computer program.

cache should be in a one-to-one correspondence to the trigger signals of manipulator controller. The manipulator controller program should ensure the correspondence between external trigger signals and scanning trajectory points and the strict synchronization of trigger time. The flow chart of data-reading loop in the automatic scanning imaging module is shown in Figure 5.11.

5.2.3 Lower Computer Software of a Dual-Manipulator Testing System

The manipulator controller program is compiled in Variable Assembly Language (VAL) on the compilation platform "Robotics Suite". VAL was first developed by Unimation in 1979 as an assembly language for minicomputer control. After two generations of improvement, the VAL3 programming language currently used by controllers has combined the advantages of assembly language and high-level programming language. Its instructions are concise and readable, suitable for both online and offline programming.

Parallel multi-task processing is an important feature of VAL3 control language. It allows parallel multi-task threads to be created in the program and the task priorities to be established. The tasks coordinate with each other in the operation process. Therefore, complex control relations such as program variable refreshing, arm state monitoring, state feedback and communication are established to constitute interdependent and mutually supportive logic control program systems. At the same time, the C# language used by upper computer can interact with VAL3 language very well. The C# language can view the global state variables of VAL3 language and call the VAL3 program functions. This provides a basis for the interaction and feedback between upper and lower computer software systems.

The program of manipulator controller can call the function method of MCP display to show the state of program execution. The interface of human-computer interaction can provide convenience for offline manipulator debugging. The main interface of lower computer program is shown in Figure 5.12.

To ensure the extension of program functions, the VAL3 presents the idea of calling global libraries. According to the characteristics of VAL3 program design, the lower computer program uses global library and multi-tasking to realize synchronous manipulator control and ensure real-time synchronous motion. The program structure of lower computer is shown in Figure 5.13.

It can be seen from the structure of lower computer program that the coordinated control of dual manipulator can be realized by multi-tasking. The contents of lower computer control program can be divided into dual-manipulator motion control, dual-manipulator communication control and manipulator trajectory file processing.

5.3 MAPPING RELATION BETWEEN DUAL-MANIPULATOR BASE COORDINATE SYSTEMS

In the testing system, the workpiece under test is automatically scanned and tested by ultrasonic transmission, as shown in Figure 5.14. The planning of workpiece trajectory is based on the workpiece coordinate system on manipulator. During testing, the workpiece coordinate system on manipulator should coincide with that in the workpiece trajectory planning. Otherwise, the actual scanning trajectory of the manipulator will deviate from the planned scanning trajectory. Therefore, in order to ensure the specified synchronous coordinated motion of the two manipulators, the settings of their workpiece coordinate systems (i.e. position and coordinate axis direction) should coincide.

The method for a manipulator to set the workpiece coordinate system is to let the manipulator move to three points and then determine the coordinates of the three points in the base coordinate system. The three points include the origin of the workpiece coordinate system, a point on the *X*-axis and a point in the *OXY* plane. The manipulator determines the position and attitude of the workpiece coordinate system relative to the base coordinate system through the internal program algorithm. The traditional method to realize the coincidence of coordinate systems on two manipulators is to let the manipulators move to the three points at the same time and align their end-effectors at those points. However,

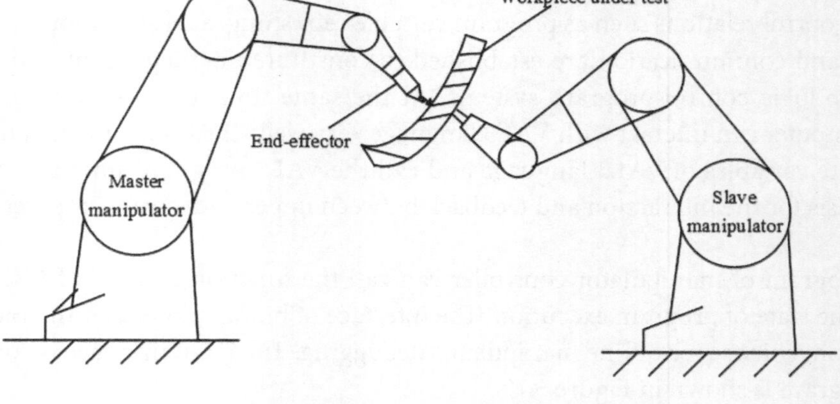

FIGURE 5.14 Ultrasonic NDT system of a dual manipulator.

for some workpieces (such as cylindrical workpieces and large curved workpieces), the two manipulators are separately located at two ends of the workpieces and are unable to reach the three points on the workpiece coordinate system at the same time. In this case, their workpiece coordinate systems can not coincide. Therefore, the method of workpiece coordinates coincidence is very necessary for automatic ultrasonic NDT. In this chapter, a method of workpiece coordinates coincidence based on the relation between base coordinate systems is proposed.

In a traditional dual-manipulator system, the calibration method of the base coordinate system relies on other equipment, or the calibration process is relatively complex to operate and is hard to be applied to the actual ultrasonic NDT system. Therefore, this chapter proposes a method of workpiece coordinates coincidence, which only needs three points in space that are not collinear. The method of matrix orthogonalization uses the exponential mapping algorithm of Lie algebra to improve the efficiency of matrix orthogonalization. The calibration results of the dual-manipulator base coordinate system are used to realize the coincidence between the workpiece coordinate systems of the two manipulators. In addition, compared with the traditional method, this method can set the workpiece coordinate system of the dual-manipulator system more conveniently. In particular, when the workpiece changes, only one manipulator needs to be driven to complete the setting of the workpiece coordinate system, and the other manipulator is used to set the three points of the workpiece coordinate system, which can be obtained through the transformation relationship between the two manipulators. Therefore, this method can simplify the operation steps and save time.

5.3.1 Transformation Relationship between Base Coordinate Systems

5.3.1.1 Definition of Parameters of a Manipulator Coordinate System

Before solving the problem of workpiece coordinates coincidence, the transformation relation between the two manipulators should be obtained at first. The transformation relation in a fixed position will not change with the position and attitude of manipulator end-effector. Therefore, when we obtain the transformation relationship between the two manipulators, we only need to know the numerical values of points in the workpiece coordinate system of one manipulator in order to obtain the values of these points in the base coordinate system of the other manipulator. Such a method can ensure the coincidence of the two workpiece coordinate systems.

The calibration process of the base-coordinates relation between the two manipulators is shown in Figure 5.15. One manipulator is defined as the master manipulator R_m, while the other cooperative manipulator is defined as the slave manipulator R_s. F_w is the world coordinate system in space, F_m is the base coordinate system of the master manipulator and F_s is the base coordinate system of the slave manipulator.

Suppose P_1 is any point in space. It is represented by ${}^{w}P_1$ in the world coordinate system, ${}^{m}P_1$ in the master base coordinate system F_m and ${}^{s}P_1$ in the slave base coordinate system F_s, namely:

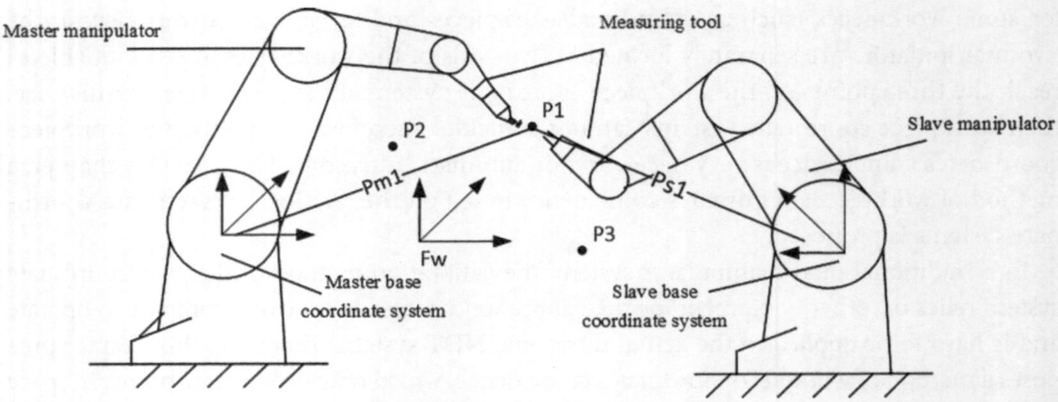

FIGURE 5.15 Transformation relationship between two manipulators.

$$\begin{aligned} {}^{w}P_1 &= \left({}^{w}X_1, {}^{w}Y_1, {}^{w}Z_1\right)^T \\ {}^{m}P_1 &= \left({}^{m}X_1, {}^{m}Y_1, {}^{m}Z_1\right)^T \\ sP_1 &= \left({}^{s}X_1, {}^{s}Y_1, {}^{s}Z_1\right)^T \end{aligned} \tag{5.3}$$

${}^{s}_{m}R$ and P_{marg} are the rotation matrix and translation matrix from the master base coordinate system F_m to the slave base coordinate system F_s. According to the general transformation mapping from one coordinate system to another based on the description of the point vector, we can obtain:

$$ {}^{s}P = {}^{s}_{m}R^{m}P + p_{morg} \tag{5.4}$$

5.3.1.2 Solution of an Unified Variable Method

Suppose there are four points in the overlapping workspace of the two manipulators. Their coordinates in the master base coordinate system are:

$$\begin{aligned} {}^{m}P_1 &= \left({}^{m}X_1, {}^{m}Y_1, {}^{m}Z_1\right)^T \\ {}^{m}P_2 &= \left({}^{m}X_2, {}^{m}Y_2, {}^{m}Z_2\right)^T \\ {}^{m}P_3 &= \left({}^{m}X_3, {}^{m}Y_3, {}^{m}Z_3\right)^T \\ {}^{m}P_4 &= \left({}^{m}X_4, {}^{m}Y_4, {}^{m}Z_4\right)^T \end{aligned} \tag{5.5}$$

Their representations in the slave base coordinate system are:

$$
\begin{aligned}
{}^{s}P_1 &= \left({}^{s}X_1, {}^{s}Y_1, {}^{s}Z_1\right)^T \\
{}^{s}P_2 &= \left({}^{s}X_2, {}^{s}Y_2, {}^{s}Z_2\right)^T \\
{}^{s}P_3 &= \left({}^{s}X_3, {}^{s}Y_3, {}^{s}Z_3\right)^T \\
{}^{s}P_4 &= \left({}^{s}X_4, {}^{s}Y_4, {}^{s}Z_4\right)^T
\end{aligned}
\tag{5.6}
$$

By substituting the coordinate values of Eqs. (5.5) and (5.6) into Eq. (5.3), we can obtain:

$$
\begin{pmatrix} {}^{s}X_1 \\ {}^{s}Y_1 \\ {}^{s}Z_1 \end{pmatrix} = {}^{s}_{m}R \begin{pmatrix} {}^{m}X_1 \\ {}^{m}Y_1 \\ {}^{m}Z_1 \end{pmatrix} + P_{morg} \tag{5.7}
$$

$$
\begin{pmatrix} {}^{s}X_2 \\ {}^{s}Y_2 \\ {}^{s}Z_2 \end{pmatrix} = {}^{s}_{m}R \begin{pmatrix} {}^{m}X_2 \\ {}^{m}Y_2 \\ {}^{m}Z_2 \end{pmatrix} + P_{morg} \tag{5.8}
$$

$$
\begin{pmatrix} {}^{s}X_3 \\ {}^{s}Y_3 \\ {}^{s}Z_3 \end{pmatrix} = {}^{s}_{m}R \begin{pmatrix} {}^{m}X_3 \\ {}^{m}Y_3 \\ {}^{m}Z_3 \end{pmatrix} + P_{morg} \tag{5.9}
$$

$$
\begin{pmatrix} {}^{s}X_4 \\ {}^{s}Y_4 \\ {}^{s}Z_4 \end{pmatrix} = {}^{s}_{m}R \begin{pmatrix} {}^{m}X_4 \\ {}^{m}Y_4 \\ {}^{m}Z_4 \end{pmatrix} + P_{morg} \tag{5.10}
$$

where ${}^{s}_{m}R$ is a 3×3 matrix describing the rotation of the slave base coordinate system relative to the master base coordinate system, and P_{morg} is a 3×1 matrix describing the vector of the origin of the slave base coordinate system relative to the master base coordinate system. Suppose that the factors in these matrices are unknown variables:

$$
{}^{s}_{m}R = \begin{pmatrix} r_{11} & r_{12} & r_{13} \\ r_{21} & r_{22} & r_{23} \\ r_{31} & r_{32} & r_{33} \end{pmatrix}, P_{morg} = \begin{pmatrix} P_{ox} \\ P_{oy} \\ P_{oc} \end{pmatrix} \tag{5.11}
$$

Eqs. (5.7)–(5.10) are all linear equations. By substituting the elements of Eq. (5.11) into Eqs. (5.7)–(5.10) and following the principle of linear equation, the target unknown variables to be solved, namely the elements in the rotation matrix and translational matrix, can be determined:

$$\begin{pmatrix} {}^mX_1 & 0 & 0 & {}^mY_1 & 0 & 0 & {}^mZ_1 & 0 & 0 & 1 & 0 & 0 \\ 0 & {}^mX_1 & 0 & 0 & {}^mY_1 & 0 & 0 & {}^mZ_1 & 0 & 0 & 1 & 0 \\ 0 & 0 & {}^mX_1 & 0 & 0 & {}^mY_1 & 0 & 0 & {}^mZ_1 & 0 & 0 & 1 \\ {}^mX_2 & 0 & 0 & {}^mY_2 & 0 & 0 & {}^mZ_2 & 0 & 0 & 1 & 0 & 0 \\ 0 & {}^mX_2 & 0 & 0 & {}^mY_2 & 0 & 0 & {}^mZ_2 & 0 & 0 & 1 & 0 \\ 0 & 0 & {}^mX_2 & 0 & 0 & {}^mY_2 & 0 & 0 & {}^mZ_2 & 0 & 0 & 1 \\ {}^mX_3 & 0 & 0 & {}^mY_3 & 0 & 0 & {}^mZ_3 & 0 & 0 & 1 & 0 & 0 \\ 0 & {}^mX_3 & 0 & 0 & {}^mY_3 & 0 & 0 & {}^mZ_3 & 0 & 0 & 1 & 0 \\ 0 & 0 & {}^mX_3 & 0 & 0 & {}^mY_3 & 0 & 0 & {}^mZ_3 & 0 & 0 & 1 \\ {}^mX_4 & 0 & 0 & {}^mY_3 & 0 & 0 & {}^mY_4 & 0 & 0 & 1 & 0 & 0 \\ 0 & {}^mX_4 & 0 & 0 & {}^mY_3 & 0 & 0 & {}^mZ_4 & 0 & 0 & 1 & 0 \\ 0 & 0 & {}^mX_4 & 0 & 0 & {}^mY_3 & 0 & 0 & {}^mX_4 & 0 & 0 & 1 \end{pmatrix}$$

$$\times \begin{pmatrix} r_{11} \\ r_{21} \\ r_{31} \\ r_{12} \\ r_{22} \\ r_{32} \\ r_{13} \\ r_{23} \\ r_{33} \\ P_{ox} \\ P_{oy} \\ P_{oz} \end{pmatrix} = \begin{pmatrix} {}^sX_1 \\ {}^sY_1 \\ {}^sZ_1 \\ {}^sX_2 \\ {}^sY_2 \\ {}^sZ_2 \\ {}^sX_3 \\ {}^sY_3 \\ {}^sZ_3 \\ {}^sX_4 \\ {}^sY_4 \\ {}^sZ_4 \end{pmatrix} \tag{5.12}$$

Eq. (5.12) can be regarded as a system of linear equations, namely $AX = B$. The necessary and sufficient condition for the unique solution of n-element linear equations is

$$r(A) = r(A, B) = n \tag{5.13}$$

In this case, n is equal to 12, which means that each column of vectors in the matrix A is linearly independent. If any four points in space are not in the same plane, any three sets of vectors consisting of two points will be linearly independent.

Suppose that there are three non-collinear points in space, namely P_1, P_2 and P_3. They satisfy the following relation:

$$P_1P_2 \neq aP_1P_3 \tag{5.14}$$

where $a \in R$ and $a \neq 0$. If the fourth point P_4 is linearly independent of them, it will not lie on the plane formed by the other three points. Their coordinates will satisfy the following relation:

$$P_1P_4 \neq aP_1P_2 + bP_1P_3 \tag{5.15}$$

where $P_1P_4 \neq 0$, $a,b \in R$ and $a,b \neq 0$.

If P_4 lies on the vector orthogonal to the vectors P_1P_2 and P_1P_3, then it must not be found on the plane composed of P_1, P_2 and P_3. Therefore, in the base coordinate system of the master manipulator, ${}^mP_4$ can be calculated by the following equation:

$$ {}^mP_1{}^mP_4 = \frac{{}^mP_1{}^mP_2 \times {}^mP_1{}^mP_3}{2\left|{}^mP_1{}^mP_2 \times {}^mP_1{}^mP_3\right|}\left(\left|{}^mP_1{}^mP_2\right| + \left|{}^mP_1{}^mP_3\right|\right) \tag{5.16}$$

Similarly, in the base coordinate system of the slave manipulator, we can obtain:

$$ {}^sP_1{}^sP_4 = \frac{{}^sP_1{}^sP_2 \times {}^sP_1{}^sP_3}{2\left|{}^sP_1{}^sP_2 \times {}^sP_1{}^sP_3\right|}\left(\left|{}^sP_1{}^sP_2\right| + \left|{}^sP_1{}^sP_3\right|\right) \tag{5.17}$$

The four points in space obtained in this way can satisfy the condition in Eq. (5.13). Because the four points are not on the same plane, the condition in Eq. (5.11) can be satisfied. The elements of the rotation matrix and translation matrix constitute the solution of Eq. (5.10).

5.3.1.3 Solving with a Homogeneous Matrix Method

Eq. (5.2) describes the general transformation mapping that transforms a vector representation from one coordinate system to another coordinate system. However, a matrix operator will be more concise and clear than Eq. (5.2) when describing this transformation mapping. We add the component "1" to the coordinate vector of the point 3×1, combine the rotation matrix 3×3 with the translation matrix 3×1 and add the component "[0, 0, 0, 1]" to the last row of the combined matrix. Then Eq. (5.2) can be rewritten as

$$\begin{bmatrix} {}^sP \\ 1 \end{bmatrix} = \begin{bmatrix} {}^s_mR & P_{morg} \\ 0,0,0, & 1 \end{bmatrix} \cdot \begin{bmatrix} {}^mP \\ 1 \end{bmatrix} \tag{5.18}$$

The matrix 4×4 in Eq. (5.16) is called homogeneous transformation matrix. It can be thought of as a simple matrix that represents the rotation and translation in a general transformation.

According to Eq. (5.16), Eq. (5.18) can be simplified as

$$ {}^{s}P = {}_{m}^{s}T \cdot {}^{m}P \tag{5.19} $$

By substituting the coordinates of the four points P_1, P_2, P_3 and P_4 in space into Eq. (5.17), we can obtain

$$ \begin{pmatrix} {}^{s}X_1 \\ {}^{s}Y_1 \\ {}^{s}Z_1 \\ 1 \end{pmatrix} = {}_{m}^{s}T \begin{pmatrix} {}^{m}X_1 \\ {}^{m}Y_1 \\ {}^{m}Z_1 \\ 1 \end{pmatrix} \tag{5.20} $$

$$ \begin{pmatrix} {}^{s}X_2 \\ {}^{s}Y_2 \\ {}^{s}Z_2 \\ 1 \end{pmatrix} = {}_{m}^{s}T \begin{pmatrix} {}^{m}X_2 \\ {}^{m}Y_2 \\ {}^{m}Z_2 \\ 1 \end{pmatrix} \tag{5.21} $$

$$ \begin{pmatrix} {}^{s}X_3 \\ {}^{s}Y_3 \\ {}^{s}Z_3 \\ 1 \end{pmatrix} = {}_{m}^{s}T \begin{pmatrix} {}^{m}X_3 \\ {}^{m}Y_3 \\ {}^{m}Z_3 \\ 1 \end{pmatrix} \tag{5.22} $$

$$ \begin{pmatrix} {}^{s}X_4 \\ {}^{s}Y_4 \\ {}^{s}Z_4 \\ 1 \end{pmatrix} = {}_{m}^{s}T \begin{pmatrix} {}^{m}X_4 \\ {}^{m}Y_4 \\ {}^{m}Z_4 \\ 1 \end{pmatrix} \tag{5.23} $$

According to the principle of linear superposition of homogeneous matrices, the elements of rotation matrix and translation matrix in Eq. (5.9) are substituted into Eq. (5.23) to obtain

$$ \begin{pmatrix} {}^{s}X_1 & {}^{s}X_2 & {}^{s}X_3 & {}^{s}X_4 \\ {}^{s}Y_1 & {}^{s}Y_2 & {}^{s}Y_3 & {}^{s}Y_4 \\ {}^{s}Z_1 & {}^{s}Z_2 & {}^{s}Z_3 & {}^{s}Z_4 \\ 1 & 1 & 1 & 1 \end{pmatrix} = \begin{pmatrix} r_{11} & r_{12} & r_{13} & P_{ox} \\ r_{21} & r_{22} & r_{23} & P_{oy} \\ r_{31} & r_{32} & r_{33} & P_{oz} \\ 0 & 0 & 0 & 1 \end{pmatrix} \begin{pmatrix} {}^{m}X_1 & {}^{m}X_2 & {}^{m}X_3 & {}^{m}X_4 \\ {}^{m}Y_1 & {}^{m}Y_2 & {}^{m}Y_3 & {}^{m}Y_4 \\ {}^{m}Z_1 & {}^{m}Z_2 & {}^{m}Z_3 & {}^{m}Z_4 \\ 1 & 1 & 1 & 1 \end{pmatrix} \tag{5.24} $$

Similarly, Eq. (5.22) can be regarded as a set of linear equations, namely $B = XA$. To ensure that the equation set has and only has a unique solution, both the matrix A and matrix B should be full-rank matrices and meet the requirement of Eq. (5.11), that is, the four points are not coplanar in public space. Then, each element of rotation matrix and translation matrix in Eq. (5.22) can be expressed as

$$\begin{pmatrix} r_{11} & r_{12} & r_{13} & P_{ox} \\ r_{21} & r_{22} & r_{23} & P_{oy} \\ r_{31} & r_{32} & r_{33} & p_{oz} \\ 0 & 0 & 0 & 1 \end{pmatrix} = \begin{pmatrix} {}^sX_1 & {}^sX_2 & {}^sX_3 & {}^sX_4 \\ {}^sY_1 & {}^sY_2 & {}^sY_3 & {}^sY_4 \\ {}^sZ_1 & {}^sZ_2 & {}^sZ_3 & {}^sZ_4 \\ 1 & 1 & 1 & 1 \end{pmatrix} \begin{pmatrix} {}^mX_1 & {}^mX_2 & {}^mX_3 & {}^mX_4 \\ {}^mY_1 & {}^mY_2 & {}^mY_3 & {}^mY_4 \\ {}^mZ_1 & {}^mZ_2 & {}^mZ_3 & {}^mZ_4 \\ 1 & 1 & 1 & 1 \end{pmatrix} \quad (5.25)$$

Like in the unified variable method, the coordinates of the fourth point in public space can be calculated by combining the coordinates of the three non-collinear points in public space with Eqs. (5.14) and (5.15).

5.3.2 Orthogonal Normalization of Rotation Matrix

Theoretically, the rotation matrix between two Cartesian coordinate systems is a unitary orthogonal matrix, but the solution of the rotation matrix may not satisfy this condition. In addition, the measurement process is limited by error and uncertainty, so the obtained rotation matrix deviates from the unitary orthogonal matrix by a small error. Therefore, it is necessary to optimize the rotation matrix to ensure that it is a unitary orthogonal matrix. This step can improve the accuracy of the transformation relation between two manipulators.

5.3.2.1 Basis of Lie Group and Lie Algebra

Lie group is applied to partial differential equations, ordinary differential equations (groups) and autonomous systems to some extent. Lie group is a smooth manifold G, as shown in Figure 5.16. It satisfies the group operation, that is, its multiplication operation and inversion operation are smooth mappings.

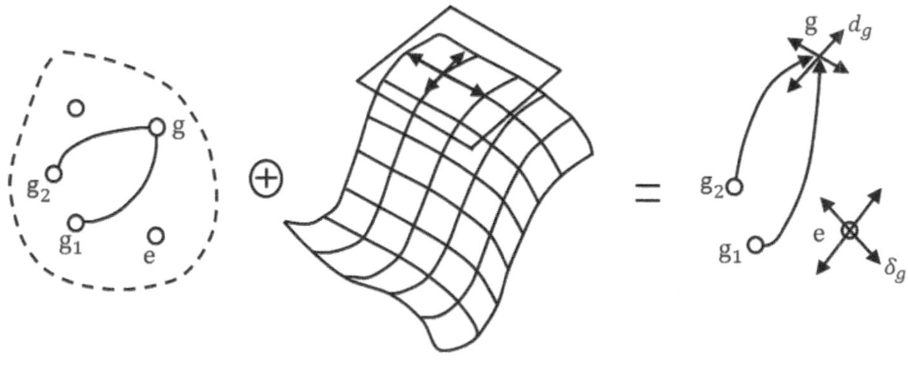

FIGURE 5.16 Basic structure of Lie group.

Multiplication operation:

$$\begin{gathered} G\times G\to G \\ (f,g)\to fg,\forall f,g\in G \end{gathered} \tag{5.26}$$

Inversion operation:

$$\begin{gathered} G\times G\to G \\ f\to f^{-1},\forall f\in G \end{gathered} \tag{5.27}$$

The multiplication operation is a two-dimensional group operation. The image of an element with respect to inverse operation is the inverse of that element. The multiplication operation satisfies the associative law in the group operation rule, while the inverse operation satisfies the group operation rule of one-to-one mapping.

The rotation transformation in Euclidean space is described by the rotation matrix R, which is an element of the special orthogonal group SO(3). SO(3) is a group of $SO(n;f)=O(n;f)\cap SL(n:f)$ matrices existing in the Lie group and is an instance of Lie group. For two elements $A,B\in G$ in a Lie group, if they satisfy the following two characteristics [1], the Lie group will be a differential manifold:

1. For the mapping, $f(A,B)=AB,\cdot f(A,B)\in G$ is required and f should be continuously differentiable;
2. The mapping $A\to A^{-1}$ must exist and must be continuously differentiable.

For $A,B\in SO(3)$, A and B must satisfy the two features of rotation matrix, namely $RR^T=R^TR=E$ and $\mathbf{det}(\boldsymbol{R})=+1$, in order to satisfy the feature (1). Since the product of rotation matrices is continuously differentiable, the element *f* is also a differentiable function [2]. In addition, if the rotation matrix satisfies $R^{-1}=R^T$, it must also satisfy the feature (2) [3].

The 4×4 homogeneous transformation matrix representing the position and attitude transformation is a Lie group and also a special group $SE(3)$ in Euclidean space. Given the rotation matrix $R\in SO(3)$ and the translation matrix $p\in SE(3)$, the matrix consisting of them will be $(R,p)\in SE(3)$, and its inverse matrix can be written as

$$\begin{bmatrix} R & p \\ 0 & 1 \end{bmatrix}^{-1}=\begin{bmatrix} R^T & -R^Tp \\ 0 & 1 \end{bmatrix} \tag{5.28}$$

An important concept associated with Lie group is Lie algebra. The tangential space of the identity element in a Lie group is called the Lie algebra of this Lie group, represented by "g". The Lie algebra, together with a bidirectional mapping, is called Lie parenthesis, which constitutes a vector space. The bilinear mapping [x, y] satisfies:

1. **Antisymmetry:** $[x,y]=-[y,x]$;
2. **Jacobi identity:** $[x,[y,z]]+[z,[x,y]]+[y,[z,x]]=0.$

where each x,y,z belongs to the Lie algebra g. The Lie parenthesis is defined as follows:

$$
\begin{gathered}
[u,v]\Big|_{t=s=0}=\frac{\partial^2}{\partial t\,\partial s}g(t)h(s)g(t)^{-1} \\
g(t),h(s)\in G, g(0)=h(0)=e, g'(0)=u, h'(0)=v
\end{gathered}
\tag{5.29}
$$

The Lie algebra so(3) related to the special orthogonal group SO(3) can be obtained by calculating the tangent vector of a smooth curve on the Lie group SO(3). The Lie algebra so(3) is a set composed of antisymmetric matrices $R^{3\times3}$, represented by

$$
[w]=\begin{bmatrix} 0 & -w_z & w_y \\ w_z & 0 & -w_x \\ -w_y & w_x & 0 \end{bmatrix}
\tag{5.30}
$$

where $w=(w_x,w_y,w_z)\in R^{3\times3}$. Z_B''' is an algorithm that converts a three-dimensional vector into the corresponding antisymmetric matrix.

Similar to the Lie algebra so(3), the Lie algebra of the Lie group $SE(3)$ is represented by $se(3)$. It is composed of 4×4 homogeneous transformation matrices, in the form of

$$
g=\begin{bmatrix} [w] & v \\ 0 & 0 \end{bmatrix}
\tag{5.31}
$$

where v is a 3×1 matrix. Its representation can be further simplified, as shown below:

$$
g=(w,v)
\tag{5.32}
$$

where w is a rotation matrix and v is a translation matrix. Such an expression contains some important geometric information about the position transformation. To better understand this information, the matrix exponential function needs to be introduced.

The Lie group $(SO(3),SE(3))$ is related to the Lie algebra $(se(3),\mathrm{se}(3))$ through the matrix exponential function e^{AT} [4]. The matrix exponential function of $se(3)$ and its elements has a closed expression. If $[w]\in so(3)$, then

$$
\exp(|w|)=I+\frac{\sin\phi}{\phi}+\frac{1-\cos\phi}{\phi^2}[w]^2
\tag{5.33}
$$

where $\phi = \|w\|$.

From $\exp([w])(\exp([w]))^T = (\exp([w]))^T \exp([w]) = I$, we can directly prove that $\exp([w])$ belongs to the special orthogonal group $SO(3)$.

The matrix exponential function of $g = (w,v) \in se(3)$ is

$$e^g = \exp\begin{bmatrix} [w] & v \\ 0 & 0 \end{bmatrix} = \exp\begin{bmatrix} \exp([w]) & Av \\ 0 & 0 \end{bmatrix} \tag{5.34}$$

where $\exp([w]) \in SO(3)$ and $Av \in R^{3\times1}$. From the definition of $SE(3)$, the result $e^g \in SE(3)$ can be easily derived. Physically speaking, $g = (w,v)$ is the spiral parameter of transformation. e^g is a spiral motion, whose rotation axis is the pointing direction of w. $\|W\|$ indicates the angle of rotation. The vector v determines the translational component of the spiral motion. For the pure rotation passing a point, we obtain $v = q \times w$.

Contrary to matrix exponent, the matrix logarithm can be used to calculate the Lie algebra corresponding to an element of a Lie group. The element R of the special orthogonal group $SO(3)$ also has a closed expression form:

$$\log R = \frac{\phi}{2\sin\phi}\left(R - R^T\right) = [w] \tag{5.35}$$

For the special group $SE(3)$:

$$\log\begin{bmatrix} R & p \\ 0 & 1 \end{bmatrix} = \begin{bmatrix} [w] & A^{-1}p \\ 0 & 0 \end{bmatrix} \tag{5.36}$$

where

$$A^{-1} = E - \frac{[w]}{2} + \frac{2\sin\phi - \phi(1-\cos\phi)}{2\phi^2 \sin\phi}[w]^2 \tag{5.37}$$

5.3.2.2 Orthogonalization of Rotation Matrix Identity

In algebra, matrix exponential is a matrix function, similar to a general exponential function. The exponent of a matrix determines the mapping relation between the Lie algebra of the matrix and the region of its Lie group. On the other hand, matrix logarithm is the inverse function of matrix exponential [5]. Not all matrices have a logarithm. Those matrices with logarithm may have more than one logarithm. The study of logarithmic matrices has been accompanied by the development of the theory of Lie group, because a matrix with logarithm must be in a Lie group, and its logarithm must be the corresponding element of Lie algebra.

The rotation matrix ${}_m^sR$ is processed into

$$J_1 = P_1^{-1} \cdot {}_m^sR \cdot P_1 \tag{5.38}$$

where J_1 is a diagonal matrix of the matrix ${}_m^sR$ and contains its characteristic root. The matrix P_1 is an invertible matrix. In linear algebra, a diagonal matrix in finite dimensional space is a matrix where all the elements outside the main diagonal are zero elements. It represents an operator to some extent.

Therefore, the logarithm of the rotation matrix ${}_m^sR$ is

$$Q = P \cdot In(J) \cdot P^{-1} \tag{5.39}$$

According to the theories of Lie group and Lie algebra, the exponential matrix of a skew symmetric matrix is an orthogonal matrix. In the inverse function, the logarithm matrix of an orthogonal matrix is a skew symmetric matrix. Skew symmetric matrix is a square matrix whose transpose plus itself is equal to null matrix. In the real number domain, a skew symmetric matrix represents an identity matrix mapped from the tangent space to the real orthogonal group $O(n)$. This is the so-called special orthogonal Lie algebra. Therefore, skew symmetric matrix can be regarded as an infinitesimal rotation, and its space constitutes the Lie algebra $O(n)$ of the Lie group $O(n)$. It is easy to prove that the addition and subtraction of two skew symmetric matrices will still produce a skew symmetric matrix.

However, due to the limitation of precision and the existence of error, the rotation matrix ${}_m^sR$ is not a standard orthogonal matrix. The matrix Q is an approximate diagonal matrix, so

$$Q + Q^T \neq 0 \tag{5.40}$$

Because of a certain error between the matrix Q and the standard diagonal matrix, the matrix Q plus its transpose is not equal to 0. In order to find the standard diagonal matrix closest to the matrix Q, we suppose that

$$Q_1 = \frac{1}{2}\left(Q^T - Q\right) \tag{5.41}$$

Then the matrix Q_1 is verified:

$$Q_1^T = \left(\frac{\left(Q^T - Q\right)}{2}\right)^T = \frac{\left(Q - Q^T\right)}{2} = -Q_1 \tag{5.42}$$

Therefore, we can ensure that the matrix Q_1 is a standard skew symmetric matrix closest to the matrix Q. By recalculating the rotation matrix ${}_m^sR$, we can obtain

$$J_2 = P_2^{-1} \cdot Q_1 \cdot P_2 \tag{5.43}$$

Furthermore, by using the exponential of the matrix, the optimized rotation matrix ${}_m^sR_1$ can be obtained:

$${}_m^sR_1 = P_2 \cdot \exp(J_2) \cdot P_2 \tag{5.44}$$

According to the above method, after the rotation matrix ${}_{m}^{s}R$ is optimized into the orthogonal matrix ${}_{m}^{s}R_1$, the translation matrix in Eq. (5.9) also needs to be recalculated. That is, the new translation matrix is

$$P_{morg} = \frac{1}{3}\left[\left({}^{s}P_1 - {}_{m}^{s}R_1 \cdot {}^{m}P_1\right) + \left({}^{s}P_2 - {}_{m}^{s}R_1 \cdot {}^{m}P_2\right) + \left({}^{s}P_3 - {}_{m}^{s}R_1 \cdot {}^{m}P_3\right)\right] \tag{5.45}$$

To sum up, we have solved the problem of the base-coordinates transformation between two manipulators. Through a series of matching operations between the two manipulators, the coordinates of three non-collinear points in space are determined. Then, a preliminary result can be obtained with the unified variable solution of transformation matrix or with the homogeneous matrix method. Next, the rotation matrix is orthonormalized. We can ensure that ${}_{m}^{s}R_1$ is an orthonormal identity matrix closest to the preliminary result. Finally, the translation matrix between the two manipulators is recalculated.

5.3.3 Experiment of Dual-Manipulator Base Coordinate Transformation Relationship

To verify the theories and conclusions in this chapter, an effective experiment was carried out in the laboratory. The test objects were two industrial manipulators. The main IPC was connected to the manipulator controllers. The two manipulators were driven by a control box in the teaching mode. At the same time, a measuring tip, called "calibration tool", was installed on each end-effector. The shape of this tool is shown in Figure 5.17. Before the experiment, the calibration tool installed on the end-effector of the master manipulator should be recorded as 66.95 mm long, and the calibration tool installed on the end-effector

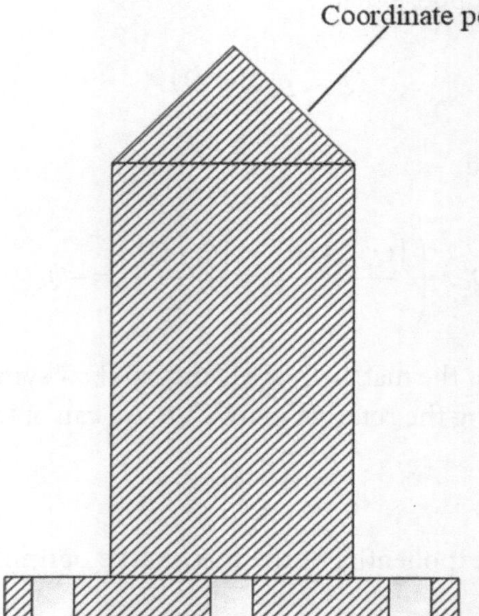

FIGURE 5.17 Real calibration tool.

of the slave manipulator should be recorded as 67.08 mm long. The two calibration tools can reduce the difficulty in the coordination between the two manipulators and improve the coordinate accuracy of the calibration points. This kind of calibration tool is widely used to calibrate the workpiece coordinate systems on manipulators. The tools are made of aluminum with a sharp end, so they're easy to machine.

The system composed of two handshaking manipulators is shown in Figure 5.18. The coordinates of the center points (TCPs) on the top of the calibration tools relative to the base coordinate systems of the manipulators can be obtained directly from the data recorded by IPC. The manipulator control software in the IPC records a 1×6 vector $(x,y,z,\alpha,\beta,\gamma)$ representing the position and attitude of calibration tool at the manipulator end. The vector $(x,y,z,)$ represents the position coordinates of calibration tool at the manipulator end. The vector (α,β,γ) is called Euler angle, which represents the direction of a rotation matrix. The manipulator position accuracy specified by manufacturer is 0.04 mm, as shown in the manipulator manual. The manipulator accuracy is much higher than the operation error, so the manipulator position coordinates recorded by IPC can be considered as true TCP coordinates.

Based on the previous discussion, the coordinates of three non-collinear points in the intersection workspace of two manipulators can be obtained through the following steps. In the acquisition of coordinate information, a series of handshakes between the two driven manipulators are completed. At first, the TCP at the end of the master manipulator is driven to the point P_1 in the intersection workspace. Then, the TCP at the end of the slave manipulator is taught to move in any attitude to the point P_1. The matching between master manipulator and slave manipulator at this time is called "handshake", as shown in Figure 5.19.

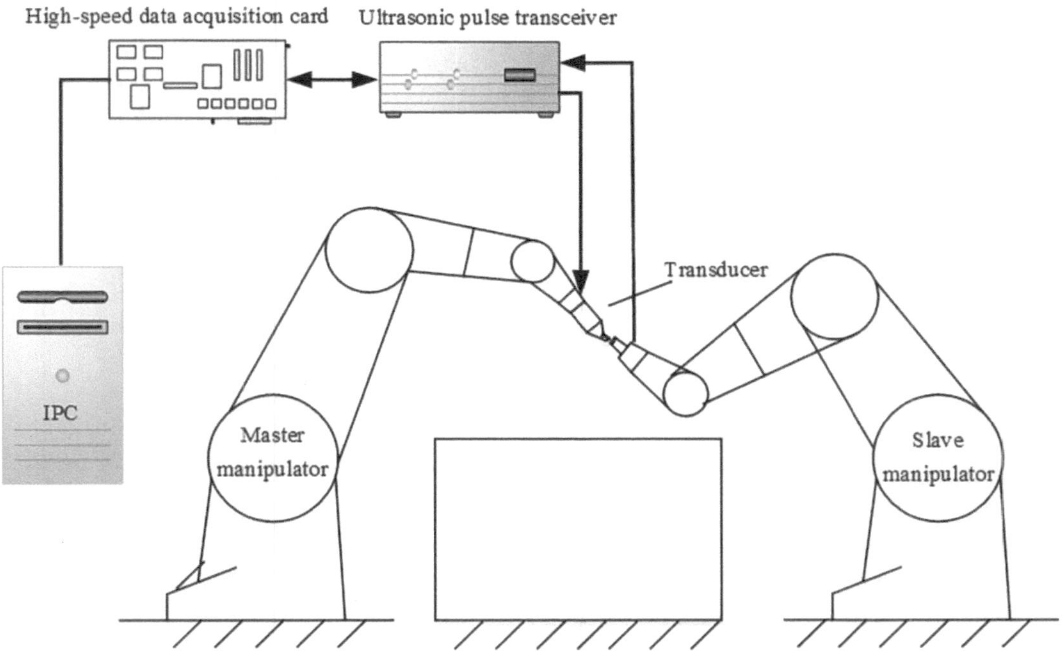

FIGURE 5.18 Structure of a dual-manipulator matching system.

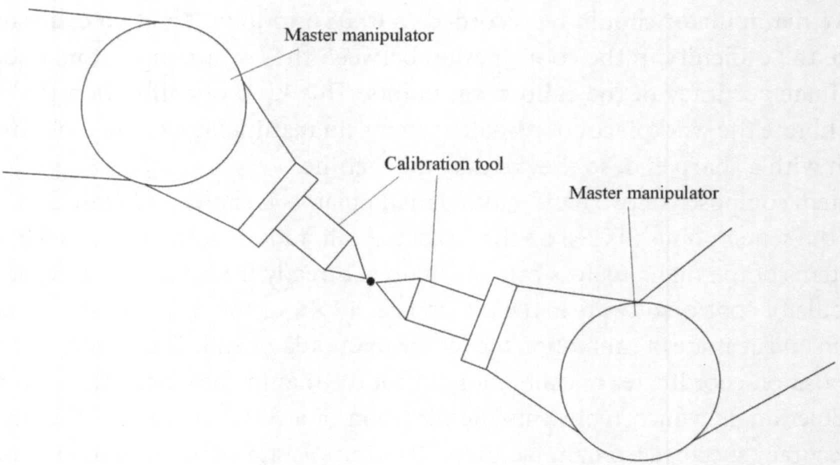

FIGURE 5.19 Dual-manipulator handshake.

The position coordinates of TCPs at the end of the two manipulators are recorded as ${}^{m}P_1$ and ${}^{s}P_1$, respectively. Then, the master manipulator is taught to move to another position so that the TCP stops at the point P_2. After that, the slave manipulator is taught to move the TCP at its end to the point P_2. Similarly, a new handshake has been achieved. In the same way, the position coordinates of TCPs at the end of the two manipulators are recorded, namely ${}^{m}P_2$ and ${}^{s}P_2$, respectively. Then, this process is repeated to obtain the coordinates of the third point P_3 in the two base coordinate systems, namely ${}^{m}P_3$ and ${}^{s}P_3$. In the whole process, the three points must not be on the same line, which should be easy to do.

The coordinate values obtained from the dual-manipulator handshake experiment are shown in Table 5.2.

${}^{m}P$ represents the coordinates of a point in the base coordinate system of the master manipulator, while ${}^{s}P$ represents the coordinates of a point in the base coordinate system of slave manipulator. By substituting these coordinates into Eqs. (5.44) and (5.45), we can obtain the coordinates of the two points not coplanar with them, namely ${}^{m}P_4$ and ${}^{s}P_4$:

$$\begin{cases} {}^{m}P_4 = (755.27\ 732.53\ 504.44)^T \\ {}^{s}P_4 = (296.89 - 835.94\ 242.10)^T \end{cases} \tag{5.46}$$

The coordinate values of the four points are substituted into Eq. (5.10) to obtain the preliminary results of the rotation matrix ${}^{s}_{m}R$ and the translation matrix P_{marg}:

TABLE 5.2 Data from the Manipulator Handshake Experiment

Point	Calibration Point of Master Manipulator (mm)			Calibration Point of Slave Manipulator (mm)		
1	830.22	276.54	534.72	752.88	−760.52	275.83
2	854.70	726.37	495.73	302.91	−736.40	234.61
3	816.62	693.94	83.06	336.99	−769.55	−178.69

$$\begin{cases} {}^{s}_{m}R = \begin{pmatrix} -0.0026 & -1.0005 & -0.0037 \\ 0.9995 & -0.0018 & -0.0116 \\ 0.0101 & -0.0055 & 1.0010 \end{pmatrix} \\ P_{m\arg} = (1033.73, -1583.5, -266.27)^{T} \end{cases} \tag{5.47}$$

The orthogonality of the rotation matrix ${}^{s}_{m}R$ is judged:

$${}^{s}_{m}R\left({}^{s}_{m}R\right) = \begin{pmatrix} 1.001 & -0.0008 & 0.0017 \\ -0.0008 & 0.9990 & -0.0017 \\ 0.0017 & -0.0017 & 1.0000 \end{pmatrix} \tag{5.48}$$

The calculation results show that the rotation matrix ${}^{s}_{m}R$ in Eq. (5.45) is not an orthogonal identity matrix but slightly deviates from the orthonormal identity matrix. Therefore, we need to use the matrix optimization method proposed in Chapter 4 to unitize and orthogonalize the rotation matrix.

By optimizing the rotation matrix ${}^{s}_{m}R$, we can get a new rotation matrix ${}^{s}_{m}R_1$:

$${}^{s}_{m}R_1 = \begin{pmatrix} -0.0022 & -1.0000 & -0.0046 \\ 0.9999 & -0.0022 & -0.0109 \\ 0.0109 & -0.0046 & 0.9999 \end{pmatrix} \tag{5.49}$$

The orthogonality of the rotation matrix ${}^{s}_{m}R_1$ is rejudged:

$${}^{s}_{m}R_1 \cdot \left({}^{s}_{m}R_1\right)' = \begin{pmatrix} 1.0000 & 0.0000 & 0.0000 \\ 0.0000 & 1.0000 & 0.0000 \\ 0.0000 & 0.0000 & 1.0000 \end{pmatrix} \tag{5.50}$$

Here, the translation matrix $P_{m\arg}$ needs to be recalculated:

$$P_{morg} = (1033.54, -1584.14, -266.92)^{T} \tag{5.51}$$

To sum up, a strict orthogonal identity rotation matrix and a new translation matrix can be obtained from the optimized results. The transformation relation between the base coordinate systems of the two manipulators can be expressed as

$$\begin{pmatrix} {}^{s}X \\ {}^{s}Y \\ {}^{s}Z \end{pmatrix} = \begin{pmatrix} -0.0022 & -1.0000 & -0.0046 \\ 0.9999 & -0.0022 & -0.0109 \\ 0.0109 & -0.0046 & 0.9999 \end{pmatrix} \begin{pmatrix} {}^{m}X \\ {}^{m}Y \\ {}^{m}Z \end{pmatrix} + \begin{pmatrix} 1033.54 \\ -1584.14 \\ -266.92 \end{pmatrix} \tag{5.52}$$

5.3.4 Analysis of Transformation Relation Error

In the previous section, we have obtained the transformation relation between the base coordinate systems of the two manipulators. To further verify the accuracy of this transformation relation and the effectiveness of calibration strategy of the base coordinate systems, the accuracy of this transformation relation needs to be further analyzed. The calibrated and measured experimental data is given in Table 5.4. X_m, Y_m and Z_m represent the measured coordinates of the calibration tool TCP in X-axis, Y-axis and Z-axis, respectively. They can be obtained in experiment in accordance with the coordinates of end-effector TCP recorded by IPC. X_c, Y_c and Z_c represent the coordinates of TCP in X-axis, Y-axis and Z-axis, respectively, which were calculated from the transformation relation. The calculated and measured results are in mm.

Although the data in Table 5.3 shows the experimental results of base-coordinates transformation relation, the error between the calculated result and the measured result can't be intuitively observed from the data in the table. In the experiment, the calculated and measured coordinates were sorted from smallest to biggest and then fitted into curves. The fitting curves of calculated and measured results are shown in Figure 5.20. It can be seen from the figure that the calculation curve and the measurement curve overlap, indicating a small error between the two results.

To accurately evaluate the error, e_x (in m) is defined as the error between x_m and x_c. Similarly, e_y (in m) is the error between y_m and y_c, and e_z (in m) is the error between z_m and z_c. Then we can obtain

TABLE 5.3 Calibration Experiment Results of the Dual-Manipulator Base Coordinates Transformation Relation

	Measured Result			Calculated Result		
Point No.	X_m (mm)	Y_m (mm)	Z_m (mm)	X_c (mm)	Y_c (mm)	Z_c (mm)
1	302.91	−736.40	234.61	302.99	−736.51	234.78
2	336.99	−769.55	−178.69	337.4	−770.01	−178.15
3	752.88	−760.52	275.83	752.69	−760.42	275.55
4	1181.29	−432.9	315.21	1181.00	−432.44	314.79
5	879.13	−947.14	57.94	879.68	−947.09	57.21
6	290.11	−1320.44	−97.93	290.71	−1320.8	−98.77
7	−284.66	−909.47	−94.56	−284.21	−910.4	−94.05
8	543.71	−1136.71	−489.76	544.28	−1137.1	−490.23
9	890.61	−598.4	153.55	891.38	−598.15	153.2
10	888.17	−375.16	154.46	887.33	−375.18	155.16
11	522.05	−375.89	152.35	522.03	−376.15	153.1
12	624.81	−664.74	344.02	625.21	−664.21	334.51
13	588.65	−858.61	192.21	588.92	−858.83	192.32
14	689.95	−566.78	−10.26	684.44	−566.31	−9.78

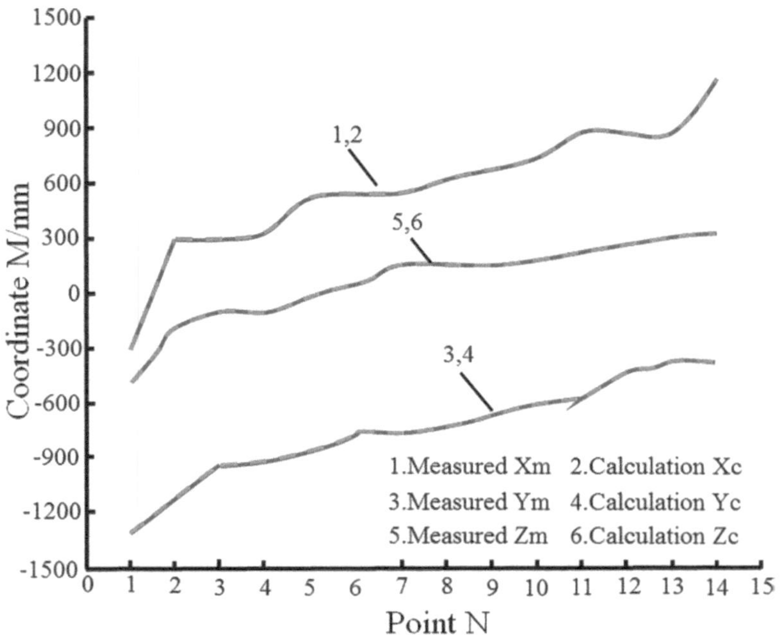

FIGURE 5.20 Fitting curves of calculated and measured results.

$$\begin{cases} e_{x\max} = \|X_m - X_c\|_\infty = 8.3\times10^{-4} \\ e_{y\max} = \|Y_m - Y_c\|_\infty = 9.3\times10^{-4} \\ e_{z\max} = \|Z_m - Z_c\|_\infty = 8.5\times10^{-4} \end{cases} \tag{5.53}$$

Eq. (3.51) confirms that the calibration error of each coordinate axis does not exceed 1 mm, which can meet the requirement for equipment error in ultrasonic NDT. The traditional normalization method of the rotation matrix is based on the optimal approximation of Frobenius norm. To compare the accuracy of the two methods under the same experimental conditions, the two methods are used to process the data in Table 5.2 and fit the curve, as shown in Figure 5.21. It can be easily seen from the figure that the maximum error of the coordinate value obtained by the traditional method is more than 5 mm, while the maximum error of the coordinate value obtained by our method is less than 1.3 mm. Therefore, it can be concluded that our method is more accurate than previous research reports.

In Figure 5.22, the calculation time of the Lie-algebra exponential mapping method is compared with that of the Frobenius-norm optimal approximation method. The main configuration of computer is: CPU Intel I5–3210 m, 4 GB memory. Because the Frobenius-norm optimal approximation method transforms the solving process with rotation matrix optimization into seven sets of independent equations, the equation sets are solved by the Levenberg-Marquardt numerical iterative method. In contrast, the method of Lie-algebra exponential mapping directly operates on the rotation matrix, using the characteristics of orthogonal-matrix exponential mapping to obtain the optimized rotation matrix.

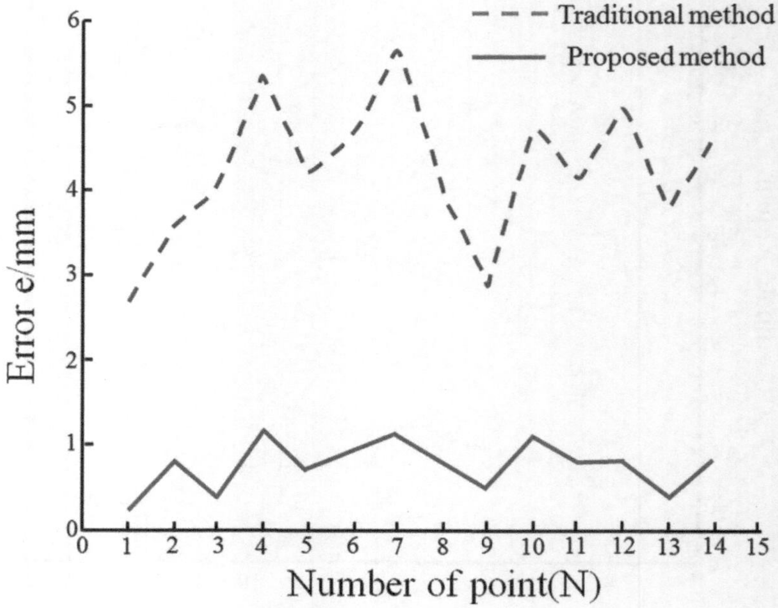

FIGURE 5.21 Accuracy comparison between two algorithms.

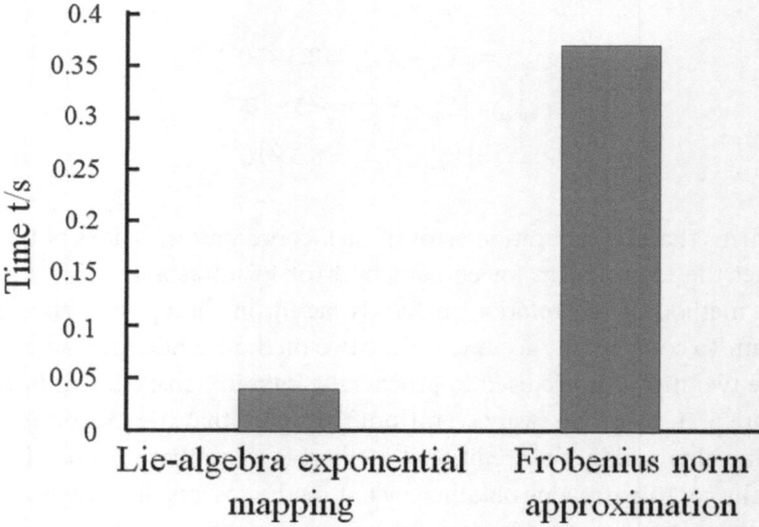

FIGURE 5.22 Efficiency comparison between the two calculation methods.

Therefore, the calculation efficiency of the Lie-algebra exponential mapping method is significantly higher than that of the Frobenius-norm optimal approximation method.

In order to verify the effectiveness of dual-manipulator base coordinates transformation relation in the automatic coincidence of workpiece coordinates, the settings of the workpiece coordinate systems of the two manipulators have been tested in practice. In the experiment, we only need to drive the master manipulator to determine its workpiece coordinate system. The point coordinates in the workpiece coordinate system of the slave manipulator will be determined by the point coordinates recorded by master manipulator and the transformation relation between the base coordinate systems of the two manipulators. It can be seen from the above that the coordinate values of three points are needed by manipulator to determine the workpiece coordinate system, including the origin of the workpiece coordinate system, a point on the *X*-axis and a point in the *OXY* plane. Table 5.5 shows the coordinate values of the three points that the master manipulator in the dual-manipulator UT system uses to set the workpiece coordinate system.

In the traditional method of dual-manipulator workpiece coordinates coincidence, the slave manipulator needs to be driven to the three points in the workpiece coordinate system, and the coordinates of the three points in the slave base coordinate system need to be recorded. According to the automatic coincidence method proposed by this chapter, as long as the coordinates of three workpiece points in the master base coordinate system (see Table 5.4) are substituted into Eq. (3.50), the coordinates of those points in the slave base coordinate system, namely the coordinates necessary for the setting of the slave workpiece coordinate system, can be obtained. Through calculation, the coordinates of the three points in the slave workpiece coordinate system are obtained, as shown in Table 5.5.

Table 5.5 lists the errors between the calculated and calibrated coordinate values of the three points necessary for slave manipulator to set the workpiece coordinate system. The deviation from the coordinate value of an actual locating point was calculated based on the coordinate relation.

It can be seen from Table 5.6 that in the method of automatic workpiece coordinate coincidence based on dual-manipulator base-coordinates transformation relation, the position error of the workpiece coordinate system is less than 1 mm. This position accuracy can meet the requirement for workpiece-coordinates coincidence error in the process of dual-manipulator ultrasonic NDT. The obtained dual-manipulator transformation relation will also play an important role in the later analysis of dual-manipulator constraints.

TABLE 5.4 Point Coordinates Necessary for Setting the Workpiece Coordinate System of Master Manipulator

Point	*X* (mm)	*Y* (mm)	*Z* (mm)
1	990.84	138.06	409.98
2	1213.82	141.61	409.52
3	1213.65	506.92	409.41

TABLE 5.5 Point Coordinates Necessary for Setting the Workpiece Coordinate System of Slave Manipulator

	Calculated Result			Calibrated Result		
Point	*X* (mm)	*Y* (mm)	*Z* (mm)	*X* (mm)	*Y* (mm)	*Z* (mm)
1	891.61	−598.60	153.55	891.38	−598.15	153.21
2	887.57	−375.16	154.46	887.32	−374.28	155.16
3	522.05	−375.89	153.35	522.03	−376.15	153.01

TABLE 5.6 Coordinate Value Errors Calculated with the Two Methods

Point	e_x (mm)	e_y (mm)	e_z (mm)
1	0.23	−0.45	0.34
2	0.25	−0.88	−0.7
3	0.02	−0.26	0.34

5.4 DUAL-MANIPULATOR MOTION CONSTRAINTS DURING TESTING

Dual-manipulator ultrasonic NDT is a complex motion process. The transmitting and receiving transducers at the end of the two manipulators are subject to not only the constraints of position and attitude but also relative motion. These constraints vary with the detection task and object. It is generally hoped that the transmitting transducer is in the normal direction of the incident surface, while the receiving transducer is on the path of sound beam propagation in order to obtain the best UT results. However, in the ultrasonic detection of a curved surface, this cooperative dual-manipulator UT system satisfies the constraints of above position and attitude at some moment. To realize automatic UT, the two manipulators need to adapt the position and attitude of the transducers at their ends to the change of surface shape, so as to ensure that more ultrasonic energy can pass through the workpiece to the receiving transducer to achieve a better detection result.

According to the test object, automatic ultrasonic NDT tasks can be mainly divided into the testing of equi-thickness workpiece and the testing of variable-thickness workpiece. The two tasks correspond to different constraint relations between the transmitting and receiving transducers. According to the constraint relation between two transducers, the dual-manipulator coordinated motions are mainly divided into coupled synchronous motion and superposed synchronous motion. This chapter mainly studies the constraint relation between the two manipulators in different coordinated motions and establishes the constraint equation, which will provide a theoretical basis for dual-manipulator scanning technology.

UT is an application of the dual-manipulator system, which needs to determine the ultrasonic constraints according to the detection task and object. Traditional testing methods are all applied to isotropic materials, rather than composite materials that show anisotropic acoustic properties due to their special material and structure. In the ultrasonic testing of curved composite workpieces, the inspector cannot simply judge the propagation law of acoustic beam in the workpieces, so the analysis of detection signal has become more and more limited. Therefore, the quantitative research of sound field in complex surfaces and composite materials based on the sound-field calculation model can help improve the detection technique and optimize the detection parameters. In this chapter, the ultrasonic measurement model of composite material is established according to the characteristics of sound field distribution in composite material obtained with the model of multiple Gaussian beams, in order to analyze the influence of different transducer settings on the amplitude of the received signal and the constraint relation between the two transducers. The analysis results will provide a basis for the application of dual-manipulator coordinated motion in ultrasonic testing.

5.4.1 Constraints on the Position and Attitude of Dual-Manipulator End-Effectors in the Testing of Equi-Thickness Workpiece

The ultrasonic testing of equi-thickness workpiece is an ultrasonic transmission method. It is a typical dual-manipulator-coupled synchronous motion where the receiving transducer moves with the transmitting transducer. During testing, the constraint relationship between the two transducers in the manipulator base coordinate systems remains unchanged. The transmitting and receiving transducers are at the ends of the two manipulators, respectively. The position and attitude of transducers relative to manipulator end-effectors remain unchanged during testing. Therefore, the constraint relationship between the two transducers can be transformed into that between the two end-effectors.

The testing process of equi-thickness workpiece in the dual-manipulator UT system is shown in Figure 5.23. The transmitting transducer moves along the equidistant lines on the workpiece surface. At each trajectory point, the normal direction of the transmitting transducer coincides with the normal direction of the workpiece surface. The transmitted ultrasonic wave passes through the workpiece to the receiving transducer. In order to obtain the optimal transmitted wave signal, the axial direction of the receiving transducer needs to coincide with that of the ultrasonic beam. Therefore, the axial direction of the transmitted wave beam is the normal direction of the workpiece surface at this point.

In the testing process, the transducer needs not only to change its position but also to change its orientation according to the shape of workpiece surface. Therefore, the position and attitude of the transducer can be expressed by a 4×4 homogeneous matrix. The homogeneous transformation matrix of the transducer is expressed as follows:

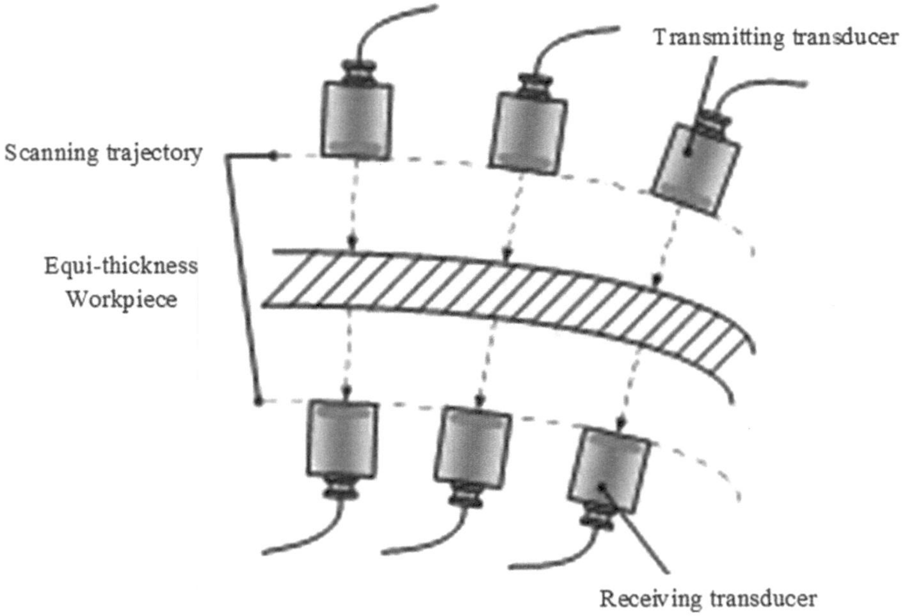

FIGURE 5.23 Constraints on two transducers in the testing of equi-thickness workpiece.

$$T_i = \begin{bmatrix} R_i & P_{oi} \\ 0 & 1 \end{bmatrix} \tag{5.54}$$

where $R_i \in R^{3\times3}$ is the selection matrix describing the attitude and sound beam direction of the transducer; and $P_{oi} \in R^{3\times1}$ indicates the position of the center point of the transducer surface. Before the use of R_i and P_{oi}, the reference coordinate system should be defined. Generally, the base coordinate system of the manipulator where the transducer is installed is selected as the reference coordinate system of the homogeneous transformation matrix of the transducer.

The testing of an equal-thickness workpiece is implemented by two manipulators in the master-slave coupling mode. Therefore, the manipulator holding the transmitting transducer can be defined as the master manipulator R_m, whose base coordinate system is $\{M\}$, while the manipulator holding the receiving transducer is the slave manipulator R_s, whose base coordinate system is $\{S\}$. The transmitting transducer is assumed to move from the point T_{a1} to the point T_{a2} on the detection trajectory. The two points are represented by ${}^mT_{a1}$ and ${}^mT_{a2}$ in the base coordinate system of the master manipulator. According to the detection constraints on equi-thickness workpiece, the receiving transducer needs to move from the point T_{b1} to the point T_{b2} on the detection trajectory at the same time. The two points are represented by ${}^sT_{b1}$ and ${}^sT_{b2}$ in the basal coordinate system of the slave manipulator. The transmitting and receiving transducers are installed on the flanges at the end of the two manipulators, respectively, so there is a transformation relationship between one transducer and the corresponding manipulator end flange. The transformation relationship between the transmitting transducer and the master manipulator end-effector is ${}^mT_{ea}$, and the transformation relationship between the receiving transducer and the slave manipulator end-effector is ${}^sT_{eb}$. Therefore, the transformation relations between the trajectory points ${}^mP_i$ and ${}^sP_i$ at the end of the two manipulators and the trajectory points ${}^mT_{ai}$ and ${}^mT_{bi}$ of the transducers are as follows:

$$\begin{cases} {}^mT_{ai} = {}^mT_{eai} \cdot {}^mP_i \\ {}^sT_{bi} = {}^sT_{ebi} \cdot {}^sP_i \end{cases} \tag{5.55}$$

In the physical sense, the transformation relations ${}^mT_{eai}$ and ${}^sT_{ebi}$ are the design parameters of transducer-holding devices, respectively. Once the device is determined, the transformation relation can also be determined. Although the transducer holder may have several parameterized designs, its design must remain unchanged once the detection task begins. When replacing the transducer holder, the transformation relations ${}^mT_{eai}$ and ${}^sT_{ebi}$ need to be recalculated (Figure 5.24).

The constraint relations for the above-mentioned dual-manipulator-coupled synchronous motion need to be established in the same reference coordinate system, which is determined to be the base coordinate system of the master manipulator. Therefore, we need to obtain the representation of the position/attitude of the receiving transducer in the master base coordinate system in the testing process. In the previous chapter, we have

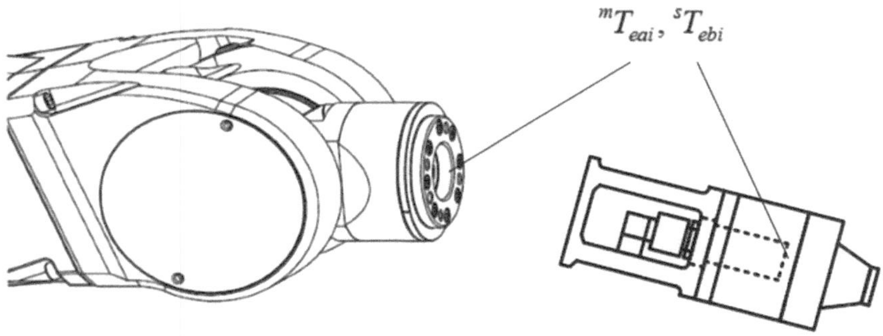

FIGURE 5.24 Position/attitude relation between the manipulator and transducer.

obtained the position/attitude transformation relation ${}^{s}_{m}T$ in the form of homogeneous transformation matrix from the slave base coordinate system to the master base coordinate system:

$$ {}^{s}_{m}T = \begin{bmatrix} R & P \\ 0 & 1 \end{bmatrix}, {}^{s}_{m}T^{-1} = \begin{bmatrix} R^T & -R^T \cdot P \\ 0 & 1 \end{bmatrix} \tag{5.56} $$

so

$$ {}^{m}T_{b1} = {}^{m}_{s}T \cdot {}^{s}T_{b1} \tag{5.57} $$

where ${}^{m}T_{b1}$ is the representation of the receiving transducer in the master base coordinate system. The representation of the transmitting transducer in the master base coordinate system is shown in Eq. (5.55). The constraint relation between the two transducers in the master base coordinate system is specified as ${}^{m}T_1$. According to the definition of constraint relation, we can obtain

$$ {}^{m}T_{b1} = {}^{m}T_1 \cdot {}^{m}T_{a1} \tag{5.58} $$

Eqs. (5.52) and (5.54) are substituted into Eq. (5.56) to obtain

$$ {}^{m}_{s}T \cdot {}^{s}T_{eb1} \cdot {}^{s}P_1 = {}^{m}T_1 \cdot {}^{m}T_{ea1} \cdot {}^{m}P_1 \tag{5.59} $$

where ${}^{m}P_1$ and ${}^{s}P_1$ represent a pair of trajectory points at the end of the two manipulators in the dual-manipulator-coupled synchronous motion. They are 4×4 homogeneous matrices describing the position and attitude of manipulator end-effectors. For the second pair of trajectory points ${}^{m}P_2$ and ${}^{s}P_2$ at the end of the manipulators, ${}^{m}P_2$ represents the position and attitude of master manipulator end-effector, and ${}^{s}P_2$ represents the position and attitude of slave manipulator end-effector. According to the constraint relations, the two points undergo the operations similar to those of ${}^{m}P_1$ and ${}^{s}P_1$ to obtain

$$ {}^{m}_{s}T \cdot {}^{s}T_{eb2} \cdot {}^{s}P_2 = {}^{m}T_2 \cdot {}^{m}T_{ea2} \cdot {}^{m}P_2 \tag{5.60} $$

In the testing process, the relative positions of the two manipulators are fixed, so their transformation relation in the form of homogeneous matrix remains unchanged. It can be derived from Eq. (5.56) that

$$ {}^{m}_{ss}{}^{m}T = {}^{m}T_{1} \cdot {}^{m}T_{ea1} \cdot {}^{m}P_1 \cdot \left({}^{s}T_{eb1} \cdot {}^{s}P_1 \right)^{-1} \tag{5.61} $$

Eq. (5.61) is substituted into Eq. (5.60) to obtain

$$ {}^{m}T_1 \cdot {}^{m}T_{ea1} \cdot {}^{m}P_1 \cdot \left({}^{s}T_{eb1} \cdot {}^{s}P_1 \right)^{-1} \cdot {}^{s}T_{eb2} \cdot {}^{s}P_2 = {}^{m}T_{2} \cdot {}^{m}T_{ea2} \cdot {}^{m}P_2 \tag{5.62} $$

In the trajectory planning, since the scanning trajectory of the master manipulator has been obtained in the third chapter, the positions/attitudes ${}^{m}P_1$ and ${}^{m}P_2$ are known. The position/attitude ${}^{s}P_2$ of the slave manipulator can be determined from Eq. (5.61):

$$ \begin{aligned} {}^{s}P_2 &= \left({}^{s}T_{eb2} \right)^{-1} \left({}^{m}T_1^{m}T_{ea1}^{m}P_1 \left({}^{s}T_{eb1}^{s}P_1 \right)^{-1} \right)^{-1} {}^{m}T_2^{m}T_{ea2}^{m}P_2 \\ &= \left({}^{s}T_{eb2} \right)^{-1} \left({}^{m}T_1^{m}T_{ea1}^{m}P_1 \left({}^{s}P_1 \right)^{-1} \left({}^{s}T_{eb1} \right)^{-1} \right)^{-1} {}^{m}T_2^{m}T_{ea2}^{m}P_2 \\ &= {}^{s}T^{-1}_{eb2} {}^{s}T_{eb1} {}^{s}P_1 {}^{m}P_1^{-1} {}^{m}T_{ea1}^{-1} {}^{m}T_2^{m}T_{ea2}^{m}P_2 \end{aligned} \tag{5.63} $$

The study in this chapter is under the condition that the transducer holders remain unchanged during testing. Therefore, when the two manipulators move from the trajectory points ${}^{m}P_1$ and ${}^{s}P_1$ to ${}^{m}P_2$ and ${}^{s}P_2$, respectively, the homogeneous transformation matrices from the transducers to the manipulator end-effectors will remain unchanged, that is,

$$ {}^{m}T_{ea1} = {}^{m}T_{ea2},\ {}^{s}T_{eb1} = {}^{s}T_{eb2} \tag{5.64} $$

According to the definition of coupled synchronous motion, the constraint relation between the transmitting and receiving transducers remains unchanged during testing, so

$$ {}^{m}T_1 = {}^{m}T_2 \tag{5.65} $$

By substituting Eqs. (5.60) and (5.61) into Eq. (5.62), the dual-manipulator motion model at any trajectory point on a workpiece with equal wall thickness can be obtained:

$$ {}^{s}P_2 = {}^{s}P_1 \cdot {}^{m}P_1^{-1} \cdot {}^{m}P_2 \tag{5.66} $$

Eq. (5.13) is just the constraint on the trajectory of the transmitting and receiving transducers scanning a workpiece with equivalent wall thickness. In the coupled cooperative motion, the coordinates of the receiving transducer at any trajectory point ${}^{s}P_2$ in the slave base coordinate system can be calculated by using the trajectory point ${}^{m}P_2$ of the receiving transducer as well as the trajectory points ${}^{m}P_1$ and ${}^{m}P_2$ of the transmitting transducer in the master base coordinate system. Generally, ${}^{m}P_1$ and ${}^{s}P_1$ are the positions and attitudes of the two manipulators measured when the detection task is initialized.

For the convenience of analysis, the coupled synchronous coordination relation can be decomposed into the coupling relation of rotation matrix and the coupling relation of position matrix. Eq. (5.63) is substituted into Eq. (5.66) to obtain

$$\begin{cases} {}^{s}R_2 = {}^{s}R_1 \cdot {}^{m}R_1^{-1} \cdot {}^{m}R_2 \\ {}^{s}P_2 = {}^{s}R_1 \cdot {}^{m}R_1^{-1}\left({}^{m}P_2 - {}^{m}P_1\right) + {}^{s}P_1 \end{cases} \tag{5.67}$$

5.4.2 Constraints on the Position and Attitude of Dual-Manipulator End-Effectors in the Testing of Variable-Thickness Workpiece

The ultrasonic NDT of a variable-thickness workpiece is characterized by a more complex constraint relation between the transmitting and receiving transducers, which includes not only their synchronous motion but also their relative motion. This constraint relation – composed of both synchronous motion and relative motion – is called superposed synchronous motion. Compared with coupled synchronous motion, the constraint relation in superimposed synchronous motion is more complicated because of the relative motion between the two transducers. In the testing process, the two transducers should arrive at the specified positions and attitudes at the same time and need to be adjusted into the corresponding attitudes. Therefore, the superposed synchronous motion is also composed of position constraint and attitude constraint. In the testing process, the transmitting and receiving transducers are respectively installed at the end of the two manipulators. This chapter studies the situation where the transducer holder at the end of the manipulator remains unchanged during testing. Therefore, the constraint relation between the two transducers is transformed into the constraint relation between the two end-effectors.

In the ultrasonic NDT of a workpiece with variable wall thickness, the trajectories of the two transducers are as shown in Figure 5.25. Like in the coupled synchronous motion, the transmitting transducer moves along the equidistant lines on the workpiece surface. Meanwhile, the normal direction of the transmitting transducer at each trajectory point should coincide with the normal direction of the corresponding workpiece surface. This motion constraint on the transmitting transducer is to obtain the optimal incident point of ultrasonic wave and to ensure the maximum energy of ultrasonic transmission. Based on the modeling of the sound field in a composite material, the best transmitted wave signal can be obtain. The specific constraints on the ultrasonic testing of composite materials will be studied in depth in the next section.

The hardware configuration in the ultrasonic NDT system of variable-thickness workpiece is similar to that in the ultrasonic NDT system of equi-thickness workpiece, that is, the transmitting and receiving transducers are installed on the flanges at the end of the two manipulators, respectively. Therefore, we can obtain

$$\begin{cases} {}^{m}T_{ai} = {}^{m}T_{eai} \cdot {}^{m}P_{ai} \\ {}^{s}T_{bi} = {}^{s}T_{ebi} \cdot {}^{s}P_{bi} \end{cases} \tag{5.68}$$

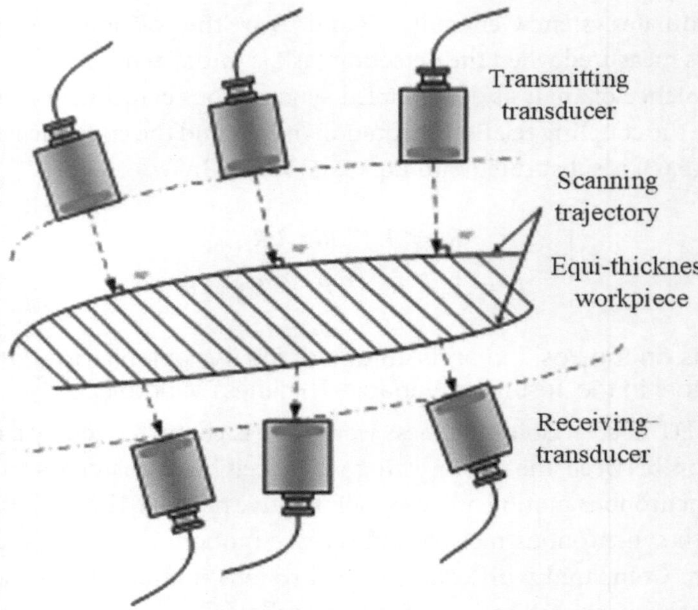

FIGURE 5.25 Constraints on two transducers in the testing of variable-thickness workpiece.

where ${}^{m}T_{ai}$: position and attitude of the surface center point of the transmitting transducer;

${}^{s}T_{bi}$: position and attitude of the surface center point of the receiving transducer;

${}^{m}T_{eai}$: transformation relation between the transmitting transducer and the end of the master manipulator;

${}^{s}T_{ebi}$: transformation relation between the receiving transducer and the end of the slave manipulator;

${}^{s}P_{ai}$: position and attitude of the end of the master manipulator at the trajectory point i;

${}^{s}P_{bi}$: position and attitude of the end of the slave manipulator at the trajectory point i.

Here, they are all 4×4 homogeneous transformation matrices. During the automatic testing of variable-thickness workpiece described in this chapter, the transducer holder remains unchanged. For the automatic UT of different variable-thickness workpieces, different transducer holders can be used. However, once the detection task starts, the transducer holder cannot be changed until the task ends.

Considering the testing process of variable-thickness workpiece as shown in Figure 5.25, the scanning trajectory of the master manipulator can be obtained with the method proposed in Chapter 3, and the scanning trajectory of the slave manipulator needs to be specified according to the scanning trajectory of the master manipulator and the law of sound beam propagation at this position. Therefore, the scanning trajectory of the slave manipulator is the superposition of both the constraint conditions of sound-beam propagation law and the motion of the master manipulator. Compared with the testing of an equi-thickness workpiece, the sound-beam propagation law of a variable-thickness workpiece at each scanning trajectory point cannot remain unchanged, and the two manipulators need to synchronously adjust their end-effector attitude to adapt to the transformation of sound-beam propagation law. Therefore, in the superposed synchronous motion, the position/attitude constraint relation between the end-effectors of two cooperative manipulators is a set of time-varying constraints.

The constraint relation in the above dual-manipulator superposed synchronous motion also needs to be established in the same reference coordinate system. Here, the base coordinate system of the master manipulator is selected as the reference coordinate system of this constraint relation. The constraint relation is a group of time-varying relations. Therefore, the homogeneous transformation matrix ${}^{m}T_{i}$ is the representation of the constraint relation between the two transducers at the i-th scanning trajectory point in the base coordinate system of the master manipulator. The homogeneous transformation matrix ${}^{m}T_{ai}$ represents the position and attitude of the transmitting transducer in the master base coordinate system at the i-th point on the motion trajectory of the transmitting transducer. The homogeneous transformation matrix ${}^{m}T_{bi}$ represents the position and attitude of the receiving transducer in the master base coordinate system at the i-th point on the motion trajectory of the receiving transducer. By means of the homogeneous transformation matrix, we obtain

$$ {}^{m}T_{bi} = {}^{m}T_{i} \cdot {}^{m}T_{ai} \tag{5.69} $$

The receiving transducer moves while being clamped by the slave manipulator. The reference coordinate system of its motion trajectory should be the base coordinate system of the slave manipulator. By means of the transformation relation ${}^{m}_{s}T$ between the base coordinate systems of the two manipulators, the motion trajectory of the receiving transducer can be transformed from the representation ${}^{m}T_{bi}$ in the master base coordinate system to the representation ${}^{s}T_{bi}$ in the slave base coordinate system. Then

$$ {}^{m}T_{bi} = {}^{m}_{s}T \cdot {}^{s}T_{bi} \tag{5.70} $$

where the homogeneous transformation matrix ${}^{m}_{s}T$ can be obtained by using the calibration method of the dual-manipulator base coordinate systems proposed in this chapter. By combining Eq. (5.68) with Eq. (5.69), we can obtain

$$ {}^{m}_{s}T \cdot {}^{s}T_{bi} = {}^{m}T_{i} \cdot {}^{m}T_{ai} \tag{5.71} $$

Eq. (5.18) represents the constraint equation of the transmitting and receiving transducers at each trajectory point during the superposed synchronous motion. The schematic diagram of this constraint relation is shown in Figure 5.26.

The constraint relationship between manipulator end-effector and transducer is shown in Eq. (5.67). The end-effector trajectory can reflect the motion law and range of the manipulator more intuitively. The motion angle of each joint can be derived from the position and attitude of manipulator end-effector, in order to judge whether each joint angle is within the motion range. Therefore, by substituting Eq. (5.67) into Eq. (5.70), we can obtain

$$ {}^{m}_{s}T \cdot {}^{s}T_{ebi} \cdot {}^{s}P_{bi} = {}^{m}T_{i} \cdot {}^{m}T_{eai} \cdot {}^{m}P_{ai} \tag{5.72} $$

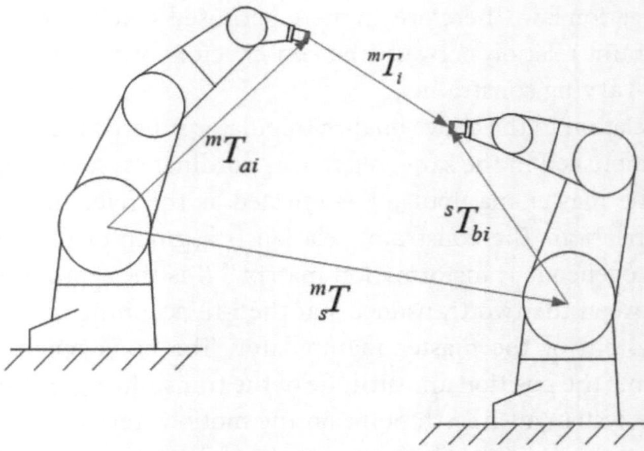

FIGURE 5.26 Constraint relation chain of superposed synchronous motion.

So

$$
\begin{aligned}
{}^{s}P_{bi} &= \left({}^{s}T_{ebi}\right)^{-1}\left({}^{m}_{s}T\right)^{-1} {}^{m}T_{i} \cdot {}^{m}T_{eai} \cdot {}^{m}P_{ai} \\
&= \left({}^{s}T_{ebi}\right)^{-1} {}^{s}_{m}T \cdot {}^{m}T_{i} \cdot {}^{m}T_{eai} \cdot {}^{m}P_{ai}
\end{aligned}
\tag{5.73}
$$

As seen from Eq. (5.73), the motion trajectory of slave end-effector, denoted as ${}^{s}p_{i}$, can be obtained from the motion trajectory of master end-effector (denoted as ${}^{m}p_{i}$), the homogeneous transformation matrix ${}^{s}_{m}T$ from the master base coordinate system to the slave base coordinate system, the constraint of sound beam propagation law ${}^{m}T_{i}$, and the homogeneous transformation matrices ${}^{m}T_{eai}$ and ${}^{s}T_{ebi}$ from the two end-effectors to the two transducers. In the testing process, the homogeneous transformation matrices ${}^{m}T_{eai}$ and ${}^{s}T_{ebi}$ remain unchanged, so they are constant matrices. Eq. (5.72) lays a theoretical foundation for the testing of variable-thickness workpieces in the dual-manipulator testing system. After planning the scanning trajectory of the master manipulator, the scanning trajectory of the slave manipulator can be derived from the constraint relation for superposition synchronous motion.

The time-varying constraint relation ${}^{m}T_{i}$ is a 4×4 variable homogeneous transformation matrix, which can be expressed as

$$
{}^{m}T_{i} = \begin{bmatrix} {}^{m}R_{i} & P_{i} \\ 0 & 1 \end{bmatrix} \tag{5.74}
$$

Similarly, with the help of Eqs. (5.1) and (5.21), the position/attitude constraint Eq. (5.20) of the two end-effectors in the testing process of variable-thickness workpiece can be decomposed into a rotation matrix constraint and a position constraint, namely

$$\begin{cases} {}^{s}R_{bi} = {}^{s}T_{ebi}^{T} \cdot {}_{s}^{m}R^{T} \cdot {}^{m}R_{i} \cdot {}^{m}R_{eai} \cdot {}^{m}R_{ai} \\ {}^{s}P_{bi} = {}^{s}R_{ebi}^{T} \cdot {}_{s}^{m}R^{T} \cdot {}^{m}R_{i} \cdot {}^{m}R_{eai} \cdot {}^{m}P_{ai} + {}^{s}R_{ebi}^{T} \cdot {}_{s}^{m}R^{T} \cdot {}^{m}R_{i} \cdot {}^{m} \\ \quad P_{eai} + {}^{s}R_{ebi}^{T} \cdot {}_{s}^{m}R^{T} \cdot {}^{m}P_{ai} - {}^{s}R_{ebi}^{T} \cdot {}_{s}^{m}R^{T} \cdot {}^{m}P_{i} - {}^{s}R_{ebi}^{T} \cdot {}^{s}P_{ebi} \end{cases} \tag{5.75}$$

The first part of Eq. (5.22) is the attitude constraint relation between the end-effectors of the two manipulators during the testing of variable-thickness workpiece. The second part of Eq. (5.22) is the position constraint relation between the end-effectors of the two manipulators in this test. In the testing process, the homogeneous transformation matrices ${}^{m}T_{eai}$ and ${}^{s}T_{ebi}$ from the two end-effectors to the two transducers are known constants. The relative positions of the bases of the two manipulators remain unchanged. The homogeneous transformation matrix ${}_{m}^{s}T$ from the master base coordinate system to the slave base coordinate system can be obtained with the method proposed in this chapter. After the detection object is identified, the scanning trajectory of the master manipulator can be obtained through trajectory planning, and the sound beam propagation constraint ${}^{m}T_{i}$ at each trajectory point can be derived in the form of homogeneous transformation matrix. By using the above parameters, the motion trajectory of slave end-effector can be determined.

REFERENCES

1. Bing S. *Research on Recursive Mechanics of Unified Open and Closed Loop Mechanical Multi-body System based on Lie Group Lie Algebra [D]*. Nanjing: Nanjing University of Aeronautics and Astronautics, 2010.
2. Serre J.P. *Lie Algebras and Lie Groups[M]*. 2nd edition. Berlin: Springer, 2006.
3. Celledoni E., Iserles A., Methods for the approximation of the matrix exponential in a Lie-algebraic setting [J], *IMA Journal of Numerical Analysis*, 2001, 21(2):463–488.
4. Lenard I.E., The matrix exponential [J], *SIAM Review*, 1996, 38(3):507–512.
5. Higham N.J. *Functions of Matrices: Theory and Computation [M]*. Philadelphia: Siam, 2008.

CHAPTER 6

Error Analysis in Robotic NDT

THE FRAME-TYPE MOTION DEVICE, which has a linear mapping relation with the Cartesian coordinate system, can realize curvilinear motion in space through linear interpolation. The mapping between the six-degree of freedom (DOF) articulated joint manipulator and Cartesian coordinate system is nonlinear. For an expected position (attitude) of end-effector in the Cartesian coordinate system, several feasible solutions may lie in the joint rotation axis, thus bringing the problems of optimal solution selection and singular position and attitude. The position error of a moving device in linear mapping relation is generally proportional and deviating, while the position error of a moving device in nonlinear mapping relationship is represented by differential [1].

6.1 KINEMATICS ANALYSIS FOR ROBOTIC TESTING PROCESS

In this section, several coordinate systems are established on the moving devices (manipulator and turntable), the position/attitude transformation relations in the testing system are expressed by homogeneous transformation matrices and the coordination between manipulator and turntable is studied.

6.1.1 Establishment of the Coordinate System in a Moving Device

Denavit and Hartenberg (DH) proposed the DH method, that is, the establishment of a coordinate system at each connecting rod of a manipulator to describe the position/attitude relationship between the connecting rods. Thus, the DH model [2], a standardized model combining geometric parameters with the position/attitude relationship between the connecting rods, was established. In this model, four geometric parameters, namely DH parameters, were used to describe the position/attitude relationship between the connecting rods. Later, scholars proposed other kinematic modeling methods, such as the CPC (Completeness and Parametric Continuity) model [3], MCPC (Modified Completeness and Parametric Continuity) model [4] and POE (Product-Of-Exponentials) model [5–7]. However, the DH model is still the most classical. As shown in Figure 6.1, a coordinate system is established on each connecting rod, the ultrasonic transducer and the tested component. The position/attitude relationship between the connecting rods is described

DOI: 10.1201/9781003212232-6

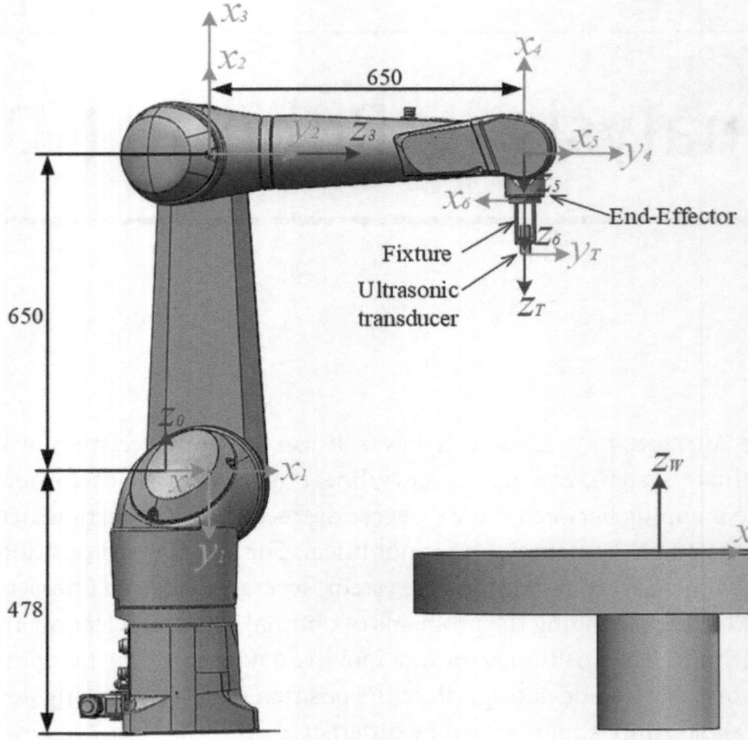

FIGURE 6.1 Distribution of coordinate systems in the moving devices.

with the DH method. {0} represents the base coordinate system; {*i*} (*i*=1… 5) represents the coordinate system of each connecting rod; {6} represents the end-effector coordinate system; {*T*} represents the tool coordinate system and {*R*} represents the turntable coordinate system. The unmarked coordinate axes can be inferred by the right-hand rule.

6.1.2 Matrix Representation of the Position/Attitude Relationship between Coordinate Systems

According to the DH method, the position/attitude relationship between any two rod coordinate system {*i*–1} and {*i*} in the manipulator can be expressed as the homogeneous transformation matrix T_{i-1}^{i}:

$$T_{i-1}^{i} = \text{Rot}(Z,\theta_i)\text{Trans}(a_i,0,d_i)\text{Rot}(X,\alpha_i)$$

$$= \begin{bmatrix} \cos\theta_i & -\sin\theta_i\cos\alpha_i & \sin\theta_i\sin\alpha_i & a_i\cos\theta_i \\ \sin\theta_i & \cos\theta_i\cos\alpha_i & -\cos\theta_i\sin\alpha_i & a_i\sin\theta_i \\ 0 & \sin\alpha_i & \cos\alpha_i & d_i \\ 0 & 0 & 0 & 1 \end{bmatrix} \tag{6.1}$$

where the four parameters θ_i, d_i, α_i and a_i are DH parameters:

θ_i: the joint angle around z_{i-1} from x_{i-1} to x_i;

d_i: the joint offset along z_{i-1} from x_{i-1} to x_i;

α_i: the deflection angle around x_i from z_{i-1} to z_i;

a_i: the offset along x_i from z_{i-1} to z_i;

where $i=0\ldots6$.

Take a six-DOF articulated manipulator as an example. Its kinematic parameters (DH parameters) are shown in Table 6.1. By substituting the DH parameters in Table 6.1 into Eq. (6.1), the position/attitude relationship between the two adjacent coordinate systems can be expressed as:

$$T_0^1=\begin{bmatrix} c_1 & 0 & -s_1 & a_1c_1 \\ s_1 & 0 & c_1 & a_1s_1 \\ 0 & -1 & 0 & 0 \\ 0 & 0 & 0 & 1 \end{bmatrix} \quad T_1^2=\begin{bmatrix} s_2 & c_2 & 0 & a_2s_2 \\ -c_2 & s_2 & 0 & -a_2c_2 \\ 0 & 0 & 1 & d_2 \\ 0 & 0 & 0 & 1 \end{bmatrix}$$

$$T_2^3=\begin{bmatrix} s_3 & 0 & c_3 & 0 \\ -c_3 & 0 & s_3 & 0 \\ 0 & -1 & 0 & 0 \\ 0 & 0 & 0 & 1 \end{bmatrix} \quad T_3^4=\begin{bmatrix} c_4 & 0 & s_4 & 0 \\ s_4 & 0 & -c_4 & 0 \\ 0 & 1 & 0 & d_4 \\ 0 & 0 & 0 & 1 \end{bmatrix}$$

$$T_4^5=\begin{bmatrix} c_5 & 0 & -s_5 & 0 \\ s_5 & 0 & c_5 & 0 \\ 0 & -1 & 0 & 0 \\ 0 & 0 & 0 & 1 \end{bmatrix} \quad T_5^6=\begin{bmatrix} -c_6 & s_6 & 0 & 0 \\ -s_6 & -c_6 & 0 & 0 \\ 0 & 0 & 1 & d_6 \\ 0 & 0 & 0 & 1 \end{bmatrix}$$

where c_i and s_i are the simplifications of $\cos\theta_i$ and $\sin\theta_i$ ($i=0\ldots6$).

TABLE 6.1 Kinematic Parameters (DH Parameters) of Manipulator

DOF	θ_i (°)	d_i (mm)	a_i (mm)	α_i (°)
1	θ_1	0	50	–90
2	θ_2–90°	50	650	0
3	θ_3–90°	0	0	–90
4	θ_4	650	0	90
5	θ_5	0	0	–90
6	θ_6–180°	100	0	0

The position/attitude relationship between any two coordinate systems can be obtained by multiplying the matrices. For example, the position/attitude relationship between the tool coordinate system $\{T\}$ in the front of the ultrasonic transducer and the base coordinate system $\{0\}$ can be expressed by:

$$T_0^T = T_0^1 T_1^2 T_2^3 T_3^4 T_4^5 T_5^6 T_6^T = \begin{bmatrix} n_x^T & o_x^T & a_x^T & p_x^T \\ n_y^T & o_y^T & a_y^T & p_y^T \\ n_z^T & o_z^T & a_z^T & p_z^T \\ 0 & 0 & 0 & 1 \end{bmatrix} \tag{6.2}$$

The Eq. (2.2) can be further simplified as:

$$T_0^T = \begin{bmatrix} \boldsymbol{R}_0^T & \boldsymbol{P}_0^T \\ \boldsymbol{0} & 1 \end{bmatrix} \tag{6.3}$$

where $\boldsymbol{P}_0^T$ represents position and $\boldsymbol{R}_0^T$ represents attitude. The position and attitude relationship between two coordinate systems in space can be represented by a homogeneous transformation matrix.

6.1.3 Coordinated Motion Relation between Manipulator and Turntable

As shown in Figure 6.2, this testing system is the combination of a manipulator and a turntable. The component under test, which is a curved rotary body, is clamped on the turntable. The whole-volume testing of the component is realized through the coordinated movement between the manipulator and the turntable. This operating mode is similar to but slightly different from that between a manipulator and a positioner in the automatic welding system. Generally speaking, in the automatic welding system, the positioner adjusts the component to an appropriate position and attitude and then stops, while the manipulator end-effector holds the welding gun to complete a series of welding tasks [8]. In this study,

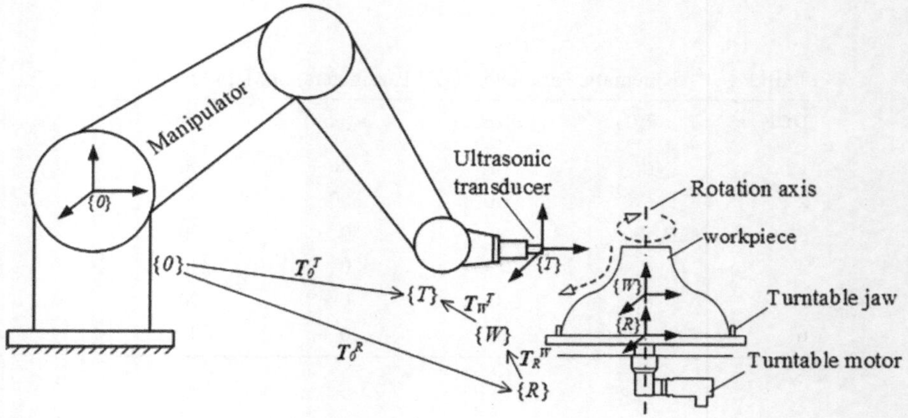

FIGURE 6.2 Coordinated motion between the manipulator and turntable.

the manipulator and the turntable need to move in coordination in order to complete the testing task. Therefore, in addition to modeling the manipulator kinematics, the coordinated motion between the manipulator and the turntable should also be considered.

On the one hand, the turntable drives the component to rotate and controls its rotation speed. On the other hand, a relative motion of the manipulator along the generatrix is superposed to the turntable motion according to a certain implicated relation. The kinematic relationships between them can be divided into two types and represented by top views, as shown in Figure 6.3.

1. The tested component in Figure 6.3a is a spatial axisymmetric geometry. The component under test is fixed by the turntable jaw, and its axis coincides with the gyration center of the turntable. Therefore, the ultrasonic transducer only needs to move along the generatrix with a certain step interval in order to cooperate with the turntable in meeting the testing requirements.
2. The tested component in Figure 6.3b is a spatial non-axisymmetric geometry. If the component under test is still fixed by the turntable jaw and the component centroid coincides with the gyration center of the turntable, the ultrasonic transducer should have a certain traction movement relation with the turntable. Whenever the turntable rotates by a stepper angle, the transducer will have a deflection ($\Delta\theta$) and an offset (Δl). Meanwhile, the manipulator needs to move along the generatrix with a certain step interval in order to meet the testing requirements.

However, due to the assembly error of turntable jaw, the axis or centroid of the component under test may deviate from the gyration center of the turntable, causing the eccentric motion of the rotating component. Thus, it can be seen that in the testing process with manipulator and turntable in coordinated motion, the clamping deviation will reduce the

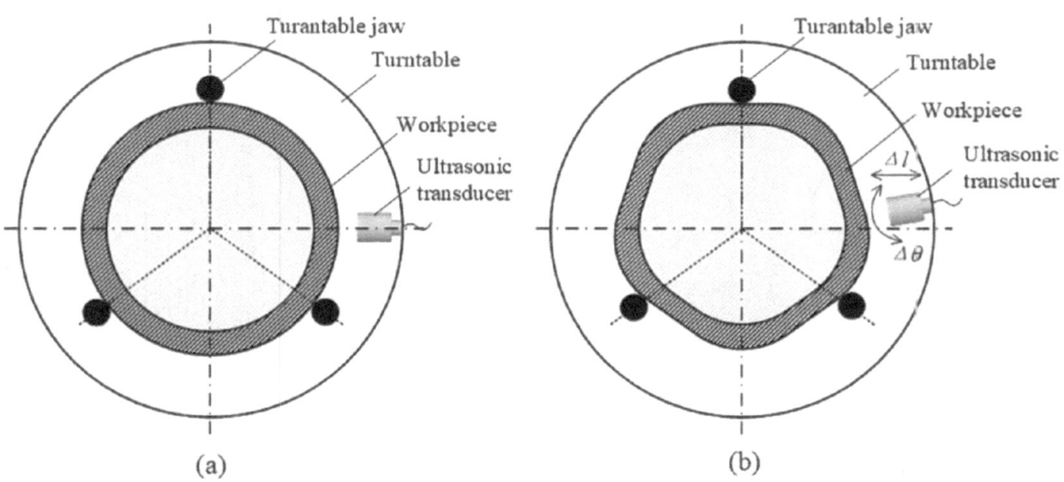

FIGURE 6.3 Different coordinated motions for the two components with different shapes. (a) Spatial axisymmetric geometry and (b) spatial non-axisymmetric geometry.

positioning accuracy and affect the detection results. Therefore, the eccentricity needs to be corrected before testing or compensated for by the coordinated manipulator movement during testing.

6.1.4 Matrix Representation of Coordinated Motion Relation

In the testing process, the manipulator drives the ultrasonic transducer to move and the turntable drives the component under test to move, so the position and attitude of the tool coordinate system relative to the workpiece coordinate system directly affect the detection results. In Figure 6.4, the base coordinate system {0}, the turntable coordinate system {*R*}, the workpiece coordinate system {*W*} and the tool coordinate system {*T*} are connected through the spatial dimension chain and expressed by the matrix multiplication ***T*** *T* 0=***T*** *R* 0***T*** *W* *R****T*** *T* *W*. Therefore, the transformation relation between the tool coordinate system and workpiece coordinate system is ***T*** *T* *W*=(***T*** *R* 0***T*** *W* *R*)⁻¹***T*** *T* 0. The matrix ***T*** *R* 0 represents the position and attitude between the turntable and the manipulator:

$$T_0^R = \begin{bmatrix} c_z & -s_z & 0 & p_x^R \\ s_z & c_z & 0 & p_y^R \\ 0 & 0 & 1 & p_z^R \\ 0 & 0 & 0 & 1 \end{bmatrix} \tag{6.4}$$

where (p_x^R, p_y^R, p_z^R) represents the position of the turntable coordinate system relative to the base coordinate system, that is, the position relation between the turntable gyration center and the manipulator; c_z and s_z are the simplifications of $\cos\theta_z$ and $\sin\theta_z$. The turntable state in Figure 6.2 is defined as the initial state, where the turntable angle is zero. θ_z $(0° \leq \theta_z < 360°)$ represents the rotation angle of the turntable. The component under test rotates with the turntable. The matrix ***T*** *W* *R* in Figure 6.2 represents the position/attitude relationship between the component and the turntable.

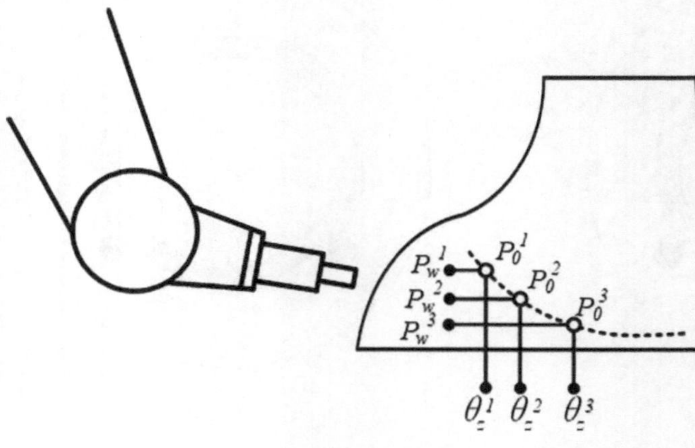

FIGURE 6.4 Schematic diagram of an implicated motion relationship.

As shown in Figure 6.4, the dashed line along the curved surface of the component contains three hollow points. If the manipulator is hoped to clamp the ultrasonic transducer and move along the dashed curve containing the three hollow points (see Figure 6.4), this curvilinear motion can be the coordinated motion composed of *P* 1 *w*, *P* 2 *w*, *P* 3 *w* and *θ* 1 *z*, *θ* 2 *z*, *θ* 3 *z*. To be specific, *P* 1 *w*, *P* 2 *w* and *P* 3 *w* represent the coordinates of the three points relative to the workpiece coordinate system, and θ 1 *z*, θ 2 *z* and θ 3 *z* represent the rotation angles of the three points relative to the turntable coordinate system. This coordinated motion relation is expressed by the matrix ***P*** ***i*** 0=***T*** *R* 0(*θ i z*) ***T*** *W R P i W* (*i*=1, 2, 3).

6.2 PLANNING OF MOTION PATH IN THE TESTING PROCESS

Motion path planning is the foundation of automatic ultrasonic testing. When using manipulator to test a complex surface or a rotary body with complex surface, the generation algorithm of detection path should be made clear so as to give play to the manipulator's flexibility and improve the surface fitting accuracy. In this section, the five-axis numerical control (NC) machining theory of complex surface [9] is based on to study the planning of detecting motion path. The NC machining uses cutter as a tool to remove the material, while the ultrasonic testing uses transducer as a tool for imaging, flaw detection and material property evaluation. Thus, it is necessary to study their differences. The motion trajectory obtained through five-axis NC machining simulation software in the computer needs to be post-processed before becoming the motion program available for the manipulator controller. The post-processing is based on the inverse kinematics algorithm. The process of generating a motion program through computer-aided machining software is called offline programming in robotics. Its ultimate goal is to ensure that the manipulator moves along the planned path.

6.2.1 Algorithm of Detection Path Generation

Before ultrasonic testing, the measured surface should be discretized into a series of paths. This process is called detection path generation. This process is generally implemented by CNC machining simulation software. The methods more commonly used in path generation include: equal parameter (isoparametric method), equal plane (isoplanar method), constant residual height, spatial filling curve and continuous curve generation [10]. The purpose of ultrasonic testing is not to remove materials but to ensure that the instruction points are distributed on the measured surface as evenly as possible, so that the scanning image can reflect the actual appearance of defects without distortion. The isoplanar method can satisfy this requirement better. The process of dividing the free-form surface *S* by a series of parallel planes (P_1, P_2, P_3...) to obtain the intersection curves (namely paths) is just isoplanar path planning, as shown in Figure 6.5. The parameters involved in path planning are mainly step interval and scanning interval:

1. Step interval

   In NC machining, the path is planned according to the cutting speed, cutting depth, feed rate and other parameters to obtain ideal machining accuracy. In

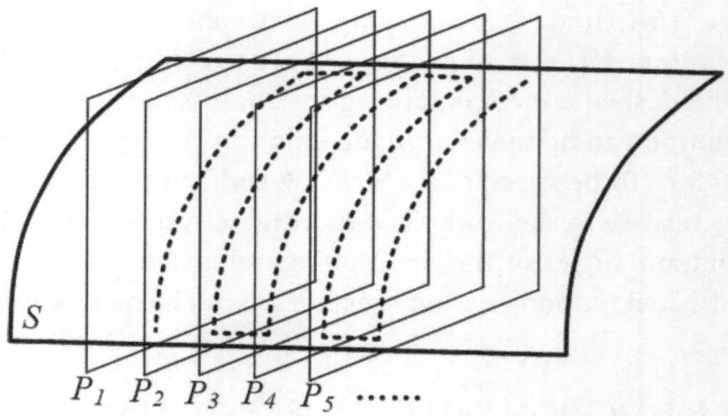

FIGURE 6.5 Isoplanar path planning.

ultrasonic testing, the path is planned according to the focal length F, focal beam diameter Φ, material sound velocity C_M and underwater sound velocity C_W, in order to obtain more ideal detection sensitivity and resolution. Take the path planning for a focusing probe in surface detection as an example. As shown in Figure 6.6, the distance between transducer surface and incident surface is the water path h, whose calculation formula is

$$h = F - t\left(C_M / C_W\right) \tag{6.5}$$

where t is the depth of focus. The water path can be calculated with Eq. (6.5) according to the requirement of detection depth, in order to locate the focus at the detection depth. For the path planning of surface detection shown in Figure 6.7, the tool coordinate system $\{T\}$ is established at the focus, the water path is h, the step interval is approximate to or slightly larger than the focal beam diameter Φ and the incident angle is 0°, in order to better cover the whole detection area.

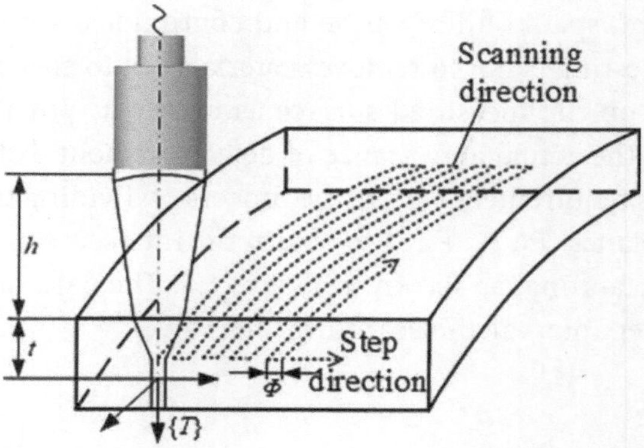

FIGURE 6.6 Surface scanning path of an ultrasonic transducer.

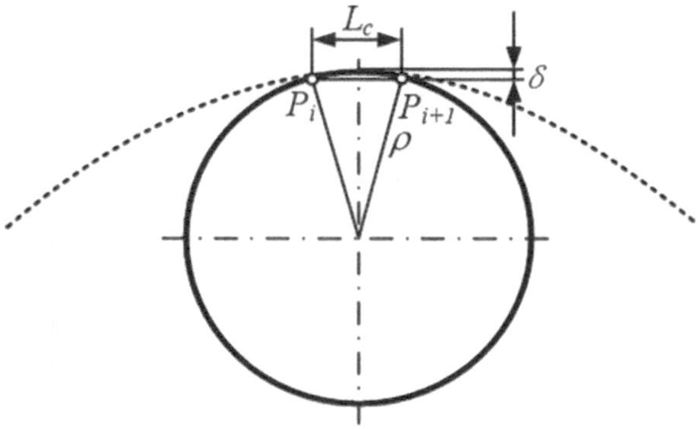

FIGURE 6.7 Linearized approximation of a scanning curve.

2. Scanning interval

In Figure 6.7, the dashed line is part of the scanning curve, and the spatial distance between the two adjacent instruction points P_i and P_{i+1} is the scanning interval. In the detection process, the tracking of the surface profile is realized by using linear interpolation to approximate the scanning curve, so the fitting error of linear interpolation will affect the tracking accuracy of the surface profile. In path planning, the scanning interval must be set reasonably according to the detection requirement of the smallest defect, so as to ensure the detection efficiency and give full play to the detection performance of the system. If the radius of curvature within a scanning interval is ρ and the scanning interval is L_c, a triangular relationship will exist:

$$\rho^2 = \left(\frac{L_c}{2}\right)^2 + (\rho - \delta)^2 \tag{6.6}$$

Then the error of linearized approximation is $\delta = \rho - \sqrt{(L_c/2)^2 - \rho^2}$. The smaller the error is, the higher the tracking accuracy of the surface profile will be.

6.2.2 Resolving of Manipulator Motion Path

The detection path generated by CNC machining simulation software takes the workpiece coordinate system as the reference. It is a neutral trajectory file that is not specific to a specified moving device and cannot be directly used to drive the manipulator motion. Instead, the path file must be converted into the command file that can be recognized by manipulator controller. During the CNC programming, the tool coordinate system and the motion direction are defined according to international standards, and the motion trajectory is converted into the machining instructions in the CNC system of a specific machine tool. In the manipulator, the implementation of this process is based on inverse operation, that is, the variables of each joint are solved according to the kinematics equation provided that the target position and attitude at a certain point in space to be reached by the tool coordinate system are known. In general, inverse operation methods are divided into analytical method and numerical method. The analytical method can obtain the solution directly and rapidly. But whether analytical solutions can be obtained depends on the manipulator structure. In general, analytic solutions can be obtained only when most of the DH parameters are 0. The numerical method is independent of the manipulator structure, but its calculation speed is slow. If the initial value is improperly selected, non-convergence or wrong convergence may occur [11]. For a six-DOF manipulator, the existence of an analytical solution must be based on the Piper rule, that is, one of the following two conditions must be satisfied:

1. The Z-axes of the three adjacent joints intersect at a point;
2. The Z-axes of the three adjacent joints are parallel to each other.

In Figure 6.8, the axes Z_3, Z_4 and Z_5 intersect at a point, so the analytical solution can be obtained for the manipulator used in this study. If the origins of the two coordinate systems {4} and {5} at the manipulator wrists are defined as the point D, it is not difficult to find that the translation of the coordinate system {6} along the $-Z$ axis by d_6 is just the point D. This translation process can be expressed by the following homogeneous coordinate transformation:

$$\begin{bmatrix} n_x^6 & o_x^6 & a_x^6 & p_x^6 \\ n_y^6 & o_y^6 & a_y^6 & p_y^6 \\ n_z^6 & o_z^6 & a_z^6 & p_z^6 \\ 0 & 0 & 0 & 1 \end{bmatrix} \begin{bmatrix} 0 \\ 0 \\ -d_6 \\ 1 \end{bmatrix} = \begin{bmatrix} p_x^6 - a_x^6 d_6 \\ p_y^6 - a_y^6 d_6 \\ p_z^6 - a_z^6 d_6 \\ 1 \end{bmatrix} \tag{6.7}$$

It can be seen that the coordinates of the point D can be deduced from the spatial position and attitude of the end-effector:

$$\begin{cases} p_x^D = p_x^6 - a_x^6 d_6 \\ p_x^D = p_y^6 - a_y^6 d_6 \\ p_x^D = p_z^6 - a_z^6 d_6 \end{cases} \tag{6.8}$$

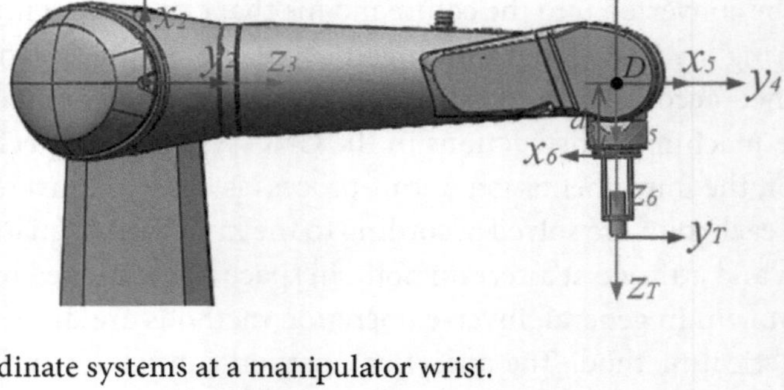

FIGURE 6.8 Coordinate systems at a manipulator wrist.

For the manipulator conforming to the Piper rule, each joint angle can be deduced from the special position relation of the point D:

1. Calculate the joint angle θ_1

   It can be seen from Eq. (6.2) that the position/attitude relationship between the coordinate systems can be obtained through matrix multiplication. In Figure 6.8, the position and attitude of the coordinate system {4} relative to the base coordinate system {0} are

$$T_0^4 = T_0^1 T_1^2 T_2^3 T_3^4 = \begin{bmatrix} n_x^D & o_x^D & a_x^D & p_x^D \\ n_y^D & o_y^D & a_y^D & p_y^D \\ n_z^D & o_z^D & a_z^D & p_z^D \\ 0 & 0 & 0 & 1 \end{bmatrix}$$

$$= \begin{bmatrix} c_1s_2s_3c_4 - c_1c_2c_3c_4 + s_1s_4 & c_1s_2c_3 + c_1c_2s_3 & c_1s_2s_3s_4 - c_1c_2c_3s_4 - s_1c_4 & c_1s_2c_3d_4 + c_1c_2s_3d_4 + c_1a_2s_2 - s_1d_2 + a_1c_1 \\ s_1s_2s_3c_4 - s_1c_2c_3c_4 - c_1s_4 & s_1s_2c_3 + s_1c_2s_3 & s_1s_2s_3s_4 - s_1c_2c_3s_4 + c_1c_4 & s_1s_2c_3d_4 + s_1c_2s_3d_4 + s_1a_2s_2 + c_1d_2 + a_1s_1 \\ c_2s_3c_4 + s_2c_3s_4 & c_2c_3 - s_2s_3 & c_2s_3s_4 + s_2c_3s_4 & c_2c_3d_4 - s_2s_3d_4 + a_2c_2 \\ 0 & 0 & 0 & 1 \end{bmatrix} \quad (6.9)$$

In Figure 6.9, the coordinates of the point D are the position coordinates of the coordinate system {4} relative to the base coordinate system:

$$\begin{cases} p_x^D = c_1s_2c_3d_4 + c_1c_2s_3d_4 + c_1a_2s_2 - s_1d_2 + a_1c_1 \\ p_y^D = s_1s_2c_3d_4 + s_1c_2s_3d_4 + s_1a_2s_2 + c_1d_2 + a_1s_1 \\ p_z^D = c_2c_3d_4 - s_2s_3d_4 + a_2c_2 \end{cases} \quad (6.10)$$

It can be found that there is a relationship between p_x^D and p_y^D in Eq. (6.10):

$$p_x^D s_1 - p_y^D c_1 = -d_2 \quad (6.11)$$

Eq. (6.11) is combined with the trigonometric function relation into:

$$\begin{cases} p_x^D s_1 - p_y^D c_1 = d_2 \\ s_1^2 + c_1^2 = 1 \end{cases} \quad (6.12)$$

Eq. (6.12) has two sets of solutions in the range of (−π, π).

2. Calculate the joint angle θ_2

The fourth column of the matrix in Eq. (6.9) follows the following relations:

$$\begin{cases} p_x^D c_1 + p_y^D s_1 = s_{23} d_4 + a_2 s_2 + a_1 \\ p_z^D = c_{23} d_4 + a_2 c_2 \end{cases} \tag{6.13}$$

where s_{23} and c_{23} are the simplifications of $\sin(\theta_2+\theta_3)$ and $\cos(\theta_2+\theta_3)$. To calculate θ_2, we need to eliminate θ_3 from Eq. (6.13). Then Eq. (6.13) can be rewritten as:

$$\begin{cases} s_{23} d_4 = p_x^D c_1 + p_y^D s_1 - a_2 s_2 - a_1 \\ c_{23} d_4 = p_z^D - a_2 c_2 \end{cases} \tag{6.14}$$

Both sides of the equal sign in Eq. (6.14) are squared and added:

$$d_4^2 = \left(p_x^D c_1 + p_y^D s_1 - a_2 s_2 - a_1\right)^2 + \left(p_z^D - a_2 c_2\right)^2 \tag{6.15}$$

Eq. (6.15) is combined with the trigonometric function relation into:

$$\begin{cases} \left(p_x^D c_1 + p_y^D s_1 - a_2 s_2 - a_1\right)^2 + \left(p_z^D - a_2 c_2\right)^2 = d_4^2 \\ s_2^2 + c_2^2 = 1 \end{cases} \tag{6.16}$$

Eq. (6.16) has two sets of solutions in the range of (−π, π).

3. Calculate the joint angle θ_3

From Eq. (6.14), the relation between θ_2 and θ_3 can be obtained:

$$\tan(\theta_2 + \theta_3) = \frac{p_x^D c_1 + p_y^D s_1 - a_2 s_2 - a_1}{p_z^D - a_2 c_2} \tag{6.17}$$

The calculation of θ_3 based on the known θ_1 and θ_2 can obtain a set of solutions.

It can be known from the above calculation process of the first three joint angles that θ_1 has two sets of solutions, each of which corresponds to two sets of solutions to θ_2 and θ_3. Therefore, the first three joints have four sets of solutions in total. Due to the limitation of joint motion range, some of the solutions are actually invalid. Before calculating the last three joint angles, the solutions that are not within the motion range can be eliminated.

4. Calculate the joint angles θ_4, θ_5 and θ_6

R_0^3 can be deduced from the determined first three joint angles. The attitude matrix R_0^6 of the manipulator end-effector has the following relationship with R_0^3:

$$R_3^6 = \left(R_0^3\right)^{-1} R_0^6 = \begin{bmatrix} n_x^w & o_x^w & a_x^w \\ n_y^w & o_y^w & a_y^w \\ n_z^w & o_z^w & a_z^w \end{bmatrix} \tag{6.18}$$

where all the matrix elements are known. R_3^6 is determined by the last three joint angles:

$$R_3^6 = R_3^4 R_4^5 R_5^6 = \begin{bmatrix} -c_4c_5c_6 + s_4s_6 & c_4c_5s_6 + s_4c_6 & -c_4s_5 \\ -s_4c_5c_6 - c_4s_6 & s_4c_5s_6 - c_4c_6 & -s_4s_5 \\ -s_5c_6 & s_5s_6 & c_5 \end{bmatrix} \tag{6.19}$$

From Eqs. (6.18) and (6.19), we obtain:

$$\begin{cases} s_5 = \pm\sqrt{\left(a_x^w\right)^2 + \left(a_y^w\right)^2} \\ c_5 = a_z^w \end{cases} \tag{6.20}$$

Then the last three joint angles are:

$$\begin{cases} \tan\theta_4 = \left(a_y^w / s_5\right) / \left(a_x^w / s_5\right) \\ \tan\theta_5 = \pm\sqrt{\left(a_x^w\right)^2 + \left(a_y^w\right)^2} \Big/ a_z^w \\ \tan\theta_6 = \left(o_z^w / s_5\right) / \left(-n_z^w / s_5\right) \end{cases} \tag{6.21}$$

When $\theta_5=0$, the joints 4 and 6 are in a singular posture on the same axis. At this point, the established angle of the joint 4 in the current posture is used to calculate the angle of the joint 6. For each set of $(\theta_1, \theta_2, \theta_3)$, two sets of $(\theta_4, \theta_5, \theta_6)$ can be obtained. Given any posture in space, a total of eight sets of solutions can be obtained without considering the limitation of joint angle. By using the joint angle limitation and the condition of shortest path, the optimal solution set (namely the joint angles obtained when achieving the target posture) can be selected from the eight sets of solutions. The calculation process of converting each target posture point into manipulator joint angles is implemented by manipulator controller.

6.3 ERROR SOURCES IN ROBOTIC UT PROCESS

Through the combination of kinematics research and motion path planning with UT engineering practice, the error sources in the robotic UT system are categorized into three types: the geometric error in path copying, the localization error in manipulator motion and the clamping error of tested component. The three error sources are relatively independent and can be analyzed separately for accuracy compensation.

6.3.1 Geometric Error in Path Copying

How to plan the path to keep the incident angle and water path always unchanged in the profiling motion of ultrasonic transducer determines the detection performance of the system. It has been found in the actual testing of various components (such as hub, blade and turbine disk) that curve fitting error, component shape error and joint interpolation error are the main factors:

1. Curve fitting error

   Theoretically, the incident angle and water path between the ultrasonic transducer and the tested surface should remain unchanged in the process of moving along the trajectory, which is a spatial curve. In the CNC simulation software, the motion program finally generated through linear interpolation and space curve fitting is composed of some discrete position/attitude control points [12]. The method of fitting the space curves through linear interpolation will lead to the direction error of a sound beam incident on the surface. The value of this error depends on the curvature change, the diameter of sound beam focus, the position of the transducer relative to the surface under test and the distance between control points. In order to identify the smallest defect, the error of curve fitting should be controlled within the lower limit of defect size, which is just the allowable error of surface fitting. Obviously, an effective way to reduce the error is to increase the interpolation points. However, it is found in production practice that the setting of too many interpolation points will reduce the detection efficiency. Therefore, reasonable curve fitting should be based on the comprehensive consideration of defect resolution and detection efficiency. A better motion path should occupy a smaller computer memory while guaranteeing the uniform distribution of scanning intervals and the reasonable distribution of step intervals.

2. Component shape error

   Some complex-shaped components under test are often composed of multiple complex surfaces in space. Shape error is the deviation between the component under test and its CAD model. In the CNC machining of a complex surface, the rigidity of machine tool and fixture is strong enough. However, the rigidity of cutter is relatively weak, so the cutter may deform easily under an external force and deviate from the ideal cutting position defined in the CAD model. For thin-walled components such as blades, their torsion and bending deformation in the machining process is also an important reason for their shape error. In addition, tool vibration, thermal deformation and stress release will cause a certain deviation between the tested component and its CAD model. For a component that has been in service for a long time, the influence of external load, temperature change, corrosion and other external factors during service should also be considered. For this reason, some methods of on-machine detection or offline measurement are introduced. On-machine detection is to replace the cutter by a high-precision probe to measure the shape of the machined component by means of contact, without removing the machined component. Offline measurement is to fix the component on special equipment such as CMM (Coordinate Measuring Machine). This method generally applies to the maintenance of the parts in service. After the shape error is determined, the precision compensation can be made during ultrasonic testing to reduce the beam incidence error to a certain extent. However, both of the two methods will increase the complexity and equipment cost of ultrasonic detection. For the robotic ultrasonic testing system, a more advantageous measurement method is to use ultrasonic measurement function to evaluate the shape error of the measured component.

3. Joint interpolation error

   After the position/attitude control points generated in the CNC machining simulation software are imported into the manipulator controller, the curved surface can be tracked and copied through the interpolation of six moving joints. For some corner areas with large curvature variation, if the same scanning interval and step interval are applied to these areas in the actual detection process, the accuracy of curve fitting will decline sharply. If the control points in these areas are increased, frequent acceleration and deceleration will occur. In some areas where the transducer moves faster, the echo signals may be lost due to the limitation of excitation frequency. In the CNC machine, the cutter control instructions supporting the spline curve format are provided. The use of the spline curves with continuous curvature instead of linear interpolation can not only avoid the error caused by motion path discontinuity but also reduce the volume of program files and improve the machining efficiency and accuracy. However, after the position/attitude control points generated by CNC machining simulation software are imported into the manipulator controller, continuous curvilinear motion can only be achieved through joint interpolation, during which a certain error will be introduced.

6.3.2 Localization Error in Manipulator Motion

In ultrasonic testing, the ultrasonic transducer is fixed on the manipulator end-effector through a fixture and then moves with the end-effector along the trajectory generated by offline programming. The analysis of manipulator localization error focuses on the error introduced by the manipulator itself. In robotics, a coordinate system is generally established on each moving part to describe the spatial posture relationship. Then, the localization error is analyzed through four steps, namely motion modeling, data measurement, parameter identification and error compensation. The base coordinate system {0} is the reference coordinate system of the whole workspace and is also considered as the world coordinate system. The connecting-rod coordinate systems {1–5} and the end-effector coordinate system {6} are initially defined by the manufacturer before the manipulator is delivered. The tool coordinate system $\{T\}$ is established at the focus of the sound beam. The manipulator's localization error is the location error of sound beam focus relative to the base coordinate system. This error is mainly affected by manipulator kinematic parameter error and tool installation error.

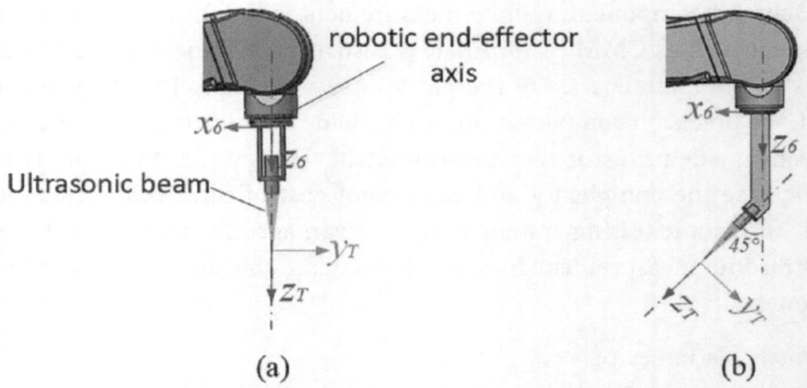

FIGURE 6.9 Installation form of an ultrasonic transducer. (a) Ultrasonic beam coincides with robotic end-effector axis and (b) ultrasonic beam and robotic end effector axis deflection 45°.

The errors in kinematic parameters include the errors in connecting rod length, joint angle, connecting rod angle and connecting rod offset. For example, in the DH motion model, the kinematic parameter errors are the deviations between the theoretical values and actual values of the four parameters θ_i, d_i, α_i and a_i; and the tool installation error is the installation error of the ultrasonic transducer fixed to the end-effector. As the detection tool, ultrasonic transducer is connected to the end-effector through fixture in a way that meets the detection requirements. For example, in Figure 6.9a, the beam axis is coaxial to the end-effector; in Figure 6.9b, there is a 45° deflection angle between them. In addition, an array detection method where multiple transducers are held can also be designed. According to the description of motion path planning in Section 2.4, more ideal resolution can be obtained by defining the tool coordinate system at the focus of the sound beam. Therefore, the deviation between the theoretical value and actual value of the position/attitude matrix ***T*** *T* 6 of the tool coordinate system {*T*} relative to the end-effector coordinate system {6} will affect the defect evaluation result.

6.3.3 Clamping Error of Tested Component

The connecting-rod coordinate systems {1–5}, the end-effector coordinate system {6} and the tool coordinate system {*T*} are fixed to the manipulator itself, while the workpiece coordinate system {*W*} and the turntable coordinate system {*R*} are outside the manipulator. It is generally recognized that the coordinate systems {*W*} and {*R*} are the main error source of this detection system:

1. Workpiece coordinate system {*W*}

   The workpiece coordinate system is of great significance in offline programming. As shown in Figure 6.10, the motion path of ultrasonic transducer in the simulation software is planned with the workpiece coordinate system as the reference to generate the corresponding coordinate points, without considering the turntable. In the actual detection, these coordinate points are converted into the coordinate points in the base coordinate system according to the relation ***T*** *T* 0 = ***T*** *W* 0***T*** *T W*. Therefore,

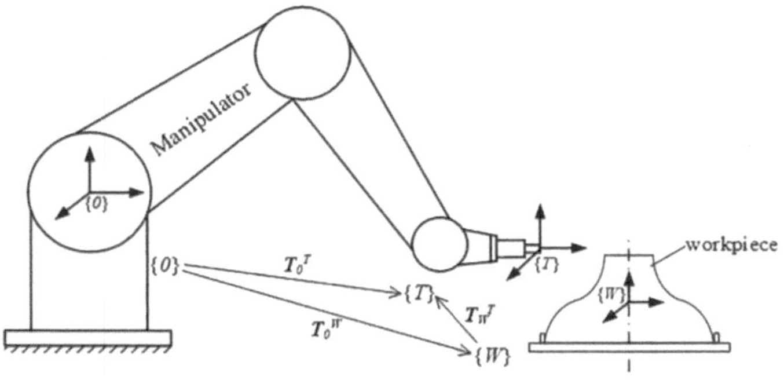

FIGURE 6.10 Coordinate transformation relation between the workpiece coordinate system and base coordinate system in space.

the deviation between the theoretical value and actual value of the position/attitude matrix ***T*** *W* 0 of the workpiece coordinate system {*W*} relative to the base coordinate system {0} will affect the detection result.

2. Turntable coordinate system {*R*}

   Turntable will play a role in the testing of rotary components. As can be seen from the description of the detection system, the manipulator is fixed, the turntable is integrated with water tank and the rotary component under test is fixed on the turntable through a three-jaw chuck. Accurate clamping is to ensure the collinearity between the center axis of the rotary component and the turntable axis. As shown in Figure 6.11, the verticality error and eccentricity error of the turntable axis will cause radial run-out in the testing of rotary components. In Figure 6.11a, the turntable axis deviates from the vertical direction by a deflection angle $\Delta\alpha$. In Figure 6.11b, the turntable axis deviates from the component center axis by an eccentricity Δl. The deflection angle $\Delta\alpha$ and the eccentricity Δl will cause the radial run-out of the component rotating with turntable. Under the influence of errors, the sound beam incident on the tested surface will deflect by an angle that will change in the form of

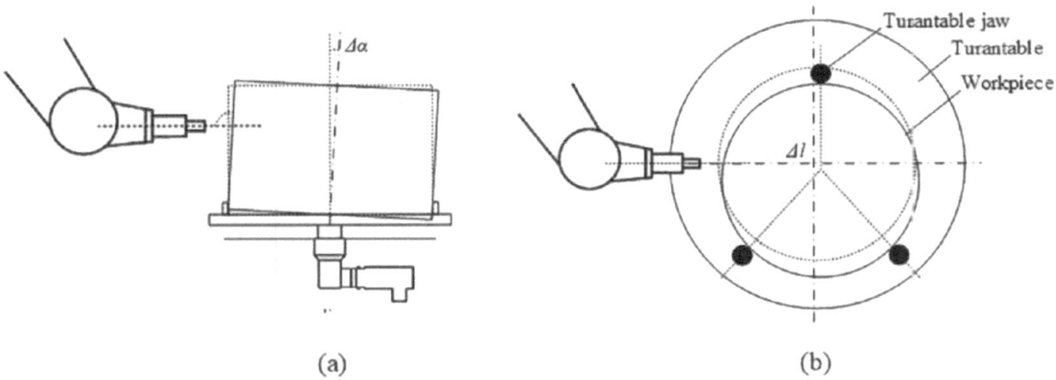

FIGURE 6.11 Perpendicularity error and eccentricity error of the turntable axis. (a) Existence of deflection angle Δα and (b) existence of eccentric distance Δl.

trigonometric function as the turntable rotates. This radial run-out is represented by unstable echo waveform in the A-scan image. A smaller radial run-out can be solved through the tracking gate. If the radial run-out is too large, the error needs to be corrected. Therefore, consideration should be given to the matrix representing the coordinated motion between the manipulator and the turntable, that is, the theoretical value of the position/attitude matrix $\boldsymbol{T}\ W\ R$ should be consistent with its actual value.

REFERENCES

1. Wang W., Wang L., Yun C., Design of a two-step calibration method of kinematic parameters for serial robots [J], *Chinese Journal of Mechanical Engineering*, 2017, 30(2):438–448.
2. Denavit J., Hartenberg R.S., A kinematic notation for lower-pair mechanisms based on matrices [J], *Journal of Applied Mechanics*, 1955, 23:215–221.
3. Zhuang H., Roth Z.S., Hamano F., A complete and parametrically continuous kinematic model for robot manipulators [J], *IEEE Transactions on Robotics and Automation*, 1992, 8(4):451–463.
4. Meng Y., Zhuang H.Q., Hamano F., Autonomous robot calibration using vision technology [J], *Robotics and Computer-Integrated Manufacturing*, 2007, 23(4):436–446.
5. Chen I.M., Yang G.L., Tan C.T., et al., Local POE model for robot kinematic calibration [J], *Mechanism and Machine Theory*, 2001, 36(11–12):1215–1239.
6. Wang H.X., Shen S.H., Lu X. et al., A screw axis identification method for serial robot calibration based on the POE model [J], *Industrial Robot: An International Journal*, 2012, 39(2):146–153.
7. Wang Y.B., Wu H.P., Handroos H., Accuracy improvement of a hybrid robot for ITER application using POE modeling method [J], *Fusion Engineering and Design*, 2013, 88(9–10):1877–1880.
8. He G., Gao H., Wu L., Inverse kinematics algorithm of positioner based on the mathematical model of welding position [J], *Journal of Mechanical Engineering*, 2006, 42(6):86–91.
9. Bi Q., Ding H., Wang Y. *Theory and Technology of Five-axis CNC Machining of Complex Curved Parts [M].* Wuhan: Wuhan University of Technology Press, 2016:16–21.
10. Zhou J., Liu G. Lou Y., Gou F. *High-efficiency Precision CNC Machining of Complex Surface [M].* Hangzhou: Zhejiang University Press, 2014:23–37.
11. Bekir K., Serkan A., An improved approach to the solution of inverse kinematics problems for robot manipulators [J], *Engineering Applications of Artificial Intelligence*, 2000, 13(2):159–164.
12. Liu X., Li Y.G., Xu X.C., A region-based tool path generation approach for machining freeform surfaces by applying machining strip width tensor [J], *International Journal of Advanced Manufacturing Technology*, 2018, 98(9–12):3191–3204.

CHAPTER 7

Error and Correction in Robotic Ultrasonic Testing

THE CHARACTERIZATION OF ULTRASONIC PROPAGATION in the coupling medium and tested components can help optimize the detection parameters and motion path, improve the detection resolution and obtain better detection results. In the research of sound field, mathematical model is widely used at home and abroad to simulate the propagation mechanism and predict the ultrasonic echo signal results. The methods of sound field simulation in computational acoustics are generally divided into analytical methods and numerical methods. The analytical methods include angular spectrum, multivariate Gaussian superposition and ray tracing. The numerical methods include finite difference, finite element and boundary element [1].

7.1 ULTRASONIC PROPAGATION MODEL

An ultrasonic transducer is a device to realize the conversion between sound energy and electric energy mainly through piezoelectric, electrostrictive or magnetostrictive effect. The ultrasonic source described in this chapter is a piezoelectric ultrasonic transducer that is mainly equipped with a normal probe or a focusing probe in the detection process. The ultrasonic transducers with other probes such as ordinary angular probe, double crystal probe and phased array probe can also be used.

1. The normal probe is made of a single circular piezoelectric wafer with both transmitting and receiving functions. Its internal structure is shown in Figure 7.1a. The wafer vibrates and excites the longitudinal wave in the thickness direction. The wafer size is directly related to the resonant frequency, directivity and near field length of sound wave. The resonant frequency depends on the wafer thickness t, that is, $t=\lambda/2$ (λ is the wavelength of sound wave in the wafer material). Given a certain wavelength range, a larger wafer diameter D indicates better beam directivity, as approximately reflected in the relational expression $\theta_0=70\lambda/D$ (θ_0 is the semi-diffusion angle of sound beam).

DOI: 10.1201/9781003212232-7

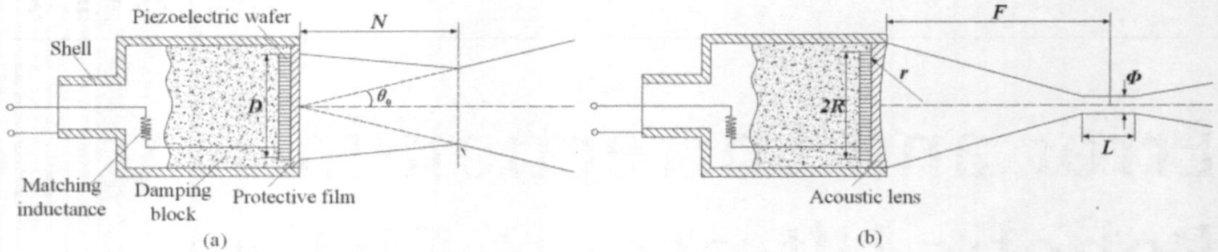

FIGURE 7.1 Internal structure of an ultrasonic transducer. (a) Normal probe and (b) focusing probe.

The area before the last sound pressure maximum on the acoustic axis is near-field area. The relationship among the near-field length N, the wafer diameter D and the wavelength λ is $N=(D^2—\lambda^2)/4\lambda$. The sound beam emitted by normal probe is treated as natural focus at the last sound pressure maximum, and the near-field length is deemed as natural focal length.

2. In the focusing probe (see Figure 7.1b for its internal structure), a focusing acoustic lens is added in front of piezoelectric wafer to converge sound beams into a narrow beam at a certain distance, so as to improve the detection sensitivity, resolution and directivity and reduce the influence of interference signals. When a focusing acoustic lens is used, the relationship between the focal length F in water and the curvature radius r of the acoustic lens is $F=r/[1—(C_W/C_L)]$ (C_L and C_W are the velocities of ultrasound in the lens material and water, respectively). After the beam is focused, a cylindrical focal zone is formed at the focus. The reflector (generally a ball) is moved outward from the focus along the radial direction of the sound beam. Twice the radial movement distance covered when the reflected sound pressure drops by 6 dB is the focal beam diameter Φ, as reflected in the relational expression $\Phi_{-6\,dB}\approx\lambda F/2R$ (R is wafer radius). The smaller the Φ is, the higher the lateral resolution will be. Now the reflector (generally a ball) is moved outward from the focus along the axial direction of the beam. Twice the axial movement distance covered when the reflected sound pressure drops by 6 dB is the focal beam length L, as reflected in the relational expression $L_{-6\,dB}\approx\lambda F/R^2$. The longer the L is, the greater the effective detection depth will be.

7.1.1 Fluctuation of Sound Pressure in an Ideal Fluid Medium

As a macroscopic mechanical vibration, ultrasonic wave satisfies three basic physical laws under ideal conditions, namely, Newton's second law, law of conservation of mass and equation of state (describing the relationship among pressure, temperature and volume). From the three basic laws, three basic equations of ideal fluid medium can be derived:

1. Equation of motion

Suppose that what is in Figure 7.2a is a volume element small enough in the ultrasonic propagation medium. P_0 is the static pressure intensity of the medium, P is sound pressure and S is the side area. The force acting on the left side of this volume element is $F_1=(P_0+p)S$ and the force acting on the right side is $F_2=(P_0+p+dp)S$, so

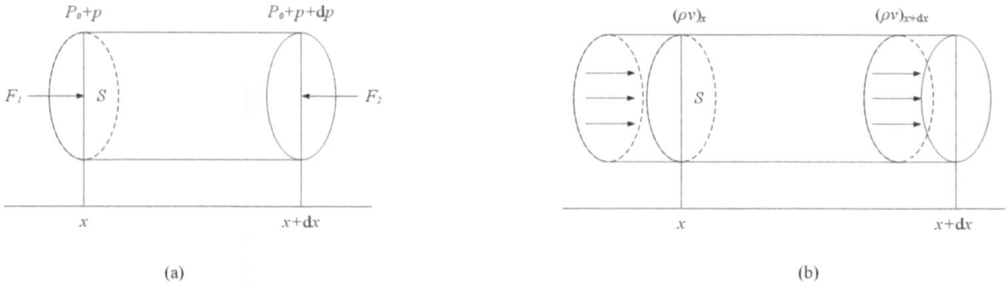

FIGURE 7.2 Fluctuation of differential element in a continuous fluid medium. (a) A sufficiently small volume element in the ultrasonic propagation medium and (b) a small volume element in the propagation medium.

the resultant force on both sides is $F = F_1 - F_2 = -S(\partial p/\partial z)\mathrm{d}z$. According to Newton's second law:

$$\rho S\mathrm{d}z\frac{\partial v}{\partial t} = -S\frac{\partial p}{\partial z}\mathrm{d}z \Rightarrow \rho\frac{\partial v}{\partial t} = -\frac{\partial p}{\partial z} \tag{7.1}$$

Eq. (3.1) describes the relationship between the sound pressure P and the particle vibration velocity v in the propagation medium.

2. Equation of continuity
Suppose that a small volume element is taken from the propagation medium, as shown in Figure 7.1b. The mass flowing through the left side into the volume element in unit time is $(\rho v)_z S$ and the mass flowing out of the right side of the volume element is $-(\rho v)_{z+dz}S$, so the mass increment of the volume element is $-[\partial(\rho v)/\partial z]Sdz$. According to the law of conservation of mass, the mass increment of the volume element in unit time must be equal to its density increment:

$$-\frac{\partial(\rho v)}{\partial z}\mathrm{S}\mathrm{d}z = \frac{\partial \rho}{\partial t}\mathrm{S}\mathrm{d}z \Rightarrow -\frac{\partial}{\partial z}(\rho v) = \frac{\partial \rho}{\partial t} \tag{7.2}$$

Eq. (7.2) describes the relationship between the particle vibration velocity v and the density ρ.

3. Equation of state
Ultrasonic propagation is considered as an adiabatic process. If the effect of temperature change is not considered, the pressure intensity P is only a function of the density ρ, namely $P = P(\rho)$. The relationship between the incremental changes of pressure intensity and density caused by acoustic wave fluctuation is:

$$dP = \left(\frac{dP}{d\rho}\right)d\rho \Rightarrow dP = c^2 d\rho \tag{7.3}$$

where the changes of pressure intensity and density are positively correlated, so the coefficient c^2 is used. Eq. (7.3) describes the relationship between the pressure intensity P and the density ρ. In this equation, c is actually the speed of sound.

In the three basic equations, the parameters p, v and ρ are all the variables that change with time and position. For example, the propagation medium density ρ includes the static density ρ_0 and the density change ρ' caused by acoustic disturbance, namely $\rho=\rho_0+\rho'$. Therefore, the relationship between the parameters is nonlinear. By ignoring the small quantities higher than the first order, the three basic equations can be linearly expressed as:

$$\rho_0\frac{\partial v}{\partial t}=-\frac{\partial p}{\partial z} \tag{7.4}$$

$$-\rho_0\frac{\partial v}{\partial z}=-\frac{\partial \rho'}{\partial t} \tag{7.5}$$

$$p=c^2\rho' \tag{7.6}$$

By eliminating v and ρ' according to Eqs. (7.4)–(7.6), the one-dimensional wave equation of the sound pressure in an ideal fluid medium can be obtained:

$$\frac{\partial^2 p}{\partial z^2}=\frac{1}{c^2}\frac{\partial^2 p}{\partial t^2} \tag{7.7}$$

Eq. (7.7) is a one-dimensional wave equation obtained by linear simplification.

By generalizing the one-dimensional Eqs. (7.1) and (7.2) to the three-dimensional case and ignoring the influence of the small quantities higher than the first order, we can obtain:

$$\rho\frac{\partial v}{\partial t}=-\nabla p \;\Rightarrow\; \rho_0\frac{\partial v}{\partial t}=-\nabla p \tag{7.8}$$

$$-\nabla\cdot(\rho v)=\frac{\partial \rho}{\partial t} \;\Rightarrow\; -\rho_0\nabla\cdot(v)=\frac{\partial \rho'}{\partial t} \tag{7.9}$$

Similarly, through the elimination of v and ρ', the linearly simplified three-dimensional wave equation can be obtained:

$$\nabla^2 p=\frac{1}{c^2}\frac{\partial^2 p}{\partial t^2} \tag{7.10}$$

The solution p of the wave equation is expressed as a time-dependent simple harmonic function $p=P\exp(j\omega t)$ and then substituted into Eq. (7.10) to obtain:

$$\nabla^2 P+k^2 P=0 \tag{7.11}$$

where $k=\omega/c$ is the wave number and ω is the circular frequency. For the acoustic wave with small amplitude in an ideal fluid medium, P satisfies Eq. (7.11), namely Helmholtz equation.

7.1.2 Expression of Sound Pressure Amplitude

The particular solutions satisfying the Helmholtz equation exist in the radiation sound field in the form of either plane wave or spherical wave. The difference of Gaussian beam solution lies in the fact that it is not the exact solution of Helmholtz equation but the approximate solution of the beam with slow amplitude change. The ultrasonic transducers involved in this study are all circular piezoelectric wafers, and the radiation sound field of circular piston (sound source) is axisymmetric. If the sound wave travels along the *Z*-axis, the radial sound field, in addition to the sound field in the *Z* direction, will also change. When the radiated sound field is defined in the cylindrical coordinate system for analysis, *P* can be expressed as:

$$P(r,z) = A(r,z)\exp(jkz) \tag{7.12}$$

By substituting Eq. (7.12) into Eq. (7.11), we can obtain:

$$\frac{\partial^2 A}{\partial z^2} + 2jk\frac{\partial A}{\partial z} + \frac{1}{r}\frac{\partial}{\partial r}\left(r\frac{\partial A}{\partial r}\right) = 0 \tag{7.13}$$

While satisfying the parabolic approximation (namely $kw_0 >> 1$), $\partial^2 A/\partial z^2$ can be ignored and Eq. (7.13) can be converted into:

$$2jk\frac{\partial A}{\partial z} + \frac{1}{r}\frac{\partial}{\partial r}\left(r\frac{\partial A}{\partial r}\right) = 0 \tag{7.14}$$

Eq. (7.14) may have many forms of solutions. Here, the solution in the form of Gaussian beam is derived. Suppose the solution of $A(r, z)$ is:

$$A(r,z) = Af_1(z)\exp\left(f_2(z)r^2\right) \tag{7.15}$$

where A is an arbitrary complex constant, and $f_1(z)$ and $f_2(z)$ are undetermined functions. If Eq. (7.15) is substituted into Eq. (7.14), then:

$$\left[2f_2^2(z) + jkf_2'(z)\right]r^2 - \left[2f_2(z) + \frac{jkf_1'(z)}{f_1(z)}\right] = 0 \tag{7.16}$$

If Eq. (7.16) is expected to be true for any r, the following conditions need to be satisfied:

$$\begin{cases} 2f_2^2(z) + jkf_2'(z) = 0 \\ 2f_1(z)f_2(z) + jkf_1'(z) = 0 \end{cases} \tag{7.17}$$

Eq. (7.17) is a system of differential equations, whose solution is:

$$\begin{cases} f_1(z)=\left[B+\dfrac{2jz}{k}\right]^{-1} \\ f_2(z)=-\left[B+\dfrac{2jz}{k}\right]^{-1} \end{cases} \tag{7.18}$$

By substituting Eq. (7.18) into Eq. (7.15) and then substituting the result into Eq. (7.12), we can obtain:

$$P(r,z)=\frac{A}{B+\dfrac{2jz}{k}}\exp\left(-\frac{r^2}{B+\dfrac{2jz}{k}}\right)\exp(jkz) \tag{7.19}$$

where B is an arbitrary complex constant. Eq. (7.19) is the solution to the P in Helmholtz equation, where both A and B are undetermined coefficients.

7.1.3 Superposition of Multiple Gaussian Beams

The values of A and B in Eq. (7.19) vary with propagation medium type and wafer size. If $P(r, z)$ is represented in the form of superposition by assigning different values to A and B, Eq. (7.19) can be rewritten as:

$$P(r,z)=\frac{i}{k}\sum_{n=1}^{N}\frac{A_n'}{B_n'+\dfrac{2jz}{k}}\times\exp\left(-\frac{r^2}{B_n'+\dfrac{2jz}{k}}+jkz\right) \tag{7.20}$$

where N is the number of superpositions. We define $\xi=r/w_0$, $z_R= kw_0^2/2$ and $\sigma=z/z_R$, and simplify the two complex terms in Eq. (7.20) as:

$$\frac{A_n'}{B_n'+\dfrac{2jz}{k}}=\frac{\dfrac{A_n'}{w_0^2}}{\dfrac{B_n'}{w_0^2}+\dfrac{2jz}{kw_0^2}}=\frac{\dfrac{A_n'}{B_n'}}{1+j\sigma\dfrac{w_0^2}{B_n'}}=\frac{A_n}{1+j\sigma B_n} \tag{7.21}$$

$$\frac{r^2}{B_n'+\dfrac{2jz}{k}}=\frac{\dfrac{r^2}{w_0^2}}{\dfrac{B_n'}{w_0^2}+\dfrac{2jz}{kw_0^2}}=\frac{\xi^2}{\dfrac{B_n'}{w_0^2}+j\sigma}=\frac{B_n\xi^2}{1+j\sigma B_n} \tag{7.22}$$

where A_n' and B_n' are complex constants. In the above two equations, we define $A_n'/B_n'=A_n$ and $w_0^2/B_n'=B_n$. Both A_n and B_n are also complex constants. Then Eq. (3.20) can be simplified as:

$$P(r,z)=\frac{i}{k}\sum_{n=1}^{N}\frac{A_n}{1+jB_n\sigma}\times\exp\left(-\frac{B_n\xi^2}{1+jB_n\sigma}+jkz_R\sigma\right) \tag{7.23}$$

Eq. (3.23) is just the representation of $P(r, z)$ in the form of superposition. The partial derivative of $P(r, z)$ at the beginning of sound beam is:

$$\begin{aligned} f_0(\xi) &= -\frac{\partial P}{\partial z}\Big|_{z=0} \\ &= -\frac{1}{z_R}\frac{\partial P}{\partial \sigma}\Big|_{\sigma=0} \\ &= \sum_{n=1}^{N} A_n\exp\left(-B_n\xi^2\right)\left(1-\frac{2B_n}{\left(kw_0\right)^2}\left(1-B_n\xi^2\right)\right) \end{aligned} \tag{7.24}$$

The condition of parabolic approximation (namely, $kw_0 >> 1$) is satisfied at the boundary between the piezoelectric wafer and the propagation medium. Then, Eq. (3.24) can be approximated as:

$$f_0(\xi)\approx\sum_{n=1}^{N}A_n\exp\left(-B_n\xi^2\right) \tag{7.25}$$

A set of coefficients A_n and B_n can be found with the minimization method, so that Eq. (3.24) can satisfy the condition of parabolic approximation at the boundary. The objective function is defined as:

$$Q=\int_0^{\infty}\left[f_0(\xi)-\sum_{n=1}^{N}A_n\exp\left(-B_n\xi^2\right)\right]^2 d\xi \tag{7.26}$$

According to the result of optimization calculation, the following requirements should be met at the extreme point:

$$\frac{\partial Q}{\partial A_i}\Big|_{A_n,B_n}=0,\quad \frac{\partial Q}{\partial B_i}\Big|_{A_n,B_n}=0 \tag{7.27}$$

where i, $n=1, 2, \ldots, N$. By using the method of Gaussian beam superposition, the sound field distribution can be numerically simulated. The numerical simulation of sound pressure amplitude along the central Z-axis of the transducer is shown in Figure 7.3. It can be seen that the sound pressure amplitude at the focus is the largest.

7.1.4 Influence of the Curved Surface on Ultrasonic Propagation

When using the multi-Gaussian-beam superposition model to calculate the sound pressure distribution along the axis of the ultrasonic transducer, we find that at the focal point, the amplitude of sound pressure is the largest and the detection sensitivity is the highest.

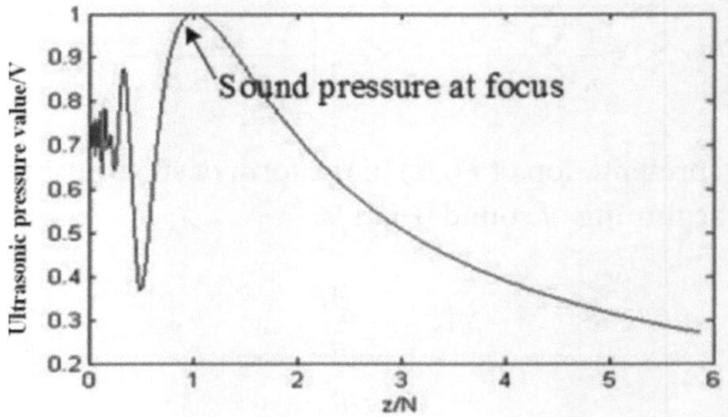

FIGURE 7.3 Axial sound pressure distribution of an ultrasonic transducer.

In the liquid-immersion water-coupling detection, the sound beam radiated by the ultrasonic transducer is transmitted through water into the material under test. At the junction of water and the material, the beam waveform is converted. The robotic UT system is mostly used for the testing of curved surfaces. Therefore, understanding the impact of the curved surface on ultrasonic propagation is of great significance for the improvement of detection resolution. For example, when a longitudinal-wave straight probe is used, the conversion of sound beam waveform at the plane should be considered first. As shown in Figure 7.4a, a longitudinal wave obliquely incident on the plane is converted into a reflected longitudinal wave, a reflected transverse wave, a refracted longitudinal wave and a refracted transverse wave.

According to the study on the expression of sound pressure amplitude in Section 7.1.2, the sound waves in various forms can be expressed as:

$$\begin{cases} L_{\text{inc}} = A_{\text{inc}} \exp\left[ik_{L1}\left(x \cdot \sin\theta_{L1} + y \cdot \cos\theta_{L1}\right) - i\omega t\right] \\ L_{\text{reflt}} = A_{\text{reflt}} \exp\left[ik_{L1}\left(x \cdot \sin\theta_{L1} + y \cdot \cos\theta_{L1}\right) - i\omega t\right] \\ S_{\text{reflt}} = B_{\text{reflt}} \exp\left[ik_{S1}\left(x \cdot \sin\theta_{S1} + y \cdot \cos\theta_{S1}\right) - i\omega t\right] \\ L_{\text{trans}} = A_{\text{trans}} \exp\left[ik_{L2}\left(x \cdot \sin\theta_{L2} + y \cdot \cos\theta_{L2}\right) - i\omega t\right] \\ S_{\text{trans}} = B_{\text{trans}} \exp\left[ik_{S2}\left(x \cdot \sin\theta_{S2} + y \cdot \cos\theta_{S2}\right) - i\omega t\right] \end{cases} \tag{7.28}$$

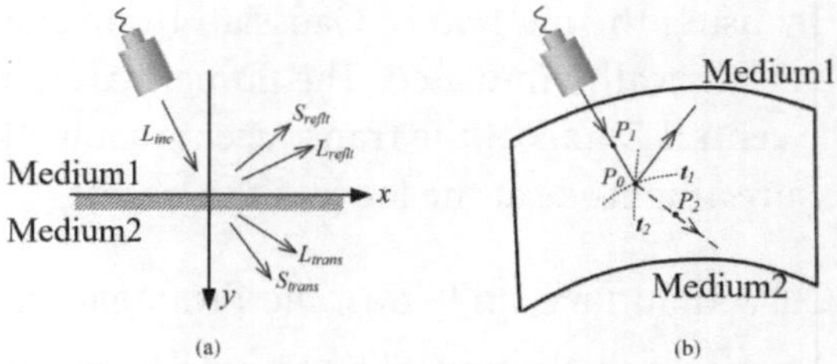

FIGURE 7.4 Propagation of transducer sound beam on the interface. (a) Beam oblique incidence to plane interface and (b) beam oblique incidence to curved interface.

where L_{inc}, L_{reflt}, S_{reflt}, L_{trans} and S_{trans} represent the incident longitudinal wave, reflected longitudinal wave, reflected transverse wave, refracted longitudinal wave and refracted transverse wave, respectively; A_{inc}, A_{reflt}, B_{reflt}, A_{trans} and B_{trans} represent the amplitudes of particle vibration velocities and θ_{L1}, θ_{S1}, θ_{L2} and θ_{S2} represent the reflection angle of the longitudinal wave, the reflection angle of the transverse wave, the refraction angle of the longitudinal wave and the refraction angle of the transverse wave, respectively. The transmission coefficient of the sound beam in waveform transformation at the plane is:

$$T_{12}^{L} = \frac{2\cos\theta_{L1}\left(1-2\sin^2\theta_{S2}\right)\left(1-2\sin^2\theta_{S1}\right)}{\Delta_1+\Delta_2} \tag{7.29}$$

$$T_{12}^{S} = \frac{-4\sin\theta_{S2}\cos\theta_{L1}\cos\theta_{L2}\left(1-2\sin^2\theta_{S1}\right)}{\Delta_1+\Delta_2} \tag{7.30}$$

where

$$\Delta_1 = \cos\theta_{L2}\left[1-4\sin^2\theta_{S1}\cos^2\theta_{S1}+4\frac{c_{S1}^2}{c_{L1}^2}\sin\theta_{S1}\cos\theta_{S1}\sin\theta_{L1}\cos\theta_{L1}\right]$$

$$\Delta_2 = \frac{\rho_2 c_{L2}}{\rho_1 c_{L1}}\cos\theta_{L1}\left[1-4\sin^2\theta_{S2}\cos^2\theta_{S2}+4\frac{c_{S2}^2}{c_{L2}^2}\sin\theta_{S2}\cos\theta_{S2}\sin\theta_{L2}\cos\theta_{L2}\right]$$

The sound beam waveform is converted into reflect and refracted waves at the plane. It can be seen from the transmission coefficient that, during waveform conversion, the amplitude of sound pressure does change, but the propagation phase of sound beam does not change much.

However, when the sound beam is incident on the surface, both the sound pressure amplitude and the propagation phase will change in a certain relationship with the radius of curvature. As shown in Figure 7.3b, P_1 and P_2 are the points in media 1 and 2, respectively, and P_0 is the point at the interface between the two media. The sound beam travels from P_1 through P_0 to P_2. The sound wave at the point P_2 can be expressed as:

$$V_{trans} = A_i T_{12}^{L}\exp[iu] \tag{7.31}$$

where $u = u_{11}+u_{12}+u_{21}+u_{22}$,

$$u_{11} = \frac{k_{L1}\cos^2\theta_1}{r_{10}}+\frac{k_{L2}\cos^2\theta_2}{r_{20}}-h_{11}\left(k_{L1}\cos\theta_1-k_{L2}\cos\theta_2\right)$$

$$u_{12} = u_{21} = -h_{12}\left(k_{L1}\cos\theta_1-k_{L2}\cos\theta_2\right) \tag{7.32}$$

$$u_{22} = -h_{22}\left(k_{L1}\cos\theta_1-k_{L2}\cos\theta_2\right)$$

where r_{10} and r_{20} represent the distances from the points P_1 and P_2 in Figure 7.3b to the point P_0, respectively; and θ_1 and θ_2 represent the incident angle and refraction angle of sound beam, respectively. Then, h_{11}, h_{12}, h_{21} and h_{22} are expressed as:

$$h_{11} = \frac{\cos^2\varphi}{R_1} + \frac{\sin^2\varphi}{R_2}$$

$$h_{12} = h_{21} = -\sin\varphi \cdot \cos\varphi \left(\frac{1}{R_1} - \frac{1}{R_2} \right)$$

$$h_{22} = \frac{\sin^2\varphi}{R_1} + \frac{\cos^2\varphi}{R_2}$$

where R_1 and R_2 represent the surface curvatures at the point P_0 along the two tangential directions $\mathbf{t}_1$ and $\mathbf{t}_2$ (see Figure 7.3b) and φ represents the incident angle of sound beam at the point P_0 on the surface. When the sound beam is incident along the normal direction of the surface (i.e. $\varphi=0$), the h_{11}, h_{12}, h_{21} and h_{22} in the above equation can be simplified as:

$$h_{11} = \frac{1}{R_1}, \quad h_{12} = h_{21} = 0, \quad h_{22} = \frac{1}{R_2} \tag{7.33}$$

In case of normal incidence, the phase variation caused by waveform conversion can be simplified. When the surface curvature at the incident point is large, the incidence of sound beam can be approximated as plane incidence. In this case, the noise waves generated by waveform conversion will also be significantly reduced and the echo signal-to-noise ratio (SNR) will become better. From the above theoretical derivation process, we can find the influence of two factors on surface detection: (1) When the pulse reflection method is used to test the curved components, the deviation of incident angle will lead to unnecessary waveform conversion and reduce the echo SNR and (2) for the curved components with larger curvature, an ultrasonic transducer with smaller sound beam diameter should be selected as far as possible. Therefore, the study on the geometric error in manipulator path copying can improve the detection resolution.

7.2 3D POINT CLOUD MATCHING ALGORITHM BASED ON NORMAL VECTOR ANGLE

One of the factors affecting the SNR of defect echo is the deviation between the actual installation position/attitude of the tested component and the theoretical position/attitude defined in the simulation environment. Three-dimensional point cloud matching is to match the measuring points on the component surface with the 3D data points in the simulation model and to calculate the deviation of the measured component relative to the simulation model. Considering that most of the tested components are curved surfaces, a matching algorithm characterized by the normal vector angle of the curved surface is proposed in this study. In this algorithm, the point-cloud spatial transformation relationship is described by means of adjacent points, and the characteristic quantity does not change with the rotation, translation and scale change of point clouds.

7.2.1 Matching Features of 3D Point Clouds

Point cloud is a set of discrete points distributed on a curved surface. It is assumed that the source point cloud and the target point cloud are two point clouds to be matched, which are denoted as S and G, respectively. The normal vector of each point relative to the surface will change with the surface position/attitude. However, the surface curvature at any point will not change with the surface position/attitude in space. Inspired by this property, we conclude that the included angle between the normal vector of any point on the surface and the normal vector of its adjacent point will not change with the surface position/attitude as well. This characteristic can be used as the reference for judging the surface alignment. Take the point cloud set S on surface as an example. As shown in Figure 7.5, s_i is any point in the point cloud S, and its normal vector is n_i. $\{s\ n\ i\}$ is a point adjacent to s_i, and its normal vector is $\{n\ n\ i\}$. The angle between n_i and $\{n\ n\ i\}$ is $\{\theta\ n\ i\}$, namely the normal vector angle. The normal vector n_i and the n normal vectors around it constitute the eigenvector $\boldsymbol{\theta}_i=(\theta\ 1\ i\ldots\ \theta\ n\ i)^T$. $\boldsymbol{\theta}_i$ will not change with the surface position/attitude. This characteristic can be applied to the matching of surface point clouds.

7.2.2 Calculation of the Normal Vector on a Curved Surface

While taking the normal vector angle as the feature of point cloud matching, we calculate the normal vector of each point on the surface through principal component analysis (PCA) and then determine the normal vector angle based on the obtained normal vector. The PCA method is to transform multiple groups of indicators in data into one or several comprehensive indicators through dimensionality reduction, so that each indicator can reflect most of the information contained in the original data. This method can simplify a problem with multiple complex factors into a problem with one or several principal components through eigenvalue decomposition, singular value decomposition or non-negative matrix decomposition, so as to obtain a more scientific and effective result. In the actual calculation process, PCA is to project the sample data from one matrix to another new space. After the matrix diagonalization, the eigenroot and eigenvector of the matrix are generated. This is just the process of projecting the matrix to the orthonormal

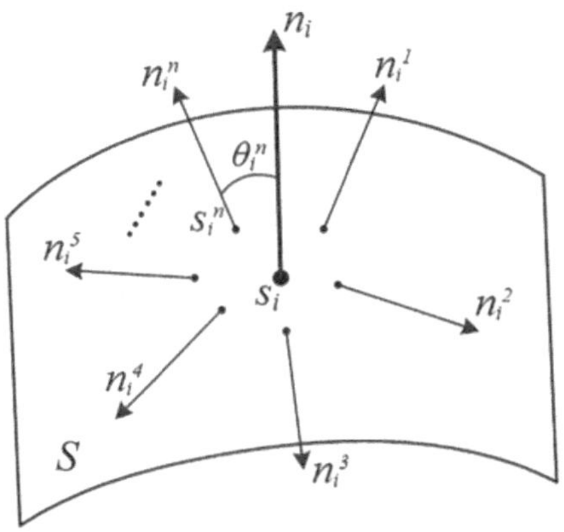

FIGURE 7.5 Normal vector angle on the surface point cloud S.

basis. Eigenvalue is the projection length along the eigenvector direction, which can reflect the features of original data more comprehensively. Therefore, through this matrix transformation, most of the original variables can be represented by few variables, namely the principal components that can be used as comprehensive indexes after the data dimension reduction.

Take the source point cloud S in Figure 7.5 as an example. Suppose that a point s_i belongs to S and its normal vector can be estimated by using k points adjacent to it. The *kd*-tree method proposed by Bentley et al. [2,3] is used to search the k nearest points around s_i, which are denoted as {*s n i*($n\in[1, k]$)}. The point s_i and its adjacent points {*s n i* ($n\in[1, k]$)} constitute a covariance matrix:

$$\boldsymbol{E}_{3\times3}=\frac{1}{k}\sum_{n=1}^{k}\left(\mathbf{s}_i^n-\overline{\mathbf{s}}_i\right)\cdot\left(\mathbf{s}_i^n-\overline{\mathbf{s}}_i\right)^T \tag{7.34}$$

where $\overline{\mathbf{s}}_i$ is the geometric center calculated by using k nearest points, and $\boldsymbol{E}_{3\times3}$ is a symmetric positive semi-definite matrix. The eigenvalue and eigenvector of the matrix $\boldsymbol{E}_{3\times3}$ are:

$$\boldsymbol{E}_{3\times3}\cdot\boldsymbol{v}_l=\lambda_l\cdot\boldsymbol{v}_l,\quad l\in\{1,2,3\} \tag{7.35}$$

where λ_1, λ_2 and λ_3 ($\lambda_1<\lambda_2<\lambda_3$) are the eigenvalues of the constructed matrix, and $\boldsymbol{v}_1$, $\boldsymbol{v}_2$ and $\boldsymbol{v}_3$ are the corresponding eigenvectors. The eigenvector $\boldsymbol{v}_1$ corresponding to the minimum eigenvalue λ_1 can be estimated as the normal vector of the point s_i on the surface.

7.2.3 Identification and Elimination of Surface Boundary Points

It is found through research that the calculation of the normal vector at the surface boundary point through PCA analysis is not accurate and will finally affect the result of point cloud matching. Therefore, the boundary points of surface point cloud need to be identified and eliminated before matching. The identification of point-cloud boundary point is widely used in reverse engineering, robot navigation and image recognition. The identification method may vary with the feature of boundary points. For example, the boundary area is extracted by using the directional curvature and subdivided and identified by using the point normal vector and octree. However, the computational processes of these methods are quite complex and are mainly applied to the algorithms that need accurate identification.

In this study, a relatively simple boundary-point identification method is proposed. Its combination with the point cloud matching method characterized by the normal vector angle can effectively improve the matching accuracy. If s_i is a boundary point of the point set S, as shown in Figure 7.6a, the side with this point should contain no data points. The modulus of the sum vector composed of this point and its k neighbor points {*s n i* ($n\in[1, k]$)} will be greater than the first critical value Δe. On the other hand, if s_i is an internal point of the point set S, as shown in Figure 7.6b, the modulus of the sum vector composed of this point and its k neighbor points {*s n i* ($n\in[1, k]$)} will be smaller than the first critical

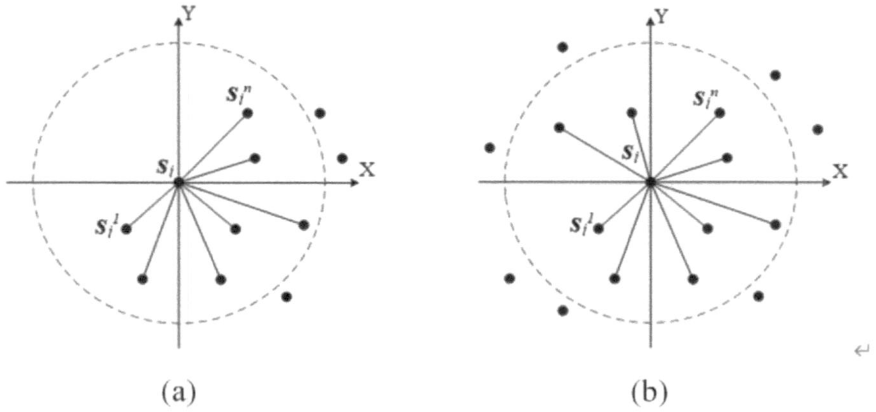

FIGURE 7.6 Normal vector angle on the surface point cloud S. (a) s_i is the boundary point on the point set S and (b) s_i is an interior point on the point set S.

value Δe. The identification of the boundary point s_i can be expressed by the following discriminant:

$$\left| \sum_{n=1}^{k} \overrightarrow{\boldsymbol{s}_i \boldsymbol{s}_i^n} \right| > \Delta e \tag{7.36}$$

When Eq. (7.36) holds true, the point s_i will be the boundary point.

In addition to the above identification method based on the critical value Δe, there are many other methods for identifying the boundary point of a point cloud on a spatial curved surface. For example, a robust and superior boundary-point identification algorithm without defining a critical value similar to Δe was proposed in Ref. [4]. To obtain better matching results, the same identification algorithm can also be used in this study.

7.2.4 Calculation of Spatial Position/Attitude Deviation of 3D Point Cloud

Suppose that S and G are two groups of point clouds reflecting the theoretical and actual positions/attitudes of the component under test, called source point cloud and target point cloud respectively. The goal of point cloud matching is to calculate the position/attitude deviation between the two groups of point clouds. Generally speaking, there is no one-to-one correspondence between the points in S and those in G, so the point pairs with the same features should be found from S and G. First, m non-boundary points $\{g_i\ (i \in [1, m])\}$ are randomly selected from G and the normal vector angle of each point in G is calculated. The normal vector angle of each point in S is calculated. By searching for the points in S whose normal vector angle matches $\{g_i\ (i \in [1, m])\}$, m point pairs with one-to-one correspondence can be determined. Finally, the transformation matrix of this point pair is calculated through PCA by calculating first the geometric centers of S and G:

$$\boldsymbol{C}_S = \frac{1}{N_S} \sum_{i=1}^{N_S} \boldsymbol{s}_i \quad (\boldsymbol{s}_i \in S) \tag{7.37}$$

$$C_g = \frac{1}{N_g}\sum_{i=1}^{N_g} g_i \ (g_i \in G) \tag{7.38}$$

where $C_s = (C_{sx}, C_{sy}, C_{sz})^T$ and $C_g = (C_{gx}, C_{gy}, C_{gz})^T$ are the three-dimensional vectors, and N_s and N_g are the total numbers of the points on source point cloud and target point cloud, respectively, after eliminating the boundary points. Then, C_s and C_g are used to construct the covariance matrix:

$$M_{3\times3} = \frac{1}{m}\sum_{i=1}^{m}(s_i - C_s)\cdot(g_i - C_g)^T \tag{7.39}$$

The singular values in $M_{3\times3}$ are decomposed to obtain $M_{3\times3}=\mathbf{U\Lambda V}^T$, where $\mathbf{\Lambda}$ is a diagonal matrix composed of the eigenvalues of $M_{3\times3}$. Finally, the rotation matrix R and the translation matrix T can be obtained as follows:

$$R = \begin{bmatrix} R_{11} & R_{12} & R_{13} \\ R_{21} & R_{22} & R_{23} \\ R_{31} & R_{32} & R_{33} \end{bmatrix} = UV^T \tag{7.40}$$

$$T = \begin{bmatrix} T_x & T_y & T_z \end{bmatrix}^T = C_g - RC_s \tag{7.41}$$

R represents the attitude deviation between the two point clouds, and T represents the position deviation between the two point clouds. The rotation matrix R can be considered as the rotation about three orthogonal axes in a sequence that follows the X–Y–Z Euler angle transformation relation. It can be expressed by matrix multiplication as:

$$\begin{aligned} R &= R(r_x)R(r_y)R(r_z) \\ &= \begin{bmatrix} 1 & 0 & 0 \\ 0 & cr_x & -sr_x \\ 0 & sr_x & cr_x \end{bmatrix} \begin{bmatrix} cr_y & 0 & sr_y \\ 0 & 1 & 0 \\ -sr_y & 0 & cr_y \end{bmatrix} \begin{bmatrix} cr_z & -sr_z & 0 \\ sr_z & cr_z & 0 \\ 0 & 0 & 1 \end{bmatrix} \\ &= \begin{bmatrix} cr_y\, sr_z & -cr_y\, sr_z & sr_y \\ sr_x\, sr_y\, cr_z + cr_x\, sr_z & -sr_x\, sr_y\, sr_z + cr_x\, sr_z & -sr_x\, cr_y \\ -cr_x\, sr_y\, cr_z + sr_x\, sr_z & cr_x\, sr_y\, sr_z + sr_x\, cr_z & cr_x\, cr_y \end{bmatrix} \end{aligned} \tag{7.42}$$

where c and s are the simplifications of the trigonometric functions cos and sin, and r_x, r_y and r_z are the rotation about three orthogonal axes, respectively. Therefore, the attitude transformation can be expressed by the following Euler angles:

$$r_x = atan2(-R_{23}, R_{33})$$
$$r_y = atan2\left(R_{13}, \sqrt{R_{11}^2 + R_{12}^2}\right) \quad (7.43)$$
$$r_z = atan2(-R_{22}, R_{11})$$

The position/attitude deviation between the source point cloud S and the target point cloud G is expressed in vector form as $(T_x, T_y, T_z, r_x, r_y, r_z)$.

7.3 CORRECTION EXPERIMENT FOR 3D POINT CLOUD COLLECTION AND INSTALLATION DEVIATION

The validity of verification algorithm in correcting the installation deviation is demonstrated through simulation and experimental test. The model point cloud mentioned in the experiment is a group of points in the motion path generated by CNC machining simulation software. Equivalent to the source point cloud in Section 7.2, the model point cloud reflects the nominal position/attitude of 3D component model defined in the simulation software. The measurement point cloud mentioned in the experiment is a group of incident points on the actually measured component. Equivalent to the target point cloud in Section 7.2, the measurement point cloud reflects the actual installation position/attitude of the component in the workspace. The strategy of this method is to calculate the deviation between the model point cloud and the measurement point cloud by using the algorithm described in Section 7.2, and then correct the deviation by modifying the coordinate values of the workpiece.

7.3.1 Steps of 3D Point Cloud Matching

In combination with the algorithm described in Section 7.2, the matching steps are described here in detail:

A. The geometric model of the tested component is imported into CNC machining simulation software. In the software, the spatial position/attitude coordinates of the manipulator relative to the geometric model are set and defined as nominal coordinates (namely the theoretical installation position/attitude of the tested component). According to the constraint conditions, scanning interval, step interval and other requirements of ultrasonic testing, the manipulator motion path is designed, and the manipulator motion program is generated and imported into the manipulator controller;

B. The component under test is installed according to the theoretical position/attitude coordinates. At this time, there is often a deviation between the actual installation position/attitude and the theoretical position/attitude. The coordinates at the actual installation position/attitude are defined as actual coordinates (that is, the actual installation coordinates of the tested component). The ultrasonic transducer held by manipulator moves along the path generated in the step A and scans the component under test. As shown in Figure 7.7, there is a deviation between the actual installation

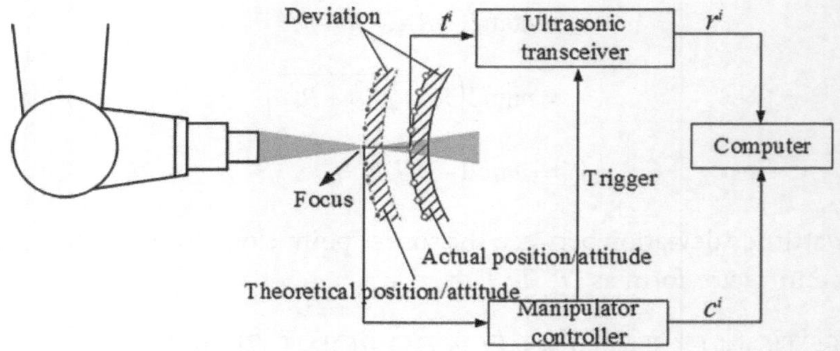

FIGURE 7.7 Principle of point cloud measurement and acquisition from a tested component surface.

position/attitude of the component and the in-model installation position/attitude set in the CNC machining simulation software. In the process of manipulator movement, the time for ultrasonic wave to propagate from the ultrasonic transducer to the component surface is measured. The distance between the transducer and component surface is calculated according to the sound velocity in the medium. The position and attitude of the ultrasonic transducer in the manipulator coordinate system are obtained from the control system of the manipulator. Then, the coordinates of the component surface points in the manipulator coordinate system are obtained. Automatic measurement is taken according to the scanning path. The position coordinates are collected once every movement distance of the manipulator (known as trigger interval) to form a point cloud that can reflect the actual installation position/attitude of the component, namely measurement point cloud;

C. The trajectory programmed and generated in the CNC machining simulation software contains a series of constraint points to ensure that the manipulator moves in accordance with the preset constraints, scanning interval and step interval. These constraint points are represented in tabular form. By importing the table into a mathematical processing software, the point cloud of the component surface in theoretical installation position/attitude, namely model point cloud, can be obtained;

D. Suppose that the model point cloud contains N_s points $\{s_i\}$, which constitute the point set S; and that the measurement point cloud contains N_g points $\{g_i\}$, which constitute the point set G. The two sets of points are imported into the mathematical calculation software;

E. According to the normal vector calculation method in Section 7.2.2, the normal vector of each point in the point sets S and G is calculated. For any point $s_i \in S$, its normal vector is denoted as n_{si}. For any point $g_i \in G$, its normal vector is denoted as n_{gi};

F. Points are randomly selected from the point set G. Each selected point is judged by the boundary point identification method described in Section 7.2.3. Boundary points are discarded, and finally m non-boundary points are identified and denoted as g_i ($i \in [1, m]$). The adjacent normal vector angles of g_i ($i \in [1, m]$) are calculated and denoted as $\boldsymbol{\theta}_{gi} = (\theta_1 g_i \ldots \theta_n g_i)^T$ ($i \in [1, m]$);

G. Like in the step F, all the non-boundary points in the point set S are identified by the boundary point identification method. It is assumed that there are n non-boundary points, denoted as s_i ($i \in [1, n]$). The adjacent normal vector angles of s_i ($i \in [1, n]$) are calculated and denoted as $\boldsymbol{\theta}_{si} = (\theta_1 s_i \ldots \theta_n s_i)^T$ ($i \in [1, n]$);

H. For each of the adjacent normal vector angles $\boldsymbol{\theta}_{gi} = (\theta_1 g_i \ldots \theta_n g_i)^T$ ($i \in [1, m]$) calculated in the step F, the point in S whose adjacent normal vector angle $\boldsymbol{\theta}_{si} = (\theta_1 s_i \ldots \theta_n s_i)^T$ ($i \in [1, n]$) is equal to or closest to $\boldsymbol{\theta}_{gi}$ is searched out to form, together with the point in G, a pair of points with the same features;

I. By using this point pair and referring to the calculation method of point-cloud position/attitude deviation in Section 7.2.4, the deviation between the model point cloud and the measurement point cloud is calculated and denoted as (ΔX, ΔY, ΔZ, ΔRX, ΔRY, ΔRZ);

J. It is assumed that the theoretical position/attitude coordinates set in the CNC machining simulation software are (X, Y, Z, RX, RY, RZ). According to the calculated deviation, the installation position/attitude of the tested component is modified into ($X+\Delta X$, $Y+\Delta Y$, $Z+\Delta Z$, $RX+\Delta RX$, $RY+\Delta RY$, $RZ+\Delta RZ$). By regenerating the motion program, the scanning path can satisfy the actual detection requirements better.

7.3.2 Simulation Verification of Position/Attitude Deviation Correction Algorithm

The validity of the algorithm is verified by CNC programming simulation software. First, an arc model with a length of 180 mm and a width of 80 mm is designed in the simulation software to simulate the actual component, as shown in Figure 7.8a. The upper surface is taken as the surface under test, where the ultrasound is incident toward the inside of the material. According to the detection resolution, the scanning interval is determined and the instruction points of manipulator motion path are generated. In Figure 7.8a, the scanning interval is defined as 1 mm. Then, the corresponding manipulator model is imported into the simulation software, and the manipulator holds an ultrasonic transducer to move above the component surface under test, as shown in Figure 7.8b. In the software, the three coordinate systems are defined to describe the position/attitude relationship, including the base coordinate system fixed on the manipulator, the tool coordinate system fixed at the focus and the workpiece coordinate system fixed on the model. The position/attitude relationship between the base coordinate system and workpiece coordinate system is given in the nominal coordinates of Table 7.1, which represent the theoretical position/attitude of the model in the workspace. The manipulator motion path generated by the software is shown in Figure 7.8c. For each instruction point on the motion path, the incident direction at that point (the Z-axis of the tool coordinate system) is a normal direction of the tested surface. The incident focus (the origin of the tool coordinate system) is located on the tested surface. Between the tool coordinate system and the top of the ultrasonic transducer

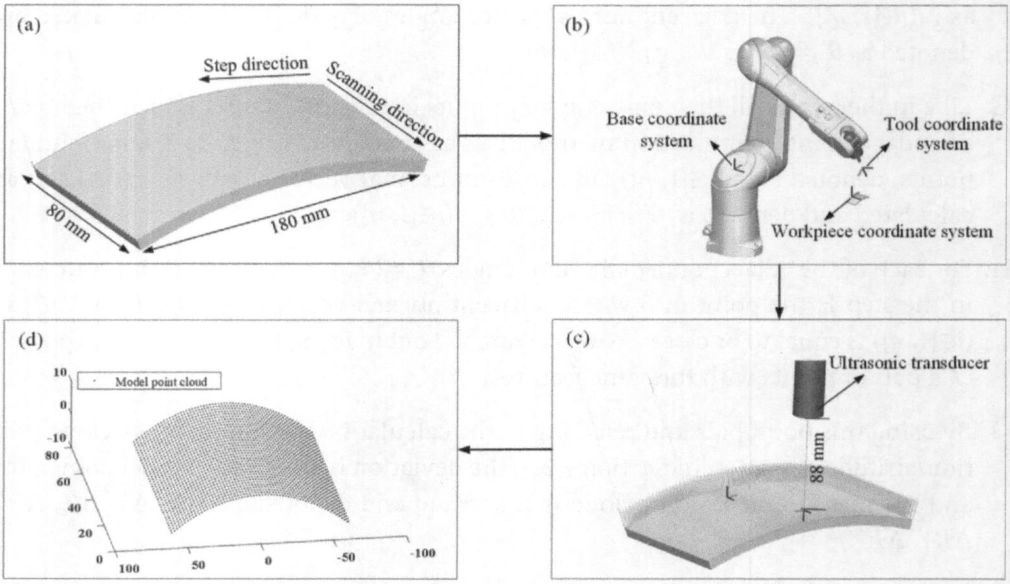

FIGURE 7.8 Acquisition of model point cloud by simulation software. (a) Detected components, (b) trajectory simulation, (c) generate motion paths and (d) 3D point cloud.

TABLE 7.1 Comparison between Assumed Deviation and Calculated Deviation

Nominal Coordinates (mm)	Assumed Deviation (mm)	Actual Coordinates (mm)	Calculated Deviation (mm)
X=700.000	ΔX=1.500	X=701.500	ΔX=1.498
Y=100.000	ΔY=1.500	Y=101.500	ΔY=1.497
Z=0.000	ΔZ=1.500	Z=1.500	ΔZ=1.499
RX=0.000	ΔRX=1.000	RX=1.000	ΔRX=0.987
RY=0.000	ΔRY=1.000	RY=1.000	ΔRY=0.985
RZ=0.000	ΔRZ=1.000	RZ=1.000	ΔRZ=1.000

is an 88 mm distance, which is the focal distance of the transducer. The instruction points generated by the software are imported into MATLAB® to form a point cloud in the 3D space, as shown in Figure 7.8d.

To simulate the deviation between the nominal coordinates and actual coordinates of the tested component in the detection process, a hypothetical deviation is added to the nominal coordinates to simulate the component coordinates in the actual installation position/attitude, as listed in Table 7.1. Similar to the generation process of model point cloud in Figure 7.8, the measurement point cloud is generated by CNC machining simulation software and MATLAB. The two point clouds are shown in Figure 7.9a, of which the blue point cloud is model point cloud and the red one is measurement point cloud. The deviation between the two point clouds is calculated and determined, as shown in Table 7.1. It can be seen that the calculated deviation is basically consistent with the assumed deviation. Then, the calculated deviation is utilized to correct the actual coordinates. As can be seen from Figure 7.9b, the two point clouds basically coincide with each other after the deviation correction.

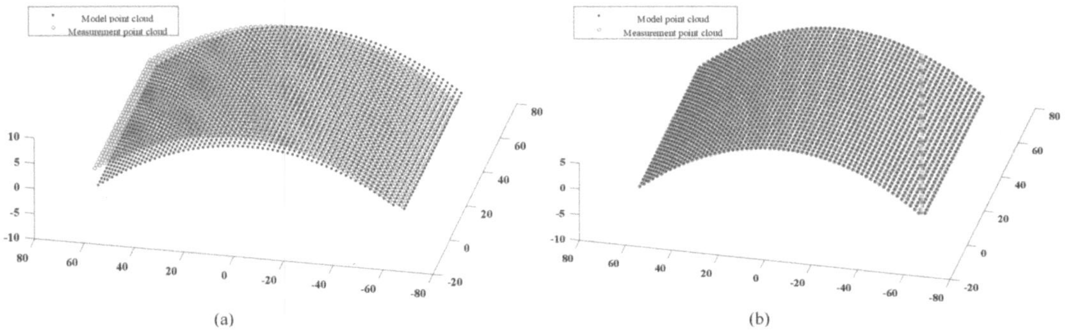

FIGURE 7.9 Two point clouds before and after deviation correction. (a) Calculate the deviation and (b) correction for deviation.

7.3.3 Experiment and Detection Verification of Curved-Component Deviation Correction

The simulation is only to verify the algorithm without considering the influence of external factors, so the algorithm needs to be further verified through actual detection. In the experiment, an arc component was tested. Four artificial defects in the form of flat bottom holes with different diameters were prefabricated on the outer wall of the component. An ultrasonic transducer was installed on the manipulator end-effector through a fixture. The manipulator moved with the transducer to scan the component surface, and the ultrasonic signal was transmitted into ultrasonic flaw detector through a coaxial cable. Before the deviation correction, an initial motion program was generated in the CNC machining simulation software according to the initial coordinate values. Then, the program was input into the manipulator controller to perform scanning. The experimental process is shown in Figure 7.10.

In the simulation process described in Section 7.3.2, point clouds are the ideal data generated by the simulation software. However, in the actual detection process, they are the real data obtained by the detection system. The focus of a beam always falls on the component under test in ideal condition but will deviate in real life due to the influence of external factors. As shown in the diagram of point-cloud acquisition principle (Figure 7.6), whenever the coordinates of the blue point c^i (the coordinates of the tool coordinate system

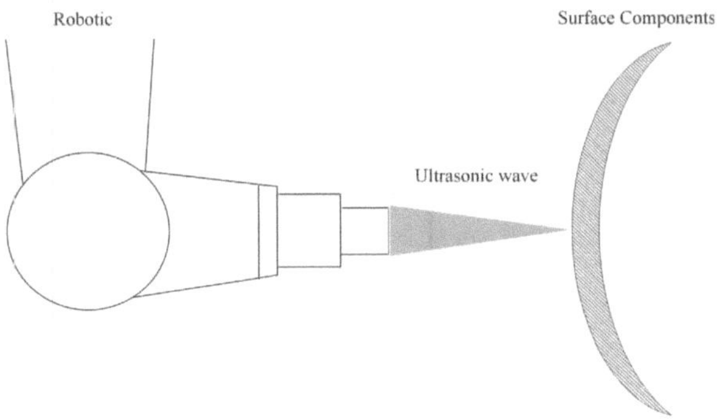

FIGURE 7.10 Tested surface.

relative to the base coordinate system) change by d mm, the manipulator controller will collect the position coordinate value of c^i and input it into the computer. All the c^i coordinates constitute a model point cloud to reflect the ideal position and attitude of the tested surface. On the other hand, while sending the position information, the manipulator controller triggers the ultrasonic transceiver through the I/O port to record the sound interval t_i between the pulse transmitted at this point and the pulse reflected from the upper surface. The coordinates of the red incident point r^i in Figure 7.6 can be calculated according to the ranging principle. All the r^i coordinates constitute a measurement point cloud to reflect the actual position and attitude of the tested surface. In the experiment shown in Figure 7.11c, a curved surface area capable of covering the hole defects was measured. Whenever the transducer focus moved for 0.2 mm, the manipulator controller input the coordinates into the computer. The obtained C-scan detection results are shown in Figure 7.11a. Meanwhile, the model point cloud and the measurement point cloud were obtained according to the point cloud acquisition principle, as shown in Figure 7.11b. The position/attitude deviation between the two point clouds in Figure 7.11b was calculated by

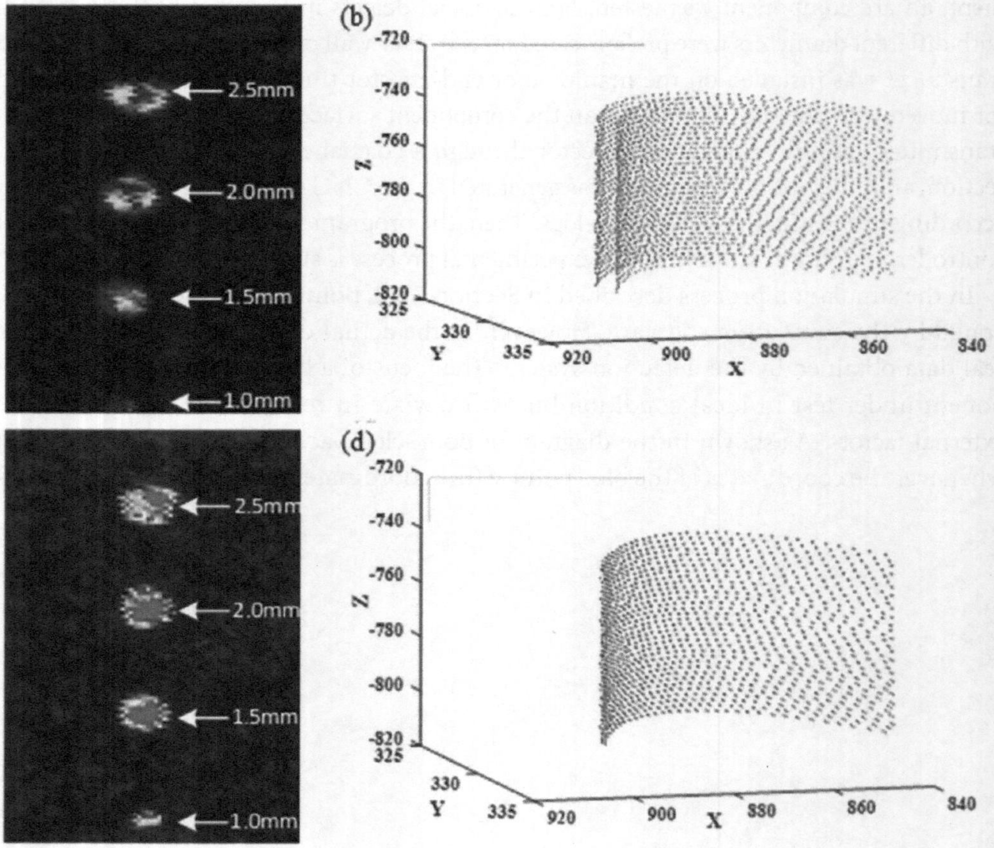

FIGURE 7.11 Comparison between the C-scan detection result and collected point cloud data of a curved component. (a and b) Trajectory deviation and; (c and d) trajectory calibration.

deviation correction algorithm and then used to compensate for the initial coordinates so as to obtain the corrected coordinates. After correction, the curved component was scanned again, and the new model point cloud and measurement point cloud were obtained during scanning. As a result, the artificial defects looked clearer, as shown in Figure 7.11c; and the two point clouds basically coincided with each other, as shown in Figure 7.11d. Thus, it can be seen that after compensation, the motion path of the manipulator holding a transducer satisfies the constraints of ultrasonic testing better and the detection resolution is improved.

REFERENCES

1. Li T. *Computational Acoustics – Equations and Calculation Methods of Sound Field [M].* Beijing: Science Press, 2005:17–33.
2. Bentley J.L., Multidimensional binary search trees used for associative searching [J], *Communications of the ACM*, 1975, 18(9):509–517.
3. Bentley J.L., Multidimensional divide-and-conquer [J], *Communications of the ACM*, 1980, 23(4):214–229.
4. Mineo C., Pierce S.G., Summan R., Novel algorithms for 3D surface point cloud boundary detection and edge reconstruction [J], *Journal of Computational Design and Engineering*, 2019, 6(1):81–91.

CHAPTER 8

Kinematic Error and Compensation in Robotic Ultrasonic Testing

IN THE NATIONAL STANDARD "Industrial Robots – Performance Criteria and Related Test Methods" (GB/T 12642-2013), the term "distance accuracy" represents the position/attitude deviation between the averaged actual trajectory distance and designed theoretical distance of a manipulator that has operated for an instructed distance. Usually, the error of manipulator trajectory is accurately measured by laser ranging and then compensated for by displacement-driven data compensation in order to improve the scanning trajectory accuracy in robotic nondestructive testing (NDT).

The change in the distance between manipulator scanning trajectory and tested component surface can be effectively detected by using either the ultrasonic sound velocity and echo time difference or the lift-off effect of electromagnetic detection, so as to determine the motion trajectory error of a manipulator in NDT. In this chapter, the manipulator trajectory error in ultrasonic testing (UT) is measured and compensated for based on the characteristics of ultrasonic ranging. By establishing the relationship between distance accuracy and localization error and deducing the kinematic parameter error from distance error, the process of error calibration can be simplified to meet the general requirements of industrial sites.

8.1 THREE-DIMENSIONAL SPATIAL DISTRIBUTION MODEL OF ROBOTIC UT ERROR

8.1.1 Model of Manipulator Localization Error

The modeling of differential manipulator motion and position/attitude error has been widely studied in the manipulator field [1–3]. The position/attitude transformation relation between the moving manipulator parts can be represented by a homogeneous

DOI: 10.1201/9781003212232-8

transformation matrix, and the error of their position and attitude can be represented by matrix differential. On this basis, the method of modeling the localization error of the robotic UT system is sketched out in this section to describe the relationships between the location error of transducer beam focus and the errors in the kinematic parameters of connecting rods and joints.

According to the described position/attitude relationship between the coordinate systems, the position/attitude relationship between the connecting-rod coordinate systems $\{i-1\}$ and $\{i\}$ in the manipulator can be expressed by the homogeneous transformation matrix $\boldsymbol{T}_{i-1}^{i}$:

$$\boldsymbol{T}_{i-1}^{i}=\begin{bmatrix} \cos\theta_i & -\sin\theta_i\cos\alpha_i & \sin\theta_i\sin\alpha_i & a_i\cos\theta_i \\ \sin\theta_i & \cos\theta_i\cos\alpha_i & -\cos\theta_i\sin\alpha_i & a_i\sin\theta_i \\ 0 & \sin\alpha_i & \cos\alpha_i & d_i \\ 0 & 0 & 0 & 1 \end{bmatrix} \tag{8.1}$$

where $a_{i,}$ α_i, d_i and θ_i are the parameters of this matrix. For the manipulator used in this system, θ_i is a variable; d_i, α_i and a_i are constants and $a_{i,}$ α_i, d_i and θ_i are the kinematic parameters for the manipulator. Considering the actual situation, the errors Δa_i, $\Delta\alpha_i$, Δd_i and $\Delta\theta_i$ are assigned to the kinematic parameters. If the four variables in the matrix $\boldsymbol{T}_{i-1}^{i}$ are written in differential form, then:

$$d\boldsymbol{T}_{i-1}^{i}=\frac{\partial \boldsymbol{T}_{i-1}^{i}}{\partial a_i}\Delta a_i+\frac{\partial \boldsymbol{T}_{i-1}^{i}}{\partial \alpha_i}\Delta\alpha_i+\frac{\partial \boldsymbol{T}_{i-1}^{i}}{\partial d_i}\Delta d_i+\frac{\partial \boldsymbol{T}_{i-1}^{i}}{\partial \theta_i}\Delta\theta_i \tag{8.2}$$

According to Eq. (8.1), the partial derivatives of the four parameters $a_{i,}$ α_i, d_i and θ_i in Eq. (8.2) are calculated to be:

$$\frac{\partial \boldsymbol{T}_{i-1}^{i}}{\partial a_i}=\boldsymbol{T}_{i-1}^{i}\cdot\boldsymbol{Q}_a,\quad \frac{\partial \boldsymbol{T}_{i-1}^{i}}{\partial \alpha_i}=\boldsymbol{T}_{i-1}^{i}\cdot\boldsymbol{Q}_\alpha,\quad \frac{\partial \boldsymbol{T}_{i-1}^{i}}{\partial d_i}=\boldsymbol{T}_{i-1}^{i}\cdot\boldsymbol{Q}_d,\quad \frac{\partial \boldsymbol{T}_{i-1}^{i}}{\partial \theta_i}=\boldsymbol{T}_{i-1}^{i}\cdot\boldsymbol{Q}_\theta \tag{8.3}$$

where

$$\boldsymbol{Q}_a=\begin{bmatrix} 0 & 0 & 0 & 1 \\ 0 & 0 & 0 & 0 \\ 0 & 0 & 0 & 0 \\ 0 & 0 & 0 & 0 \end{bmatrix}\quad \boldsymbol{Q}_\alpha=\begin{bmatrix} 0 & 0 & 0 & 0 \\ 0 & 0 & -1 & 0 \\ 0 & 1 & 0 & 0 \\ 0 & 0 & 0 & 0 \end{bmatrix}$$

$$\boldsymbol{Q}_d=\begin{bmatrix} 0 & 0 & 0 & 0 \\ 0 & 0 & 0 & \sin\alpha_i \\ 0 & 0 & 0 & \cos\alpha_i \\ 0 & 0 & 0 & 0 \end{bmatrix}\quad \boldsymbol{Q}_\theta=\begin{bmatrix} 0 & -\cos\alpha_i & \sin\alpha_i & 0 \\ \cos\alpha_i & 0 & 0 & a_i\cos\alpha_i \\ -\sin\alpha_i & 0 & 0 & -a_i\sin\alpha_i \\ 0 & 0 & 0 & 0 \end{bmatrix}$$

According to the calculation equations of all the partial derivatives in Eq. (8.3), Eq. (8.2) can be further rewritten as:

$$dT_{i-1}^{i} = T_{i-1}^{i} \cdot \left(Q_a \Delta a_i + Q_a \Delta \alpha_i + Q_a \Delta d_i + Q_a \Delta \theta_i\right) = T_{i-1}^{i} \cdot \delta T_{i-1}^{i} \tag{8.4}$$

where δT_{i-1}^{i} is the homogeneous transformation matrix representing the differential translation and differential rotation between the two adjacent coordinate systems $\{i-1\}$ and $\{i\}$:

$$\delta T_{i-1}^{i} = \begin{bmatrix} 0 & -\delta z_{i-1}^{i} & \delta y_{i-1}^{i} & dx_{i-1}^{i} \\ \delta z_{i-1}^{i} & 0 & -\delta x_{i-1}^{i} & dy_{i-1}^{i} \\ -\delta y_{i-1}^{i} & \delta x_{i-1}^{i} & 0 & dz_{i-1}^{i} \\ 0 & 0 & 0 & 0 \end{bmatrix}$$
$$\times \begin{bmatrix} 0 & -\cos\alpha_i \Delta\theta_i & \sin\alpha_i \Delta\theta_i & \Delta a_i \\ \cos\alpha_i \Delta\theta_i & 0 & -\Delta\alpha_i & a_i \cos\alpha_i \Delta\theta_i + \sin\alpha_i \Delta d_i \\ -\sin\alpha_i \Delta\theta_i & -\Delta\alpha_i & 0 & -a_i \sin\alpha_i \Delta\theta_i + \cos\alpha_i \Delta d_i \\ 0 & 0 & 0 & 0 \end{bmatrix} \tag{8.5}$$

where

$$\delta_{i-1}^{i} = \begin{bmatrix} \delta x_{i-1}^{i} \\ \delta y_{i-1}^{i} \\ \delta z_{i-1}^{i} \end{bmatrix} = \begin{bmatrix} \Delta\alpha_i \\ \sin\alpha_i \Delta\theta_i \\ \cos\alpha_i \Delta\theta_i \end{bmatrix} \tag{8.6}$$

$$d_{i-1}^{i} = \begin{bmatrix} dx_{i-1}^{i} \\ dy_{i-1}^{i} \\ dz_{i-1}^{i} \end{bmatrix} = \begin{bmatrix} \Delta a_i \\ a_i \cos\alpha_i \Delta\theta_i + \sin\alpha_i \Delta d_i \\ -a_i \sin\alpha_i \Delta\theta_i + \cos\alpha_i \Delta d_i \end{bmatrix} \tag{8.7}$$

are, respectively, the differential rotation vector and differential translation vector between the coordinate systems $\{i-1\}$ and $\{i\}$.

For the robotic UT system, the calculation of localization error considers the position/attitude error of sound beam focus relative to the base coordinate system. According to the relationship of matrix multiplication, we can obtain:

$$T_0^T + dT_0^T = \left(T_0^1 + dT_0^1\right)\left(T_1^2 + dT_1^2\right)\left(T_2^3 + dT_2^3\right)\left(T_3^4 + dT_3^4\right)\left(T_4^5 + dT_4^5\right)\left(T_6^T + dT_6^T\right) \tag{8.8}$$

The right term of Eq. (8.8) is expanded to contain the first-order and second-order error terms of the beam focus error introduced by kinematic parameter errors. The rest can be done in the same manner. The low-order error of each kinematic parameter is much larger than its high-order error. By ignoring the differential terms higher than the first order, dT_0^T can be expressed as:

$$\begin{aligned} dT_0^T &\cong T_0^1\delta T_0^1 \cdot T_1^T + T_0^2\delta T_1^2 \cdot T_2^T + T_0^3\delta T_2^3 \cdot T_3^T + T_0^4\delta T_3^4 \cdot T_4^T + T_0^5\delta T_4^5 \cdot T_5^T + T_0^6\delta T_5^6 \cdot T_6^T + T_0^T\delta T_6^T \\ &\cong T_0^T \cdot \Big[\left(T_1^T\right)^{-1} \delta T_0^1 \cdot T_1^T + \left(T_2^T\right)^{-1} \delta T_1^2 \cdot T_2^T + \left(T_3^T\right)^{-1} \delta T_2^3 \cdot T_3^T \\ &\quad + \left(T_4^T\right)^{-1} \delta T_3^4 \cdot T_4^T + \left(T_5^T\right)^{-1} \delta T_4^5 \cdot T_5^T + \left(T_6^T\right)^{-1} \delta T_5^6 \cdot T_6^T + \delta T_6^T \Big] \cong T_0^T \cdot \delta T_0^T \end{aligned} \tag{8.9}$$

where δT_0^T is the error matrix of sound beam focus relative to the base coordinate system and is expressed as:

$$\delta T_0^T = \begin{bmatrix} 0 & -\delta z_0^T & \delta y_0^T & dx_0^T \\ \delta z_0^T & 0 & -\delta x_0^T & dy_0^T \\ -\delta y_0^T & \delta x_0^T & 0 & dz_0^T \\ 0 & 0 & 0 & 0 \end{bmatrix} \tag{8.10}$$

where $\boldsymbol{d}_0^T = \begin{bmatrix} dx_0^T & dy_0^T & dz_0^T \end{bmatrix}$ and $\delta_0^T = \begin{bmatrix} \delta x_0^T & \delta y_0^T & \delta z_0^T \end{bmatrix}$ are the position error and attitude error of sound beam focus, respectively. ($\angle a_i$, $\angle \alpha_i$, $\angle d_i$, $\angle \theta_i$) are the parameters in $\boldsymbol{d}_0^T$. For the sake of simplicity, $\boldsymbol{d}_0^T$ is represented by:

$$\begin{aligned} \boldsymbol{d}_0^T &= \begin{bmatrix} dx_0^T \\ dy_0^T \\ dz_0^T \end{bmatrix} = \boldsymbol{M}_1\Delta \boldsymbol{a} + \boldsymbol{M}_2\Delta \boldsymbol{\alpha} + \boldsymbol{M}_3\Delta \boldsymbol{d} + \boldsymbol{M}_4\Delta \boldsymbol{\theta} \\ &= \begin{bmatrix} \boldsymbol{M}_1 & \boldsymbol{M}_2 & \boldsymbol{M}_3 & \boldsymbol{M}_4 \end{bmatrix} \begin{bmatrix} \Delta \boldsymbol{a} \\ \Delta \boldsymbol{\alpha} \\ \Delta \boldsymbol{d} \\ \Delta \boldsymbol{\theta} \end{bmatrix} = \boldsymbol{M}\Delta \boldsymbol{q} \end{aligned} \tag{8.11}$$

where

$\angle \mathbf{a} = (\angle a_1\ \angle a_2\ \angle a_3\ \angle a_4\ \angle a_5\ \angle a_6)^T$
$\angle \boldsymbol{\alpha} = (\angle \alpha_1\ \angle \alpha_2\ \angle \alpha_3\ \angle \alpha_4\ \angle \alpha_5\ \angle \alpha_6)^T$
$\angle \mathbf{d} = (\angle d_1\ \angle d_2\ \angle d_3\ \angle d_4\ \angle d_5\ \angle d_6)^T$
$\angle \boldsymbol{\theta} = (\angle \theta_1\ \angle \theta_2\ \angle \theta_3\ \angle \theta_4\ \angle \theta_5\ \angle \theta_6)^T$

In the detection system, the manipulator localization error model in Eq. (8.11) is represented by the location error of sound beam focus.

8.1.2 Relationship between Distance Error and Kinematic Parameter Error

Figure 8.1 shows the motion trajectory covered within an instruction period in workspace and the corresponding actual motion trajectory. $\boldsymbol{p}_c^i$ and $\boldsymbol{p}_c^{i+1}$ are the two instruction position points on the trajectory, and $\boldsymbol{p}_r^i$ and $\boldsymbol{p}_r^{i+1}$ are the corresponding actual position points. The vector of the distance between two points on the instructed trajectory is expressed as:

$$\boldsymbol{l}_c(i+1,i)=\left[x_c^{i+1}-x_c^i \; y_c^{i+1}-y_c^i \; z_c^{i+1}-z_c^i\right]^T \tag{8.12}$$

where (x_c^i, y_c^i, z_c^i) and $(x_c^{i+1}, y_c^{i+1}, z_c^{i+1})$ are the Cartesian coordinates at two instruction position points. The vector of the distance between the corresponding two points on the actual motion trajectory is expressed as:

$$\boldsymbol{l}_r(i+1,i)=\left[x_r^{i+1}-x_r^i \; y_r^{i+1}-y_r^i \; z_r^{i+1}-z_r^i\right]^T \tag{8.13}$$

where (x_r^i, y_r^i, z_r^i) and $(x_r^{i+1}, y_r^{i+1}, z_r^{i+1})$ are the Cartesian coordinates of actual position points. According to Eqs. (8.12) and (8.13), the vector of the distance error between two points in the workspace can be expressed as:

$$\Delta\boldsymbol{l}(i+1,i)=\boldsymbol{l}_c(i+1,i)-\boldsymbol{l}_r(i+1,i) \tag{8.14}$$

To eliminate the influence of random error, the above calculation formula for the distance error between two points is generalized to N sampling points. The distance error is represented by the arithmetic mean value of the distance errors among N sampling points in the target area:

$$\overline{\Delta\boldsymbol{l}}=\frac{1}{N}\sum_{i=0}^{N}\Delta\boldsymbol{l}(i+1,i) \tag{8.15}$$

For the point $\boldsymbol{P}^i$ on the motion trajectory in Figure 8.1, the location error $\boldsymbol{d}_0^T$ of sound beam focus in Eq. (8.11) can be referred to in order to define the location error of the point $\boldsymbol{P}^i$ as:

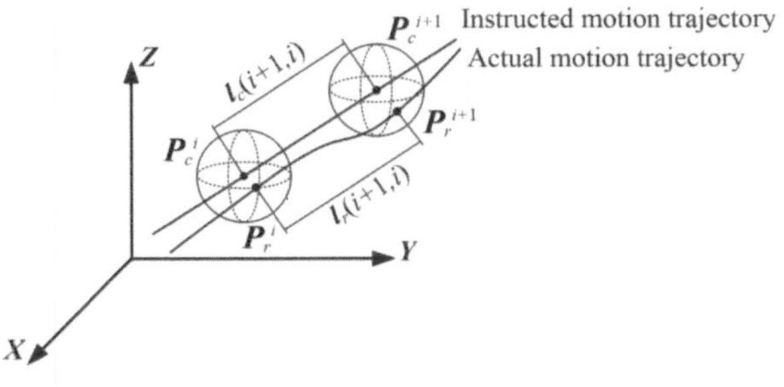

FIGURE 8.1 Distance error in workspace.

$$\Delta \boldsymbol{P}^i = \begin{bmatrix} \Delta x^i \\ \Delta y^i \\ \Delta z^i \end{bmatrix} = \begin{bmatrix} x_c^i - x_r^i \\ y_c^i - y_r^i \\ z_c^i - z_r^i \end{bmatrix} = \boldsymbol{P}_c^i - \boldsymbol{P}_r^i \tag{8.16}$$

Then it can be proved by Eq. (8.14) that:

$$\begin{aligned} \Delta \boldsymbol{l}(i+1,i) &= \begin{bmatrix} x_c^{i+1} - x_c^i \\ y_c^{i+1} - y_c^i \\ z_c^{i+1} - z_c^i \end{bmatrix} - \begin{bmatrix} x_r^{i+1} - x_r^i \\ y_r^{i+1} - y_r^i \\ z_r^{i+1} - z_r^i \end{bmatrix} = \begin{bmatrix} x_c^{i+1} - x_r^{i+1} \\ y_c^{i+1} - y_r^{i+1} \\ z_c^{i+1} - z_r^{i+1} \end{bmatrix} - \begin{bmatrix} x_c^i - x_r^i \\ y_c^i - y_r^i \\ z_c^i - z_r^i \end{bmatrix} \\ &= \Delta \boldsymbol{P}^{i+1} - \Delta \boldsymbol{P}^i \end{aligned} \tag{8.17}$$

According to Eq. (8.11), the relationship between the location error of the point $\boldsymbol{P}^i$ and the kinematic parameter error is defined as:

$$\Delta \boldsymbol{P}^i = \boldsymbol{M}^i \cdot \Delta \boldsymbol{q} \tag{8.18}$$

where each element in the matrix $\boldsymbol{M}^i$ is a function containing the kinematic parameters a, α, d, θ. The joint angle θ is a variable, while the joint offset d, the deflection angle α and the distance offset a are constants. $\boldsymbol{M}^i$ will change with the manipulator position/attitude. Eq. (8.18) is substituted into Eq. (8.17) to obtain:

$$\Delta \boldsymbol{l}(i+1,i) = \Delta \boldsymbol{P}^{i+1} - \Delta \boldsymbol{P}^i = \left(\boldsymbol{M}^{i+1} - \boldsymbol{M}^i\right) \cdot \Delta \boldsymbol{q} \tag{8.19}$$

where the relationship between distance error and kinematic parameter error is established; $\angle \mathbf{l}\ (I+1, i)$ is the vector of the distance error between two points and can be obtained by ultrasonic ranging; $\boldsymbol{M}^i$ and $\boldsymbol{M}^{i+1}$ are related to the coordinate values of instruction points and the nominal values of kinematic parameters and can be obtained by calculation and Δq is an error vector composed of kinematic parameter errors (containing $4n$ parameters, that is, 24 parameters for six-degree of freedom [DOF] manipulator) and is also an unknown quantity to be solved. In Eq. (8.19), $\angle \mathbf{l}\ (I+1, i)$ is a three-dimensional vector, that is, the distance change along three orthogonal directions. However, the transducer can only measure the distance change along a single axis of the beam, so the modulus of this 3D vector should be taken. According to Eqs. (8.12) and (8.19), we can obtain:

$$l_c(i+1,i) \cdot \Delta l(i+1,i) \cong \begin{bmatrix} x_c^{i+1} - x_c^i & y_c^{i+1} - y_c^i & z_c^{i+1} - z_c^i \end{bmatrix} \cdot \left(\boldsymbol{M}^{i+1} - \boldsymbol{M}^i\right) \Delta \boldsymbol{q} \tag{8.20}$$

where $l_c\ (i+1, i)$ and $\angle l\ (i+1, i)$ are the moduli of the 3D vector, namely $l_c\ (i+1, i) = |\mathbf{l}_c\ (i+1, i)|$ and $\angle l\ (i+1, i) = |\angle \mathbf{l}\ (i+1, i)|$. Then $l_c\ (i+1, i)$ is moved to the right-hand side of Eq. (8.20) to change the equation into:

$$\Delta l(i+1,i)=\begin{bmatrix} \dfrac{x_c^{i+1}-x_c^i}{l_c(i+1,i)} & \dfrac{y_c^{i+1}-y_c^i}{l_c(i+1,i)} & \dfrac{z_c^{i+1}-z_c^i}{l_c(i+1,i)} \end{bmatrix}\cdot\left(\boldsymbol{M}^{i+1}-\boldsymbol{M}^i\right)\Delta\boldsymbol{q}$$

$$=\begin{bmatrix} L_x(i+1,i) & L_y(i+1,i) & L_z(i+1,i) \end{bmatrix}\cdot\left(\boldsymbol{M}^{i+1}-\boldsymbol{M}^i\right)\Delta\boldsymbol{q} \tag{8.21}$$

To solve the parameter in Δq, at least 24 distance errors should be measured, and 24 equations like Eq. (8.21) should be established into a system of equations. Statistically speaking, in order to reduce the impact of random factors, more than 24 distance errors should be actually measured.

8.1.3 Three-Dimensional Spatial Distribution of Errors

The kinematic parameter errors of the manipulator can be deduced from distance errors. The distribution of distance errors in the three-dimensional space can reflect the influence of kinematic parameter errors on the working performance of the robotic UT system to some extent. According to the analysis of manipulator kinematics, each DOF of the manipulator contains four kinematic parameters (a, α, d, θ), each of which has a different influence on distance error. The base coordinate system of the manipulator is taken as the reference coordinate system. Several instructed distances with the same length are selected from X- and Y-axes, respectively. The coordinates of each instructed distance change in a single direction but do not change in the other two directions. By adding the kinematic parameter errors, the variation of distance error distribution in the three-dimensional space can be analyzed.

Four instructed distances are set along the X direction. Each instructed distance is 50 mm long, extending from one instruction point to another instruction point (refer to Table 8.1 for their coordinates). The errors $\Delta\theta$ (0.1°, 0.3°, 0.5°, 1.0°) are separately added to the theoretical value of the joint angle θ. As shown in Figure 8.2a, the distance error in the +X-axis increases with the X-coordinate value or with the joint angle error in general. Similarly, the same trend can also be seen when adding errors to other kinematic parameters. The variation tendencies of distance error caused by the variation of the joint offset d, the deflection angle α and the distance offset a are shown in Figure 8.2b–d, respectively.

Five instructed distances are set along the Y direction. Each instructed distance is 50 mm long, extending from one instruction point to another instruction point (refer to Table 8.2 for their coordinates). The errors $\Delta\theta$, Δd, $\Delta\alpha$, Δa (0.1, 0.3, 0.5, 1.0 for each error) are, respectively, added to the theoretical values of the joint angle θ, the joint offset d, the

TABLE 8.1 Instructed Distance Coordinates in X Direction

No.	Position Coordinates
1	(700, 0, 230) → (750, 0, 230)
2	(750, 0, 230) → (800, 0, 230)
3	(800, 0, 230) → (850, 0, 230)
4	(850, 0, 230) → (900, 0, 230)
5	(900, 0, 230) → (950, 0, 230)

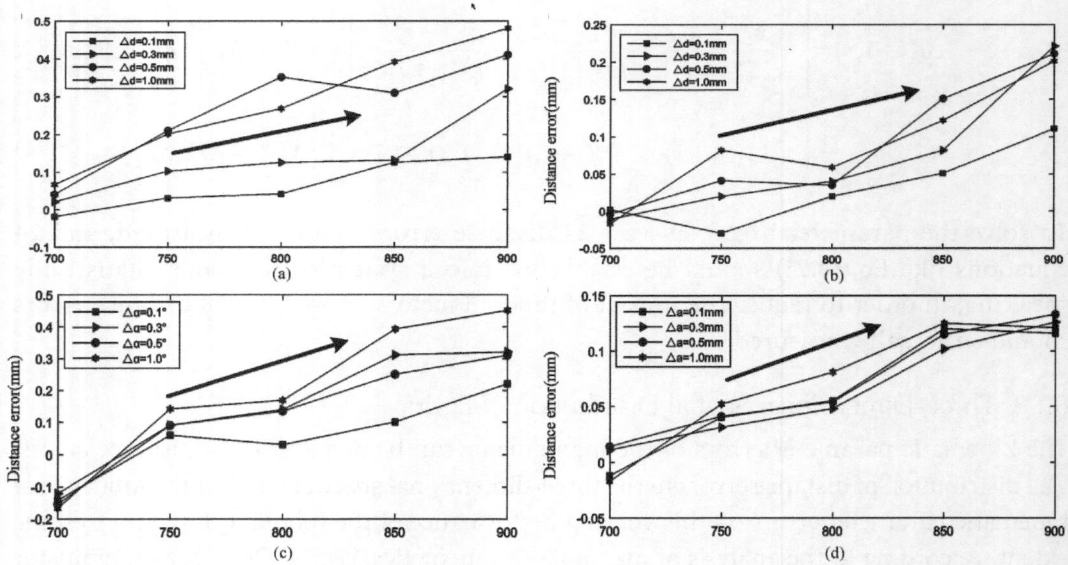

FIGURE 8.2 Distance error change in X direction. (a) The distance error in the +X axis increases with the X-coordinate value or with the joint angle error in general, (b) the variation tendencies of distance error caused by the variation of the joint offset d, (c) the variation tendencies of distance error caused by the variation of the deflection angle α and (d) the variation tendencies of distance error caused by the variation of the distance offset a.

TABLE 8.2 Instructed Distance Coordinates in *Y* Direction

No.	Position Coordinates
1	(800, −150, 180) → (800, −100, 180)
2	(800, −100, 180) → (800, −50, 180)
3	(800, −50, 180) → (800, 0, 180)
4	(800, 0, 180) → (800, 50, 180)
5	(800, 50, 180) → (800, 100, 180)
6	(800, 100, 180) → (800, 150, 180)

deflection angle α and the distance offset a. Different from the trend in Figure 8.2 that the distance error increases along the +*X*-axis, it can be seen in Figure 8.3 that the distance error increases in both +*Y* and −*Y* directions but is relatively small near the $Y = 0$ point. The analysis of distance error distribution in *X* and *Y* directions shows that the localization error of the manipulator will increase as it gets closer to the workspace boundary, that is, the manipulator near the workspace boundary has a large localization error.

8.2 FEEDBACK COMPENSATION MODEL OF ROBOTIC UT ERROR

Affected by the manipulator localization error, the motion path of the detection system will deviate from the ideal path planned according to the principle of ultrasonic propagation, thus resulting in the decrease of the echo signal-to-noise ratio (SNR). As a result, the echo signals from some small defects may be covered by the clutter signals generated in

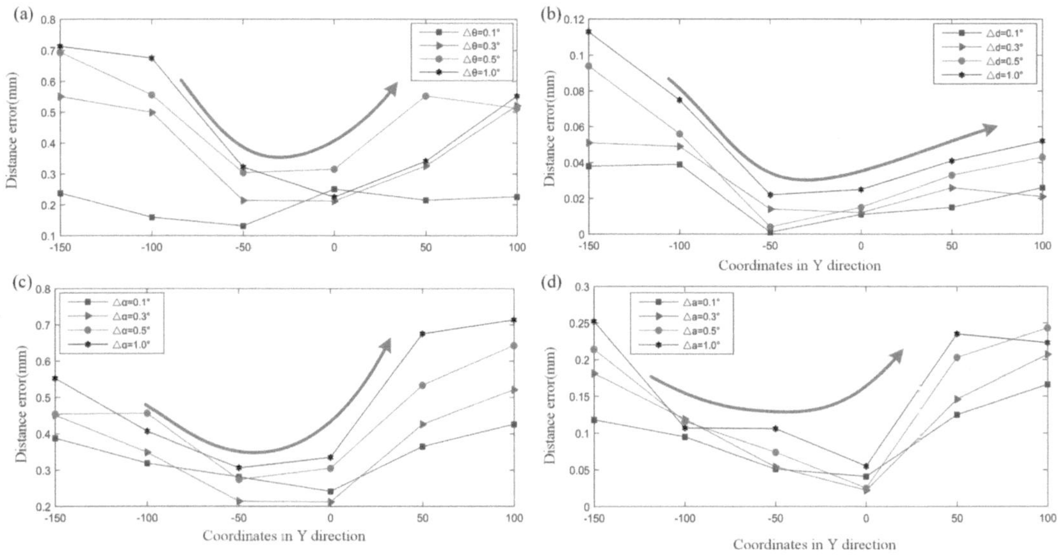

FIGURE 8.3 Distance error change in *Y* direction.

waveform conversion. Due to the low SNR of defect echo, the data gate may find it hard to capture the peak value of defect echo. Therefore, some refracted and diffracted wave signals generated by waveform conversion may be misjudged as defect waves, causing false or missed detection in the detection process and reducing the defect resolution of the detection system. For the NDT system, a greater concern is to compensate for these errors in some ways so as to reduce the impact of errors on the detection resolution.

8.2.1 Principle of Error Feedback Compensation

At present, two compensation methods are commonly used. One method is to measure the position/attitude error of manipulator end-effector in real time with the aid of external measuring equipment and compensate for the position/attitude error by adjusting the joint angle in order to improve the location accuracy of the manipulator. This method has high requirements for data transmission speed and computation load, thus leading to the decrease of manipulator speed to some extent. The other method is to use the kinematic parameter errors obtained by calibration to compensate for the theoretical values of kinematic parameters. The latter method requires the completion of calibration before the manipulator operation. Since it has no requirements for data transmission speed and computation load, the manipulator location accuracy can be improved to a certain degree but is still inferior to that obtained by external real-time compensation.

The schematic diagram of external real-time compensation method is shown in Figure 8.4. The instructed coordinate value of a point on the motion path is P^n. In the manipulator controller, P^n is converted into the motion with six joint angles $(\theta_1 \ldots \theta_6)$ through inverse kinematics transformation. Due to the existence of kinematic parameter errors $(\Delta\theta, \Delta d, \Delta\alpha, \Delta a)$, the actual position/attitude of the end-effector arriving at the point is $P^n + \Delta P$. The error $-\Delta P$ is measured by external measuring device in the process of

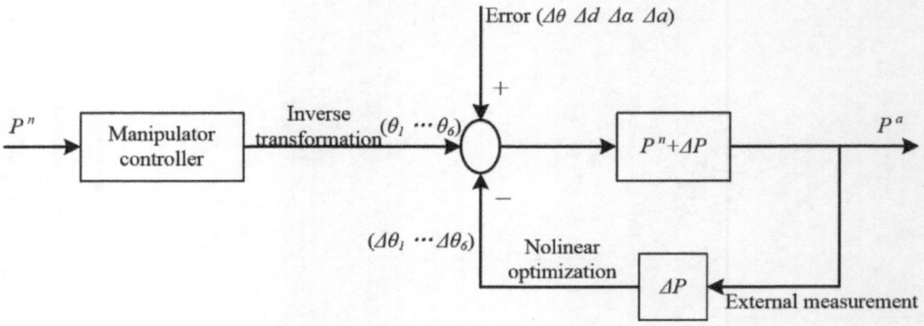

FIGURE 8.4 External real-time compensation method.

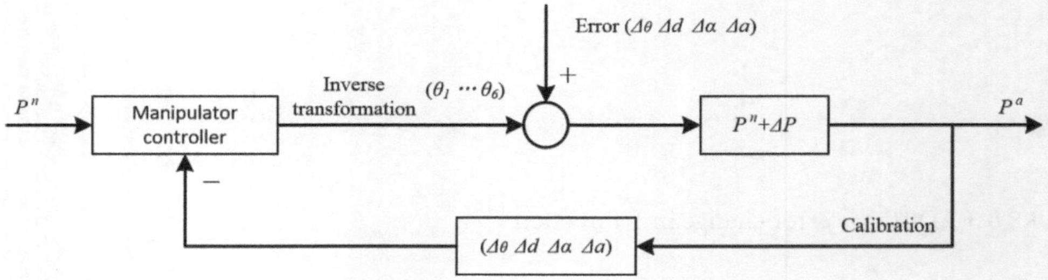

FIGURE 8.5 Method of kinematic parameter error calibration and compensation.

manipulator movement, so an offset should be preset between the actual position/attitude P^a and the theoretical position/attitude P^n. Through the offsets of joint angle variables, the offset of end-effector position/attitude is realized. In essence, this process is to solve the problem of nonlinear optimization. The solved joint angle offsets are $(\Delta\theta_1 \ldots \Delta\theta_6)$. Then the joint angle offsets are added to the joint angle motion in manipulator controller. At this point, the error ΔP can be compensated for to a certain extent, and the actual position/attitude P^a of end-effector is more in line with the instructed position/attitude P^n, so the location accuracy is improved. The schematic diagram of kinematic parameter error calibration and compensation is shown in Figure 8.5. The purpose of this calibration is to obtain the kinematic parameter errors ($\Delta\theta$, Δd, $\Delta\alpha$, Δa) and then use them as compensations in the manipulator controller, so as to offset the kinematic parameter errors introduced by external factors. This method requires the completion of calibration before testing.

8.2.2 Calculation of Kinematic Parameter Errors

Calibration is a way to improve the location accuracy of a manipulator by modifying its own kinematic parameters. The accurate measurement and calculation of kinematic parameter errors is the premise to ensure the correction accuracy. In addition to a commonly used three-dimensional position/attitude coordinate measurement device, a measurement method suitable for the robotic UT system is needed. According to the relationship between distance error and kinematic parameter error, a feasible method for distance error measurement and kinematic parameter error calculation is proposed in this section.

As shown in Figure 8.6, on the hemisphere (radius: R) with the point $\mathbf{o}^1$ as the center and the other N points distributed on its surface, there are N distances with the equal length relative to the point $\mathbf{o}^1$. N independent equations can be established according to Eq. (8.21). For the manipulator with six-DOF, N distances ($N = 24$) should be measured to calculate all the elements of $\angle\mathbf{q}$. In the actual solving process, two opposite hemispherical surfaces are used and N distances ($N = 48$) are measured to reduce the impact of random errors on the measurement results. For the distance $\mathbf{o}^1\mathbf{p}^1$ in Figure 8.6, it can be substituted into Eq. (8.21) and simplified as $\angle l(\mathbf{o}^1, \mathbf{p}^1) = \mathbf{C}_1 \angle\mathbf{q}$. Similarly, for 48 measured distances, they can be expressed by the following matrix:

$$\begin{bmatrix} \Delta l(\boldsymbol{o}^1,\boldsymbol{p}^1) \\ \Delta l(\boldsymbol{o}^1,\boldsymbol{p}^2) \\ \vdots \\ \Delta l(\boldsymbol{o}^1,\boldsymbol{p}^{23}) \\ \Delta l(\boldsymbol{o}^1,\boldsymbol{p}^{24}) \\ \Delta l(\boldsymbol{o}^1,\boldsymbol{p}^{25}) \\ \Delta l(\boldsymbol{o}^2,\boldsymbol{p}^{26}) \\ \vdots \\ \Delta l(\boldsymbol{o}^2,\boldsymbol{p}^{47}) \\ \Delta l(\boldsymbol{o}^2,\boldsymbol{p}^{48}) \end{bmatrix} \begin{bmatrix} \begin{bmatrix} L_x(\boldsymbol{o}^1,\boldsymbol{p}^1) & L_y(\boldsymbol{o}^1,\boldsymbol{p}^1) & L_z(\boldsymbol{o}^1,\boldsymbol{p}^1) \end{bmatrix} \cdot \left(\boldsymbol{M}^{o^1} - \boldsymbol{M}^{p^1}\right) \\ \begin{bmatrix} L_x(\boldsymbol{o}^1,\boldsymbol{p}^2) & L_y(\boldsymbol{o}^1,\boldsymbol{p}^2) & L_z(\boldsymbol{o}^1,\boldsymbol{p}^2) \end{bmatrix} \cdot \left(\boldsymbol{M}^{o^1} - \boldsymbol{M}^{p^2}\right) \\ \vdots \\ \begin{bmatrix} L_x(\boldsymbol{o}^1,\boldsymbol{p}^{23}) & L_y(\boldsymbol{o}^1,\boldsymbol{p}^{23}) & L_z(\boldsymbol{o}^1,\boldsymbol{p}^{23}) \end{bmatrix} \cdot \left(\boldsymbol{M}^{o^1} - \boldsymbol{M}^{p^{23}}\right) \\ \begin{bmatrix} L_x(\boldsymbol{o}^1,\boldsymbol{p}^{24}) & L_y(\boldsymbol{o}^1,\boldsymbol{p}^{24}) & L_z(\boldsymbol{o}^1,\boldsymbol{p}^{24}) \end{bmatrix} \cdot \left(\boldsymbol{M}^{o^1} - \boldsymbol{M}^{p^{24}}\right) \\ \begin{bmatrix} L_x(\boldsymbol{o}^2,\boldsymbol{p}^{25}) & L_y(\boldsymbol{o}^2,\boldsymbol{p}^{25}) & L_z(\boldsymbol{o}^2,\boldsymbol{p}^{25}) \end{bmatrix} \cdot \left(\boldsymbol{M}^{o^1} - \boldsymbol{M}^{p^{25}}\right) \\ \begin{bmatrix} L_x(\boldsymbol{o}^2,\boldsymbol{p}^{26}) & L_y(\boldsymbol{o}^2,\boldsymbol{p}^{26}) & L_z(\boldsymbol{o}^2,\boldsymbol{p}^{26}) \end{bmatrix} \cdot \left(\boldsymbol{M}^{o^1} - \boldsymbol{M}^{p^{26}}\right) \\ \vdots \\ \begin{bmatrix} L_x(\boldsymbol{o}^2,\boldsymbol{p}^{47}) & L_y(\boldsymbol{o}^2,\boldsymbol{p}^{47}) & L_z(\boldsymbol{o}^2,\boldsymbol{p}^{47}) \end{bmatrix} \cdot \left(\boldsymbol{M}^{o^2} - \boldsymbol{M}^{p^{47}}\right) \\ \begin{bmatrix} L_x(\boldsymbol{o}^2,\boldsymbol{p}^{48}) & L_y(\boldsymbol{o}^2,\boldsymbol{p}^{48}) & L_z(\boldsymbol{o}^2,\boldsymbol{p}^{48}) \end{bmatrix} \cdot \left(\boldsymbol{M}^{o^2} - \boldsymbol{M}^{p^{48}}\right) \end{bmatrix}$$

$$\cdot \Delta \boldsymbol{q} = \begin{bmatrix} \boldsymbol{C}_1 \\ \boldsymbol{C}_2 \\ \vdots \\ \boldsymbol{C}_{23} \\ \boldsymbol{C}_{24} \\ \boldsymbol{C}_{25} \\ \boldsymbol{C}_{26} \\ \vdots \\ \boldsymbol{C}_{47} \\ \boldsymbol{C}_{48} \end{bmatrix} \cdot \Delta \boldsymbol{q} \tag{8.22}$$

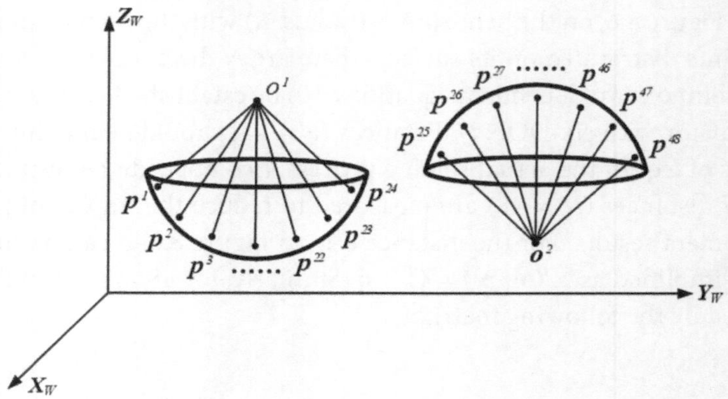

FIGURE 8.6 Measurement method of error in *N* distances.

According to the optimization method (such as the theory of least squares), the optimal solution of Eq. (8.22) is:

$$\Delta \boldsymbol{q} = \left(\boldsymbol{C}^T \boldsymbol{C}\right)^{-1} \boldsymbol{C}^T \left[\Delta l\left(\boldsymbol{o}^1, \boldsymbol{p}^1\right) \ldots \Delta l\left(\boldsymbol{o}^2, \boldsymbol{p}^{48}\right)\right] \tag{8.23}$$

where $\mathbf{C} = [\mathbf{C}_1, \mathbf{C}_2, \ldots, \mathbf{C}_{47}, \mathbf{C}_{48}]^T$. According to Eq. (8.23), $\boldsymbol{\Delta}\mathbf{q}$, namely kinematic parameter error, is solved. Based on this relation, the values of elements in $\boldsymbol{\Delta}\mathbf{q}$ are introduced to the theoretical values of kinematic parameters to offset the influence of external factors on kinematic parameters and thus reduce the localization error of the manipulator.

8.2.3 Step of Feedback Compensation of Kinematic Parameter Error

According to the derivation process of the relation between kinematic parameter error and distance error, the kinematic parameter error can be calculated and compensated for by the following steps:

A. The motion path of the manipulator is planned by referring to Figure 8.6, so that the manipulator clamping an ultrasonic transducer can make an arc movement on the hemisphere $\mathbf{o}^1$ and an arc movement on the hemisphere $\mathbf{o}^2$ in the opposite direction. The radii of both $\mathbf{o}^1$ and $\mathbf{o}^2$ are R;

B. During the movement, the sound interval between the transmitting surface of the transducer and the reflective surface of the hemisphere is measured at each instruction point, and the distance is calculated accordingly. The distance from the reflective surface of one hemisphere is denoted as l_r $(\mathbf{o}^1, \mathbf{p}^i)$, while the distance from the reflective surface of the other hemisphere is denoted as l_r $(\mathbf{o}^2, \mathbf{p}^i)$;

C. The position coordinates at each instruction point are known. The coordinates of the instruction point $\mathbf{p}^i$ are denoted as **p** i c=(x i c, y i c, z i c). The distance between two instruction points is known. The distance between {**o** 1 c} and {**p** i c} is $\{l_c\ (\mathbf{o}^1, \mathbf{p}^i) = R\}$, and the distance between {**o** 2 c} and {**p** i c} is $\{l_c\ (\mathbf{o}^2, \mathbf{p}^i) = R\}$;

D. The distance error Δl is calculated according to Eq. (8.14). The distance error measured on the hemisphere $\mathbf{o}^1$ is $\Delta l\ (\mathbf{o}^1, \mathbf{p}^i) = R - l_r\ (\mathbf{o}^1, \mathbf{p}^i)$, and the distance error measured on the hemispheric $\mathbf{o}^2$ is $\Delta l\ (\mathbf{o}^2, \mathbf{p}^i) = R - l_r\ (\mathbf{o}^2, \mathbf{p}^i)$;

E. $\angle\mathbf{q}$, namely kinematic parameter error, is calculated according to Eq. (8.23);

F. The kinematic parameter values in the manipulator controller are compensated for by the obtained kinematic parameter errors, as shown in Figure 8.5.

According to the above steps, the kinematic parameter error in a certain area can be calculated and compensated for. Obtaining the optimal solution of $\angle\mathbf{q}$ in Eq. (8.23) is the key to calculate the kinematic parameter error. Some optimization methods have been widely used in error calibration and accuracy compensation. For example, the recursive least squares (RLS) algorithm is used to identify the geometric parameters of four-DOF SCARA (Selective Compliance Assembly Robot Arm) manipulator and five-DOF tree-typed modular manipulator. The methods of nonlinear optimization, recursive linear optimization and extended Kalman filtering are used to identify the kinematic parameter error of SCARA manipulator and are compared in terms of accuracy through experiment. The method of minimum linear combination is used to analyze the manipulator localization error, and then the method of linear prediction and real-time error compensation is used to reduce the error. Similar optimization methods can also be used to calculate the kinematic parameter error $\angle\mathbf{q}$ in Eq. (8.23).

8.3 DESIGN AND APPLICATION OF BI-HEMISPHERIC CALIBRATION BLOCK

It is found from the robotic UT process that with the change of the scanning area, the location error of sound beam focus will change accordingly. The location error is smaller in some areas but larger in other areas. Therefore, the evaluation of system error has become an urgent problem to be solved. The error of the detection system must be quantified, but currently no method is available for the comprehensive evaluation of this system. A set of standardized application processes and technical plans can be established to measure the manipulator motion accuracy and verify the trajectory error. Bi-hemispheric calibration blocks can effectively and comprehensively evaluate the characteristics of the UT system and improve the detection accuracy and reliability, thus improving the manufacturing quality and service reliability of tested components and reaping the best technical and economic benefits.

8.3.1 Design of Bi-hemispheric Calibration Block

In the above sections, the processes of measuring, identifying and compensating for the kinematic parameter errors are presented. As mentioned above, the manipulator with ultrasonic transducer needs to move along the paths on two opposite hemispheres, while measuring the distances between the transmitting front end of the transducer and the reflective surfaces of the hemispheres in order to obtain the distance error as a data source for the subsequent calculation of kinematic parameter errors. The distance measurement in this study is based on the principle of ultrasonic propagation time measurement through pulse reflection. As shown in Figure 8.7, the pulse transceiver excites the piezoelectric wafer of ultrasonic transducer by means of fixed-frequency narrow pulses so that the wafer vibrates and generates an ultrasonic wave. The sound wave propagates forward in the medium (water) and generates the pulsed reflection echoes on the upper surface and back side of the reflector [4]. By comparing Figure 8.7a and b, it can be found that there is a relationship between the time interval from the emission of sound wave to the reception of reflected wave and the distance from the emitting surface of the transducer to the reflector surface, that is, $l_r = (C_w \times t)/2$, where C_w is the sound velocity in water and t is the propagation time of sound wave in water. The comparison between Figure 8.7a and c shows that, when the axis of the sound beam emitted by transducer deviates from the normal direction of the reflecting surface by a small angle, the amplitude of the reflected wave will decrease to a certain extent. However, as long as the deflection angle is not large, the sound interval can still reflect the distance between the emitting surface and the reflecting surface.

According to the above ranging principle, a standard test block with two opposite hemispherical reflectors was designed and used as an auxiliary tool for the ultrasonic measurement of distance error. The test block was designed by referring to the theoretical model shown in Figure 8.6. Its CAD model is shown in Figure 8.8. The test block is composed of

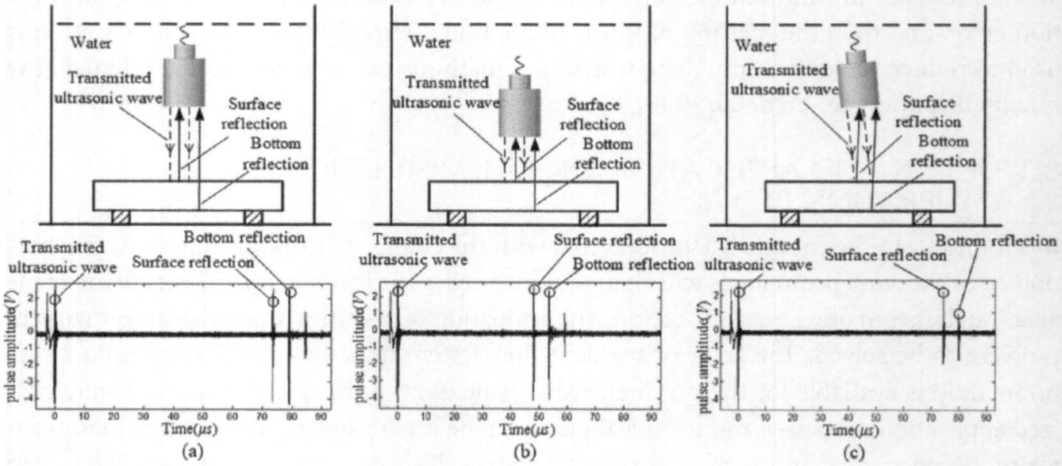

FIGURE 8.7 Ranging principle of ultrasonic transducer based on pulse reflection method. (a) Normal incidence - long water range, (b) normal incidence - short water range and (c) oblique incidence - long water range.

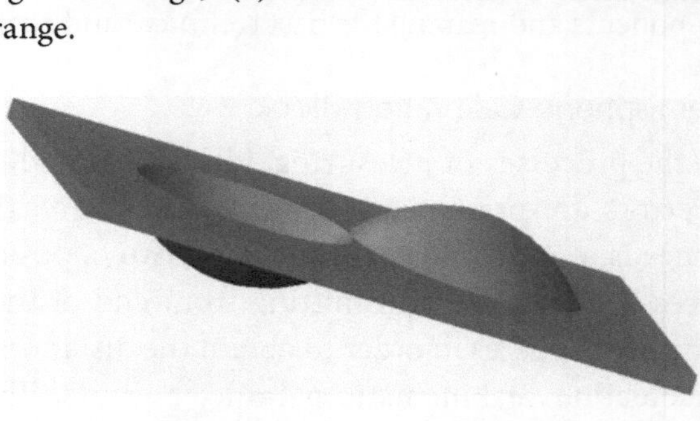

FIGURE 8.8 Oblique view of bi-hemispheric calibration block.

a flat plate and two opposite hemispherical shells. It is usually made of aluminum alloy, stainless steel, titanium alloy and other metal materials that are not susceptible to corrosion and rust. In order to reduce weight, the test block can also be made of resin, carbon fiber and other non-metallic materials with high hardness.

8.3.2 Method of UT System Compensation with Bi-hemispheric Calibration Error

Suppose that the manipulator with ultrasonic transducer moves from the point P^i to another point P^{i+1}. In this process, the distance error of beam focus is $\Delta\mathbf{l}(i+1, i)$. Then the manipulator moves from the point P^j to another point P^{j+1}. In this process, the distance error of beam focus is $\Delta\mathbf{l}\ (j+1, j)$. Therefore, the deviation between the two distance errors is expressed as the vector $\Delta\mathbf{l}(i+1, i)-\Delta\mathbf{l}(j+1, j)$, and the deviation degree is expressed as the scalar $\mathrm{e}=|\Delta\mathbf{l}(i+1, i)-\Delta\mathbf{l}(j+1, j)|$. The relationship between the two distance errors is shown in Figure 8.9. For two distance errors close to each other in the workspace, the closer they are, the smaller the deviation degree e will be and vice versa. If the deviation degree between the distance errors $\Delta\mathbf{l}(i+1, i)$ and $\Delta\mathbf{l}(j+1, j)$ is small, they can be interchanged. Thus it can be seen that when the deviation degree between two distance errors is small, the replacement of one distance error by the other distance error will not significantly affect the evaluation result of kinematic parameter error.

When the case of two distance errors is generalized to multiple distance errors, the same conclusion can also be drawn, that is, the distance errors in a certain area of the workspace tend to be the same. A vector with statistical significance can be used to represent the distance error in that area, and then the localization error can be determined without significant deviation. When the manipulator joint is confined to a small motion range, the location error of sound beam focus will change slightly. The localization errors in different areas of manipulator workspace will be different. The localization errors in small areas are similar to each other. If the manipulator deviates greatly from its initial position and attitude, the localization error will also be large.

If the method of measuring the distance error in an area and then calculating and compensating for the kinematic parameter error, as described in Section 8.2.3, is generalized to the whole workspace, then the kinematic parameter error of the manipulator at a certain position can be determined through the linear fitting of the errors in multiple areas. When the unknown error at a certain position in the space is linearly represented by several known errors, the weight of influence of each known error will be different. The closer the distance is, the greater the weight will be. This weight is just the coefficient in linear fitting. If the reciprocal of the distance between two points is taken as the coefficient, a closer distance will indicate a larger coefficient and a greater weight of influence in linear fitting. On the contrary, a farther distance implies a smaller weight of influence. When the measuring

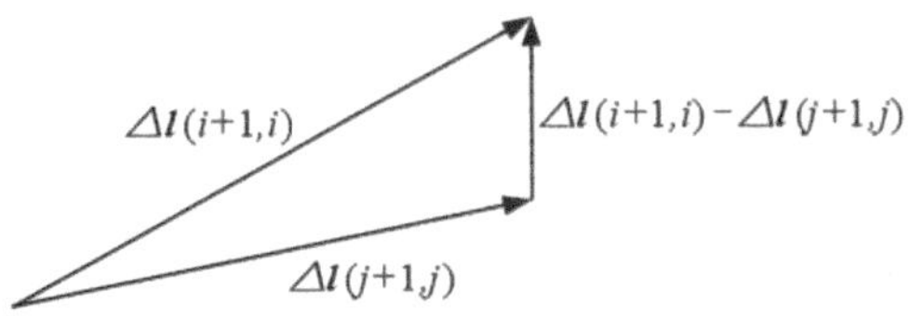

FIGURE 8.9 Deviation degree between two distance errors.

points are uniformly distributed at the boundary and center of the workspace, the linear fitting will be more accurate and the computing speed will be faster.

In order to distribute the measurement points more evenly to achieve higher fitting accuracy, and to evaluate the localization error distribution in different areas of manipulator workspace, the workspace can be divided into several three-dimensional grids. When the manipulator is in any position/attitude, its kinematic parameter error can be determined through the linear fitting of kinematic parameter errors in the grids. Take Figure 8.10 as an example. In Figure 8.10a, the manipulator workspace is divided into six cubic areas *A–F*. Then the kinematic parameter error in each area is determined and all of them are defined as the vectors $\boldsymbol{\Delta q}^A$–$\boldsymbol{\Delta q}^F$. In Figure 8.10b, $\boldsymbol{\Delta q}^A$–$\boldsymbol{\Delta q}^F$, respectively, represent the kinematic parameter errors at the centers of the six grids. The kinematic parameter error $\boldsymbol{\Delta q}^i$ at any point $\boldsymbol{P}^i$ in space is obtained through the linear fitting of $\boldsymbol{\Delta q}^A$–$\boldsymbol{\Delta q}^F$. Thus, the steps of gridded measurement and compensation of kinematic parameter error are as follows:

A. According to the method of measurement and identification of kinematic parameter error, the kinematic parameter errors in the 3D grids A–F are determined, respectively, denoted as $\boldsymbol{\Delta q}^A$–$\boldsymbol{\Delta q}^F$;

B. The linear distances between the center of each grid and the measured point are calculated and denoted as d^A–d^F;

C. The coefficient for the linear fitting to the kinematic parameter error at the measured point is calculated. For example, the linear fitting coefficient of the area *A* is:

$$k^A = \left(\frac{1}{d^A}\right)^r \bigg/ \left(\frac{1}{d^A}+\frac{1}{d^B}+\frac{1}{d^C}+\frac{1}{d^D}+\frac{1}{d^E}+\frac{1}{d^F}\right)^r \tag{8.24}$$

where the power exponent r is usually 1. The value of r can be adapted appropriately to a larger or smaller k^A.

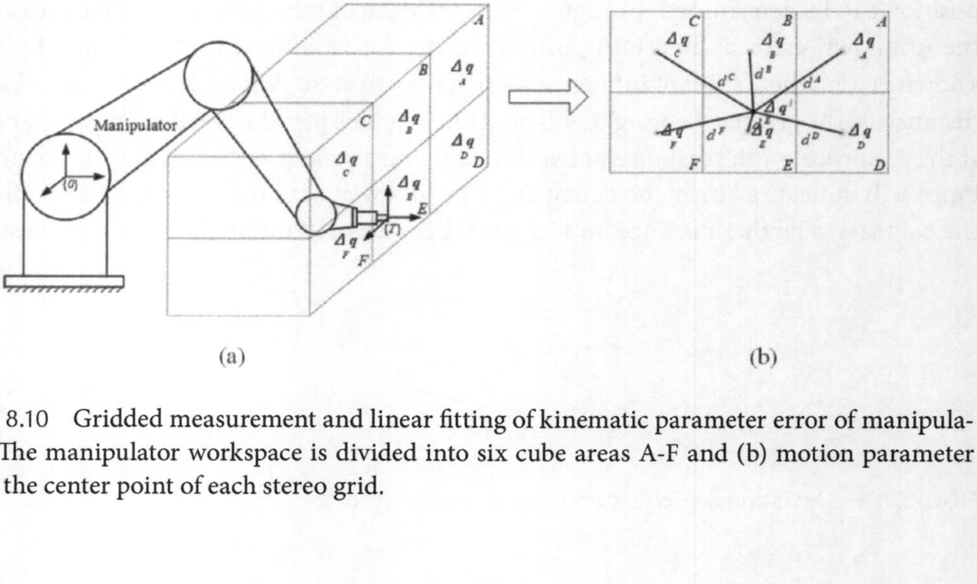

FIGURE 8.10 Gridded measurement and linear fitting of kinematic parameter error of manipulator. (a) The manipulator workspace is divided into six cube areas A-F and (b) motion parameter error at the center point of each stereo grid.

D. The kinematic parameter error at any point P^i in space is linearly represented by:

$$\Delta q^i = k^A \Delta q^A + k^B \Delta q^B + k^C \Delta q^C + k^D \Delta q^D + k^E \Delta q^E + k^F \Delta q^F \tag{8.25}$$

E. The kinematic parameter error at the point P^i is determined. The sum of this error and the actual value of the kinematic parameter is just the compensated kinematic parameter.

Based on the above steps, the kinematic parameter error of the manipulator at any position and attitude can be obtained and compensated for. Similarly, the theoretical values of kinematic parameters at each position point can be compensated for during the manipulator movement. After compensation, the localization error of the manipulator is reduced and the resolution of the detection system is improved accordingly.

8.3.3 Application of Calibration Block in Kinematic Parameter Error Compensation

The bi-hemispheric calibration block shown in Figure 8.8 is combined with the feedback compensation of kinematic parameter error to improve the manipulator location accuracy. The front and top views of bi-hemispheric calibration block and its dimensions are shown in Figure 8.11. The block is formed by two hemispheric shells welded on a flat plate. Either of the two hemispheric shells has a through hole at the bottom as the drain hole.

In the CNC machining simulation software, the three-dimensional model of calibration block illustrated in Figure 8.12a is used to plan the motion trajectory. The zig-zag (reciprocating) motion process of the manipulator holding ultrasonic transducer above the hemispheric surface is simulated by CNC five-axis milling method. The five-axis machining method is mainly used for machining complex curved surfaces, curved contours and a series of holes distributed on curved surface. The machine tool should contain at least five axes (three for linear motion and two for rotational motion) in coordinated movement under the control of the numerical control system to implement a complex motion process. Therefore, this method can simulate the flexible movement of a manipulator. The tool path

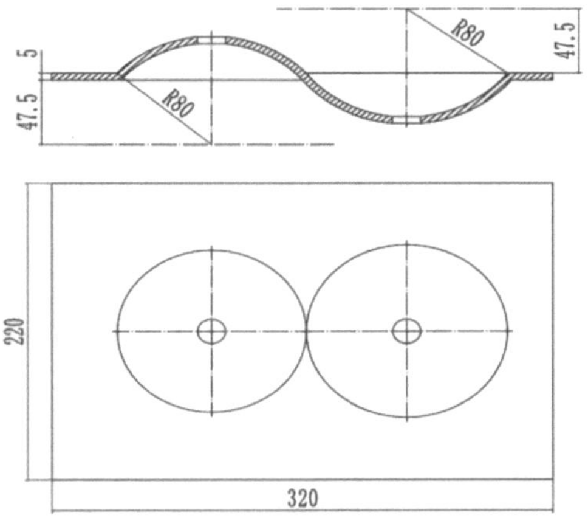

FIGURE 8.11 Size of bi-hemispheric calibration block.

FIGURE 8.12 Generation of manipulator scanning path by CNC machining simulation software. (a) Trajectory planning, (b) trajectory simulation and (c) scan path.

planned and generated in CNC machining simulation software, as shown in Figure 8.12b, is used as the surface-scanning path of the manipulator with transducer. The step size in the truncation direction and that in the guide direction during NC machining correspond to the scanning interval and step interval, respectively, during ultrasonic testing. On the motion path generated by CNC machining simulation software, each instruction point is expressed as Cartesian coordinates, and the workpiece coordinate system is taken as the reference coordinate system. The scanning path of manipulator shown in Figure 8.12c is just the tool machining path that has been converted to the manipulator's working space.

The motion path generated in the manipulator workspace shown in Figure 8.12c takes the base coordinate system of the manipulator as the reference coordinate system. A calibration block is placed in each of the six cubic grids A–F with equal dimensions in the workspace, in a way that the block center coincides with the grid center. Thus, six sets of motion programs specific to the bi-hemisphere are generated. The spatial distribution of test blocks in the six cubic grids of manipulator workspace is shown in Figure 8.13a. The coordinates of test block relative to the base coordinate system are shown in Table 8.3.

In the actual measurement process, the calibration block is placed in the areas *A*–*F* in turn by referring to the position coordinates shown in Table 8.3, and the motion programs are run in the corresponding areas, respectively. Then the manipulator with ultrasonic transducer

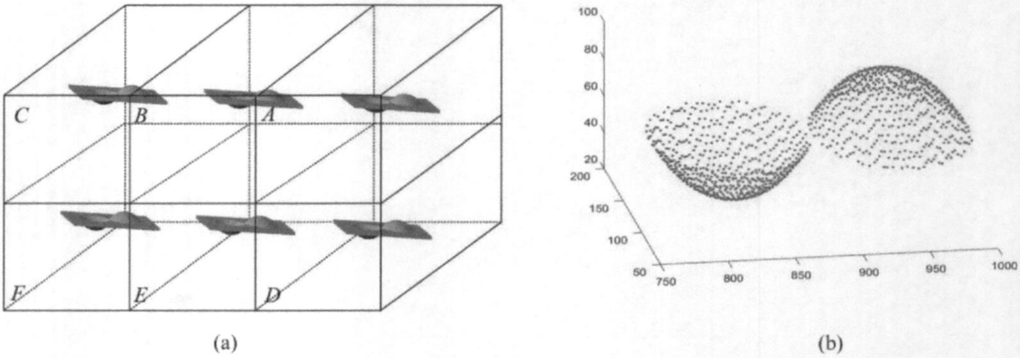

FIGURE 8.13 Distribution of calibration blocks in workspace and distance measuring process. (a) The spatial pose of the calibration block and (b) 3D point cloud.

TABLE 8.3 Position Coordinates of Calibration Block Relative to Base Coordinate Systems

Area	Position Coordinates		
A	$X = 1200$	$Y = 300$	$Z = 500$
B	$X = 1200$	$Y = 0$	$Z = 500$
C	$X = 1200$	$Y = -300$	$Z = 500$
D	$X = 1200$	$Y = 300$	$Z = 300$
E	$X = 1200$	$Y = 0$	$Z = 300$
F	$X = 1200$	$Y = -300$	$Z = 300$

moves above the two hemispheres, while collecting the manipulator position coordinates and the echo waveform data received by transducer in fixed motion intervals. In the experiment process when the calibration block is placed in the area *B*, the focus position coordinates and the echo waveform data are collected once every 2 mm shift of the sound beam focus until the end of the motion programs designed for the two hemispheres. After the acquired focus position coordinates are imported into MATLAB®, the point cloud map as shown in Figure 8.13 can be drawn to reflect the actual movement trajectory of the focus in the workspace.

The position coordinates of 1236 points (618 points for each hemisphere) and the corresponding echo waveform data were measured on two hemispheres, as shown in Figure 8.13.

TABLE 8.4 Measurement Data of Calibration Block in Area B

Focus Position Coordinates (mm)	Ultrasonic Propagation Distance (mm)	Distance Error (mm)
(1196.92, −30.34, 529.18)	36.42	−1.26
(1185.51, −31.51, 530.82)	36.57	−0.11
(1189.04, −28.73, 529.30)	36.77	1.19
(1196.53, −26.02, 525.47)	36.66	−0.92
(1200.22, −25.16, 522.40)	36.51	−1.17
(1201.09, −23.75, 520.36)	36.34	−1.34
(1199.78, −21.56, 519.17)	36.28	−1.43
(1196.18, −19.30, 518.62)	36.19	−1.59
(1190.51, −18.18, 518.30)	36.19	−0.99
(1183.70, −19.27, 517.68)	36.42	−0.86
(1178.22, −21.96, 516.18)	36.57	−1.11
(1178.23, −21.17, 515.33)	36.60	−1.08
(1184.70, −14.90, 512.86)	36.56	0.98
(1192.73, −11.22, 509.42)	36.86	0.85
(1200.24, −10.46, 505.44)	36.83	1.02
(1205.15, −11.53, 501.93)	36.69	−0.95
(1204.58, −11.49, 502.70)	36.63	−0.91
(1204.56, −9.78, 500.03)	36.57	−1.37
(1204.09, −8.59, 498.69)	36.31	−1.57
(1203.22, −7.18, 497.33)	36.11	−1.52
(1197.42, −2.93, 494.03)	36.16	−0.95
(1180.82, −5.86, 495.01)	36.13	−1.40
(1195.65, −1.11, 489.92)	36.28	−1.05
(1195.09, −1.28, 490.93)	36.54	0.92

TABLE 8.5 Kinematic Parameter Errors in Areas *A*–*C*

Area	Kinematic Parameter Error		
A	$\angle\mathbf{q}^1 = (0.20, 0.03, 2.00, 0.02)$	$\angle\mathbf{q}^2 = (-0.10, -2.06, 1.33, -0.06)$	$\angle\mathbf{q}^3 = (-0.29, 0.10, 2.94, -0.01)$
	$\angle\mathbf{q}^4 = (0.07, 0.38, -0.01, 0.01)$	$\angle\mathbf{q}^5 = (-0.15, 0.15, 0.00, 0.00)$	$\angle\mathbf{q}^6 = (1.20, -0.12, 0.21, 0.84)$
B	$\angle\mathbf{q}^1 = (0.00, 0.00, 1.12, 0.00)$	$\angle\mathbf{q}^2 = (-0.10, -1.12, 1.13, -0.06)$	$\angle\mathbf{q}^3 = (-0.19, 0.00, 1.38, 0.20)$
	$\angle\mathbf{q}^4 = (-0.06, 0.00, 0.00, 0.02)$	$\angle\mathbf{q}^5 = (-0.11, 0.45, 0.20, 0.01)$	$\angle\mathbf{q}^6 = (0.00, 0.02, 0.00, 0.00)$
C	$\angle\mathbf{q}^1 = (0.01, 0.10, 2.10, 0.01)$	$\angle\mathbf{q}^2 = (-0.11, -2.06, 1.33, -0.06)$	$\angle\mathbf{q}^3 = (-0.10, -2.06, 1.33, -0.06)$
	$\angle\mathbf{q}^4 = (0.05, 0.21, 1.02, 0.11)$	$\angle\mathbf{q}^5 = (-0.26, -2.26, 1.33, -0.06)$	$\angle\mathbf{q}^6 = (0.64, -0.32, 0.31, 1.03)$
D	$\angle\mathbf{q}^1 = (0.21, 0.45, 1.42, 0.01)$	$\angle\mathbf{q}^2 = (-0.21, -2.15, 1.25, -0.13)$	$\angle\mathbf{q}^3 = (-0.10, -2.06, 1.33, -0.06)$
	$\angle\mathbf{q}^4 = (0.16, 0.23, 1.18, 0.53)$	$\angle\mathbf{q}^5 = (-0.33, -2.53, 1.82, -0.56)$	$\angle\mathbf{q}^6 = (1.24, -1.02, 0.52, 1.13)$
E	$\angle\mathbf{q}^1 = (0.11, 0.15, 1.14, 0.02)$	$\angle\mathbf{q}^2 = (-0.11, -2.06, 1.33, -0.02)$	$\angle\mathbf{q}^3 = (-0.10, -2.06, 1.33, -0.06)$
	$\angle\mathbf{q}^4 = (0.11, 0.21, 1.02, 0.21)$	$\angle\mathbf{q}^5 = (-0.14, -1.56, 1.02, -0.01)$	$\angle\mathbf{q}^6 = (1.01, -0.32, 0.11, 0.43)$
F	$\angle\mathbf{q}^1 = (0.25, 0.51, 2.03, 0.51)$	$\angle\mathbf{q}^2 = (-0.26, -1.06, 1.21, -0.16)$	$\angle\mathbf{q}^3 = (-0.11, -2.20, 1.11, -0.18)$
	$\angle\mathbf{q}^4 = (0.15, 0.26, 1.26, 0.91)$	$\angle\mathbf{q}^5 = (-0.26, -2.26, 1.53, -0.13)$	$\angle\mathbf{q}^6 = (1.14, -1.32, 0.71, 1.05)$

Then 48 points, 24 for each hemisphere, were selected to calculate the kinematic parameter error. Based on the echo waveforms of the 48 points, the sound interval between the emitted wave and the wave reflected from the upper surface was extracted. Based on the extracted sound interval and the sound velocity in the propagating medium, the propagation distance was calculated. Table 8.4 shows the position coordinates of 24 points in a hemisphere, the ultrasonic propagation distance and the distance error deduced from the

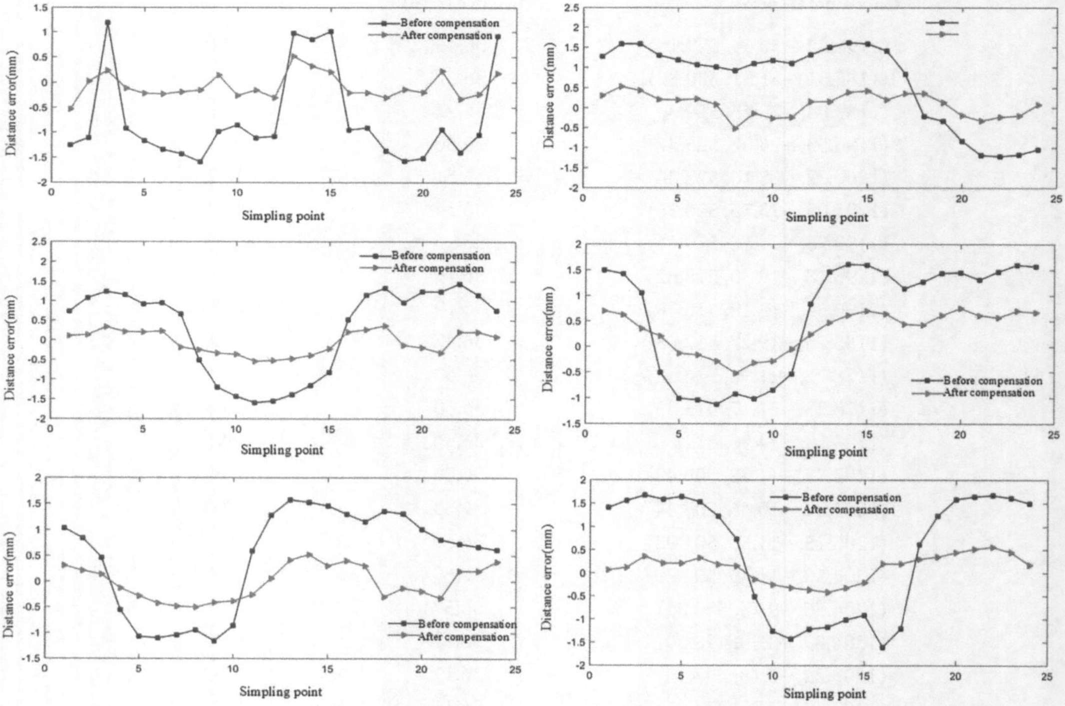

FIGURE 8.14 Comparison of distance errors of calibration blocks in areas *A*–*F* before and after compensation.

hemispherical radius when the test block is in the area *B*. According to the distance error measured on the two hemispheres, the kinematic parameter error in the area *B* can be calculated by using Eq. (8.22). Similarly, the kinematic parameter errors in the six 3D grid areas *A–F* can be obtained by repeating the above measurement process in the area *B*. The kinematic parameter errors in the six areas are listed in Table 8.5. It can be seen that the kinematic parameter errors in the areas far from the initial position are relatively larger.

During the manipulator movement, the joint rotation angles were read. By using the data in Table 8.5 and Eqs. (8.24) and (8.25), the kinematic parameter values in the manipulator controller were compensated for, so as to obtain a more accurate position of sound beam focus and reduce the localization error. After compensation, the ultrasonic propagation distance between the transmitting end of ultrasonic transducer and the hemispherical surface was measured again in each of the areas *A–F* and the distance errors at 24 points were collected. The comparison of the distance errors before and after compensation is shown in Figure 8.14. It can be seen that the distance error in each area is reduced, and that the localization accuracy of the manipulator is improved. Finally, the error measurement and compensation process for bi-hemispheric calibration block was summarized, as shown in Figure 8.15.

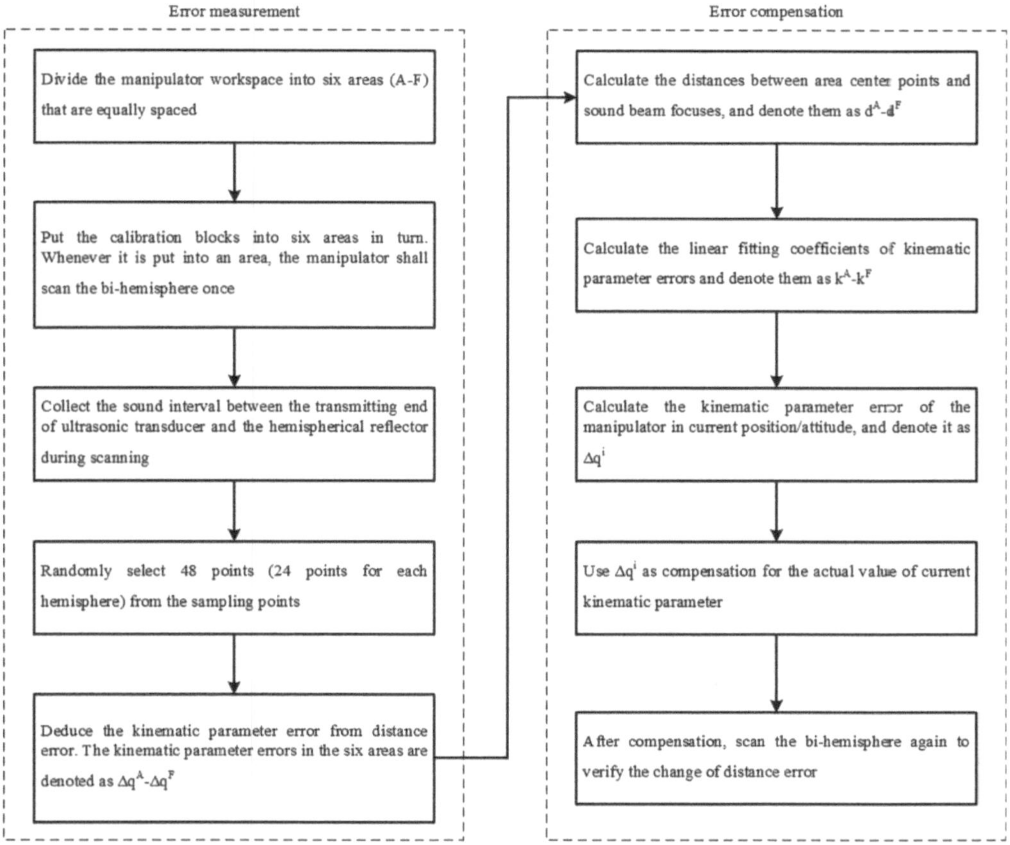

FIGURE 8.15 Measurement and compensation process for kinematic parameter errors.

REFERENCES

1. Zeng Y., Tian W., Liao W., Precision compensation technology of industrial robot oriented to aircraft-specific automatic riveting system [J], *Aeronautical Manufacturing Technology*, 2016, 18:46–52.
2. Renders J.M., Rossignol E., Becquet M., et al., Kinematic calibration and geometrical parameter identification for robots [J], *IEEE Transactions on Robotics and Automation*, 1991, 7(6):721–732.
3. Driels M.R., Swayze W.E., Automated partial pose measurement system for manipulator calibration experiments [J], *IEEE Transactions on Robotics and Automation*, 1994, 10(4):430–440.
4. Schmerr L.W. *Fundamentals of Ultrasonic Nondestructive Evaluation [M]*. New York: Plenum Press, 1998:199–238.

CHAPTER 9

Dual-Manipulator Ultrasonic Testing Method for Semi-Closed Components

THE MANIPULATOR NONDESTRUCTIVE TESTING (NDT) techniques, especially dual-manipulator NDT technique, can significantly improve the flexibility, accuracy, and precision and detection speed of the system [1]. However, it is still a challenging task to test semi-closed composite components (such as the wave-transmitting cowling at the front end of an aircraft) with the aid of manipulator.

9.1 PROBLEMS FACED BY THE ULTRASONIC AUTOMATIC TESTING OF SEMI-CLOSED CURVED COMPOSITE COMPONENTS

A semi-closed composite component has neither a small sound attenuation rate to allow the application of single-sided ultrasonic reflection detection method nor enough inner cavity space for a manipulator arm to move inside and perform a complete transmission C-scan detection process. To meet this challenge, we propose the dual-manipulator NDT strategy as shown in Figure 9.1c and d. In this strategy, the manipulator near the inner side of a workpiece holds a special-shaped extension arm tool, which has a water nozzle and an ultrasonic probe at its end. Therefore, the probe can move inside the semi-closed workpiece to perform the detection task.

According to the inverse kinematics of manipulator, if the special-shaped extension arm tool is used as manipulator end-effector without additional constraints, the system will be prone to collision in the detection task. In the simulation environment SRS2016, the typical collision situations of the dual-manipulator ultrasonic NDT system under two working conditions (with extension arm tool and without extension arm tool) were simulated, as shown in Figure 9.1.

Figure 9.1a and b shows the collision between the manipulator without special-shaped extension arm tool and the semi-closed workpiece. Figure 9.1c and d shows the collision between

DOI: 10.1201/9781003212232-9

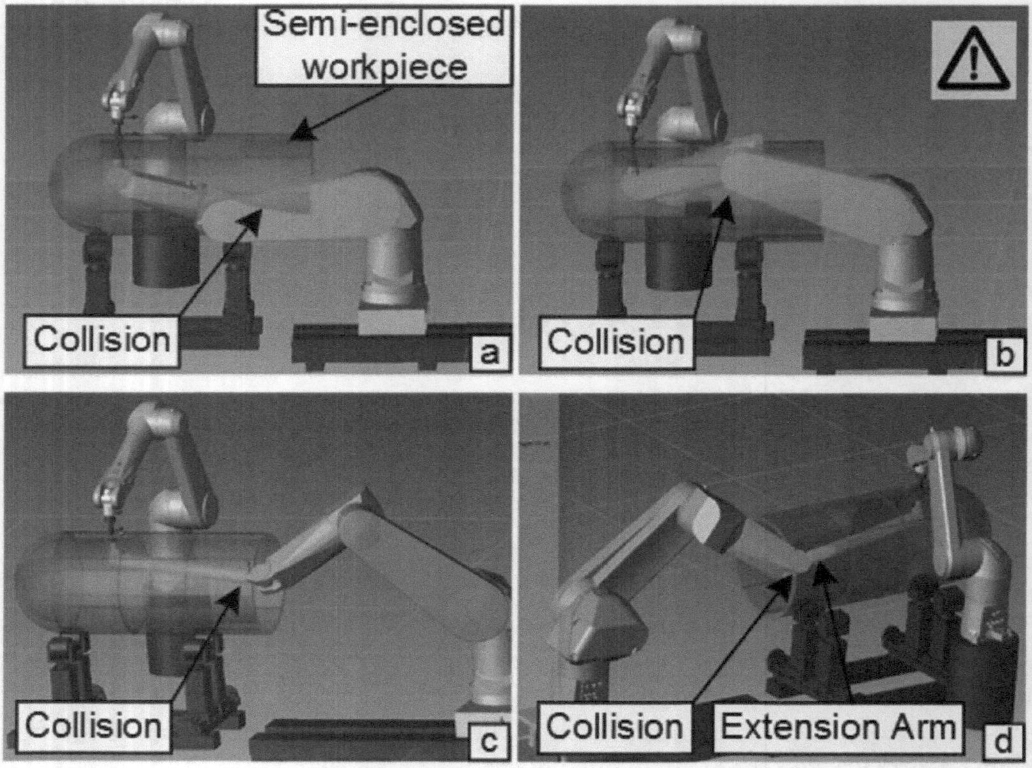

FIGURE 9.1 Typical collision situations of a dual-manipulator ultrasonic NDT system. (a and b) Collision between robot and semi-closed workpiece; and (c and d) no collision between robot and semi-closed workpiece.

the manipulator (or extension arm) with extension arm and the semi-closed workpiece. To ensure the collision-free NDT of semi-closed workpieces, a method of manipulator trajectory calculation based on coordinate system constraints is mainly presented in this chapter.

9.2 METHOD OF PLANNING THE DUAL-MANIPULATOR TRAJECTORY IN THE ULTRASONIC TESTING OF SEMI-CLOSED COMPONENTS

9.2.1 Coordinate Systems in Dual-Manipulator and Their Relations

A key point in synchronous dual-manipulator motion is to determine the position and attitude of ultrasonic probe at the same detection point and to describe them correctly in the base coordinate systems of two different manipulators.

Suppose that R_m and R_s refer to the master and slave manipulators, respectively. As shown in Figure 9.2, $\{B_M\}$ and $\{B_S\}$, respectively, represent the base coordinate systems of the master and slave manipulators. $\{W\}$ represents the workpiece coordinate system. $\{F_i\}$ represents the coordinate system at any discrete point on the scanning trajectory. $\{F_M\}$ and $\{F_S\}$, respectively, represent the flange coordinate systems of the main and

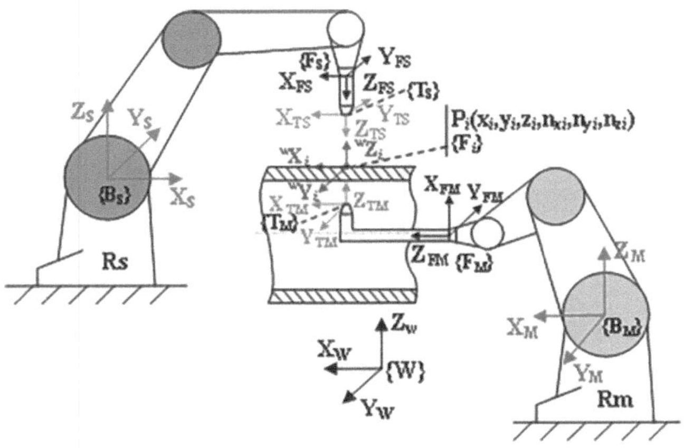

FIGURE 9.2 Coordinate systems in a dual-manipulator ultrasonic NDT system.

slave manipulators. Similarly, $\{T_M\}$ and $\{T_S\}$, respectively, represent the tool center frames (TCFs) of the master and slave manipulators. The coordinates of any point in $\{B_M\}$ are expressed as ${}^{M}P_i = \left[{}^{M}X_i, {}^{M}Y_i, {}^{M}Z_i\right]^T$. The coordinates of any point in $\{B_S\}$ are expressed as ${}^{S}P_i = \left[{}^{S}X_i, {}^{S}Y_i, {}^{S}Z_i\right]^T$.

1. Direct calibration method of workpiece coordinate system

   The calibration of the workpiece coordinate system is the basis of robotic testing. In trajectory planning, the scanning trajectory is usually relative to the workpiece coordinate system $\{W\}$. However, for the purpose of manipulator control, the discrete points on the trajectory should be described in the base coordinate systems of the manipulators. Therefore, it is necessary to map discrete trajectory points to the base coordinate systems of the two manipulators according to the position/attitude relationships between the workpiece coordinate system $\{W\}$ and the base coordinate systems $\{B_M\}$ and $\{B_S\}$. Please note that the Euler angle transformation sequence of manipulator attitude used in this system is *X–Y–Z*.

   The transformation relationships between the coordinate system $\{W\}$ and the coordinate systems $\{B_M\}$ and $\{B_S\}$ are the link connecting the discrete point cloud on trajectory and the manipulator controller and must be accurately determined. Generally, the positions of pre-selected three workpiece points in the manipulator base coordinate systems are measured by moving the two manipulators holding standard measuring tools. Then, from the coordinate values of these three points, the positions and attitudes of the workpiece coordinate system in the manipulator base coordinate systems are derived. That is, the transformation matrices between $\{W\}$ and $\{B_M\}/\{B_S\}$ are determined. The selected three points are generally the origin O_w $\{x_O, y_O, z_O\}$ of the workpiece coordinate system $\{W\}$; a point x_w $\{x_x, y_x, z_x\}$ on the +*X*-axis and a point Y_w $\{x_y, y_y, z_y\}$ on the +*Y*-axis in the *XOY* plane.

In the base coordinate systems $\{B_M\}$ and $\{B_S\}$ of the manipulators, the three points are, respectively, expressed as ${}^{M}P_0=\{{}^{M}x_0, {}^{M}y_0, {}^{M}z_0\}$, ${}^{M}P_x=\{{}^{M}x_x, {}^{M}y_x, {}^{M}z_x\}$, ${}^{M}P_y=\{{}^{M}x_y, {}^{M}y_y, {}^{M}z_y\}$ and ${}^{0}P_y=\{{}^{0}x_0, {}^{0}y_0, {}^{0}z_0\}$, ${}^{0}P_x=\{{}^{0}x_x, {}^{0}y_x, {}^{0}z_x\}$, ${}^{0}P_y=\{{}^{0}x_y, {}^{0}y_y, {}^{0}z_y\}$.

Then, the transformation relationship between the coordinate system $\{W\}$ and the coordinate system $\{B_M\}$ can be determined according to the following equation:

$$ {}^{M}_{W}T=\begin{bmatrix} {}^{W}X_W^T & {}^{W}Y_W^T & {}^{W}Z_W^T & {}^{W}P_0 \\ 0 & 0 & 0 & 1 \end{bmatrix} \tag{9.1} $$

where ${}^{M}_{W}T$ is a 4×4 square matrix.

${}^{W}X_W^T$, ${}^{W}Y_W^T$ and ${}^{W}Z_W^T$ are calculated by the following equation:

$$ \begin{cases} {}^{M}\mathbf{X}_W=\dfrac{\left({}^{M}x_x-{}^{M}x_o,\ {}^{M}y_x-{}^{M}y_o,\ {}^{M}z_x-{}^{M}z_o\right)}{\sqrt{\left({}^{M}x_x-{}^{M}x_o\right)^2+\left({}^{M}y_x-{}^{M}y_o\right)^2+\left({}^{M}z_x-{}^{M}z_o\right)^2}} \\ {}^{M}\mathbf{Y}_W'=\dfrac{\left({}^{M}x_y-{}^{M}x_o,\ {}^{M}y_y-{}^{M}y_o,\ {}^{M}z_y-{}^{M}z_o\right)}{\sqrt{\left({}^{M}x_y-{}^{M}x_o\right)^2+\left({}^{M}y_y-{}^{M}y_o\right)^2+\left({}^{M}z_y-{}^{M}z_o\right)^2}} \\ {}^{M}\boldsymbol{Z}_W={}^{M}\mathbf{X}_W\times{}^{M}\mathbf{Y}_W' \\ {}^{M}\mathbf{Y}_W={}^{M}\boldsymbol{Z}_W\times{}^{M}\mathbf{X}_W \end{cases} \tag{9.2} $$

where ${}^{W}X_W$, ${}^{W}Y_W$ and ${}^{W}Z_W$ represent the projections of three principal unit vectors of the workpiece coordinate system $\{W\}$ in the x, y and z directions of the base coordinate system $\{B_M\}$. For any vector on the $+y$-axis in the XOY plane of the workpiece coordinate system $\{W\}$, ${}^{W}Y_W$ is the representation of this vector in the base coordinate system $\{B_M\}$.

Similarly, the transformation relationship between the coordinate system $\{W\}$ and the coordinate system $\{B_S\}$ is:

$$ {}^{S}_{W}T=\begin{bmatrix} {}^{S}X_W^T & {}^{S}Y_W^T & {}^{S}Z_W^T & {}^{S}P_0 \\ 0 & 0 & 0 & 1 \end{bmatrix} \tag{9.3} $$

According to Eqs. (9.1) and (9.3), the discrete trajectory points with respect to the workpiece coordinate system can be mapped to the base coordinate systems of the two manipulators in order to accurately control the motion trajectories of manipulator end-effectors. The typical calibration method of the workpiece coordinate system is shown in Figure 9.3.

2. Indirect calibration method of workpiece coordinate system

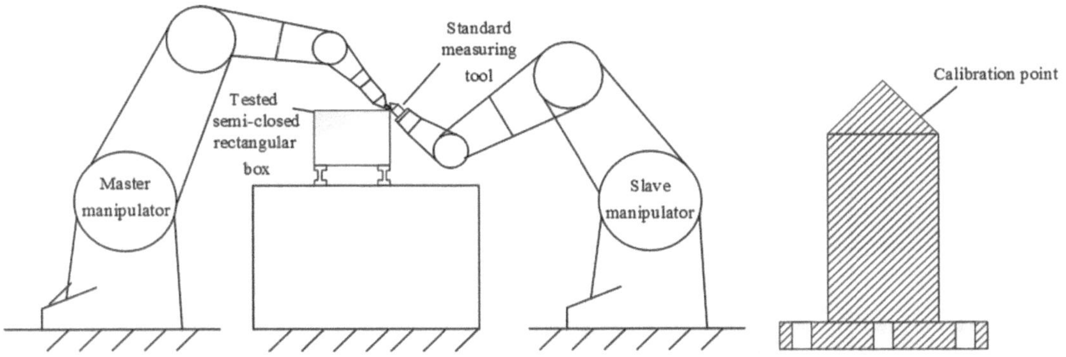

FIGURE 9.3 Typical calibration method of a workpiece coordinate system.

The direct calibration of the workpiece coordinate system requires two manipulators to arrive at the calibration points in the workpiece coordinate system at the same time. For large components, complex curved components and other special components, the two manipulators with calibration tools sometimes cannot arrive at the selected calibration points on the workpiece at the same time to define the workpiece coordinate system. Therefore, the indirect calibration method is proposed in this section. At first, the transformation relationship between the base coordinate systems of the two manipulators is determined. Then, according to this transformation relationship, the three calibration points measured by one manipulator are mapped to the base coordinate system of another manipulator. Finally, the workpiece coordinate system is determined with the method presented in the previous section.

Suppose that there are four non-coplanar points (three of which are not collinear) in the common workspace of the two manipulators. They are represented by ${}^{M}P_i = \left\{ {}^{M}x_i, {}^{M}y_i, {}^{M}z_i \right\}$ and ${}^{S}P_i = \left\{ {}^{S}x_i, {}^{S}y_i, {}^{S}z_i \right\}(i=1-4)$, respectively, in the master and slave base coordinate systems $\{B_M\}$ and $\{B_S\}$. According to the coordinate transformation theory, we can obtain

$$\begin{bmatrix} {}^{S}P_i \\ 1 \end{bmatrix} = {}^{S}_{W}P_i \cdot \begin{bmatrix} {}^{W}P_i \\ 1 \end{bmatrix} (i=1-4) \tag{9.4}$$

Eq. (9.4) is expanded into

$$\begin{bmatrix} {}^{S}X_1 & {}^{S}X_2 & {}^{S}X_3 & {}^{S}X_4 \\ {}^{S}Y_1 & {}^{S}X_2 & {}^{S}X_3 & {}^{S}X_4 \\ {}^{S}Z_1 & {}^{S}X_2 & {}^{S}X_3 & {}^{S}X_4 \\ 1 & 1 & 1 & 1 \end{bmatrix} = {}^{S}_{M}T \cdot \begin{bmatrix} {}^{M}X_1 & {}^{M}X_2 & {}^{M}X_3 & {}^{M}X_4 \\ {}^{M}Y_1 & {}^{M}X_2 & {}^{M}X_3 & {}^{M}X_4 \\ {}^{M}Z_1 & {}^{M}X_2 & {}^{M}X_3 & {}^{M}X_4 \\ 1 & 1 & 1 & 1 \end{bmatrix} \tag{9.5}$$

If the four points P_1, P_2, P_3 and P_4 are selected according to the above rules, then the vector formed by any two of them will be linearly independent, that is, $\begin{vmatrix} Pi \\ 1 \end{vmatrix} \neq 0(I=1-4)$. Then Eq. (9.5) has a unique solution, i.e.

$$
{}_M^ST = \begin{bmatrix} {}^sX_1 & {}^sX_2 & {}^sX_3 & {}^sX_4 \\ {}^sY_1 & {}^sX_2 & {}^sX_3 & {}^sX_4 \\ {}^sZ_1 & {}^sX_2 & {}^sX_3 & {}^sX_4 \\ 1 & 1 & 1 & 1 \end{bmatrix} \cdot \begin{bmatrix} {}^MX_1 & {}^MX_2 & {}^MX_3 & {}^MX_4 \\ {}^MY_1 & {}^MX_2 & {}^MX_3 & {}^MX_4 \\ {}^MZ_1 & {}^MX_2 & {}^MX_3 & {}^MX_4 \\ 1 & 1 & 1 & 1 \end{bmatrix} \tag{9.6}
$$

Eq. (9.6) is the representation of the master base coordinate system $\{B_M\}$ in the slave base coordinate system $\{B_s\}$. The above four points can be measured directly using the manipulator holding standard measuring tool, without the need for any external measuring equipment. More specifically, ${}_M^ST$ is composed of the rotation matrix ${}_M^SR$ and the translation vector $\left({}^SP_{\text{Mon}}\right)$, as shown in Eq. (9.7). $\left({}^SP_{\text{Mon}}\right)$ is the representation of the origin of the coordinate system $\{B_M\}$ in the coordinate system $\{B_s\}$.

$$
{}_M^ST = \begin{bmatrix} {}_M^SR & {}^SP_{\text{Mon}} \\ 0 & 1 \end{bmatrix} \tag{9.7}
$$

It can be known easily that ${}_M^ST^{-1}$ is just the transformation relationship between the slave base coordinate system and master base coordinate system. After determining the coordinate transformation relationship between master and slave manipulators, we can transfer the three points calibrated by one manipulator and used for calculating the workpiece coordinate system to another manipulator and then use Eq. (9.8) to determine the representation of the workpiece coordinate system in the base coordinate systems of the two manipulators (namely master and slave base coordinate systems).

$$
{}_W^ST = {}_M^ST \cdot {}_W^MT
$$

$$
{}_W^MT = {}_M^ST^{-1} \cdot {}_W^ST \tag{9.8}
$$

This simple calibration method solves the problem that two manipulators cannot complete the calibration of the workpiece coordinate system at the same time. Its second advantage is that the calibration can be done by only one manipulator after determining the transformation relationship between the two manipulators. Compared with the direct calibration method, the time consumed by this method is almost halved. Theoretically, the rotation matrix R between two Cartesian coordinate systems is an orthogonal identity matrix. However, due to the limitation of joint angle sensor accuracy, the truncation error in numerical calculation, the calibration error in calibration process and other factors, the calculated calibration rotation matrix generally cannot satisfy the orthonormalization constraints. By using the traditional rotation-matrix orthonormalization method based on Frobenius norm approximation [2,3] or the Lie algebra exponential mapping method [4], we have normalized the calibrated rotation matrix.

9.2.2 Method of Planning the *X*-Axis Constrained Trajectory in the Ultrasonic Testing of Semi-Closed Component

In an era with accelerated global integration, the manufacturing sector must improve its capacity of indigenous innovation, shorten the product life cycle and diversify the product categories. The automation system based on industrial manipulator provides the best solution for enterprises to improve the innovation strength and increase the flexibility of production facilities. However, the trajectory programming specific to the industrial manipulator systems in special applications is still very difficult, time-consuming and expensive. For example, in traditional trajectory planning, it takes more than 8 months to manually program the trajectory of a robotic arc welding system used for manufacturing large vehicle hulls but only a few hours to complete the whole welding process [5]. With the continuous development of computer graphics, the CAD/CAM technology has been widely used in industrial production. At present, the manipulator trajectories are mainly planned with two methods, namely online programming (OP) and offline programming (OLP).

The OP method has strong intuitiveness, low programming requirements and low cost. However, it is only applicable to the workpieces with simple geometry, and its program quality is limited by the operator's skills. Currently, the OLP trajectory planning method based on the 3D model of the manipulator's complete work unit is becoming more and more popular, as it shifts the programming burden from manipulator operators to engineers. Compared with the OP method, OLP is more reliable and advantageous in the trajectory programming of complex systems, but it generally relies on the CAD models of workpieces. For a workpiece without CAD model, its surface model can be easily converted from its point cloud data captured by advanced 3D scanner [6].

As the inverse kinematics of articulated six-degree of freedom (DOF) industrial manipulator has many solutions in the joint space, the problems such as accessibility, minimum moving distance and collision avoidance need to be considered in the manipulator configuration. Because most of the existing OLP software can not automatically provide optimal solutions, the manual configuration or the secondary software development based on API interface must be carried out. After the completion of trajectory planning, necessary I/O control signals should be added to the equipment in the manipulator's work unit during post-processing. If necessary, the path should be smoothed, fine-tuned and refined, and even convert into the target programming language specific to a manipulator. For commercial trajectory planning software, this problem is more highlighted, because it is difficult for commercial software to customize the functions while considering the compatibility with different manipulator brands [7]. In addition, high copyright cost is also a problem that the commercial software has to face.

This paper proposes a trajectory planning method that first obtains the data point cloud of machining process from the multi-axis machining unit of commercial CAD/CAM software and then post-processes it in CAD model into the trajectory in the format of manipulator kinematics data according to the actual manipulator configuration.

1. Extraction of point cloud in CAD model

   The method of surface trajectory planning is related to surface representation method, scanning type and scanning parameter. In modern CAD/CAM integrated environment, non-uniform rational B-splines (NURBS) function is generally used to describe the shape of curves and surfaces. Due to the introduction of a weight factor to flexibly control and adjust the shape of curves and surfaces, NURBS has been applied more and more widely in the trajectory planning on curves and surfaces. It has become the main way to write the trajectory planning software.

   The local trajectory planned for a certain type of composite torpedo shell is shown in Figure 9.4. During the trajectory planning, the CAD/CAM software such as Dassault CATIA was used to obtain the discrete point cloud of tool path on the machined component surface. For clarity, large line spacing was set.

2. Trajectory post-processing algorithm of semi-closed component

   Although high-quality zig-zag scanning trajectory can be easily and quickly generated by CAD/CAM software, this tooling trajectory is used for five-axis CNC machine tool, rather than directly for manipulator motion. Therefore, discrete point cloud must be post-processed before applying the trajectory to the ultrasonic NDT system with six-DOF manipulator, especially before testing semi-closed irregular workpieces [1]. The data format of discrete point cloud trajectory generated by CAD/CAM software is $\left[x,y,z,t_x,t_y,t_z\right]$, which is composed of the position parameters x,y,z and the unit normal vector t_x,t_y,t_z of discrete points.

   According to the principle of ultrasonic testing (UT), the ultrasonic beam should always be perpendicular to the workpiece surface during the testing. The motion control principle of industrial manipulators is to completely align the TCFs $\{TM\}$ and

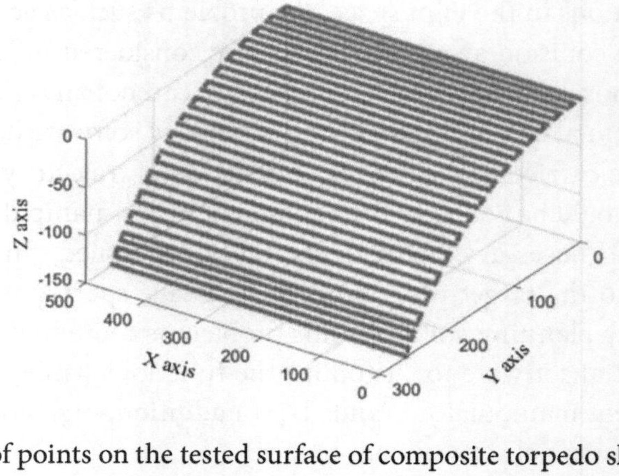

FIGURE 9.4 A cloud of points on the tested surface of composite torpedo shell.

$\{TS\}$ of master and slave manipulators with the auxiliary coordinate systems $\{Fi\}$ of discrete points, as shown in Figure 9.5.

To ensure that the manipulator moves in the correct position and direction without colliding with the workpiece, an *X*-axis constraint method is proposed to establish the auxiliary coordinate systems of discrete points *Pi*. The rules for establishing auxiliary coordinate systems are as follows:

- First, each line of discrete points is extracted from the trajectory. If the current line is odd-numbered, the *X*-axis direction of its auxiliary coordinate system should be from the current point to the next point. If the current line is even-numbered, the *X*-axis direction of its auxiliary coordinate system should be from the next point to the current point. In this way, the *X*-axes of all the auxiliary coordinate systems $\{Fi\}$ of discrete points will be parallel to the axis of the semi-closed workpiece and tangent to the trajectory;
- Second, the *Z*-axis direction of an auxiliary coordinate system is assumed to be the direction of external normal vector of curved surface at the corresponding discrete point;
- Finally, the *Y*-axis is automatically generated by the right-hand rule, that is, by the cross product of *Z* and *X*, as shown in Figure 9.5.

However, when the trajectory is not a straight line, the *X*-axis obtained with the above method will not be the tangent of the trajectory and needs to be corrected, as shown in Figure 9.6.

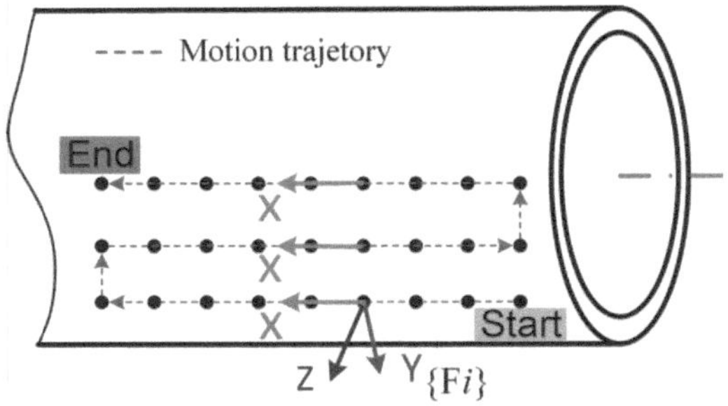

FIGURE 9.5 A method for defining the auxiliary coordinate systems $\{Fi\}$ of discrete trajectory points.

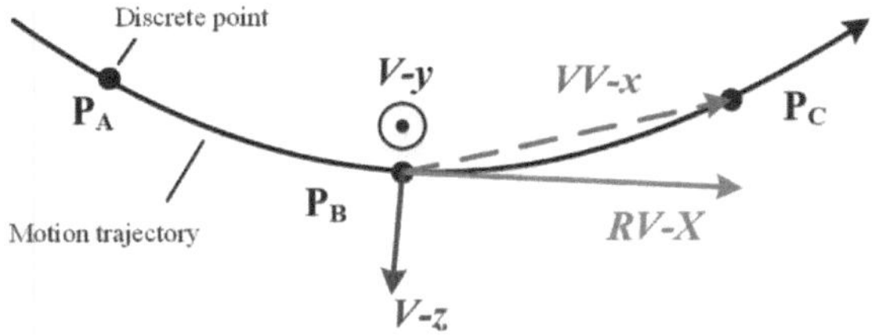

FIGURE 9.6 Calibration process of *X*-axis of an auxiliary coordinate system at a discrete point.

VV-x is the uncorrected *X*-axis of auxiliary coordinate system, and *RV-x* is the corrected *X*-axis of auxiliary coordinate system.

In addition, to constrain the TCF, the *X*-axis of the TCF $\{T_M\}$ should coincide with the axis of special-shaped extension arm tool, as shown in Figure 9.6. This method can minimize the swinging amplitude of the extension arm and avoid its collision with the workpiece when the coordinate systems $\{T_M\}$ and $\{T_S\}$ coincide completely with $\{Fi\}$ in the detection process. This method can also ensure that the ultrasonic probe on extension arm moves along a predefined trajectory and is perpendicular to the workpiece surface under test in real time.

The specific algorithm of *X*-axis constraint method is as follows:

The *X*-axis (*VV-x*) of the initial discrete-point coordinate system is called uncorrected *X*-axis, denoted as ox^t, i.e.

$$ox^t = \left(x_c - x_b, y_c - y_b, z_c - z_b\right) \tag{9.9}$$

The *Z*-axis of the discrete-point coordinate system is the tool axis vector generated during the trajectory planning, denoted as *oz*, i.e.

$$oz = \left(t_x, t_y, t_z\right) \tag{9.10}$$

The *Y*-axis of the discrete-point coordinate system is the cross product of *Z*-axis and *X*-axis, denoted as *oy*, i.e.

$$oy = oz \times ox' \tag{9.11}$$

The corrected *X*-axis is the cross product of the corrected *Y*-axis and *Z*-axis, denoted as *ox*, i.e.

$$ox = oy \times oz \tag{9.12}$$

The combination of these vectors constitutes the rotation matrix *R* of the discrete-point coordinate system $\{Fi\}$ relative to the workpiece coordinate system $\{W\}$, i.e.:

$$R = \begin{bmatrix} n_x & o_x & t_x \\ n_y & 0_y & t_y \\ n_z & 0_y & n_z \end{bmatrix} \tag{9.13}$$

In Eq. (9.13), the columns 1, 2 and 3, respectively, represent the projections of the *X*, *Y* and *Z* axes of the coordinate system $\{Fi\}$ to the workpiece coordinate system $\{W\}$.

To ensure that the manipulator moves smoothly with no risk of collision in the testing process, all the *X*-axis directions of the coordinate systems at discrete points should be constrained according to the X-axis constraint method. In this way, the minimum change of manipulator attitude can be ensured in the whole testing process.

In addition, to satisfy the control algorithm of manipulator system, the unit vector of the coordinate system $\{Fi\}$ must be converted into the Euler angle relative to the workpiece coordinate system $\{W\}$ according to the X–Y–Z Euler angle transformation rule.

The rotation matrix expressed by Euler angles is as shown in Eq. (9.14). For clarity, cosα and sinα are abbreviated as $c\alpha$ and $s\alpha$. In addition, α, β and γ are measured in degrees.

$$R_{XYZ} = R_{X(\alpha)}R_{Y(\beta)}R_{z(\gamma)}$$

$$= \begin{bmatrix} 1 & 0 & 0 \\ 0 & c\alpha & -s\alpha \\ 0 & s\alpha & c\alpha \end{bmatrix} \begin{bmatrix} c\beta & o_x & s\beta \\ 0 & 1 & 0 \\ -s\beta & 0 & c\beta \end{bmatrix} \begin{bmatrix} c\gamma & -s\gamma & 0 \\ s\gamma & c\gamma & 0 \\ 0 & 0 & 1 \end{bmatrix}$$

$$= \begin{bmatrix} c\beta c\gamma & -c\beta s\gamma & s\beta \\ s\alpha s\beta c\gamma + c\alpha s\gamma & -s\alpha s\beta c\gamma + c\alpha s\gamma & -s\alpha c\beta \\ -c\alpha s\beta c\gamma + s\alpha s\gamma & c\alpha s\beta s\gamma + s\alpha c\gamma & c\alpha c\beta \end{bmatrix} \begin{bmatrix} r_{11} & r_{12} & r_{13} \\ r_{21} & r_{22} & r_{23} \\ r_{31} & r_{32} & r_{33} \end{bmatrix} \tag{9.14}$$

Eqs. (9.14) and (9.13) have an equivalent relationship, that is, their corresponding elements are equal:

$$R_{XYZ} = \begin{bmatrix} r_{11} & r_{12} & r_{13} \\ r_{21} & r_{22} & r_{23} \\ r_{31} & r_{32} & r_{33} \end{bmatrix} = R = \begin{bmatrix} n_x & o_x & t_x \\ n_y & 0_y & t_y \\ n_z & 0_y & n_z \end{bmatrix} \tag{9.15}$$

From Eqs. (9.14) and (9.15), we can derive

$$\cos\beta = \pm\sqrt{r_{23}^2 + r_{33}^2} = \pm\sqrt{t_y^2 + t_z^2} \tag{9.16}$$

In Eq. (9.16), $\cos\beta$ is positive. In other words, the value range of β is set as (–90, 90), i.e.

$$\cos\beta = \sqrt{r_{23}^2 + r_{33}^2} = \sqrt{t_y^2 + t_z^2} \quad \alpha = A\tan 2(-k_3, r_{33})X_{180}\pi c \tag{9.17}$$

If $\cos\beta \neq 0$, then

$$\begin{aligned} \alpha &= A\tan 2(-r_{23}, r_{33}) \times 180/\pi \\ \beta &= A\tan 2\left(-r_{13}, \sqrt{r_{23}{}^2 + r_{33}{}^2}\right) \times 180/\pi \\ \gamma &= A\tan 2(-r_{21}, r_{11}) \times 180/\pi \end{aligned} \tag{9.18}$$

If $\cos\beta = 0$, the rotation matrix will be reduced to the following form:

$$R_{XYZ} = \begin{bmatrix} 0 & 0 & s\beta \\ s(\alpha+\gamma) & c(\alpha+\gamma) & 0 \\ -c(\alpha+\gamma) & s(\alpha+\gamma) & 0 \end{bmatrix} \quad (9.19)$$

Then, let $\alpha = 0$, so

$$\begin{aligned} &\alpha = 0 \\ &\beta = A\tan 2(r_{13}, 0) \times 180/\pi \\ &\gamma = A\tan 2(r_{21}, r_{22}) \times 180/\pi \end{aligned} \quad (9.20)$$

The trajectory data obtained from CAM software is post-processed by Eq. (9.20). The constraint conditions used in the algorithm enable the manipulator system to test the semi-closed workpiece smoothly and nondestructively.

9.2.3 Experimental Verification of the Trajectory Planning Method with *X*-Axis Constraint

To verify the feasibility and correctness of the proposed trajectory processing method, the rectangular semi-closed box (as shown in Figure 9.3) and the revolving composite torpedo shell were used as prototypes for trajectory planning and experimental verification. In addition, the master manipulator is equipped with a special-shaped extension arm tool where a water nozzle and an ultrasonic probe are installed.

1. Rectangular semi-closed box

   The trajectory is planned on one side of the rectangular semi-closed box (workpiece). The trajectory area is 100 × 200 mm. The thickness of the tested workpiece is 12 mm. The planned trajectory is zig-zag. To improve the readability of the trajectory image, the step interval (between two lines) is set as 10 mm, and the sampling interval (between two discrete points) is set as 0.75 mm.

   During the detection motion of dual manipulators, 1360 groups of data (including the Cartesian coordinates of end-effector trajectories and the corresponding manipulator joint angles) were collected. It can be seen from this data relative to the trajectory curve of the workpiece coordinate system {*W*} (see the Figure 9.7) that the trajectory curves of the two manipulators are parallel raster-scanning lines and are consistent with the expected scanning trajectories and attitudes without colliding with the workpiece. The joint angle curves (see the Figure 9.8) of the two manipulators are smooth and continuous, indicating that the two manipulators run smoothly without collision. The angle curves of the joints J1, J3, J4 and J5 of the master manipulator and those of the joints J1, J3 and J4 of the slave manipulator contain about eight

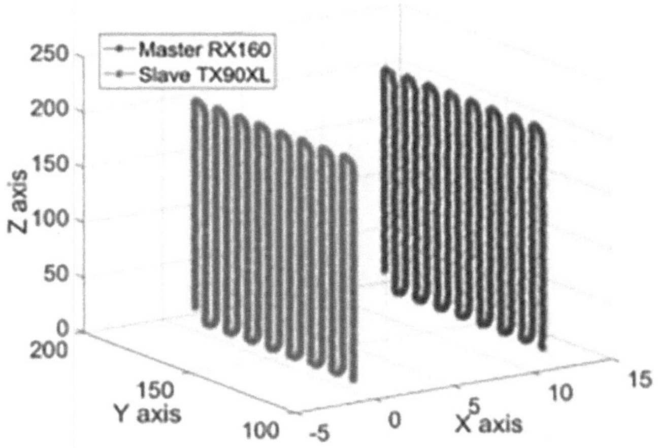

FIGURE 9.7 Motion trajectories of manipulator end-effectors.

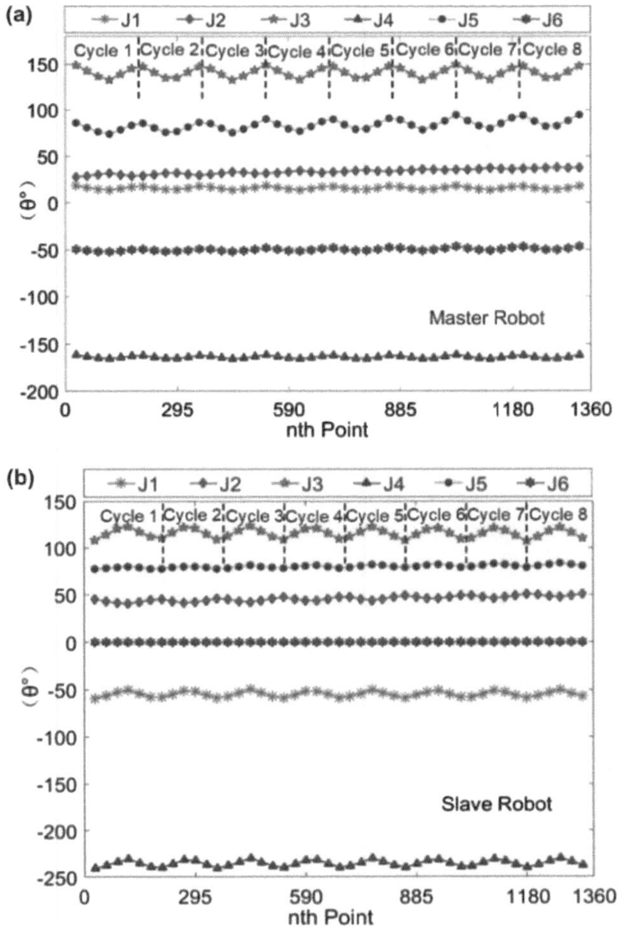

FIGURE 9.8 Joint angle curves of two manipulators. (a) The joint angle curves of the master robot and (b) the joint angle curves of the slave robot.

reciprocating motions. This is because the planned trajectories are 16-line zig-zag grating paths and the joint angle curves accurately correspond to the end-effector motion trajectories.

2. Verification of trajectory of small-diameter semi-closed rotating workpiece

The *X*-axis constraint method applies to not only the testing of the planes of a rectangular semi-closed box but also the trajectory planning for other types of semi-closed workpieces and open-type curved/flat workpieces. To further verify the correctness of this method, a semi-closed composite torpedo shell was used as the prototype. This rotary workpiece has an outer diameter of 250 mm and a thickness of 25 mm. The step interval (between two lines) is set as 15 mm, and the sampling interval is set as 2 mm.

Considering the small detection area and large sampling interval, 661 groups of Cartesian coordinate data and joint angle data were collected in this experiment. The Cartesian trajectory curves of the two end-effectors relative to the workpiece coordinate system {*W*} are shown in Figure 9.9. The joint angle curves of the two manipulators are shown in Figure 9.10. The test results are very similar to those of rectangular semi-closed box. It can be seen that the motion trajectories of the manipulator end-effectors are parallel raster-scanning lines. The motion along the trajectories was correct, and the axis of extension arm was parallel to that of the semi-closed rotating workpiece in the motion process. Except for the large rotation angle of the fourth axis of the slave manipulator caused by the workpiece position, no other influences were found on the joint angle curves.

The trajectory verification experiments of the two workpieces in the dual-manipulator NDT system demonstrate the practicability of special-shaped extension arm in

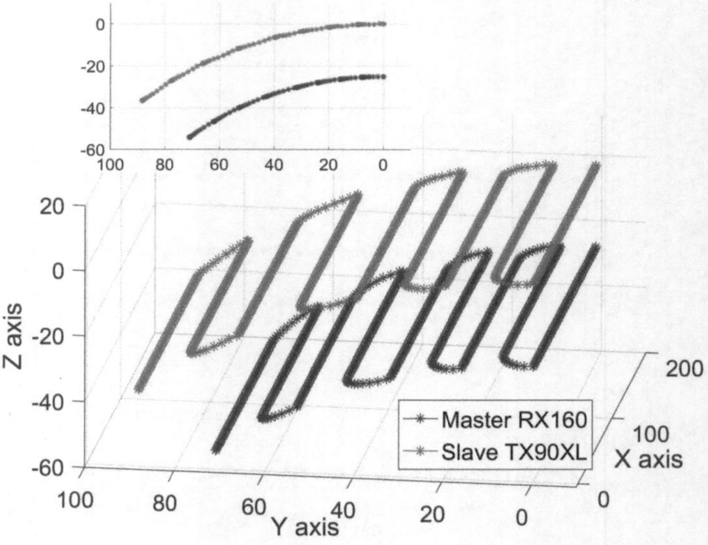

FIGURE 9.9 Motion trajectories of manipulator end-effectors.

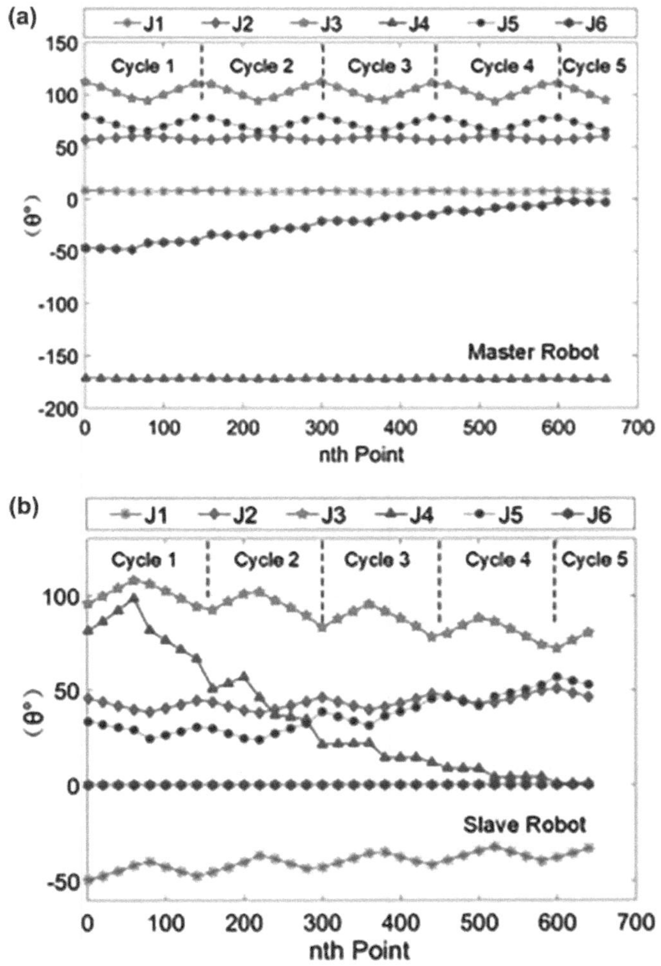

FIGURE 9.10 Joint angle curves of the two manipulators.

the NDT test of a semi-closed workpiece as well as the correctness of its combination with the *X*-axis-constrained trajectory planning method.

The above experiments show that the robotic NDT concept with special-shaped extension arm proposed for semi-closed workpieces is feasible. In this testing concept, the *X*-axis-constrained trajectory planning method is correct and reasonable.

9.3 ANALYSIS AND OPTIMIZATION OF VIBRATION CHARACTERISTICS OF SPECIAL-SHAPED EXTENSION ARM TOOL

In the last section, the dual-manipulator motion trajectory planning method suitable for the NDT of special-shaped components is introduced and experimentally verified, and the motion mode required for the UT of semi-closed components is realized. However, because the motion time between two discrete points during the synchronous motion of the two manipulators is always equal (generally a machine cycle, namely 4 ms), the manipulator

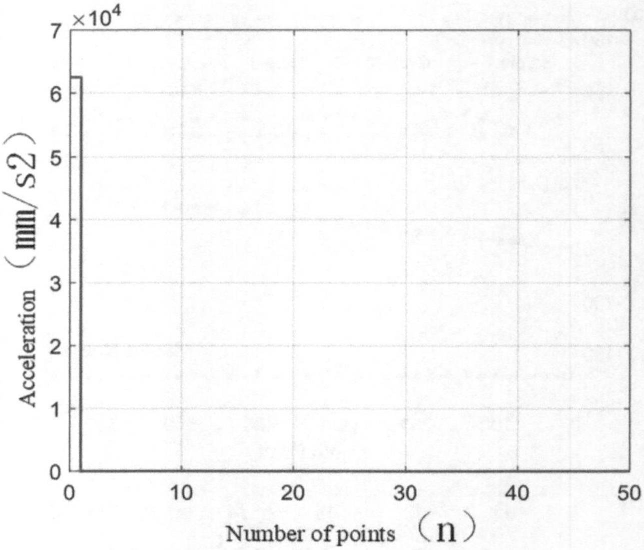

FIGURE 9.11 Acceleration curve of the manipulator at the start of motion.

holding the extension arm always starts with a large acceleration at the beginning of each line of the trajectory. Its typical initial acceleration curve is shown in Figure 9.11. The same phenomenon also occurs at the end of each line of the trajectory. Thus, the special-shaped extension arm held by the manipulator will vibrate at the beginning and end of each line of the trajectory, which will affect the reliability of ultrasonic detection results in severe case.

Proceeding from the calibration of static parameters of extension arm, this section proposes the improved S-curve acceleration (deceleration) control algorithm and linearly interpolates it as control function into the Cartesian space of the corresponding manipulator and then optimizes the manipulator trajectory planned in the last section to improve the motion smoothness of manipulator system.

9.3.1 Calibration of Static Characteristics of Special-Shaped Extension Arm Tool

The introduction of special-shaped extension arm greatly expands the detection range of manipulator system and solves the problem that special-shaped complex curved components (such as semi-closed hollow components) cannot be automatically tested. However, it increases the load on the manipulator (especially on the fifth and sixth axes, the weakest points of the manipulator). Therefore, before attaching a special-shaped extension arm tool to the end of manipulator, its torque and rotational inertia should be analyzed and calculated so that a manipulator with appropriate load capacity can be selected.

The way that the manipulator holds the extension arm is shown in Figure 9.12. One end of the extension arm is connected to the end of the manipulator's sixth axis through a flange and moves along with the manipulator in all directions in space. The other end is composed of a steering gear, an ultrasonic transducer, a coupled water-spraying device and other devices. The ultrasonic transducer can rotate around the seventh axis at the end of the extension arm. The structure, length and weight of the extension arm will have a

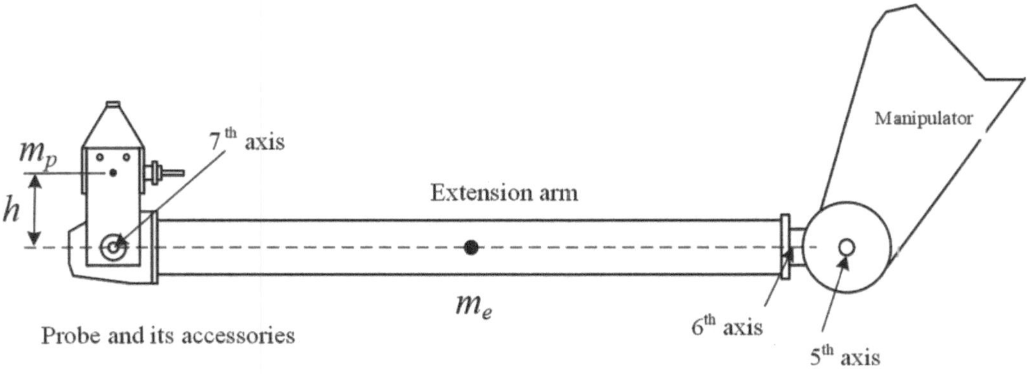

FIGURE 9.12 Structural diagram of special-shaped extension arm system.

direct impact on the motion characteristics of the manipulator. Therefore, while meeting the motion range (according to the size of the tested workpiece, the length of extension arm is designed to be $L = 1000$ mm), the design of extension arm should optimize its structure form, reduce its weight and guarantee its stable movement, so as to avoid the vibration of the manipulator and extension arm.

Calculation and analysis of equivalent rotational inertia and torque of special-shaped extension arm tool at the fifth and sixth axes of the manipulator:

If the extension arm is a long homogeneous rod, the equivalent rotational inertia of the extension arm at the fifth axis of the manipulator can be expressed as:

$$J_{51} = \frac{1}{3} m_e L^2 \tag{9.21}$$

where m_e refers to the weight of extension arm and L is the length of extension arm.

According to the parallel-axis theorem, the equivalent rotational inertia of the probe and swing mechanism held by the extension arm at the fifth axis of the manipulator can be simplified as:

$$J_{52} = I_P + m_p L^2 \tag{9.22}$$

where I_p is the rotational inertia of the ultrasonic probe and its accessories, $I_P \le 0.005\ \mathrm{kg/m^2}$; and m_p represents the weight of the ultrasonic probe and its accessories.

According to Eqs. (9.21) and (9.22), the equivalent rotational inertia of the extension arm at the fifth axis of the manipulator can be expressed as:

$$\text{Inertial}/Jt_5 = J_{51} + J_{52} = \frac{1}{3} m_e L^2 + m_p L^2 + L_r \tag{9.23}$$

The equivalent torque of the extension arm at the fifth axis of the manipulator is:

$$T_5 = m_p g L + \frac{1}{2} m_e g L \tag{9.24}$$

The equivalent rotational inertia of the extension arm at the sixth axis of the manipulator can be expressed as:

$$\text{Inertial}/Jt_6 = m_e R^2 + \frac{1}{3} m_p h^2 \tag{9.25}$$

where R is the radius of the thin-walled extension arm, $R = 25\,\text{mm}$; h is the distance between the center of gravity of the probe and the sixth axis of rotation, $h = 110\,\text{mm}$; $g = 9.8\,\text{m/s}^2$.

The equivalent torque of the extension arm at the sixth axis of the manipulator is:

$$T_6 = m_p g h \tag{9.26}$$

The carbon fiber tube with an outer diameter of 50 mm and a wall thickness of 1 mm is selected as the main structure of extension arm. The mass of extension arm is $m_e = 0.277\,\text{kg}$ and the mass of probe and swing mechanism is $m_p = 1.2\,\text{kg}$, so the equivalent mass of extension arm at the sixth axis is $m = 1.477\,\text{kg}$. By substituting these parameters into Eqs. (9.21)–(9.26), the parameters of extension arm can be calculated. The comparison between manipulator load parameters and extension arm parameters is shown in Table 9.1.

According to the parameter comparison results in the table, the extension arm can meet the load capacity requirements of the manipulator better.

Bending deformation analysis and finite element simulation for extension arm:

As one end of the extension arm is fixed on the end flange of the manipulator and the other end is free, the extension arm can be viewed as a cantilever beam structure. Under the gravity of the extension arm itself and the force F ($m_p g$) applied by the end probe and its accessories, an unreasonable design of extension arm will cause a certain displacement in the gravity direction between the probe at the end of extension arm and the flange at the end of manipulator. The excessive deformation of extension arm will have a certain influence on its location accuracy during detection. Therefore, it is required to analyze the bending deformation of extension arm, and if necessary, to redesign the extension arm or compensate for its location error.

The force on the extension arm can be decomposed, as shown in Figure 9.13a and b. w_{B1} is the maximum deflection of the extension arm under the force F, w_{B2} is the maximum deflection of the extension arm under its gravity, and θ_1 and θ_2 are the rotation angles of the arm end under the force F and its own gravity.

TABLE 9.1 Comparison between the Load Parameters of RX160 Manipulator and the Parameters of Special-Shaped Extension Arm

Rated parameters of manipulator	Maximum load	Maximum torque (fifth axis)	Maximum inertia (fifth axis)	Maximum torque (sixth axis)	Maximum inertia (sixth axis)
	34 kg	58 N×m	4 kg/m²	29 N×m	1 kg/m²
Equivalent parameters of extension arm system	Equivalent load	Equivalent torque (fifth axis)	Equivalent inertia (fifth axis)	Equivalent torque (sixth axis)	Equivalent inertia (sixth axis)
	1.477 kg	13.12 N×m	1.3 kg/m²	1.3 N×m	0.005 kg/m²

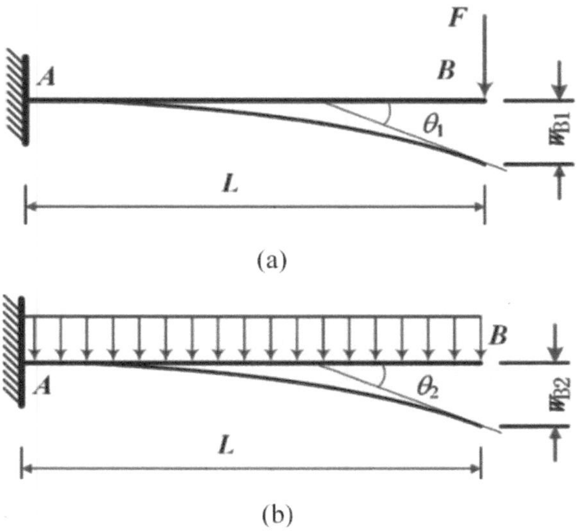

FIGURE 9.13 Deformation of extension arm under load. (a) The force F acts on the extension rod and (b) self-weight of the extension rod.

According to the relevant deflection curve equation, the maximum deflection of extension arm at its end, denoted as $w = w_{B1} + w_{B2}$, can be determined, namely:

$$w = -\frac{FL^3}{3EI} - \frac{m_e g L^3}{8EI} \tag{9.27}$$

The maximum rotation angle of its end face is $\theta = \theta_1 + \theta_2$, i.e.

$$\theta = -\frac{FL^2}{2EI} - \frac{m_e g L^2}{6EI} \tag{9.28}$$

The material of extension arm is T700-series carbon fiber. Its parameter values are shown in Table 9.2, where the elastic modulus is $E = 2.3 \times 10^{11}$ Pa. I is the inertia moment of extension arm, $I = \frac{\pi D^4}{32}\left(1 - \alpha^4\right)$ (where $\alpha = \frac{d}{D}$, $D = 0.05$ m, $d = 0.048$ m). The maximum deflection of extension arm at its end is calculated to be $w = 0.2$ mm and the rotation angle of its end face is $\theta = 0.3°$.

In the typical environment of ANSYS 15.0, the extension arm was simulated and analyzed, and its 3D model was established. A "Solid 186" hexahedral unit with 20 nodes was selected. The material properties were set according to Table 9.2. Then the hexahedral unit was divided into multiple grids in an appropriate size. To set the boundary conditions, the flange end of extension arm was fully constrained (by regarding the manipulator system as a rigid body), and a vertically downward force of 11.76 N was applied to the other end of extension arm. Finally, the model was solved while considering the gravity.

TABLE 9.2 Material Parameters of Carbon-Fiber Extension Arm

Material Properties of T700 Series	Density	Elastic Modulus	Poisson's Ratio
Value	1800 kg/m³	230 GPa	0.307

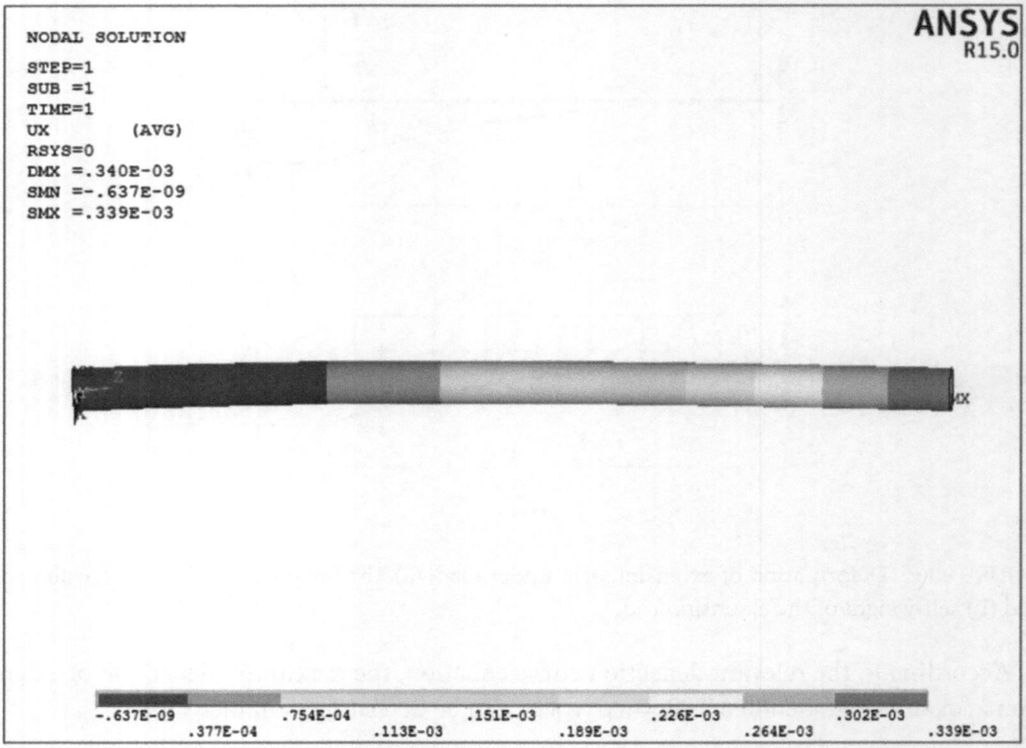

FIGURE 9.14 Displacement diagram of extension arm.

In the post-processing module, the total displacement of extension arm is as shown in Figure 9.14. The maximum displacement of extension arm at its probe end is 0.34 mm, very close to the above-mentioned calculation result. According to the calculation and simulation results, the stiffness of extension arm is enough, and the deformation and rotation angle of extension arm at its end are very small and can be ignored.

9.3.2 Improved S-Curve Acceleration Control Algorithm

As the key of CNC technology, interpolation technology can directly reflect the performance of a CNC device. Interpolation is a calculation method that determines some intermediate points between the known discrete points on a contour curve by using the selected interpolation algorithm, so as to obtain the expected velocity and acceleration. The acceleration (deceleration) control methods commonly used in CNC systems include trapezoidal acceleration (deceleration) control method [8], exponential acceleration (deceleration) control method [9], trigonometric function acceleration (deceleration) control method [10,11] and S-curve acceleration (deceleration) control method [12–16]. In the S-curve acceleration (deceleration) control method, the acceleration, velocity and displacement are the time-dependent linear, quadratic and cubic functions, respectively. This method is widely used, but its acceleration jerk changes abruptly during acceleration and deceleration. The curves of its displacement, velocity, acceleration and jerk J are shown in Figure 9.15.

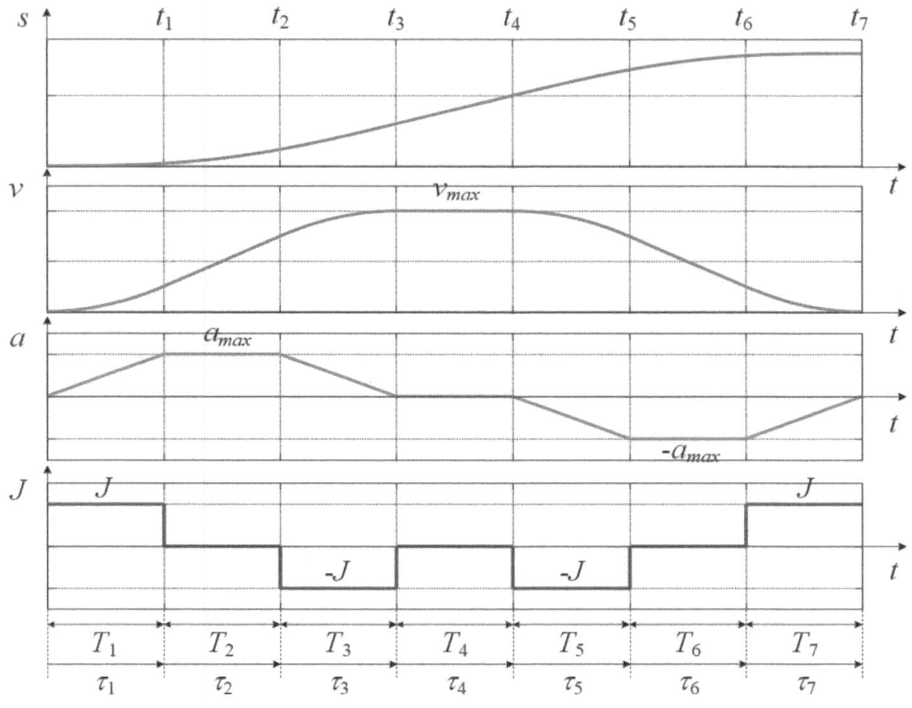

FIGURE 9.15 S-curve acceleration (deceleration) process.

In the figure, $V_{\max}$ is the maximum speed of machine tool, t_k (k=1–7) is the turning point of each stage, τ_k (k=1–7) is the time coordinate of each stage, T_k (k=1–7) is the continuous running time of each stage, $\alpha_{\max}$ is the maximum acceleration and J is acceleration jerk.

As can be seen from the above figure, a complete S-curve motion is composed of seven stages. Stage 1 is to accelerate with a linearly increasing acceleration until the acceleration reaches its maximum $\alpha_{\max}$. Stage 2 is to continue the accelerated motion with the maximum acceleration till the time point t_2. Stage 3 is to continue the accelerated motion with a linearly decreasing acceleration until the acceleration drops to 0, at which point the velocity reaches its maximum $v_{\max}$. Stage 4 is to keep moving at a constant speed till the time point t_4. The deceleration stages are symmetric with the acceleration stages 1, 2 and 3.

In this process, the jerk (J) curve shows a step change. This sudden change causes the non-smoothness of acceleration curve, the existence of an inflection point at the sudden jerk change and then the vibration of robotic testing system. Especially when the manipulator holds an extension arm, this vibration will be amplified to affect the detection reliability. Therefore, in order to improve the motion stability of the system and avoid the step change of the jerk J, we improved the S-curve acceleration (deceleration) control method. The observations show that, if the acceleration curve is a second-order or higher-order curve, the continuity of acceleration and jerk can be guaranteed to avoid the step change of the jerk J. This method is called improved S-curve acceleration (deceleration) control

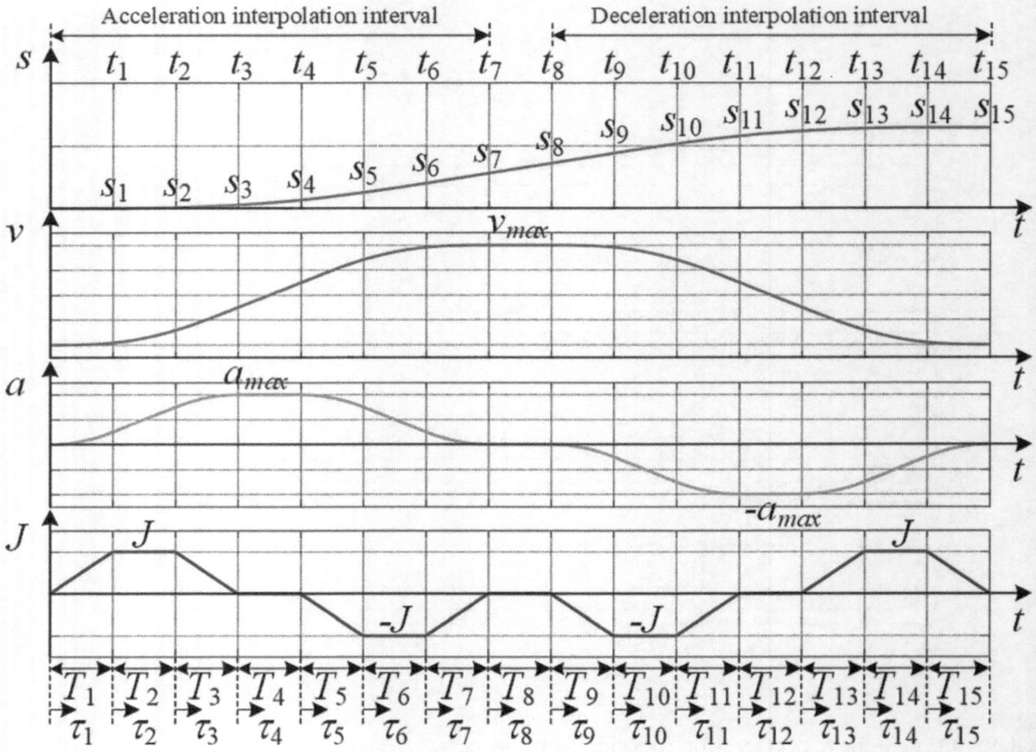

FIGURE 9.16 Improved S-curve acceleration (deceleration) process.

algorithm. The curves of its displacement, velocity, acceleration and jerk J are shown in Figure 9.16.

According to the actual situation of manipulator movement, the starting and ending velocities in the acceleration (deceleration) process are both set as zero. In this case, the movement process is completely symmetrical, as shown in the above figure. Therefore, $T_1 = T_3 = T_5 = T_7 = T_9 = T_{11} = T_{13} = T_{15}, T_4 = T_{12}, T_2 = T_6 = T_{10} = T_{14}$. For the convenience of explanation, the time intervals $T_1 \sim T_{15}$ in this paper and in Figure 9.16 are assumed to be equal. From Figure 9.16, we can observe the variable accelerated motion with increased acceleration in the period $T_0 \sim T_3$, the uniformly accelerated motion in $t_3 \sim t_4$ and the variable acceleration motion with decreased acceleration in $t_4 \sim t_7$. The velocity at t_7 is the maximum. The velocity and displacement are nonlinearly increasing before t_7. In $t_7 \sim t_8$, the acceleration is 0, the velocity is constant and the displacement increases linearly. The changes of acceleration and velocity in the second half of the curve are the inverse of those in the first half.

In the improved S-curve acceleration (deceleration) control algorithm, the equations for the jerk J, the acceleration a, the velocity v and the displacement s are Eqs. (9.29)–(9.32), respectively:

$$J(t)=\begin{cases} k\tau_1 & 0\le t<t_1 \\ kT_1 & t_1\le t<t_2 \\ kT_1-k\tau_3 & t_2\le t<t_3 \\ 0 & t_3\le t<t_4 \\ -k\tau_5 & t_4\le t<t_5 \\ -kT_1 & t_5\le t<t_6 \\ -kT_1+k\tau_7 & t_6\le t<t_7 \\ 0 & t_7\le t<t_8 \\ -k\tau_9 & t_8\le t<t_9 \\ -kT_1 & t_9\le t<t_{10} \\ -kT_1+k\tau_{11} & t_{10}\le t<t_{11} \\ 0 & t_{11}\le t<t_{12} \\ k\tau_{13} & t_{12}\le t<t_{13} \\ kT_1 & t_{13}\le t<t_{14} \\ kT_1-k\tau_{15} & t_{14}\le t<t_{15} \end{cases} \tag{9.29}$$

where k is jerk factor, $\tau_k=t-t_{k-1}$.

By integrating Eq. (9.29), the acceleration function can be easily obtained:

$$J(t)=\begin{cases} 0.5k\tau_1^2 & 0\le t<t_1 \\ 0.5kT_1^2+kT_1\tau_2 & t_1\le t<t_2 \\ 0.5kT_1^2+kT_1T_2+kT_1\tau_3-0.5k\tau_3^2 & t_2\le t<t_3 \\ kT_1(T_1+T_2) & t_3\le t<t_4 \\ kT_1(T_1+T_2)-0.5k\tau_5^2 & t_4\le t<t_5 \\ 0.5kT_1^2+kT_1T_2-kT_1\tau_6 & t_5\le t<t_6 \\ 0.5kT_1^2-kT_1\tau_7+0.5k\tau_7^2 & t_6\le t<t_7 \\ 0 & t_7\le t<t_8 \\ -0.5k\tau_9^2 & t_8\le t<t_9 \\ -0.5kT_1^2-kT_1\tau_{10} & t_9\le t<t_{10} \\ -0.5kT_1^2-kT_1T_2-kT_1\tau_{11}+0.5k\tau_{11}^2 & t_{10}\le t<t_{11} \\ -kT_1(T_1+T_2) & t_{11}\le t<t_{12} \\ -kT_1(T_1+T_2)+0.5k\tau_{13}^2 & t_{12}\le t<t_{13} \\ -kT_1(T_1+T_2)+0.5kT_1^2+kT_1\tau_{14} & t_{13}\le t<t_{14} \\ -0.5kT_1^2+kT_1\tau_{15}-0.5k\tau_{15}^2 & t_{14}\le t<t_{15} \end{cases} \tag{9.30}$$

Similarly,

$$
v(t)=\begin{cases}
\frac{1}{6}kc & 0\le t\le t_1 \\
v_1+0.5kT_1\tau_2\left(T_1+\tau_2\right) & t_1\le t\le t_2 \\
v_2+0.5kT_1^2\tau_3+kT_1\ \tau_3+0.5kT_1\tau_3^2-\frac{1}{6}k\tau_3^3) & t_2\le t\le t_3 \\
v_3+kT_1\left(T_1+T_2\ \right)\tau_4 & t_3\le t\le t_4 \\
v_4+kT_1\left(T_1+T_2\ \right)\tau_5-\frac{1}{6}k\tau_5^3 & t_4\le t\le t_5 \\
v_5+0.5kT_1^2\tau_6+kT_1\ T_2\,\tau_6-0.5kT_1\tau_6^2 & t_5\le t\le t_6 \\
v_6+0.5kT_1^2\tau_7-0.5T_1\ \tau_7^2+\frac{1}{6}kT_1\tau_6^2 & t_6\le t\le t_7 \\
v_7 & t_7\le t\le t_8 \\
v_8-\frac{1}{6}k\tau_9^3 & t_8\le t\le t_9 \\
v_9-0.5kT_1\ \tau_{10}\left(T_1\ +\tau_{10}\right) & t_9\le t\le t_{10} \\
v_{10}-0.5kT_1^2\tau_{11}-kT_1\ T_2\,\tau_{11}-0.5kT_1\tau_{11}^2+\frac{1}{6}k\tau_{11}^3 & t_{10}\le t\le t_{11} \\
v_{11}-kT_1\ \left(T_1\ +T_2\ \right)\tau_{12} & t_{11}\le t\le t_{12} \\
v_{12}-kT_1\ \left(T_1\ +T_2\ \right)\tau_{13}+\frac{1}{6}k\tau_{13}^3 & t_{12}\le t\le t_{13} \\
v_{13}-0.5kT_1^2\tau_{14}-kT_1\ T_2\,\tau_{14}+0.5kT_1\tau_{14}^2 & t_{13}\le t\le t_{14} \\
v_{14}-0.5kT_1^2\tau_{15}+0.5kT_1\ \tau_{15}^2-\frac{1}{6}k\tau_{15}^3 & t_{14}\le t\le t_{15}
\end{cases}
\tag{9.31}
$$

where

$$
v_1=\frac{1}{6}kT_1^3
$$

$$v_2 = v_1 + 0.5kT_1T_2(T_1 + T_2)$$

$$v_3 = v_2 + \frac{5}{6}kT_1^3 + kT_1^2T_2$$

$$v_4 = v_3 + kT_1T_2(T_1 + T_2)$$

$$v_5 = v_4 + \frac{5}{6}kT_1^3 + kT_1^2T_2$$

$$v_6 = v_5 + 0.5kT_1T_2(T_1 + T_2)$$

$$v_7 = v_6 + \frac{1}{6}kT_1^3$$

$$v_8 = v_7$$

$$v_9 = v_8 - \frac{1}{6}kT_1^3$$

$$v_{10} = v_9 - 0.5kT_1T_2(T_1 + T_2)$$

$$v_{11} = v_{10} - \frac{5}{6}kT_1^3 - kT_1^2T_2$$

$$v_{12} = v_{11} - kT_1T_2(T_1 + T_2)$$

$$v_{13} = v_{12} - \frac{5}{6}kT_1^3 - kT_1^2T_2$$

$$v_{14} = v_{13} - 0.5kT_1T_2(T_1 + T_2)$$

Similarly,

$$s(t)=\begin{cases} \dfrac{1}{24}k\tau_1^4 & 0 \le t < t_1 \\ s_1 + \dfrac{1}{12}kT_1\tau_2\left(2T_1^2 + 3T_1\tau_2 + 2\tau_2^2\right) & t_1 \le t < t_2 \\ s_{21} + \dfrac{1}{24}k\tau_3\left(s_{22} + s_{23}\tau_3 + 4T_1\tau_3^2 - \tau_3^3\right) & t_2 \le t < t_3 \\ s_3 + 0.5kT_1\tau_4\left(T_1 + T_2\right)\left(2T_1 + T_2 + \tau_4\right) & t_3 \le t < t_4 \\ s_{41} + \dfrac{1}{24}k\tau_5\left(s_{42} + 12T_1\tau_5\left(T_1 + T_2\right) - \tau_5^3\right) & t_4 \le t < t_5 \\ s_{51} + \dfrac{1}{12}kT_1\tau_6\left(s_{52} + 3\left(T_1 + 2T_2\right)\tau_6 - 2\tau_6^2\right) & t_5 \le t < t_6 \\ s_{61} + \dfrac{1}{24}k\tau_7\left(s_{62} + 6T_1^2\tau_7 - 4T_1\tau_7^2 + \tau_7^3\right) & t_6 \le t < t_7 \\ s_7 + kT_1\left(T_1 + T_2\right)\left(2T_1 + T_2 + T_3\right)\tau_8 & t_7 \le t < t_8 \\ s_{81} + \dfrac{1}{24}\tau_9\left(s_{82} - \tau_9^3\right) & t_8 \le t < t_9 \\ s_{91} + \dfrac{1}{12}kT_1\tau_{10}\left(s_{92} - 3T_1\tau_{10} - 2\tau_{10}^2\right) & t_9 \le t < t_{10} \\ s_{101} + \dfrac{1}{24}kT_1\tau_{10}\left(s_{102} - s_{23}\tau_{11} - 4T_1\tau_{11}^2 + \tau_{11}^3\right) & t_{10} \le t < t_{11} \\ s_{11} + 0.5kT_1\tau_{12}\left(T_1 + T_2\right)\left(2T_1 + T_2 + 2T_3 - \tau_{12}\right) & t_{11} \le t < t_{12} \\ s_{121} + \dfrac{1}{24}k\tau_{13}\left(s_{122} - 12T_1\left(T_1 + T_2\right)\tau_{13} + \tau_{13}^3\right) & t_{12} \le t < t_{13} \\ s_{131} + \dfrac{1}{12}kT_1\tau_{14}\left(s_{132} + 2\tau_{14}^2 - 3\left(T_1 + 2T_2\right)\tau_{14}\right) & t_{13} \le t < t_{14} \\ s_{14} + \dfrac{1}{24}k\tau_{15}\left(4T_1^3 - 6T_1^2\tau_{15} + 4T_1\tau_{15}^2 - \tau_{15}^3\right) & t_{14} \le t < t_{15} \end{cases} \tag{9.32}$$

where

$$s_1 = \frac{1}{24}kT_1^4$$

$$s_{21} = s_1 + \frac{1}{12} kT_1T_2\left(2T_1^2 + 3T_1T_2 + 2T_1^2\right)$$

$$s_{22} = 4T_1^3 + 12T_1T_2\left(T_1 + T_2\right)$$

$$s_{23} = 6T_1\left(T_1 + 2T_2\right)$$

$$s_3 = s_{21} + \frac{1}{24} kT_1^2\left(13T_1^2 + 24T_1T_2 + 12T_2^2\right)$$

$$s_{41} = s_3 + 0.5kT_1T_3\left(T_1 + T_2\right)\left(2T_1 + T_2 + T_3\right)$$

$$s_{42} = 12T_1\left(2T_1^2 + 3T_1T_2 + T_2^2 + 2T_3\left(T_1 + T_2\right)\right)$$

$$s_{51} = s_{41} + \frac{1}{24} kT_1^2\left(35T_1^2 + 48T_1T_2 + 24T_1T_3 + 24T_2T_3 + 12T_2^2\right)$$

$$s_{52} = 22T_1^2 + 30T_1T_2 + 6T_2^2 + 12T_3\left(T_1 + T_2\right)$$

$$s_{61} = s_{51} + \frac{1}{12} kT_1T_2\left(22T_1^2 + 33T_1T_2 + 12T_1T_3 + 12T_2T_3 + 10T_2^2\right)$$

$$s_{62} = 44T_1^3 + 72T_1^2T_2 + 24T_1T_2^2 + 24T_1\left(T_1 + T_2\right)T_3$$

$$s_7 = s_{61} + \frac{1}{24} kT_1^2\left(47T_1^2 + 72T_1T_2 + 24T_1T_3 + 24T_2T_3 + 24T_2^2\right)$$

$$s_{81} = s_7 + kT_1T_4\left(T_1 + T_2\right)\left(2T_1 + T_2 + T_3\right)$$

$$s_{82} = 48T_1^3 + 72T_1^2T_2 + 24T_1T_2^2 + 24T_1\left(T_1 + T_2\right)T_3$$

$$s_{91} = s_{81} + \frac{1}{24} kT_1^2\left(47T_1^2 + 72T_1T_2 + 24T_1T_3 + 24T_2T_3 + 24T_2^2\right)$$

$$s_{92} = 22T_1^2 + 36T_1T_2 + 12T_2^2 + 12T_3\left(T_1 + T_2\right)$$

$$s_{101} = s_{91} + \frac{1}{12} kT_1T_2\left(22T_1^2 + 33T_1T_2 + 12T_1T_3 + 12T_2T_3 + 10T_2^2\right)$$

$$s_{102} = 44T_1^3 + 60T_1^2T_2 + 12T_1T_2^2 + 24T_1(T_1 + T_2)T_3$$

$$s_{11} = s_{101} + \frac{1}{24}kT_1^2\left(35T_1^2 + 48T_1T_2 + 24T_1T_3 + 24T_2T_3 + 12T_2^2\right)$$

$$s_{121} = s_{11} + 0.5kT_1T_3(T_1 + T_2)(2T_1 + T_2 + T_3)$$

$$s_{122} = 24T_1^3 + 36T_1^2T_2 + 12T_1T_2^2$$

$$s_{131} = s_{121} + \frac{1}{24}kT_1^2\left(13T_1^2 + 24T_1T_2 + 12T_2^2\right)$$

$$s_{132} = 2T_1^2 + 6T_1T_2 + 6T_2^2$$

$$s_{14} = s_{131} + \frac{1}{12}kT_1T_2\left(2T_1^2 + 3T_1T_2 + 2T_2^2\right)$$

9.3.3 Trajectory Interpolation Based on Improved S-Curve Acceleration Control

To stabilize the movement process of dual-manipulator UT system, avoid the impact on this system during its start or stop, implement flexible deceleration and improve the detection reliability, the manipulator trajectory interpolation in Cartesian space is implemented in this section based on the improved S-curve acceleration control.

The trajectory interpolation methods in Cartesian space mainly include linear interpolation [17], circular interpolation [18] and NURBS curve interpolation [19]. Robotic UT is different from ordinary CNC machining. The latter's machining accuracy should meet the design tolerance of a workpiece, while the former's detection accuracy is lower than the latter's machining accuracy. Especially, the detection accuracy for defects in composite workpieces is generally at the millimeter level. Therefore, the discrete points on the planned trajectory are sufficient for profiling detection during robotic UT, while the trajectory interpolation focuses more on the control of acceleration and deceleration. For the zig-zag scanning trajectory planned in robotic UT, the trajectory point curve in each line always degenerates from a space curve into a plane curve or a plane line, so the interpolations at a distance from the starting and ending points in each line of the trajectory will be improved in this section based on the basic linear interpolation algorithm. Since the interpolation rules for the starting and ending points in each line are symmetric, the linear interpolation based on improved S-curve acceleration control during starting will be highlighted.

1. Interpolation algorithm for linear trajectory

   Linear trajectory interpolation is to insert some intermediate points (interpolation points) into a linear trajectory by a certain rule and determine the position and attitude of those interpolation points when the position and attitude of the starting and ending trajectory points P_1 and P_2 are known. The position and attitude of each interpolation point can be expressed as:

$$\begin{cases} x_n = X_1 + \kappa_n (X_2 - X_1) \\ y_n = Y_1 + \kappa_n (Y_2 - Y_1) \\ z_n = Z_1 + \kappa_n (Z_2 - Z_1) \\ \alpha_n = A_1 + \kappa_n (A_2 - A_1) \\ \beta_n = B_1 + \kappa_n (B_2 - B_1) \\ \gamma_n = \Gamma_1 + \kappa_n (\Gamma_2 - \Gamma_1) \end{cases} \tag{9.33}$$

where $x_n, y_n, y_n, \alpha_n, \beta_n, \gamma_n$ represent the position and attitude (in Euler angle) of the n-th interpolation point; $X_1, Y_1, Z_1, A_1, B_1, \Gamma_1, X_2, Y_2, Z_2, A_2, B_2, \Gamma_2$ represent the positions and attitudes (in Euler angles) of two adjacent points P_1, P_2 on the known trajectory; and κ_n is the ratio between the distance from the n-th interpolation point to the first interpolation point and the distance between the first and last interpolation points.

When using the improved S-curve acceleration (deceleration) control method for interpolation on the built dual-manipulator motion trajectory, the first thing to make clear is that the cycle time of the manipulator controller system is 4 ms, that is, the distance covered by the manipulators per second in the synchronous motion is composed of 250 points. Second, interpolation is mainly used in the displacement traveled before the improved S-curve arrives at the point of uniform motion. The main steps of this interpolation method are as follows:

2. Determine the length of each period. Because of $T_1 = T_3 = T_5 = T_7 = T_9 = T_{11} = T_{13} = T_{15}$, $T_2 = T_6 = T_{10} = T_{14}$, $T_4 = T_{12}$, we only need to determine, T_1, T_2, T_4;
3. Determine the manipulator speed (i.e. the $v_{\max}$ in Figure 9.16) in normal testing according to the trajectory;
4. Determine the jerk factor k according to $v_{\max}$;
5. Calculate the maximum acceleration from Eq. (9.30), which should be less than the maximum acceleration of manipulator end under the load;
6. Determine the final displacement S_7 at t_7 according to the length of each period determined in the Step 1 and the jerk factor k;
7. Determine the number of points to be interpolated in each period and the total number of points to be interpolated in the seven periods before t_7 according to the length of each period determined in Step 1 and the cycle time of manipulator controller;
8. Distribute the interpolation points for each period into the current period, and then calculate the distance ΔL between the expected interpolation position and the first interpolation point according to the relevant equation;
9. Deduce the position and attitude of the expected interpolation point from the positions and attitudes of trajectory points P_1, P_2 on both ends of the distance ΔL from the first interpolation point according to the relevant equation.

After applying interpolation to the original trajectory with the improved S-curve acceleration (deceleration) control method, the initial acceleration of the manipulator will become the acceleration curve shown in Figure 9.16, rather than the abrupt initial acceleration curve before interpolation shown in Figure 9.15.

REFERENCES

1. Mineo C., Morozov M., Pierce G., et al., Computer-aided tool path generation for robotic nondestructive inspection [J]. Conference of the British Institute for Non-Destructive Testing, 2013.
2. Gan Y., Dai X., Base frame calibration for coordinated industrial robots [J], *Robotics and Autonomous Systems*, 2011, 59(7–8):563–570.
3. Gan Y. *Research on the Multi-robot Cooperative Control in Flexible Welding System [D].* Nanjing: Southeast University, 2014.
4. Lu Z., Xu C., Pan Q., et al., Automatic method for synchronizing workpiece frames in twin-robot nondestructive testing system [J], *Chinese Journal of Mechanical Engineering*, 2015, 28(4):860–868.
5. Online R., Pan Z., Polden J., et al., Recent progress on programming methods for industrial robots [J], *Robotics and Computer Integrated Manufacturing*, 2012, 28(2):87–94.
6. Bi Z.M., Lang S.Y., A framework for CAD- and sensor-based robotic coating automation [J], *IEEE Transactions on Industrial Informatics*, 2007, 3(1):84–91.
7. Bruccoleri M., D'ONOFRIO C., La Commare U., Off-line programming and simulation for automatic robot control software generation[C]. 5th International Conference on Industrial Informatics.
8. Yu D. *Research on the High-speed High-precision Machining Technology of Complex Curves and Surfaces based on NURBS [D].* Hefei University of Technology, 2014.
9. Yu J. *Research on the Acceleration and Deceleration Control Method in High-speed High-precision Machining [D].* Graduate School of Chinese Academy of Sciences, 2009.
10. Lee A.C., Lin M.T., Pan Y.R., et al., The feedrate scheduling of NURBS interpolator for CNC machine tools [J], *CAD Computer Aided Design*, 2011, 43(6):612–628.
11. Guo X., Li C., A new flexible acceleration and deceleration algorithm [J], *Journal of Shanghai Jiaotong University*, 2003(02):205–207+212.
12. Erkorkmaz K., Altintas Y., High speed CNC system design. Part I: jerk limited trajectory generation and quintic spline interpolation [J], *International Journal of Machine Tools and Manufacture*, 2001, 41(9):1323–1345.
13. Liu P., Yang M., Ke S., et al., Research on the application of acceleration and deceleration S-curve in robotic trajectory interpolation algorithm [J], *Manufacturing Automation*, 2012, 34(10):4–11.
14. Li Z., Liu Z., Cai L., S-type acceleration and deceleration planning algorithm with continuous jerk [J], *Computer Integrated Manufacturing Systems*, 2019, 25(5):1192–1200.
15. Yue Q., Lin M., Research on the trajectory planning based on polynomial acceleration and deceleration control [J], *Computer Engineering and Science*, 2019, 41(12):2255–2260.
16. Yue Q. *Research on the Trajectory Planning of Six-axis Industrial Robots [D].* Jiangsu University of Science and Technology, 2019.
17. Zhang J. *Research on the Trajectory Planning and Simulation Experiment of Industrial Robots_ by Zhang Jian [D].* Zhejiang University of Technology, 2014.
18. Hu Y. *Research on the Robotic Trajectory Planning Method Suitable for Ultrasonic Nondestructive Testing [D].* Beijing Institute of Technology, 2019.
19. Ling G. *Research on the Position and Attitude Planning Algorithm of Six-joint Spraying Robot [D].* Guangdong University of Technology, 2018.

CHAPTER 10

Calibration Method of Tool Center Frame on Manipulator

As described in the previous chapter, the introduction of a special-shaped extension arm tool has greatly expanded the scope of detectable workpieces and provided an alternative scheme for the automatic detection of special-shaped complex curved components, such as semi-closed hollow components. However, before the dual-manipulator testing system with a special-shaped extension arm tool is officially put into use, the parameters of tool center frame (TCF) of the extension arm must be accurately determined. Accurate TCF parameters are crucial to the high-precision offline programming of manipulator trajectory [1].

TCF calibration is to determine the origin and attitude of TCF relative to the flange coordinate system at the end of the manipulator arm. Accurate TCF parameters are necessary for offline trajectory programming to accurately control the manipulator motion. It is only based on accurate TCF parameters that the unique mapping relationship between the Cartesian points in space and the joint angles in the manipulator joint space can be established and the detection task can be successfully completed [2]. TCF parameter calibration is a means to compensate for the machining, assembling and installation errors of special-shaped tool, and is also an important procedure to resume production after the manipulator collision in industrial production [3–9]. Considering the special requirements for TCF attitude in the dual-manipulator UT of semi-closed components, a method of TCF parameter self-calibration is proposed in this chapter. This calibration method is simple and reliable without the need for any external measuring tool. Moreover, the calibration result of the proposed method is verified through the dual-manipulator motion testing.

10.1 REPRESENTATION METHOD OF TOOL PARAMETERS

In robotics, the tool parameters are generally represented in the transformation relation between the TCF $\{T\}$ and the manipulator flange coordinate system $\{F\}$. This transformation

DOI: 10.1201/9781003212232-10

relation includes the position translation {P} and the attitude rotation {R} and can be expressed as the following transformation matrix:

$$ {}_T^F\boldsymbol{T} = \left[\frac{\boldsymbol{R} \quad \boldsymbol{P}}{0\,0\,0 \quad 1}\right] = \begin{bmatrix} a_x & o_x & n_x & p_x \\ a_y & o_y & n_y & p_y \\ a_z & o_z & n_z & p_z \\ 0 & 0 & 0 & 1 \end{bmatrix} \tag{10.1} $$

where $\boldsymbol{a},\boldsymbol{o},\boldsymbol{n}$ are the three-dimensional unit column vectors that are orthogonal to each other. The combination of the three vectors represents the rotation of {T} relative to {F}. P is a three-dimensional column vector that represents the position of {T} relative to {F}. However, for Staubli manipulator, the rotation relationship is represented by the Euler angle, namely the position value x,y,z and the Euler angle value rx,ry,rz:

$$ {}_T^F T = \{x,\ y,z,rx,ry,rz\} \tag{10.2} $$

Eq. (10.2) is rewritten into a matrix:

$$ {}_T^F T = \begin{bmatrix} 1 & 0 & 0 & F_x \\ 0 & 1 & 0 & F_y \\ 0 & 0 & 1 & F_z \\ 0 & 0 & 0 & 1 \end{bmatrix} \begin{bmatrix} 1 & 0 & 0 & 0 \\ 0 & c\alpha & -s\alpha & 0 \\ 0 & s\alpha & c\alpha & 0 \\ 0 & 0 & 0 & 1 \end{bmatrix} $$

$$ \times \begin{bmatrix} c\beta & 0 & s\beta & 0 \\ 0 & 1 & 0 & 0 \\ -s\beta & 0 & c\beta & 0 \\ 0 & 0 & 0 & 1 \end{bmatrix} \begin{bmatrix} c\gamma & -s\gamma & 0 & 0 \\ s\gamma & c\gamma & 0 & 0 \\ 0 & 0 & 1 & 0 \\ 0 & 0 & 0 & 1 \end{bmatrix} $$

$$ \begin{bmatrix} c\beta c\gamma & -c\beta s\gamma & s\beta & F_x \\ s\alpha s\beta c\gamma + c\alpha s\gamma & -s\alpha s\beta s\gamma + c\alpha c\gamma & -s\alpha c\beta & F_y \\ -c\alpha s\beta c\gamma + s\alpha s\gamma & c\alpha s\beta s\gamma + s\alpha c\gamma & c\alpha c\beta & F_z \\ 0 & 0 & 0 & 1 \end{bmatrix} \tag{10.3} $$

TCF calibration is generally to determine the transformation matrix of Eq. (10.2) with the TCF calibration method and then turn the transformation matrix ${}_T^F T$ of the calibrated {T} relative to {F} into the Euler angle expression rx,ry,rz recognizable for manipulator according to the transformation relation between the Euler angle and rotation matrix by referring to the relevant equation in Chapter 9.

10.2 FOUR-ATTITUDE CALIBRATION METHOD IN TCF

According to the analysis in the previous section, the TCF calibration can be divided into two steps. The first step is to calibrate the position of tool-end center point, and the second step is to calibrate the attitude of tool-end center point.

10.2.1 Calibration of the Position of Tool-End Center Point

In the field of signal analysis, the received signal strength indicator (RSSI) is an indicator of received signal strength. It is a localization technique that determines the distance between the transmitting point and the receiving point according to the strength of the received signal, and then performs the location calculation based on the corresponding data. According to the RSSI principle, a four-attitude calibration method for calibrating the TCF of the special-shaped extension arm is proposed. This method does not need any external measuring device. In four different attitudes, the center point of the end of the special-shaped tool held by the manipulator was moved to the auxiliary tool center point ${}^{B}\mathbf{P}_{\text{ATCP}}$ (ATCP), as shown in Figure 10.1.

The coordinate system $\{B\}$ in Figure 10.1 is the base coordinate system of the manipulator. The coordinate system $\{F\}$ is the flange coordinate system of the manipulator. The coordinate system $\{A\}$ is the coordinate system of the auxiliary calibration tool itself. If the position ${}^{B}\mathbf{P}_{\text{ATCP}}$ of the point P_{ATCP} relative to $\{B\}$ is known, the position value ${}^{F}\mathbf{P}_{\text{ATCP}}$ of end center point of special-shaped tool relative to $\{F\}$ can be determined by the following equation:

$$\begin{bmatrix} {}^{F}\mathbf{P}_{\text{ATCP}} \\ 1 \end{bmatrix} = {}^{B}_{F}T^{-1} \times \begin{bmatrix} {}^{B}\mathbf{P}_{\text{ATCP}} \\ 1 \end{bmatrix} \tag{10.4}$$

For a commercial manipulator, ${}^{B}_{F}T$ is usually a known matrix. ${}^{F}\mathbf{P}_{\text{ATCP}}$ is a three-dimensional column vector. During calibration, the auxiliary calibration tool is usually placed at will in the motion space of the manipulator (that is, ${}^{B}\mathbf{P}_{\text{ATCP}}$ is unknown). Without introducing an additional measuring tool or device, the parameter of the position vector ${}^{F}\mathbf{P}_{\text{ATCP}}$, denoted

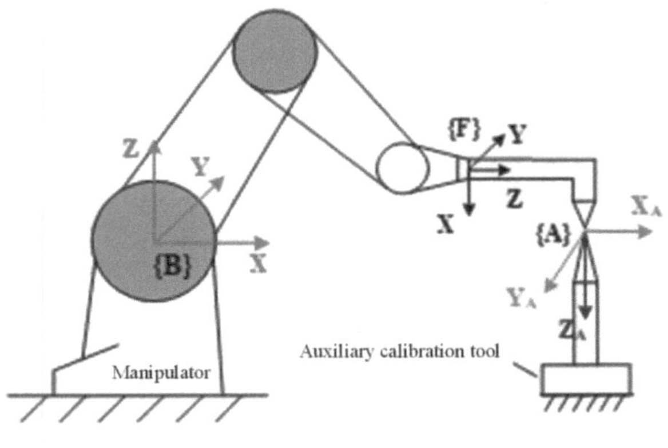

FIGURE 10.1 Calibration principle of the end center point of special-shaped tool.

as $\left({}^F x, {}^F y, {}^F z\right)^T$, can be calculated by using the four-attitude calibration method proposed in this paper to record the value $\left({}^B_F T_i,\ i=1,2,3,4\right)$ of $\{F\}$ relative to $\{B\}$ when the end center point of special-shaped tool is aligned with the center point of auxiliary calibration tool in four attitudes. In the first calibration experiment, a customized irregular tip with unknown parameters was taken as the object to be calibrated. Its four-attitude calibration process is shown in Figure 10.2. In the default setting, the manipulator is installed on the ground, and the auxiliary calibration tool is vertical to the ground.

According to Eq. (10.1), we can obtain

$$\begin{bmatrix} {}^F x \\ {}^F y \\ {}^F z \\ 1 \end{bmatrix} = \begin{bmatrix} a_{xi} & o_{xi} & n_{xi} & {}^B x_i \\ a_{yi} & o_{yi} & n_{yi} & {}^B y_i \\ a_{zi} & 0_{zi} & n_{zi} & {}^B x_i \\ 0 & 0 & 0 & 1 \end{bmatrix}^{-1} \times \begin{bmatrix} {}^B x \\ {}^B y \\ {}^B z \\ 1 \end{bmatrix} (i=1,2,3,4) \tag{10.5}$$

where $\left({}^B x_i, {}^B y_i, {}^B z_i\right)^T$ is the position vector of the coordinate system $\{F\}$ relative to the coordinate system $\{B\}$ in four attitudes; $\left({}^B x, {}^B y, {}^B z\right)^T$ is the position vector of the end

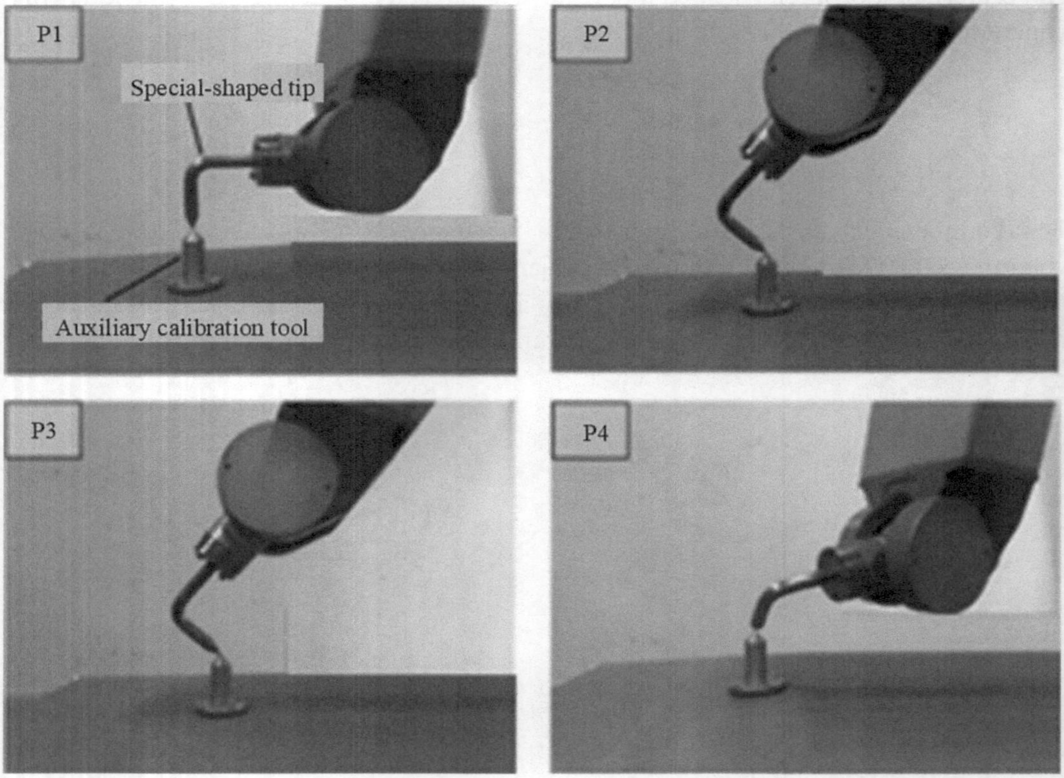

FIGURE 10.2 Four-attitude calibration process of special-shaped tip.

center point of special-shaped tool relative to $\{B\}$ and $\left({}^{F}x, {}^{F}y, {}^{F}z\right)^{T}$ is the position vector of the end center point of special-shaped tool relative to $\{F\}$.

According to the RSSI principle, we can obtain

$$\begin{cases} \left({}^{B}x-{}^{B}x_1\right)^2+\left({}^{B}y-{}^{B}y_1\right)^2+\left({}^{B}z-{}^{B}z_1\right)^2=R^2 \\ \left({}^{B}x-{}^{B}x_2\right)^2+\left({}^{B}y-{}^{B}y_2\right)^2+\left({}^{B}z-{}^{B}z_2\right)^2=R^2 \\ \left({}^{B}x-{}^{B}x_3\right)^2+\left({}^{B}y-{}^{B}y_3\right)^2+\left({}^{B}z-{}^{B}z_3\right)^2=R^2 \\ \left({}^{B}x-{}^{B}x_4\right)^2+\left({}^{B}y-{}^{B}y_4\right)^2+\left({}^{B}z-{}^{B}z_4\right)^2=R^2 \end{cases} \tag{10.6}$$

The mutual subtraction of the equations in Eq. (10.6) results in

$$\begin{cases} 2\left({}^{B}x_2-{}^{B}x_1\right)^{B}x+2\left({}^{B}y_2-{}^{B}y_1\right)^{B}y+2\left({}^{B}z_2-{}^{B}z_1\right)^{B}z=_1 \\ 2\left({}^{B}x_3-{}^{B}x_1\right)^{B}x+2\left({}^{B}y_3-{}^{B}y_1\right)^{B}y+2\left({}^{B}z_3-{}^{B}z_1\right)^{B}z=_2 \\ 2\left({}^{B}x_4-{}^{B}x_1\right)^{B}x+2\left({}^{B}y_4-{}^{B}y_1\right)^{B}y+2\left({}^{B}z_4-{}^{B}z_1\right)^{B}z=_3 \end{cases} \tag{10.7}$$

where $\Delta_1 = {}^{B}x_2^2 - {}^{B}x_1^2 + {}^{B}y_2^2 - {}^{B}y_1^2 + {}^{B}z_2^2 - {}^{B}z_1^2$;

$$\Delta_2 = {}^{B}x_3^2 - {}^{B}x_1^2 + {}^{B}y_3^2 - {}^{B}y_1^2 + {}^{B}z_3^2 - {}^{B}z_1^2;$$

$$\Delta_3 = {}^{B}x_4^2 - {}^{B}x_1^2 + {}^{B}y_4^2 - {}^{B}y_1^2 + {}^{B}z_4^2 - {}^{B}z_1^2.$$

Eq. (10.7) is rewritten into a matrix:

$$\begin{bmatrix} 2\left({}^{B}x_2-{}^{B}x_1\right) & 2\left({}^{B}y_2-{}^{B}y_1\right) & 2\left({}^{B}z_2-{}^{B}z_1\right) \\ 2\left({}^{B}x_3-{}^{B}x_1\right) & 2\left({}^{B}y_3-{}^{B}y_1\right)y & 2\left({}^{B}z_3-{}^{B}z_1\right) \\ 2\left({}^{B}x_4-{}^{B}x_1\right) & 2\left({}^{B}y_4-{}^{B}y_1\right) & 2\left({}^{B}z_4-{}^{B}z_1\right) \end{bmatrix}$$

$$\times\begin{bmatrix} {}^{B}x \\ {}^{B}y \\ {}^{B}z \end{bmatrix}=[\boldsymbol{C}]\times\begin{bmatrix} {}^{B}x \\ {}^{B}y \\ {}^{B}z \end{bmatrix}=\begin{bmatrix} \Delta_1 \\ \Delta_2 \\ \Delta_3 \end{bmatrix} \tag{10.8}$$

where $[C]=\left[\begin{bmatrix} 2\left({}^{B}x_2-{}^{B}x_1\right) & 2\left({}^{B}y_2-{}^{B}y_1\right) & 2\left({}^{B}z_2-{}^{B}z_1\right) \\ 2\left({}^{B}x_3-{}^{B}x_1\right) & 2\left({}^{B}y_3-{}^{B}y_1\right)y & 2\left({}^{B}z_3-{}^{B}z_1\right) \\ 2\left({}^{B}x_4-{}^{B}x_1\right) & 2\left({}^{B}y_4-{}^{B}y_1\right) & 2\left({}^{B}z_4-{}^{B}z_1\right) \end{bmatrix}\right]$

Then

$$\begin{bmatrix} {}^{B}x \\ {}^{B}y \\ {}^{B}z \end{bmatrix} = [\boldsymbol{C}]^{-1} \times \begin{bmatrix} 1 \\ 2 \\ 3 \end{bmatrix} \tag{10.9}$$

Eq. (10.5) is combined with Eq. (10.9) into

$$\begin{bmatrix} {}^{F}x \\ {}^{F}y \\ {}^{F}z \\ 1 \end{bmatrix} = \begin{bmatrix} a_{xi} & o_{xi} & n_{xi} & {}^{B}x_i \\ a_{yi} & o_{yi} & n_{yi} & {}^{B}y_i \\ a_{zi} & 0_{zi} & n_{zi} & {}^{B}x_i \\ 0 & 0 & 0 & 1 \end{bmatrix}^{-1} \times \left(\begin{bmatrix} & \boldsymbol{C}^{-1} & & \begin{matrix} 0 \\ 0 \\ 0 \end{matrix} \\ 0 & 0 & 0 & 1 \end{bmatrix} \begin{bmatrix} \Delta_1 \\ \Delta_2 \\ \Delta_3 \\ 1 \end{bmatrix} \right) \tag{10.10}$$

where ${}^{F}x$, ${}^{F}y$, ${}^{F}z$ are the calibrated position coordinates of end center point of special-shaped tool relative to the manipulator flange coordinate system.

10.2.2 Calibration of the Attitude of End Center Point of Special-Shaped Tool

According to the definition of the discrete trajectory point coordinate system $\{F_i\}$ and the constraint relationship between $\{F_i\}$ and TCF, the X-axis direction of TCF is defined as pointing from the manipulator flange to the end of the extension arm along the extension arm. In order to meet the above requirement, the end of special-shaped tool is required to be coaxial with auxiliary calibration tool in one of the four attitudes (i.e. the attitude $P1$ shown in Figure 10.2) when calibrating the attitude of the end of special-shaped tool. Suppose that the first calibrated attitude $\left({}^{B}_{F}T_1\right)$ meets this requirement. When the manipulator is in the $P1$ attitude (see Figure 10.2), the X-axis of TCF is assumed to be in the plane consisting of the Z-axis of the flange coordinate system $\{F\}$ and the Z-axis of the auxiliary calibration TCF $\{A\}$.

The unit vector of Z-axis of the coordinate system $\{F\}$ is represented by ${}^{B}\boldsymbol{oz}_F$; and the unit vectors of X-axis, Y-axis and Z-axis of the coordinate system $\{A\}$ are represented by ${}^{B}\boldsymbol{ox}_A$, ${}^{B}\boldsymbol{oy}_A$, ${}^{B}\boldsymbol{oz}_A$. If the constraint condition is satisfied, then

$${}^{B}\boldsymbol{oy}_A = {}^{B}\boldsymbol{oz}_A \times {}^{B}\boldsymbol{oz}_F \tag{10.11}$$

Then

$${}^{B}\boldsymbol{ox}_A = {}^{B}\boldsymbol{oy}_A \times {}^{B}\boldsymbol{oz}_A \tag{10.12}$$

If the attitude of the coordinate system $\{T\}$ is consistent with that of the coordinate system $\{T\}$ at the calibration point P_1 (note: the attitudes of the two coordinate systems are described in the manipulator base coordinate system $\{B\}$), then

$$ {}^{B}_{T}R=\left[{}^{B}\mathbf{ox}_A \ {}^{B}\mathbf{oy}_A \ {}^{B}\mathbf{oz}_A\right] \tag{10.13} $$

${}^{B}\mathbf{ox}_A$, ${}^{B}\mathbf{oy}_A$, ${}^{B}\mathbf{oz}_A$ are the three-dimensional unit column vectors. Their parameters are the projections of the unit vectors of all axes of the TCF $\{T\}$ to the manipulator base coordinate system $\{B\}$. According to the coordinate transformation theory, we obtain

$$ {}^{F}_{T}T=\begin{bmatrix} & {}^{F}_{T}R & & {}^{F}\mathbf{P}_{\text{ATCP}} \\ 0 & 0 & 0 & 1 \end{bmatrix}={}^{B}_{F}T^{-1}\times\begin{bmatrix} & {}^{B}_{T}R & & {}^{B}\mathbf{P}_{\text{ATCP}} \\ 0 & 0 & 0 & 1 \end{bmatrix} \tag{10.14} $$

where ${}^{F}_{T}R$ is the rotation matrix of $\{T\}$ relative to the flange coordinate system $\{F\}$.

It can be seen from Eq. (10.14) that the position calibration result ${}^{F}P_{\text{ATCP}}$ and direction calibration result ${}^{F}_{T}R$ of the TCF $\{T\}$ constitute the transformation relationship ${}^{F}_{T}T$ between $\{T\}$ and $\{F\}$. ${}^{F}_{T}T$ represents the unique transformation relation between $\{T\}$ and $\{F\}$. In this case, to ensure that this transformation relationship can be identified by Staubli manipulator, ${}^{F}_{T}R$ must be converted into X–Y–Z Euler angle based on Eqs. (10.13) and (10.14).

10.3 CORRECTION OF FOUR-ATTITUDE CALIBRATION ERROR OF TOOL CENTER FRAME

The four-attitude TCF calibration method is simple and easy to operate with fewer operation steps. However, the fault tolerance rate of this calibration method is zero. The four points recorded during calibration must not be wrong; otherwise, the calibration results will be completely inconsistent with actual TCF parameters. According to the analysis of the reflection and transmission of ultrasonic wave at interface in Chapter 3 (for example, the reflection coefficient and transmission coefficient of ultrasonic wave at the water-stainless steel interface shown in Figure 2.5), the transmissivity of ultrasonic wave at the water-steel interface remains almost unchanged when the incident angle is ≤13. Therefore, the calibration error of position parameter has a greater impact on ultrasonic testing than the calibration error of attitude parameter. To solve this problem, the spherical fitting method based on least square method is proposed to correct the calibration error of TCF position parameter caused by the insufficiency of calibration points, the mistake in the recorded individual calibration point and the unreasonable selection of calibration points.

Suppose that a set of data x_i, y_i, z_i $i=n$ is calibrated in any attitude with the TCF calibration method described in Section 10.2. Since the manipulator holding the same special-shaped tip moves to the same point in space but in different attitudes, the recorded $\{F\}$ origins should be scattered on the same sphere under ideal condition. Of course, due to the existence of error in each calibration, a certain data deviation is unavoidable. Suppose that the spherical equation is:

$$ (x-a)^2+(y-b)^2+(z-c)^2=R^2 \tag{10.15} $$

where (a, b, c) and R are the center of sphere and the radius, respectively.

Spherical fitting is carried out with the least square method to construct the residual function:

$$E(a,b,c,R)=\sum_{i=1}^{n}\left((x_i-a)^2+(y_i-b)^2+(z_i-c)^2-R^2\right)^2 \tag{10.16}$$

According to the least square principle, when E is the minimum, the following results can be obtained:

$$\frac{\partial E}{\partial a}=-4\sum_{i=1}^{n}(x_i-a)\left((x_i-a)^2+(y_i-b)^2+(z_i-c)^2-R^2\right)=0 \tag{10.17}$$

$$\frac{\partial E}{\partial b}=-4\sum_{i=1}^{n}(y_i-b)\left((x_i-a)^2+(y_i-b)^2+(z_i-c)^2-R^2\right)=0 \tag{10.18}$$

$$\frac{\partial E}{\partial c}=-4\sum_{i=1}^{n}(z_i-c)\left((x_i-a)^2+(y_i-b)^2+(z_i-c)^2-R^2\right)=0 \tag{10.19}$$

$$\frac{\partial E}{\partial R}=-4\sum_{i=1}^{n}R\left((x_i-a)^2+(y_i-b)^2+(z_i-c)^2-R^2\right)=0 \tag{10.20}$$

Because $R\neq 0$, we can derive

$$\sum_{i=1}^{n}\left((x_i-a)^2+(y_i-b)^2+(z_i-c)^2-R^2\right)=0 \tag{10.21}$$

To simplify Eqs. (10.17)–(10.20), we suppose

$$\begin{aligned} u_i &= x_i-\overline{x} \\ v_i &= y_i-\overline{y} \\ w_i &= z_i-\overline{z} \\ u_a &= a-\overline{x} \\ v_b &= b-\overline{y} \\ w_c &= c-\overline{z} \end{aligned} \tag{10.22}$$

where $\overline{x}=\left(\sum x_i\right)/n$, $\overline{y}=\left(\sum y_i\right)/n$, $\overline{z}=\left(\sum z_i\right)/n$. The substitution of these values into Eqs. (10.17)–(10.20) can yield

$$\sum_{i=1}^{n} u_i\left((u_i-u_a)^2+(v_i-v_b)^2+(w_i-w_c)^2-R^2\right)=0 \tag{10.23}$$

$$\sum_{i=1}^{n} v_i\left((u_i-u_a)^2+(v_i-v_b)^2+(w_i-w_c)^2-R^2\right)=0 \tag{10.24}$$

$$\sum_{i=1}^{n} w_i\left((u_i-u_a)^2+(v_i-v_b)^2+(w_i-w_c)^2-R^2\right)=0 \tag{10.25}$$

$$\sum_{i=1}^{n}\left((u_i-u_a)^2+(v_i-v_b)^2+(w_i-w_c)^2-R^2\right)=0 \tag{10.26}$$

Because $\sum u_i=\sum v_i=\sum w_i=0$, the equations can be simplified as

$$\left(\Sigma u_i^2\right)u_a+(\Sigma u_i v_i)v_b+(\Sigma u_i w_i)w_c=\frac{\sum\left(u_i^3+u_i v_i^2+u_i w_i^2\right)}{2}$$

$$(\Sigma u_i v_i)u_a+\left(\Sigma v_i^2\right)v_b+(\Sigma v_i w_i)w_c=\frac{\sum\left(u_i^2 v_i+v_i^3+v_i w_i^2\right)}{2}$$

$$(\Sigma u_i w_i)u_a+(\Sigma v_i w_i)v_b+\left(\Sigma w_i^2\right)w_c=\frac{\sum\left(u_i^2 w_i+v_i^2 w_i+w_i^3\right)}{2} \tag{10.27}$$

so

$$\begin{bmatrix} u_a \\ v_b \\ w_c \end{bmatrix}=\begin{bmatrix} \Sigma u_i^2 & \Sigma u_i v_i & \Sigma u_i w_i \\ \Sigma u_i v_i & \Sigma v_i^2 & \Sigma v_i w_i \\ \Sigma u_i w_i & \Sigma v_i w_i & \Sigma w_i^2 \end{bmatrix}^{-1}\begin{bmatrix} \dfrac{\sum\left(u_i^3+u_i v_i^2+u_i w_i^2\right)}{2} \\ \dfrac{\sum\left(u_i^2 v_i+v_i^3+v_i w_i^2\right)}{2} \\ \dfrac{\sum\left(u_i^2 w_i+v_i^2 w_i+w_i^3\right)}{2} \end{bmatrix} \tag{10.28}$$

By combining Eq. (10.22) with Eq. (10.28), the coordinates of the center of the fitted sphere are determined:

$$\begin{bmatrix} u_a \\ v_b \\ w_c \end{bmatrix} = \begin{bmatrix} \Sigma u_i^2 & \Sigma u_i v_i & \Sigma u_i w_i \\ \Sigma u_i v_i & \Sigma v_i^2 & \Sigma v_i w_i \\ \Sigma u_i w_i & \Sigma v_i w_i & \Sigma w_i^2 \end{bmatrix}^{-1} \begin{bmatrix} \dfrac{\sum\left(u_i^3 + u_i v_i^2 + u_i w_i^2\right)}{2} \\ \dfrac{\sum\left(u_i^2 v_i + v_i^3 + v_i w_i^2\right)}{2} \\ \dfrac{\sum\left(u_i^2 w_i + v_i^2 w_i + w_i^3\right)}{2} \end{bmatrix} \tag{10.29}$$

By substituting Eq. (10.29) into Eq. (10.21), the spherical radius can be determined:

$$R = \sqrt{\frac{\sum_{i=1}^{n}\left((u_i - u_a)^2 + (v_i - v_b)^2 + (w_i - w_c)^2\right)}{n}} \tag{10.30}$$

To verify the correctness of the fitting results, the special-shaped tip with a known size (as shown in Section 10.2.1) was established as the calibration object in the Creo 3D modeling software. The verification test was carried out in the SRS2016 simulation environment. The calibration tip illustrated in Figure 10.3 was bent by 90° from *Z*-axis around *Y*-axis to +*X*-axis. Its center is 80 mm long in the *X* direction and 107 mm long in the *Z* direction. According to the Pythagorean theorem, the distance between the tip center and the origin of the tip coordinate system is 133.6 mm. When the tip tool was installed on the manipulator flange, its reference coordinate system coincided with the manipulator flange coordinate system. Therefore, the distance between the center of the installed tip and the origin of the flange coordinate system is also 133.6 mm.

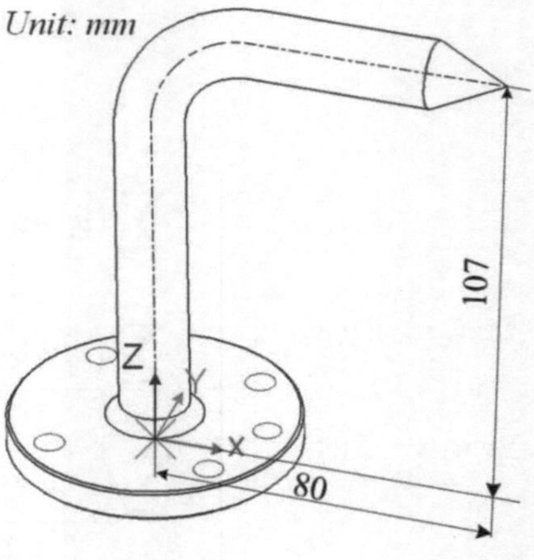

FIGURE 10.3 Special-shaped tip with a known size.

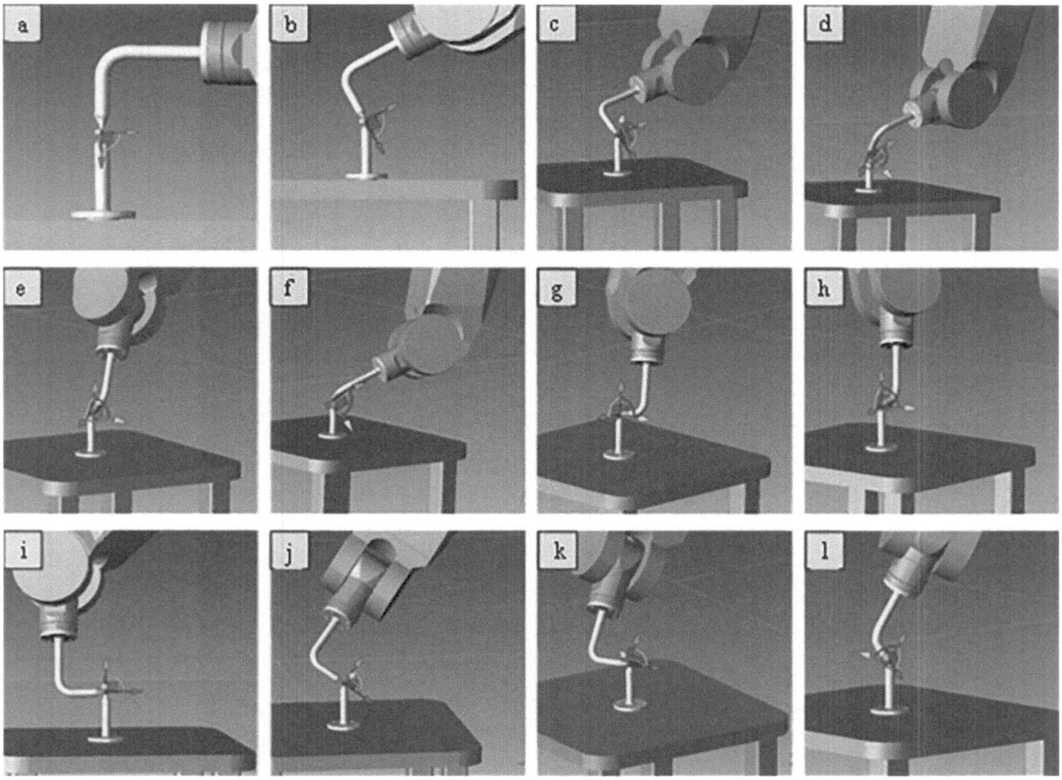

FIGURE 10.4 Twelve calibration attitudes of special-shaped tip.

A simulation platform was established in SRS2016 based on the above special-shaped calibration tip and RX160 manipulator. A total of 12 calibrations (a–i) were made in different attitudes (see Figure 10.4) by using the TCF calibration method described in Section 10.2. Meanwhile, 12 sets of data were collected.

The 12 sets of position data (x, y, z) in Table 10.1 were fitted to obtain the fitting results, as shown in Figure 10.5. The position of sphere center is $(1133.8, 80.52, -141.83)$, and the spherical radius is 133.58 mm. The position of the auxiliary calibration tool is $(1133.5, 80.16, -141.38)$. The theoretical fitted spherical radius is 133.6 mm, the fitted spherical radius error is 0.02 mm, the error of radius fitting is 0.015% and the one-way maximum position error is 0.45 mm. It can be seen that the fitting correction results of calibration data meet the requirement that the alignment accuracy of ultrasonic probe should be less than 1 mm [10].

To observe the fitting state of calibration point data, any four sets of data were selected from the above 12 sets of calibration point data and then fitted. A total of 495 combinations (C_{12}^4) were fitted. Then the R-change curve graph was drawn with the number of combinations (n) as the abscissa and the fitted spherical radius R as the ordinate, as shown in Figure 10.6.

As seen from the figure, the fitting results of any four calibration points tended to have large errors, which were even 3–4 times the size of the special-shaped calibration tool itself.

TABLE 10.1 Position Data and Euler Angles of Special-Shaped Tip in 12 Calibrations

Attitude	*x*	*y*	*z*	*rx*	*ry*	*rz*
1	1027.12	96.53	−63.2	87.19	80.62	−86.44
2	1078.93	87.45	−20.6	−15.7	120.12	18.91
3	1057.15	13.77	−54.78	170.87	53.01	−132.73
4	1043.73	121.86	−52.53	22.37	109.42	−65.27
5	1101.1	101.24	−13.93	81.41	23.98	−144.86
6	1078.62	173.59	−63.61	−34.23	120.64	−24.54
7	1111.43	128.75	−18.26	15.35	161.62	−81.49
8	1137.16	117.28	−13.89	15.35	161.62	−60.65
9	1205.86	23.43	−45.05	19.74	177.35	14.99
10	1132.72	11.27	−27.62	37.04	143.29	−10.06
11	1121.39	−24.04	−59.46	26.05	146.68	45.8
12	1084.36	57.67	−20.11	153.86	10.41	−67.69

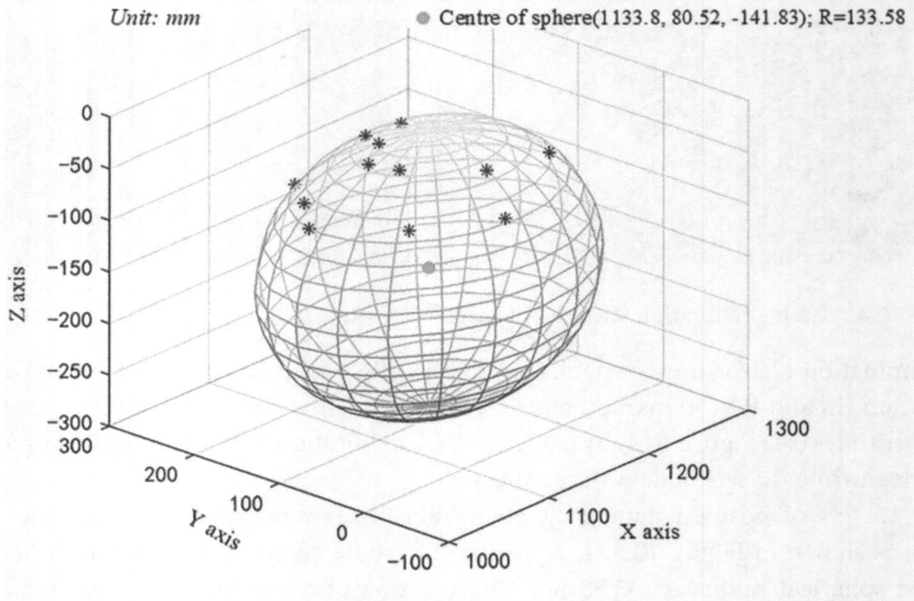

FIGURE 10.5 Fitting results of 12 calibrations of special-shaped tip.

This is unreasonable. Then the two combinations with the largest errors, namely the 25th and 406th combinations, were picked. The former combination was composed of the four data sets for attitude 1, attitude 2, attitude 6 and attitude 7, while the latter was composed of the four data sets for attitude 4, attitude 7, attitude 8 and attitude 9. The fitting results of the two combinations are shown in Figures 10.7 and 10.8, respectively.

The incorrect fitting results of the two combinations have a common feature, that is, the four calibration points are all concentrated in one orientation of the sphere and are close to each other, which is also in agreement with the theory. If they are closer to each other, larger errors may be caused. Therefore, the four points should be evenly distributed on the sphere during the calibration of special-shaped tool. In other words, a larger change of

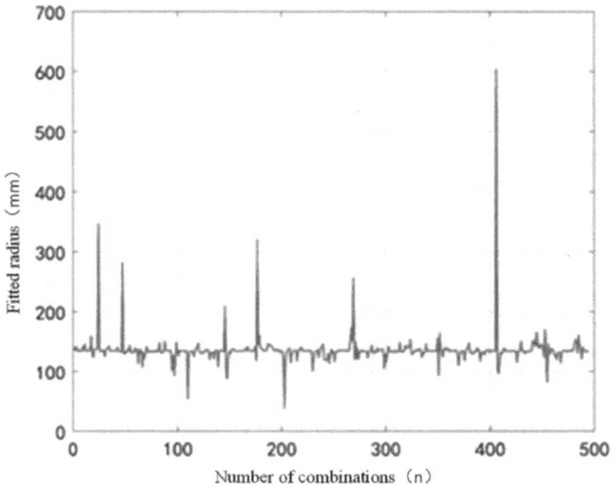

FIGURE 10.6 Spherical radii obtained after the fitting of 495 sets of data.

Note: The fitting results of any four sets of data cannot represent the actual calibration results, as there are large errors in the combinations. Next, we will analyze this problem.

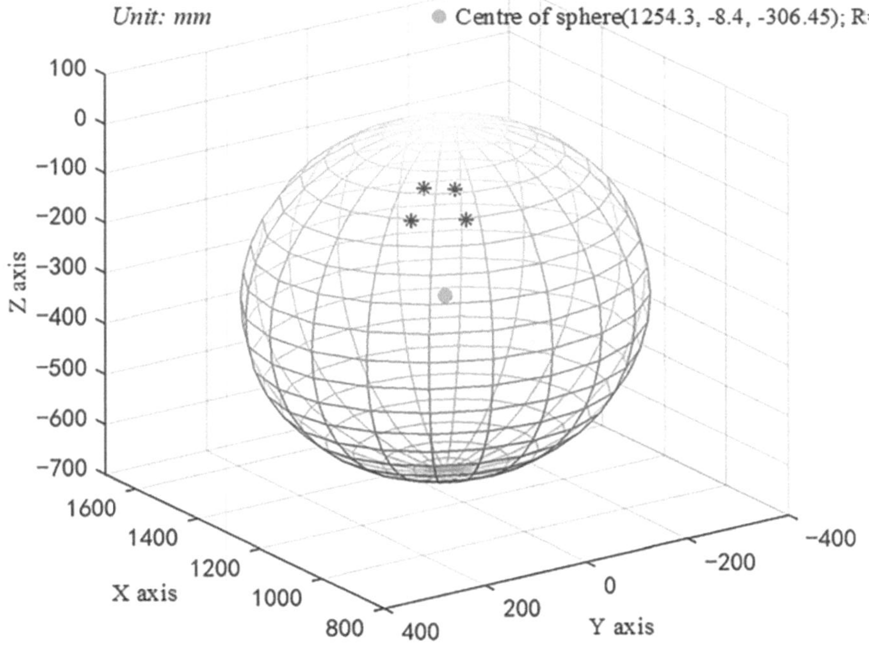

FIGURE 10.7 Spherical fitting of the 25th combination.

manipulator attitude during calibration will contribute to a more correct calibration result and higher calibration accuracy. To this end, one set of relatively decentralized calibration points (such as the set consisting of attitude 1, attitude 8, and attitude 9 and attitude 11) was randomly selected from the 12 sets of data. The fitting result is shown below (Figure 10.9).

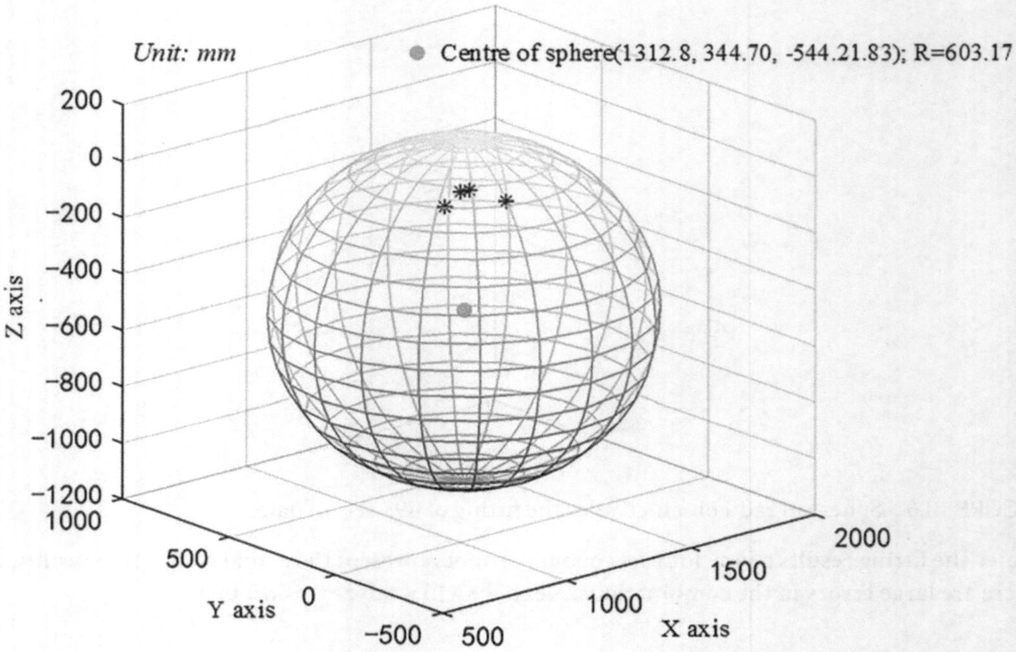

FIGURE 10.8 Spherical fitting of the 406th combination.

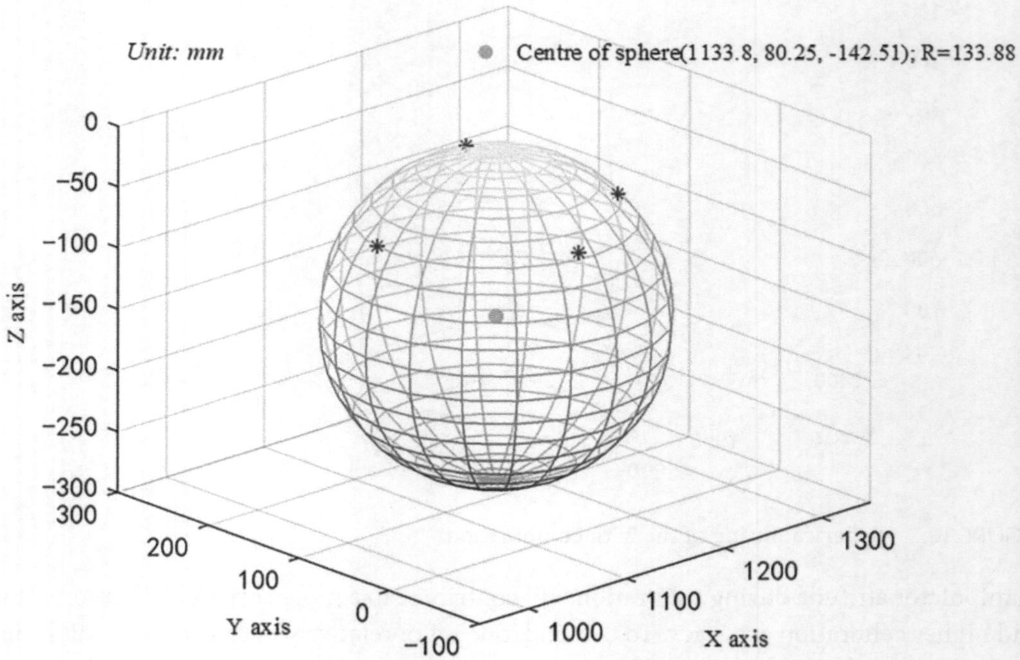

FIGURE 10.9 Spherical fitting of decentralized data points.

According to the fitting result of this set of data, both the sphere center position (1133.8, 80.25, −142.51) and the spherical radius $R = 133.88$ mm are slightly different from the actual values and can fit the actual situation well. Therefore, as long as the locations of calibration points are appropriate, an accurate result can be obtained with only four points.

Therefore, the following principle should be followed during the four-point calibration of special-shaped tool:

The selected four calibration points should be distributed on the sphere as evenly as possible. In practical operation (i.e. during the tool calibration), the manipulator attitude should change as greatly as possible to help improve the calibration accuracy.

The R curve in the fitting result of any four sets of data listed in Table 10.1 cannot represent the actual calibration result, as such a random combination may contain four concentrated points and then a large fitting error may be caused. This situation should be avoided in the actual calibration.

As for the attitude correction, it can be found from the calibration method that there is no error in the X-axis coordinate of TCF attitude. The small errors in the Y-axis and Z-axis coordinates can be corrected by minor adjustment according to the signal strength received by ultrasonic transducer during the actual detection.

10.4 FOUR-ATTITUDE TCF CALIBRATION EXPERIMENT

10.4.1 TCF Calibration Experiment of Special-Shaped Tip Tool

To verify the correctness of the proposed four-attitude calibration method, a calibration experiment was carried out with the special-shaped tip shown in Figure 10.3. The manipulator used in the calibration experiment is just the RX160 manipulator holding a special-shaped extension arm in the dual-manipulator UT system. The tip is installed on the manipulator flange, and the auxiliary calibration tool is randomly put in a place in the movement space of the manipulator.

According to the above calibration steps, the position and attitude of the flange coordinate system in the four attitudes P_1, P_2, P_3, P_4, were recorded, as shown in Table 10.2.

According to the X–Y–Z Euler-angle transformation relation and the Euler angle of the point P_1 we can obtain

$$
{}_F^BT_{P1} = \begin{bmatrix} 0.012 & 0 & 0.99993 & 1070.08 \\ 0.019 & 0.99983 & 0 & 29.56 \\ -0.99976 & 0.019 & 0.012 & -79.18 \\ 0 & 0 & 0 & 1 \end{bmatrix} \tag{10.31}
$$

TABLE 10.2 Position and Attitude of the Flange Coordinate System in Four Attitudes

Point\Coordinate	*x*	*y*	*z*	*rx*	*ry*	*rz*
*P*1	1070.08	29.56	−79.18	0	89.32	1.07
*P*2	1162.08	29.42	−41.72	0	138.99	1.07
*P*3	1113.08	−35.42	−70.09	27.45	115.17	12.84
*P*4	1063.65	64.44	−96	−119.58	66.9	65.25

According to Eq. (10.31) and the coordinates of the points P_1–P_4, we can obtain

$$\left({}^{B}x,\ {}^{B}y,\ {}^{B}z\right)^{T}=(1157,\ 28.275,\ -160.86)^{T} \tag{10.32}$$

$$\left({}^{F}x,\ {}^{F}y,\ {}^{F}z\right)^{T}=(82.671,\ -2.83,\ 85.909)^{T} \tag{10.33}$$

According to Eqs. (10.32) and (10.33), we can obtain

$$ {}^{F}_{T}\boldsymbol{T}=\begin{bmatrix} 0.012 & -0.019 & 0.99976 & 82.671 \\ 0 & -0.99976 & -0.019 & -2.83 \\ 0.99986 & 0 & -0.012 & 85.909 \\ 0 & 0 & 0 & 1 \end{bmatrix} \tag{10.34}$$

According to the above equations, we can obtain

$$\begin{cases} rx=\alpha=122.44 \\ ry=\beta=88.732 \\ rz=\gamma=57.567 \end{cases} \tag{10.35}$$

Therefore, the Euler-angle transformation relation between the special-shaped tip and flange coordinate system is

$$ {}^{F}_{T}T=\{82.671,-283,85.909,122.44,88.732,57.567\} \tag{10.36}$$

The parameters in Eq. (10.36) can be downloaded to the manipulator controller, so that the manipulator can clamp the special-shaped tip to move accurately in accordance with the predetermined trajectory.

10.4.2 Verification Experiment of TCF Calibration Result of Special-Shaped Tip Tool

Based on the calibration result given in Section 10.4.1, a verification experiment was carried out on the dual-manipulator UT system, as shown in Figure 10.10. The scanning trajectory is based on part of a semi-closed cylindrical composite torpedo shell with the diameter of 600 mm and the thickness of 25 mm.

The typical zig-zag trajectory was planned. For the convenience of explanation, the line spacing of the trajectory was set as 10 mm, and the point spacing was set as 1.5 mm. During the movement, about 650 sets of Cartesian coordinate data were collected, including the data for the end center point of special-shaped tip and for the end center point of water nozzle. The trajectories of the ends of the two tools relative to the coordinate system {W} (shown in Figure 10.11) show that the trajectories of the master and slave manipulators are parallel raster-scanning lines with a constant spacing between them. This experiment has basically verified the correctness of the calibration method and its practical application value.

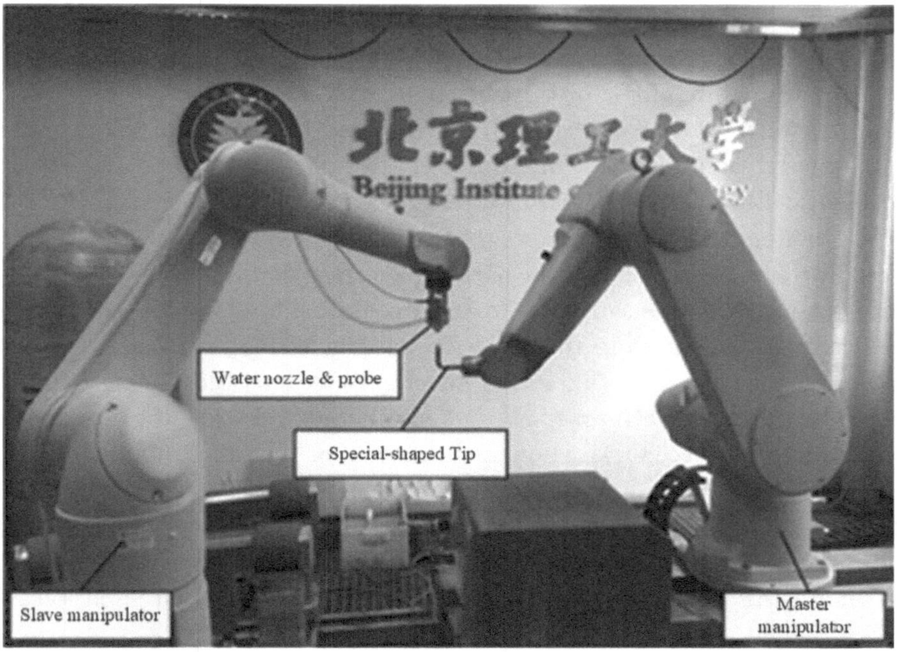

FIGURE 10.10 Verification experiment of dual-manipulator synchronous motion with a special-shaped tip on one manipulator.

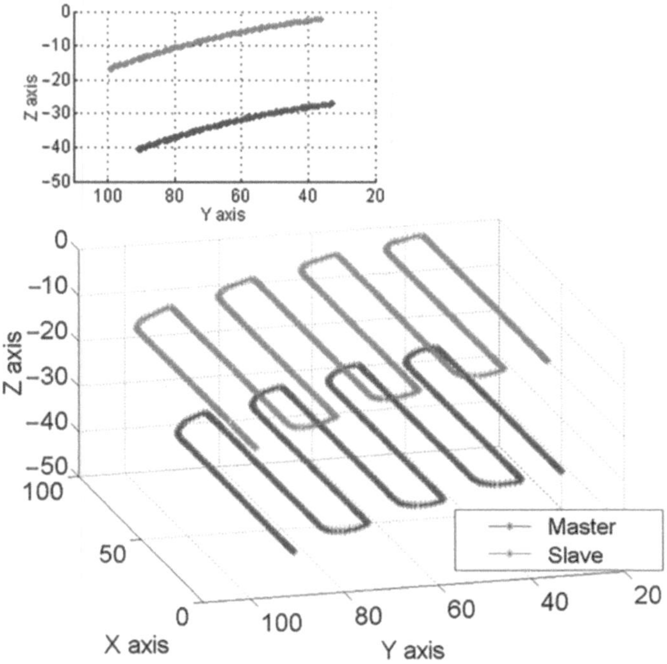

FIGURE 10.11 Actual TCP trajectories of the two manipulators.

10.5 FOUR-ATTITUDE TCF CALIBRATION EXPERIMENT OF SPECIAL-SHAPED EXTENSION ARM

According to the dual-manipulator UT concept for semi-closed workpiece, one ultrasonic probe and the water nozzle coupled with it were fixed to the end of the extension arm held by the master manipulator. The extension arm, ultrasonic probe and water nozzle are collectively called the special-shaped extension arm tool. This is a large special-shaped tool. If it cannot be precisely calibrated, the error will be amplified at its end. Therefore, it is necessary to precisely calibrate the parameters of the coordinate system at the end of the special-shaped extension arm tool. To improve the calibration accuracy and simulate the underwater acoustic path in the detection process, a small tip was fixed to the end of the nozzle during the TCF calibration and then was removed during the ultrasonic detetion. The distance between tip endpoint and water nozzle is just the distance between water nozzle and workpiece surface during the detection. The calibration process is shown in Figure 10.12. The specific position and attitude of the flange coordinate system in the four attitudes P_1, P_2, P_3, P_4 are shown in Table 10.3.

According to the calibration method of special-shaped tool described in Section 10.2, the parameter of the coordinate system at the end of the special-shaped extension arm tool,

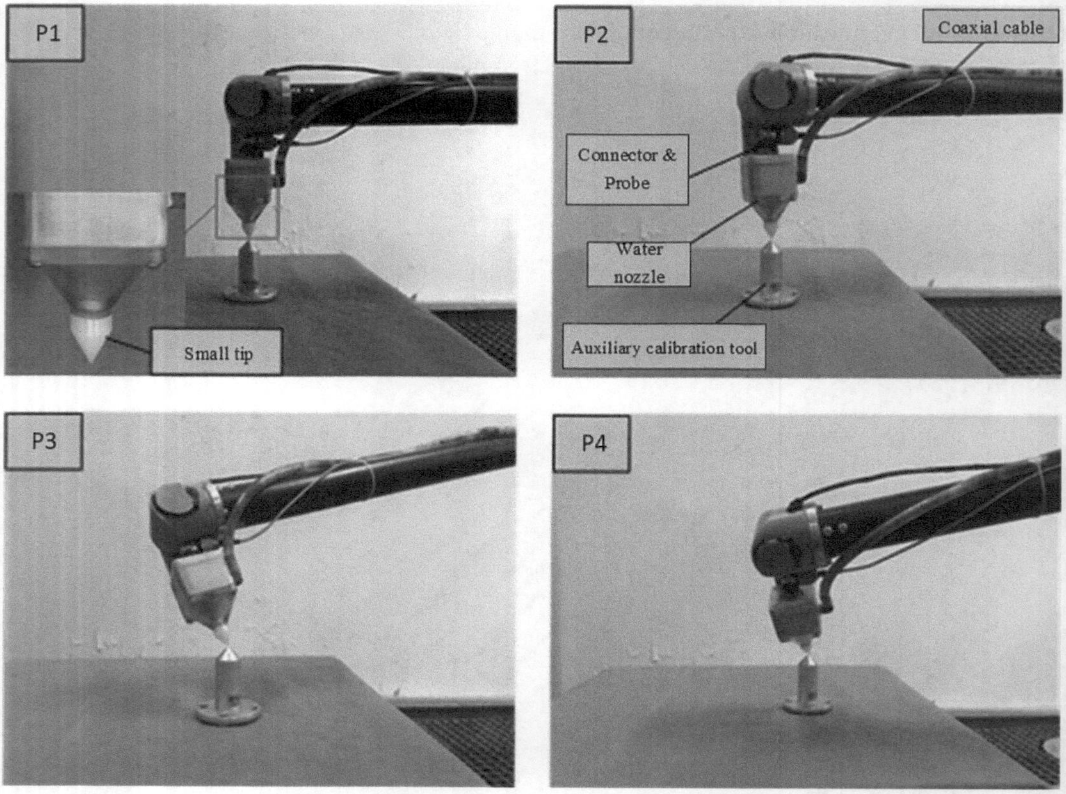

FIGURE 10.12 Calibration process of an end coordinate system of a special-shaped extension arm tool.

TABLE 10.3 Position and Attitude of the Flange Coordinate System in the Four Attitudes of the Special-Shaped Extension Arm

Point\Coordinate	x	y	z	rx	ry	rz
$P1$	1022.45	158.71	−100.71	106.33	89.14	−60.42
$P2$	1133.54	−329.68	−100.86	89.46	116.45	−43.61
$P3$	1228.96	−262.39	243.2	45.48	119.57	16.72
$P4$	1261.28	645.4	211.42	−53.18	122.32	74.92

namely the position and Euler-angle transformation relation between the TCF $\{T\}$ and the flange coordinate system $\{F\}$, is shown as follows:

$$ {}_{T}^{F}T = \{112.72, -115.11, 1065.96, 89.66, 44.09, 90.23\} $$

REFERENCES

1. Cheng F.S., The method of recovering robot TCP positions in industrial robot application programs [J]. Proceedings of the 2007 IEEE International Conference on Mechatronics and Automation, ICMA 2007, 2007:805–810.
2. Borrmann C., Wollnack J., Enhanced calibration of robot tool centre point using analytical algorithm [J], *International Journal of Materials Science and Engineering*, 2015, 3(1):12–18.
3. Gordić Z., Ongaro C. Development and Implementation of Orthogonal Planes Images Method for Calibration of Tool Centre Point[C]. Rodić A., Borangiu T. *Advances in Robot Design and Intelligent Control.* Cham: Springer International Publishing, 2017:105–115.
4. Yin S., Guo Y., Ren Y., et al., A novel TCF calibration method for robotic visual measurement system [J], *Optik*, 2014, 125(23):6920–6925.
5. Liao W.U., Xiangdong Y., Shanqing L.A.N., et al., Robotic TCF calibration based on a planar template [J], *Robot*, 2012, 34(1):98–103.
6. Xu X., Zhu D., Zhang H., et al., TCP-based calibration in robot-assisted belt grinding of aero-engine blades using scanner measurements [J], *International Journal of Advanced Manufacturing Technology*, 2017, 90(1–4):635–647.
7. Cunfeng K., Research and implementation of TCF calibration of welding robot [J], *Journal of Beijing University of Technology*, 2016, 42(01):30–34.
8. Huajun Z. *Research on Trajectory Planning and System Calibration for the Laser Cutting Application of Industrial Robots [D].* Zhejiang: Zhejiang Sci-Tech University, 2013.
9. Yongjie P., Calibration of TCF parameters of arc welding robot [J], *Robot*, 2001, 0446:109–112.
10. Glass S., Gripp S., Hasse W., et al., Robots for Automated Non-destructive Examination of Complex Shapes[C]. Asia Pacific Conference on Non-Destructive Testing (14th APCNDT), 2013.

[illegible]

TABLE 10.3 [illegible]

[illegible]	[illegible]	[illegible]	[illegible]	[illegible]	[illegible]

[illegible]

[illegible]ENCES

1. [illegible]
2. [illegible]
3. [illegible]
4. [illegible]
5. [illegible]
6. [illegible]
7. [illegible]
8. [illegible]
9. [illegible]
10. [illegible]

CHAPTER 11

Robotic Radiographic Testing Technique

THE RADIOGRAPHIC TESTING (RT) process has a certain radioactivity. The unmanned automatic robotic nondestructive testing (NDT) process can meet the requirements of on-site RT. Especially for a component with complex shape, the manipulator can hold the radiation source with one arm and the DR receiver screen with the other arm, and then detect the existence, position, shape and size of a defect in the component under test in accordance with the change in the position of X-ray transmitter relative to X-ray receiver and the attenuation rate of X-ray energy in the component. Alternatively, the manipulator can hold the component with one arm and move it in the radiation field to accomplish the transmission detection. The latter technique will be the focus of this chapter.

11.1 BASIC PRINCIPLE OF X-RAY CT TESTING

11.1.1 Theory of X-ray Attenuation

When the X-ray passes through the workpiece with a certain thickness, its intensity will attenuate to some extent. This phenomenon is called X-ray attenuation. It is found through study that the attenuation of narrow-beam monoenergetic X-ray obeys the Lambert–Beer's law (Figure 11.1), that is,

$$I = I_0 e^{-\mu \Delta x} \tag{11.1}$$

where Δx is the thickness of the workpiece under test, that is, the thickness of material; I is the X-ray intensity after attenuation; I_0 is the original X-ray intensity and μ is the attenuation coefficient of X-ray, which is mainly related to the material properties. Eq. (11.1) shows that the attenuation degree of X-ray in a specific material is exponentially related to the workpiece thickness [1].

However, in practical applications, the rays in the CT system are usually continuous wide-beam rays, namely polychromatic rays. In other words, the X-ray beams contain

DOI: 10.1201/9781003212232-11

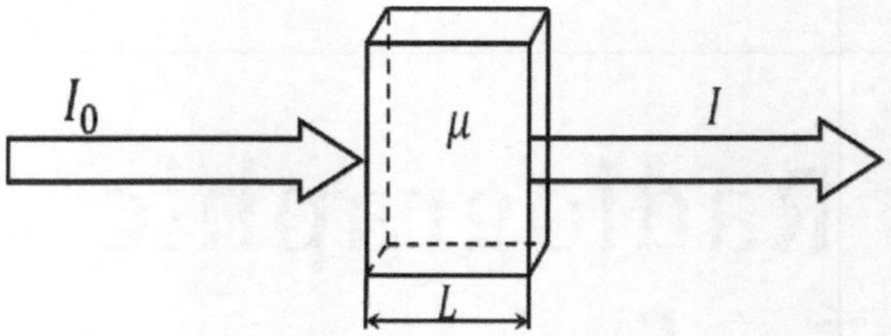

FIGURE 11.1 X-ray passing through an object with the thickness of L.

some scattered photons and show different attenuation degrees when passing through the material with a certain thickness. Therefore, the detector receives not only the attenuated beams but also some scattered photons (Figure 11.2). If the workpiece is homogeneous, the attenuation law of rays can be described by Eq. (11.2) [2].

$$I(l) = \int_0 (1+n) \cdot I_0(E) e^{-\mu_1(E)x} dE \tag{11.2}$$

where n is the ray's scattering ratio, that is, $n = I_s - I_D$, where I_s is the scattered ray intensity and I_D is the initial ray intensity;

$\mu(E)$ is the photon attenuation coefficient of the ray with the energy E;

$I_0(E)$ is the initial photon intensity of the ray with the energy E;

$I(l)$ is the ray intensity after passing through the workpiece.

What is presented above is the attenuation law of X-ray in homogeneous objects. However, the actually tested workpiece is often made of several materials with different densities and atomic compositions, so the X-ray will show different attenuation coefficients $(\mu_1\mu_2 \cdots \mu_{n-1}\mu_n)$ in different materials after entering the workpiece. In view of this phenomenon, the workpiece under test is usually divided into a number of small units, as shown in Figure 11.3. If each unit is small enough, it can be approximately viewed as homogeneous material and thus can be characterized by Eq. (11.1). Hence, the whole material can be

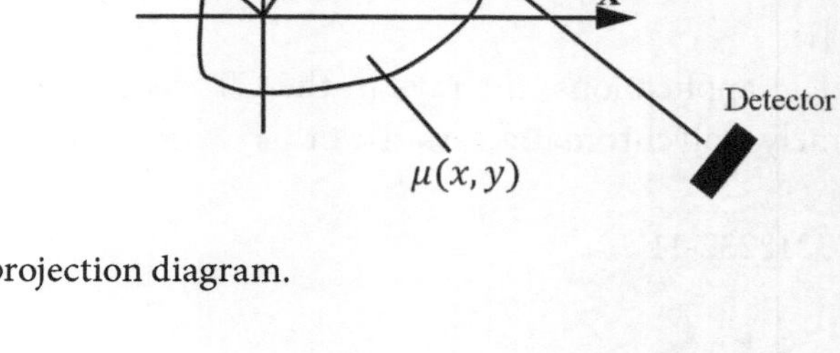

FIGURE 11.2 X-ray projection diagram.

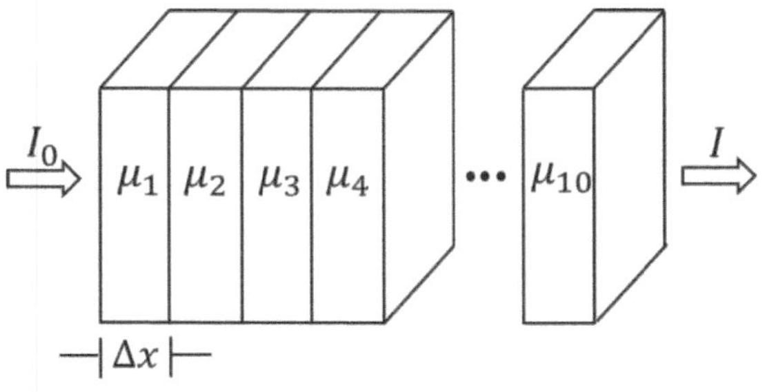

FIGURE 11.3 X-ray passing through all the units of composite material.

characterized by Eq. (11.3) [3], which is just the attenuation law of X-ray in heterogeneous materials.

$$I = I_0 e^{-\mu_1 \Delta x} e^{-\mu_2 \Delta x} e^{-\mu_3 \Delta x} \cdots e^{-\mu_n \Delta x} = I_0 e^{-\sum_{n=1}^{N} \mu_n \Delta x} \tag{11.3}$$

11.1.2 Mathematical Basis of Industrial CT Imaging

Industrial CT imaging is based on the basic attenuation law of X-ray in objects. It can be known from the above description that the attenuation process of X-ray in heterogeneous materials can be characterized by Eq. (11.3). However, in practical applications, Eq. (11.3) usually needs to be standardized. At first, it should be converted into Eq. (11.4):

$$\frac{I}{I_0} = e^{-\mu_1 x} e^{-\mu_2 x} e^{-\mu_3 x} \cdots e^{-\mu_n x} = e^{-\sum_{n=1}^{N} \mu_n a x} \tag{11.4}$$

Then the negative logarithm of Eq. (11.4) is:

$$p = -\ln\left(\frac{I}{I_0}\right) = \ln\left(\frac{I_0}{I}\right) = \sum_{n=1}^{N} \mu_n \Delta x \tag{11.5}$$

Suppose that Δx is infinitely decreasing, that is, the units constituting the object are infinitely scaled down. Then, from Eqs. (11.4) and (11.5), we can obtain

$$p = -\ln\left(\frac{I}{I_0}\right) = n\frac{I}{I_0} = \int_L \mu_n \, dx \tag{11.6}$$

where p is the projection data obtained when the X-ray passes through the material μ_n [4]. It can be concluded from the above equations that, when the units constituting the object under test are small enough, the value of object projection is equal to the linear integral of attenuation coefficients of X-ray on its path. However, in the actual calculation, p can be approximately calculated and processed by using a set of sampled values and then μ_n can be calculated by reverse solution. This is the so-called CT image reconstruction.

Currently, the reconstruction algorithms of CT images can be divided into two main categories: analytical reconstruction algorithm and iterative reconstruction algorithm. FDK is a mature algorithm that has been widely used in the society, so it will not be described again. However, we must understand that, the mathematical basis of these algorithms is a very important transformation – the Radon transformation proposed by Radon in 1917 [5,6], whose inverse transformation is the basic principle of CT image reconstruction. Next, the inverse of Radon transformation will be sketched out:

Suppose that the two-dimensional X-ray attenuation coefficient function of a point on a line in two-dimensional plane is expressed as I/I_0, as shown in Figure 11.4. Eq. (11.6) can be transformed into

$$p = \int_L f(x,y)\,dL \tag{11.7}$$

On this basis, the Dirac-delta function sampling function $\delta(x)$ is introduced to obtain

$$p(s,\theta) = \int_L f(x,y)\,dl = \int_{+\infty}^{-\infty}\int f(x,y)\,\delta(x\cos\theta + y\sin\theta - s)\,dx\,dy \tag{11.8}$$

Eq. (11.8) is the so-called Radon transformation. Then it can be expressed by polar coordinates:

$$f(x,y) = \frac{1}{2\pi^2}\int_0^{\pi}\int_{+\infty}^{-\infty} \frac{1}{r\cos(\beta-\theta)-s}\frac{\partial p}{\partial s}\,dr\,d\theta \tag{11.9}$$

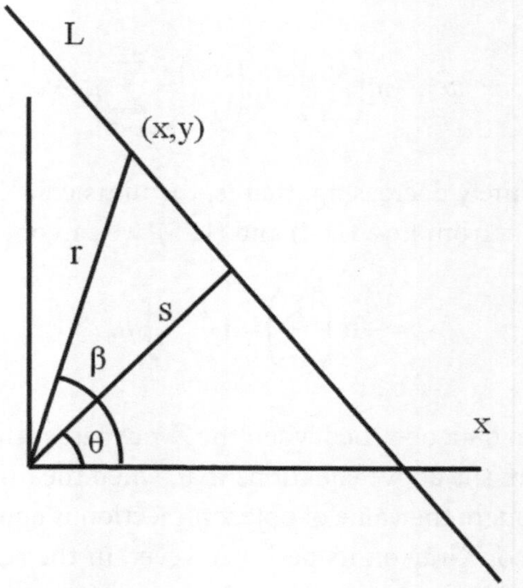

FIGURE 11.4 Radon transformation diagram.

Eq. (11.9) is just the corresponding inverse Radon transformation. In other words, if the projection data of the workpiece at all angles is obtained, the corresponding attenuation coefficient distribution in the scanned object can be obtained based on inverse Radon transformation. Meanwhile, the projection data can be obtained from the corresponding projection diagram. Therefore, inverse Radon transformation has provided a mathematical basis for CT image reconstruction.

11.2 COMPOSITION OF A ROBOTIC X-RAY CT TESTING SYSTEM

An industrial X-ray CT testing system is mainly composed of a high-energy X-ray source and its controller system, a flat panel detector (FPD), an image processing control system, a mechanical motion scanning and control system, auxiliary equipment, an industrial personal computer (IPC) and other subsystems [7]. Its structure is shown in Figure 11.5.

Although each X-ray CT system is specific to a certain type of detection object, their overall structures are basically the same. In general, the mechanical scanning system only has the most basic rotary function and lacks the activity to flexibly scan the workpiece under test. This will create serious inconvenience when testing the components with special size or complex shape and measuring the geometric parameters of the system. Therefore, if traditional rotary worktable is replaced by six-degree of freedom (DOF) industrial manipulator, the manipulator with stronger flexibility and more spatial DOFs can effortlessly scan the workpieces in a variety of positions and attitudes. In this way, not only the detection flexibility of the whole system but also the detection efficiency of workpieces can be improved. The following diagram is the basic scheme of the robotic X-ray CT testing system (Figure 11.6).

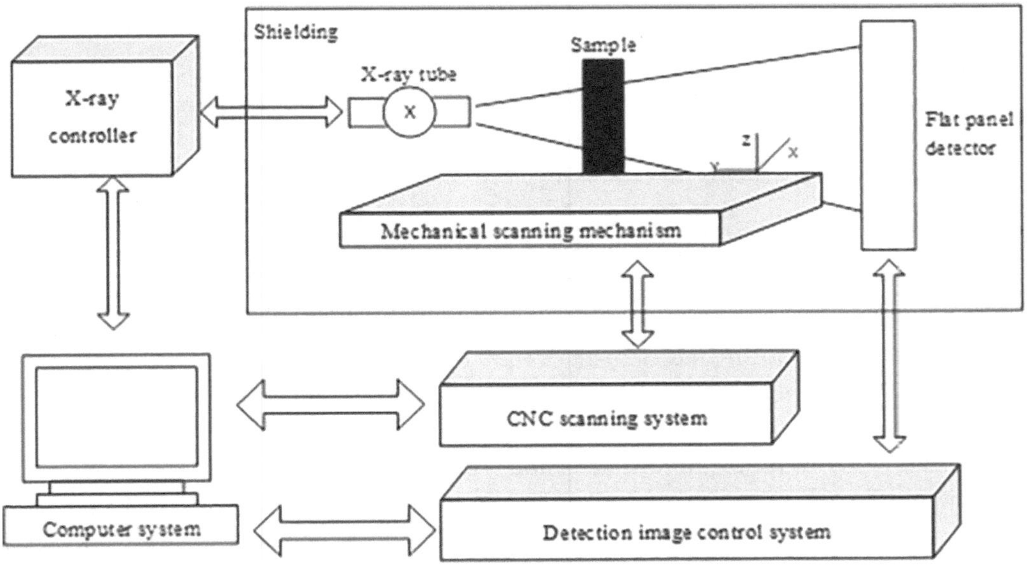

FIGURE 11.5 Schematic diagram of an industrial X-ray CT system.

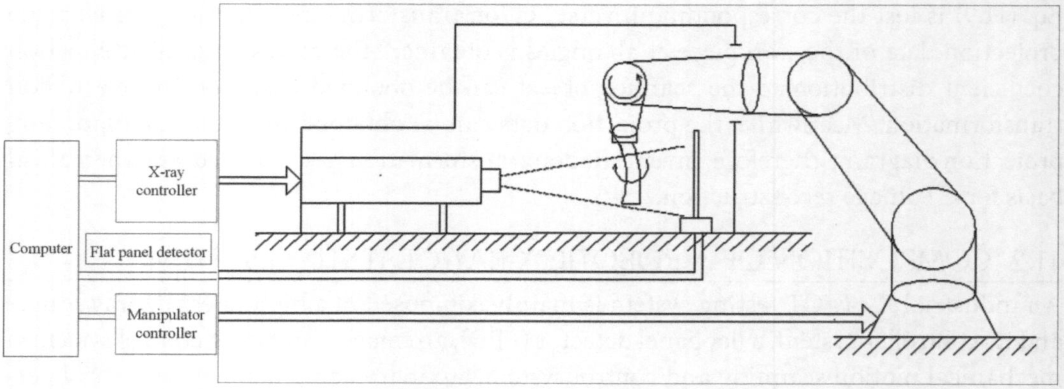

FIGURE 11.6 Schematic diagram of a robotic X-ray CT testing system.

In the X-ray CT detection system of this manipulator, the scanning mechanism is the MOTOMAN-SV3X manipulator itself, supported by its home-made control system (with PMAC card). Its main task is to clamp, spatially locate and rotate the workpiece under test. The data acquisition system is a SIX-650HD-E FPD with 960×786 pixels. The size of each pixel is 0.15 mm, and the pixel gray depth is 14 bpp. The highest data acquisition speed can reach 30 fp/s, meeting the requirement for high-speed X-ray image acquisition performance. The high-energy X-ray system is a XYG-1302 X-ray machine, which has a double-focus X-ray exposure mode and the maximum X-ray intensity of 130 kV to penetrate the workpiece to some degree. As its core control unit is S7–200PLC, its coordinated control with other subsystems can be realized through effective programmable logic controller (PLC) programming. At the same time, a dual-core multithreaded IPC is used to implement the big data computation in the process of image acquisition and CT image reconstruction.

The main working process of this system is as follows. The manipulator clamping the workpiece under test enters the conical X-ray field in an appropriate position attitude and synchronously collects the workpiece data through the coordination among X-ray source, FPD and manipulator. On this basis, the 3D workpiece image is reconstructed with the help of VGStudio software so that the workpiece can be tested in all directions.

11.3 ACQUISITION, DISPLAY AND CORRECTION OF X-RAY PROJECTION DATA

11.3.1 Principle and Working Mode of a Flat Panel Detector

1. Structure and working mechanism of the FPD

The FPD is the most widely used projection-data acquisition device in X-ray CT testing systems. According to the data acquisition principle, FPDs can be roughly divided into two types: direct conversion type and indirect conversion type [8], which convert the X-rays in different ways. The direct conversion type can directly convert X-rays into electrical signals via a photoelectric conversion material (such as α-Se), so its imaging quality is good. However, due to the repeated effects of high pressure, the imaging stability in direct conversion is poor, the image acquisition efficiency is low and the image can be damaged easily. In contrast, an indirect detector can effectively solve these problems, so it is widely used in X-ray CT testing systems. Next, the SIX-650HD-E detector used in this system will be taken as an example to emphatically describe the structure and working mechanism of an indirect FPD.

The top layer of the FPD is a scintillator, which is usually made of Gd_2O_2S or Gsl. On the layer below it are an amorphous silicon array and thin film transistors. Each amorphous pixel corresponds to a thin film transistor, and they are connected through an amorphous photodiode. Pixel array is the basic image acquisition unit of the FPD. The whole detector consists of several basic conversion units. On the bottom layer are a glass substrate and other auxiliary circuits and control units. The circuit structure of pixel array is shown in Figure 11.7. Of course, a complete FPD also includes the power supply system, image processing and storage unit, and other auxiliary devices. Among them, X-ray conversion unit is the core of the detector and directly affects the quality of the collected images.

SIX-650HD-E is an indirect FPD, whose basic working principle is shown in Figure 11.8. When being exposed to X-rays, it can convert X-rays into the visible light with a wavelength of 550 nm, which is exactly the optimal wavelength to be converted into an electrical signal by photodiode. Then, under the control of line-drive circuit control unit, the TFT (Thin Film Transistor) circuit reads the electrical signal line by line and inputs it into the corresponding amplifier, where the voltage signal suitable

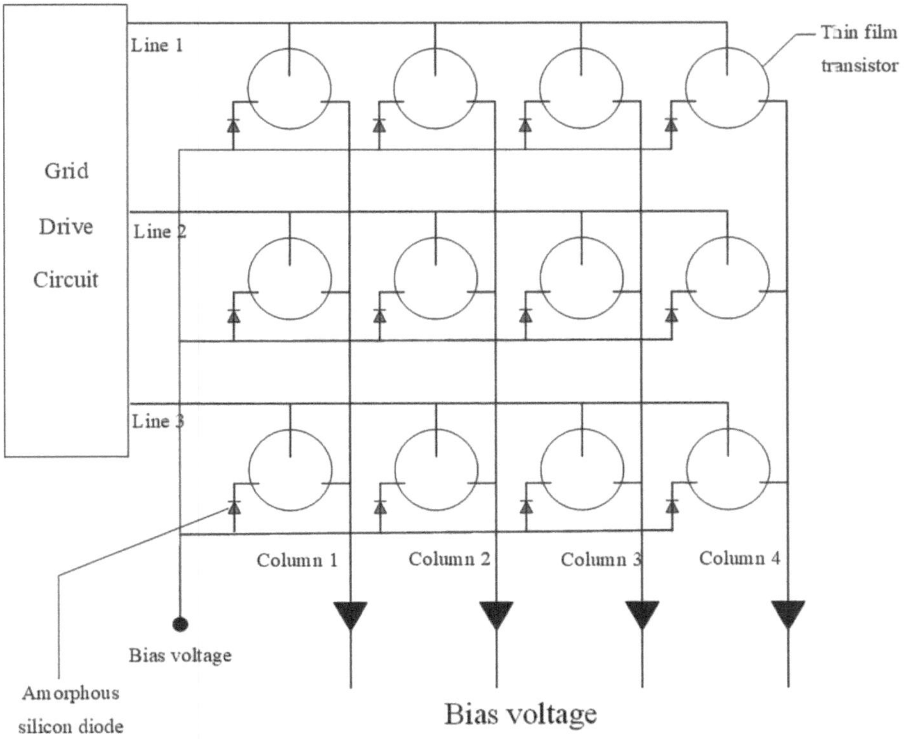

FIGURE 11.7 Circuit structure of FPD pixel array.

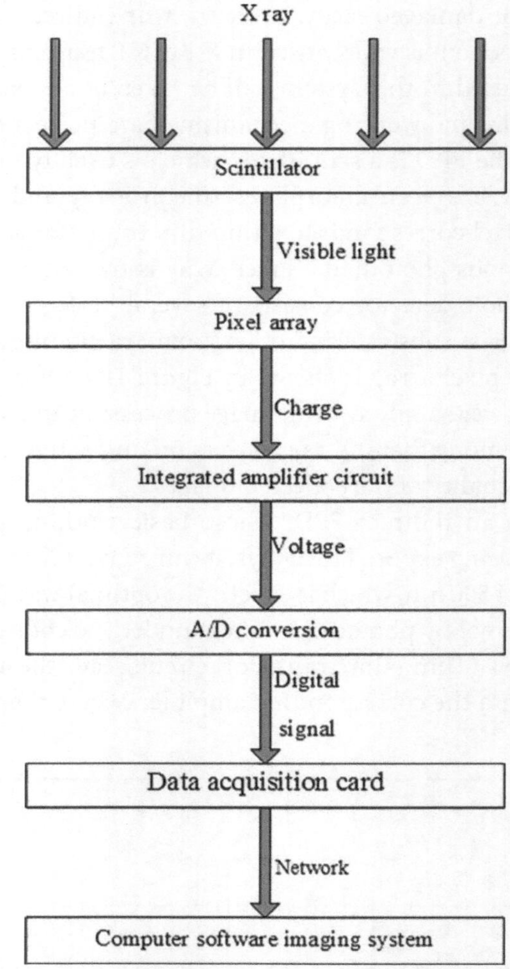

FIGURE 11.8 Imaging principle of a flat panel detector.

for A/D converter is generated. The voltage signal is then converted into a digital signal through A/D converter. Then, the data is collected by a data acquisition card and transmitted at high speed through network to the cache developed by upper-computer software. Finally, the X-ray projection image is generated.

2. Analysis of X-ray image acquisition mode

Generally, the FPDs acquire the images in two modes: internal trigger and external trigger. However, in either trigger mode, the time for acquiring an image is the integral time plus image readout time, as shown in Figure 11.9. For the SIX-650HD-E FPD, the readout time is basically 21 ms.

In the internal trigger mode, the software sends out a signal to urge the detector to acquire the image. In this system, internal trigger is applied only during the image calibration process discussed later. During the acquisition of synchronous CT projection data, the external trigger mode is selected because the projection data in the

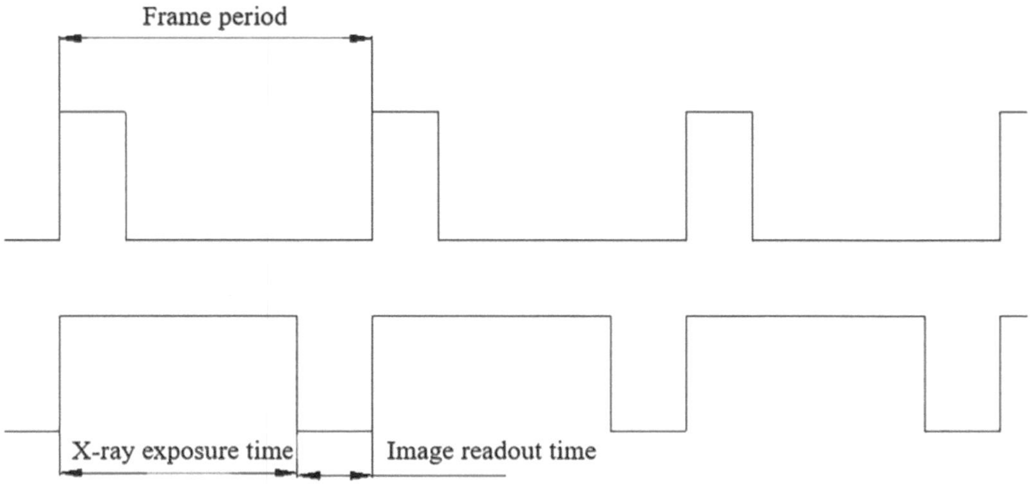

FIGURE 11.9 Schematic diagram of image acquisition process.

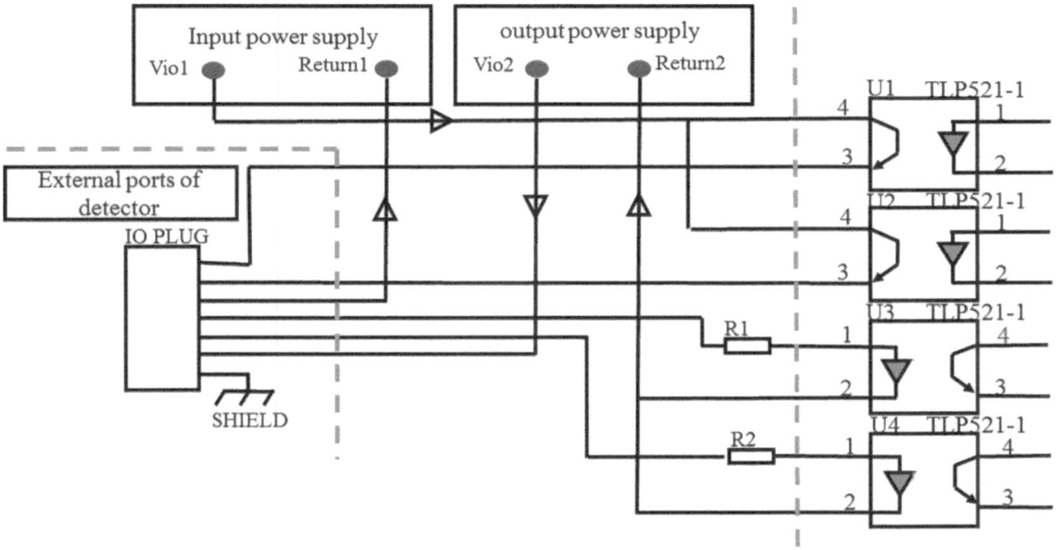

FIGURE 11.10 External communication interfaces of the SIX-650HD-E flat panel detector.

corresponding position needs to be collected. In this case, the trigger interface of the FPD will be utilized. Only when the interface receives a certain voltage trigger signal, the image can be collected. The SIX-650HD-E FPD has two sets of external ports for signal communication with other external devices, as shown in Figure 11.10.

As can be seen in Figure 11.10, the SIX-650HD-E FPD has two input signal interfaces and two output signal interfaces. Among them, In0 is the acquisition enabling port. When the enabling signal it receives is at high voltage, the acquisition enabling will be activated. In this case, if the acquisition mode is set as external trigger and the In1 port receives the trigger signal of a certain frequency, the detector will perform the projection data acquisition at this frequency. It is worth noting that the frequency of trigger acquisition signal cannot exceed the frequency set in the software acquisition mode, that is to say, the acquisition frequency set in the software is the upper limit of image acquisition frequency.

Out0 and Out1 are the two output signals of the SIX-650HD-E detector, namely X-ray exposure signal and image readout signal. Generally, the X-ray image acquisition can be realized with two methods of radiation exposure: continuous exposure and intermittent exposure. As the intermittent exposure can be guaranteed through the two external trigger signals Out0 and Out1, the X-ray exposure will be activated in the image integral time. When the image is read out, the X-ray exposure will be deactivated. In this way, the same integral time for each pixel in the FPD can be effectively ensured so as to obtain the projected image with higher quality.

11.3.2 Implementation of X-ray Image Acquisition and Real-Time Display Software

1. Method of real-time display of X-ray projection data

   Take the SIX-650HD-E FPD as an example. Its maximum acquisition frequency can reach 30 frames per second, so the acquisition time of each frame may be quite short. On the other hand, the data collected by this detector is 14 bpp, that is, the collected image is 14-digit data. However, in a high-level programming language, the smallest operable unit is generally byte, that is, the operable bits are at least 8 bits. Therefore, in the actual image acquisition, the first thing to do is to put the acquired 14-bit data into two bytes, namely the 14th bit in 16 bytes. However, in the actual image display, the 14-bit images cannot be directly displayed, that is, the screen is black now and white then so that the images can't be displayed clearly and continuously [9]. For this reason, the 14-bit data needs to be converted into 8-bit data to ensure that the computer can properly display the projection data collected by FPD. Then how to convert the 14-bit projection data with the gray level of 0–16,383, into the image with the gray level of 0–255 that can be displayed by computer? Next, this problem will be analyzed with two methods:

   The first method is threshold method, whose basic idea is to assume a reference gray value. If the gray value is X, only the images within X~$X+255$ will be displayed, as shown in Figure 11.11. As long as the gray value is beyond this range, the corresponding image will not be displayed. In other words, the gray level higher than $Y=X+255$ is considered to be the brightest, while the gray level lower than X is considered to be the darkest.

   In Figure 11.11, the first green line is X, and the second green line is Y. Since the gray values of images are mainly concentrated in this area, X can be set nearby for further study.

   The second method is mapping method [10], whose basic idea is to divide the 14-bit gray value data into three intervals, namely interval 1 ≤ minimum gray value, minimum gray value ≤ interval 2 ≤ maximum gray value and interval 3 ≥ maximum gray value. As shown in Figure 11.12, a certain percentage of gray value data in the interval 2 is mapped to the interval 0–255 and then displayed.

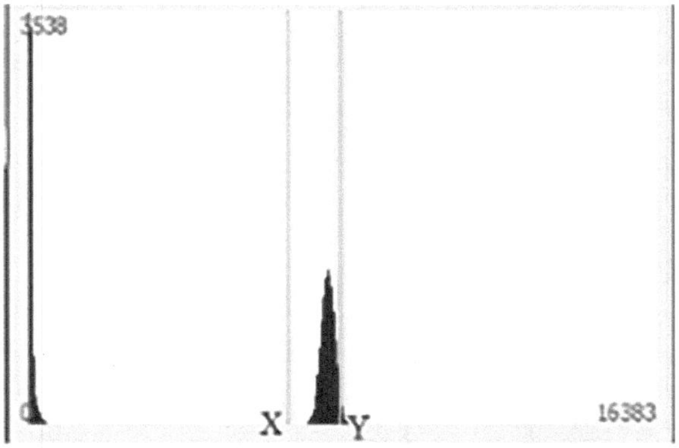

FIGURE 11.11 Representation of threshold method.

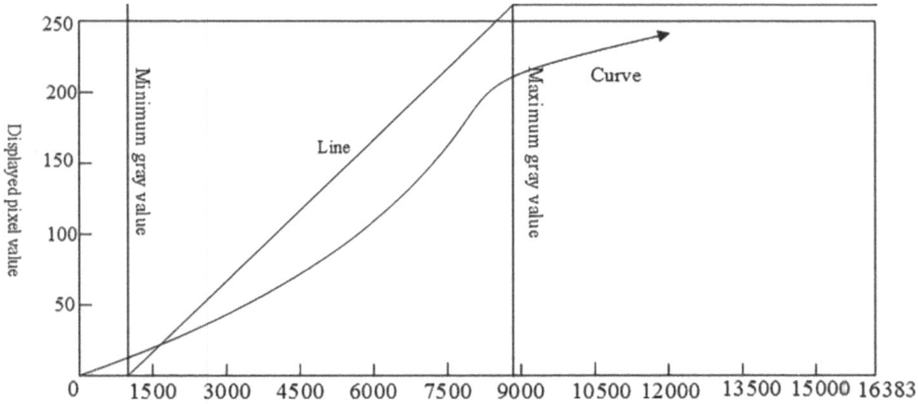

FIGURE 11.12 Segmentation and mapping of original image.

Two methods for mapping the image gray value to the interval between the minimum and maximum gray values are shown in the figure above. One method is line mapping, which can provide the image with a better contrast. The other method is curve mapping, which can soften the image. The latter is commonly used in image mapping and conversion.

2. Implementation of real-time display software of X-ray projection data

   Through the above two methods, the 14 bpp image data can be converted into the 8 bpp image data, and then the image can be displayed. In this system, the second method, namely mapping method, is adopted to realize the image display. However, in the actual software program design, the images need to be displayed in real time, so the functions of image acquisition, image gray depth conversion and image display must be implemented simultaneously. Therefore, in order to complete the above tasks at the same time, the idea of multi-threaded programming can be introduced to display the X-ray projection data in real time. The allocation of basic threaded tasks is shown in Figure 11.13.

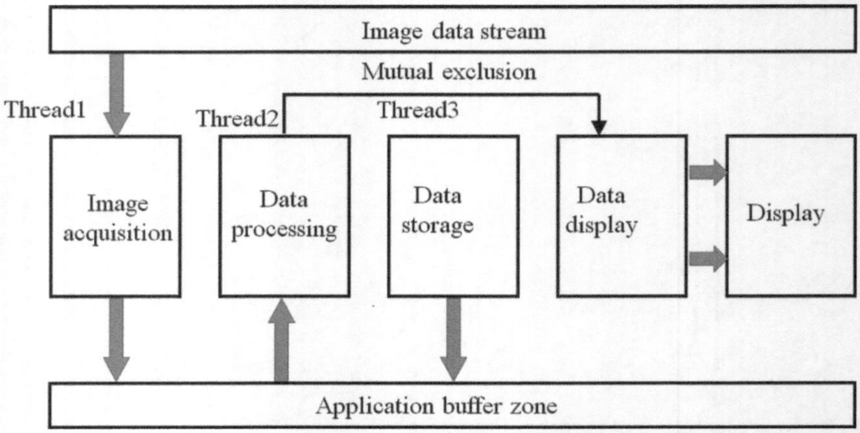

FIGURE 11.13 Block diagram of software design multithreading.

In Figure 11.13, the image acquisition is completed by the thread 1, and the image processing and display is completed by the thread 2. Then the processed image is saved by the thread 3. The image-saving thread can be controlled by the quantity of saved images, that is, when the quantity of saved images reaches the preset limit, this thread function can be blocked.

In the implementation of program design, the X-ray projection image acquired by upper computer software is a memory block, so it is displayed in the form of bitmap in this program. Therefore, we need to create a default bitmap before the program operation and then update it constantly with the images acquired by the acquisition thread. At first, the Cbitmip provided by MFC is created as a bitmap, and the CreateCompatibleDC is created as a compatible DC. Then, the Draw () function is designed to set new bitmap information and refresh the bitmap. Finally, the BitBlt function is used to paste the bitmap in the compatible DC into the current DC so that the image can be displayed. Figure 11.14 shows the real-time image display program of turbine blades.

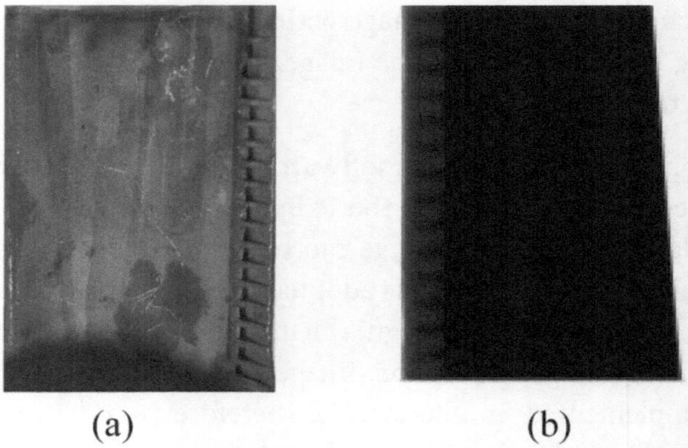

FIGURE 11.14 Real-time image display interface of turbine blade. (a) Trubine blade and (b) X-ray projection image.

```
CDC memDC;
CBitmap memBitmap;

memDC.CreateCompatibleDC(&dc);
memBitmap.CreateCompatibleBitmap(&dc, nWidth, nHeight);
memDC.SelectObject(&memBitmap);

Draw(&memDC, nWidth, nHeight);

dc.BitBlt(0, 0, nWidth, nHeight, &memDC, 0, 0, SRCCOPY);
```

11.3.3 Analysis of the Factors Affecting the Quality of X-ray Projection Images

In the actual image acquisition process, the image will inevitably be affected by dark current, inconsistent gain and bad pixels due to the manufacturing complexity of the FPD and the instability of its internal components. As a result, the projection image quality will decline to result in the artifacts in reconstruction results [11]. Therefore, this section will highlight the effects of dark current, inconsistent gain and bad pixels on image quality, as well as the implementation of software calibration.

1. Analysis of noise sources in X-ray projection images

   Generally speaking, the noise that affects the image quality of the FPD is divided into working noise and structural noise. The working noise refers to the effect of the system's mechanical motion, electronic motion and light quantum factor on the detector. The structural noise is mainly the effect of the detector structure on the image, including the effect of dark current, inconsistent response and bad pixels. Next, the structural noise will be mainly analyzed:

   1. Effect of dark current

      The image acquisition of the FPD is implemented by photoelectric conversion circuit under the control of thin film transistor. The on–off characteristics of thin film transistor are excellent. However, in practical application, a certain bias current will be generated in its output process due to its nonlinear characteristics. This results in the existence of a gray value in the detector that is not exposed to X-rays but is exposed to bias current or dark current. Thus, a certain deviation from the real dark field and an error in the calibration of dark field will be caused, and the image quality will decline. In general, in order to reduce the influence of dark current, the method of image superposition and averaging is also adopted to calibrate the images in actual image acquisition.

   2. Effect of inconsistent gain

      In an ideal state, the gray value output by the FPD is linearly related to its X-ray intensity in the acquisition of X-ray projection images. The gray value output by a single pixel can be expressed by the following equation:

$$I(x,y)=k\cdot Nd(x,y) \tag{11.10}$$

In theory, the scintillator in the detector exposed to the X-ray radiation in a certain range will convert the received X-ray photons into the corresponding output gray values in accordance with a certain conversion coefficient, also known as gain coefficient. However, the unavoidable difference in the structure of each pixel found in the manufacturing process of the FPD will lead to the difference in gain coefficient, thus causing serious non-uniformity in the collected projection data. As a result, the image quality is seriously degraded, so gain calibration is generally necessary for the acquired images.

3. Effect of bad pixels on the FPD

In the manufacturing process of the FPD, the space between the corresponding pixels is small and the number of pixels is large. For example, in the SIX-650HD-E detector used by this system, the space between pixels is 0.18 mm, and over 700,000 pixels are available. So, the defect pixels will inevitably be generated during the detector manufacturing. In general, the defect pixels are classified into too sensitive pixels or too insensitive pixels, which lead to the existence of bright or dark spots or even complete bright or dark lines scattered in the image acquired by the detector. This will directly result in the quality deterioration of a sliced image in 3D reconstruction. The acquired defect image with a defect pixel marked with a red cross is shown in Figure 11.15.

2. Comparison of corrected image and uncorrected image

In the experiment, a dark field image is first collected without any calibration, as shown in Figure 11.16, where the dark field image without correction is represented by (a), and its horizontal gray value distribution is represented by (b).

FIGURE 11.15 Image with a defect pixel. (a) The existence of bright or dark spots in the image and (b) the complete bright or dark lines scattered in the image

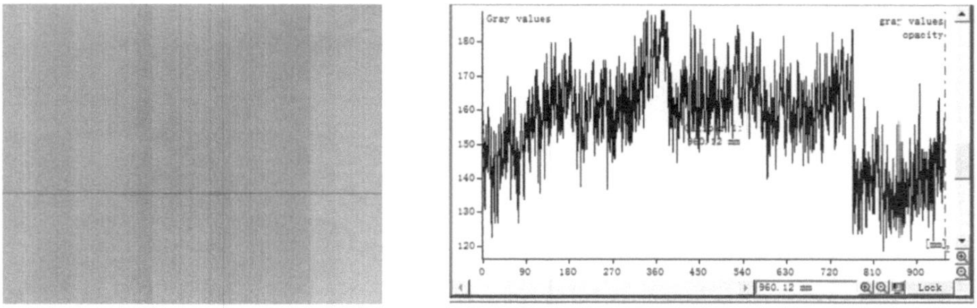

(a) (b)

FIGURE 11.16 Dark field image without calibration and its gray value distribution. (a) Uncorrected dark field images and (b) horizontal grayscale value distribution.

It can be seen from Figure 11.16 that in the dark field image without correction, the minimum gray value is generally not zero and the gray values fluctuate greatly, resulting in obvious fringe inconsistency in the projected image. It can be seen that there is a different dark current in each pixel of the detector. Then the image is corrected in the calibration process of dark current, as shown in Figure 11.17, where the corrected dark field image is represented by (a) and its horizontal gray value distribution is represented by (b).

As can be seen from Figure 11.17, the minimum gray value of the calibrated image is basically zero, and the range of its gray value fluctuation is basically controlled.

On the other hand, to calibrate the detector gain, the working voltage of X-ray source is set as 80 kev, the working current is set as 0.8 mA and the working mode of the detector is set as small-focus mode. Then, an image is acquired as shown in Figure 11.18, where its bright field image is represented by (a), its grayscale histogram is represented by (b) and its grayscale distribution is represented by (c).

It can be seen from Figure 11.18 that the responses of image pixels are obviously inconsistent and need to be corrected, as shown in Figure 11.19, where the corrected image is represented by (a), the grayscale histogram is represented by (b) and the corrected grayscale image is represented by (c).

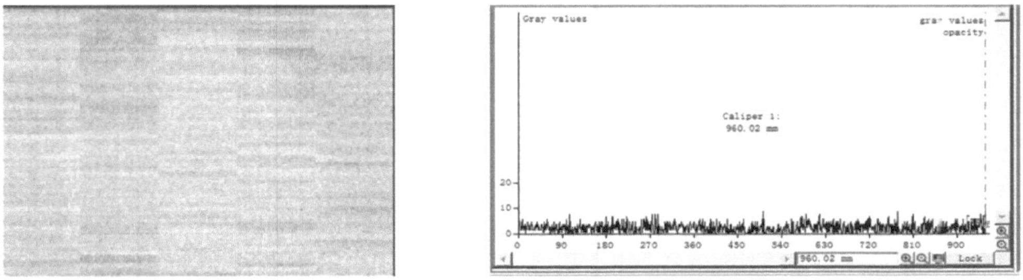

(a) (b)

FIGURE 11.17 Dark field image after bias correction. (a) Corrected dark field image and (b) horizontal grayscale value distribution.

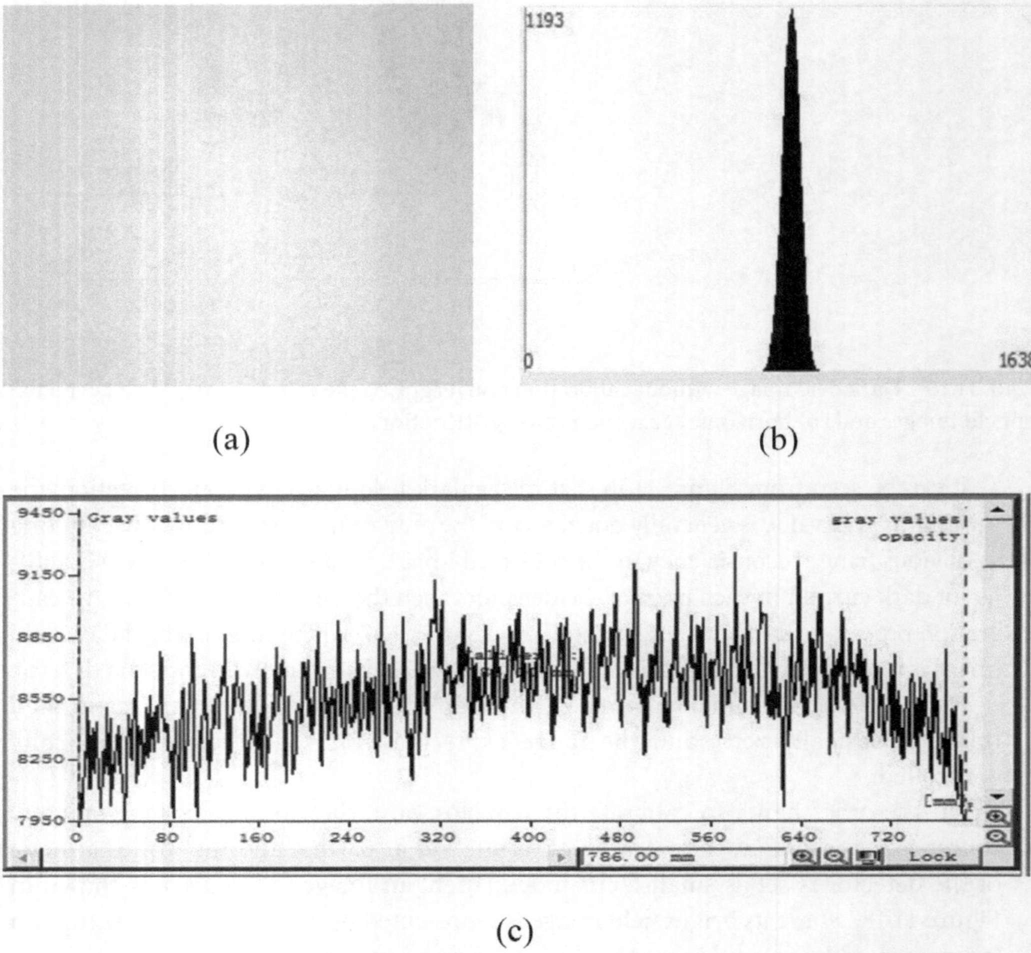

FIGURE 11.18 Uncorrected bright field image. (a) Corrected dark field image, (b) grayscale histogram and (c) horizontal grayscale value distribution.

It can be seen from Figure 11.19 that the gray values of the corrected bright field image are obviously concentrated. Although the ideal linear state isn't achieved, the calibrated image has been significantly improved compared with the uncalibrated image.

The appearance of a whole white line on the image with defect pixel is caused by excessive pixel response. This defect pixel falls into the type of too sensitive pixel. It can be clearly seen from the graph of longitudinal gray value distribution that the gray value on the white line is significantly higher, as shown in Figure 11.20.

The image whose defect pixel has been calibrated is shown in Figure 11.21. It can be seen that the curve of longitudinal gray value distribution is obviously flattened. Thus, the bad pixel that is too sensitive has been corrected to some extent.

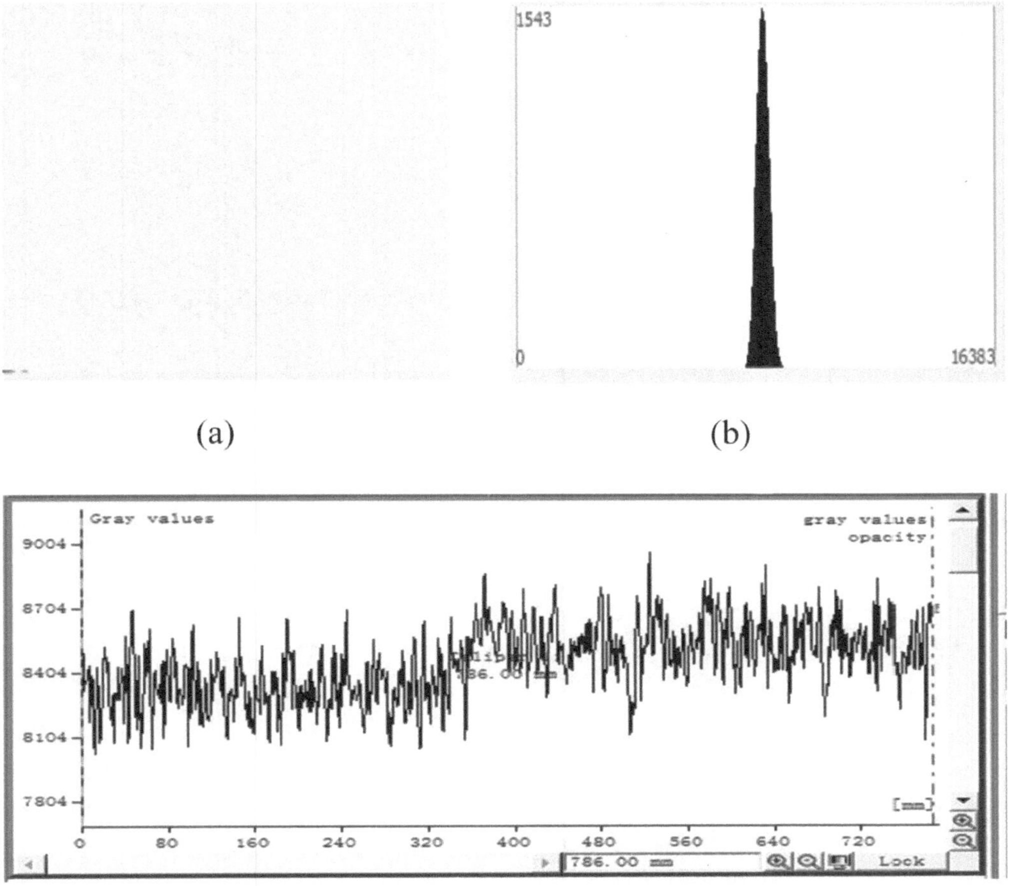

FIGURE 11.19 Corrected bright field image. (a) Corrected dark filed image, (b) grayscale histogram and (c) horizontal grayscale value distribution.

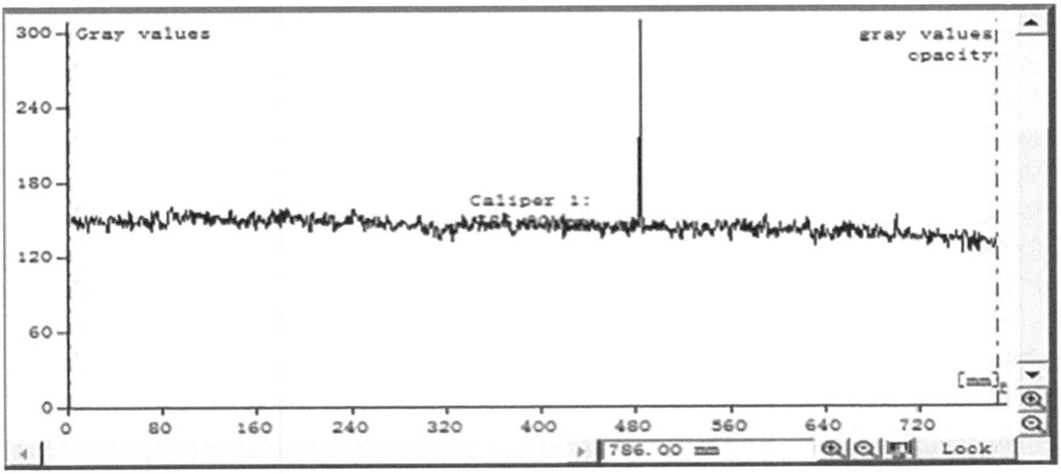

FIGURE 11.20 Longitudinal gray value distribution curve for Figure 6.16a.

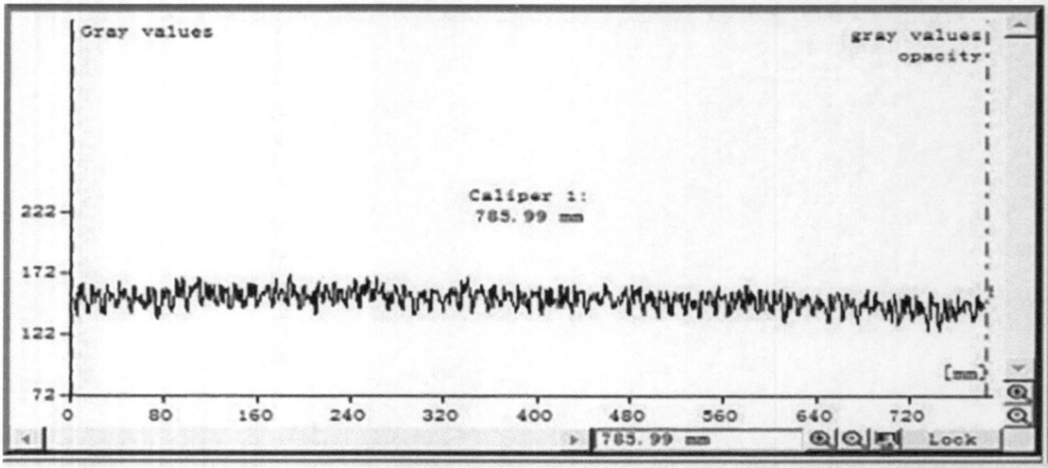

FIGURE 11.21 Longitudinal gray value distribution of corrected defect-pixel dark field image.

11.4 COOPERATIVE CONTROL OF X-RAY DETECTION DATA AND MANIPULATOR POSITION AND ATTITUDE

11.4.1 Design of Collaborative Control Concept

To complete the 3D reconstruction of the tested workpiece, its projection data at each circumferential angle should be acquired, and the acquired projection data should be consistent with the workpiece position and angle in space. Therefore, the design and completion of collaborative system control is particularly important and can directly affect the quality and accuracy of reconstructed image slices.

Through the above analysis, it can be known that collaborative data acquisition is mainly based on the spatial angle and position of the workpiece and thus should be accomplished through the external trigger function of the FPD. On the other hand, due to the line-by-line data reading of the detector, intermittent X-ray irradiation can be used in the projection data acquisition in order to avoid the image quality degradation caused by different exposure time of each line of projection data. Therefore, the on–off control of high-energy X-rays should be based on the integral time set by the detector. The core control unit of X-ray source used in this system is Siemens S7–200PLC. The coordinated and synchronous lower-level control concept of this system is designed as shown in Figure 11.22.

In the above control system, cooperative data acquisition signal or manipulator position/angle trigger signal is used as the system drive signal and is given by PMAC card (manipulator control unit) through its I/O port according to the workpiece angle and position in space. The FPD is triggered by this signal to collect data from the corresponding position, while sending an X-ray exposure signal to the X-ray controller PLC to trigger its X-ray exposure. After the preset integral time is reached, the image readout signal will be sent to the X-ray source to trigger the X-ray shutdown so as to ensure the consistency of image exposure time. Figure 11.23 describes the control timing sequences of the FPD, manipulator and high-energy X-ray source under the externally triggered intermittent exposure.

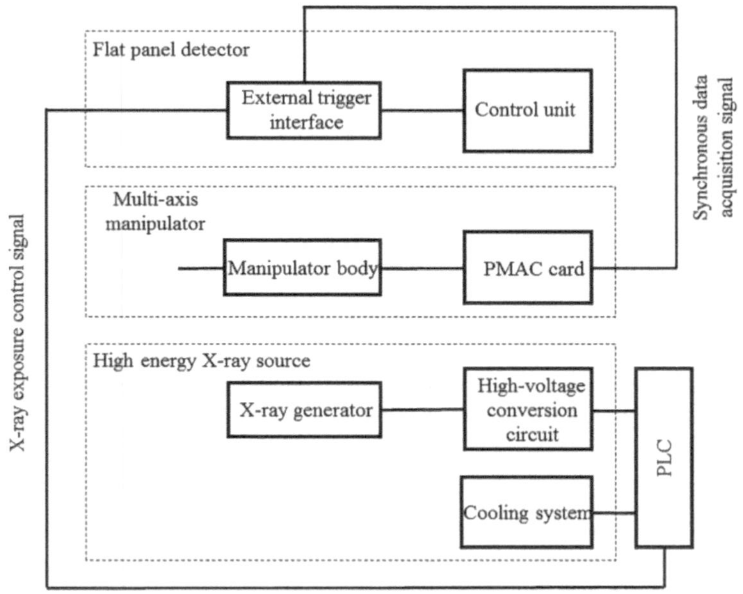

FIGURE 11.22 Coordinated lower-level control concept of a synchronous acquisition system.

As can be seen in Figure 11.23, the synchronous control of external trigger signal and X-ray exposure signal is implemented by the detector, while the synchronous control of image readout signal and X-ray shutdown signal can be implemented through an external circuit. That is to say, the X-ray shutdown can be externally triggered by image readout signal. Such a synchronization mechanism can not only accurately collect the projection data at the corresponding position and angle but also effectively ensure accurate X-ray integral time of the image.

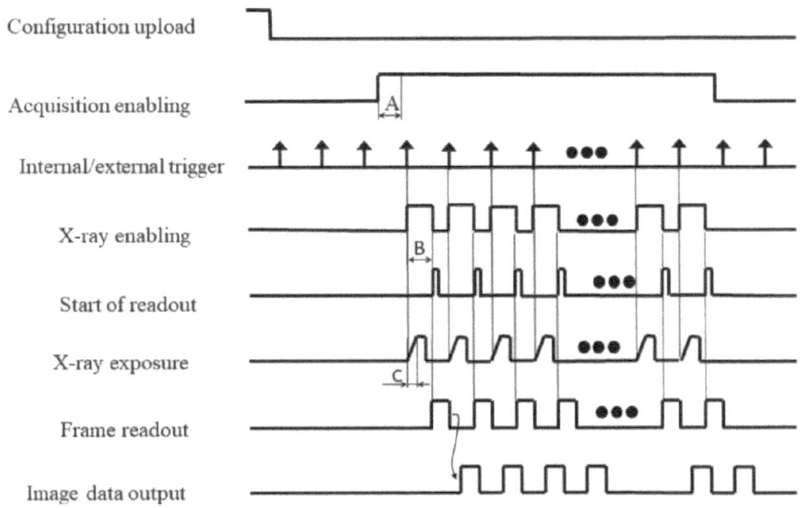

FIGURE 11.23 Acquisition timing sequences under the externally triggered intermittent X-ray exposure. (a) Acquisition delay time, (b) X-ray exposure time and (c) X-ray stabilization time.

11.4.2 Method of Manipulator Motion Control Programming in Lower Computer

The bottom program of manipulator is mainly to enable the spatial rotation, side-to-side movement, localization and homing of the tested component. Specifically, the spatial rotation includes ordinary spatial rotation and spiral spatial rotation. The movement mainly includes leftward movement and rightward movement. The localization motion is to move the workpiece under test to the center of X-ray field and ensure that its axis of rotation is coplanar with and vertical to the center line of X-ray, as shown in Figure 11.24. The homing motion is to enable the manipulator to return to its initial position.

The motion control of the manipulator starts from its initial motion position, as shown in Figure 11.25. Therefore, synchronous data acquisition should be preceded by the localization motion, that is, the rotation axis at the end of the manipulator should coincide with the Z-axis in Figure 11.24. Spatial rotation or horizontal movement should be based on the manipulator position after localization.

The motion control system of this manipulator has a core control unit, namely PMAC card, and a cache for storing multiple motion control programs (PROGs) and a PLC. Therefore, during actual programming, each motion should be implemented through a different motion control program, and each motion control program should be stored in a different buffer zone [12]. In this system, multiple motion control programs are designed to implement the localization, ordinary rotation scanning, spiral rotation scanning, horizontal leftward movement, horizontal rightward movement and homing. In addition, a logical control program (namely PLC program) is designed to switch on/off the above six motions, that is, a control variable is set for each motion program. Since the PLC program in PMAC card has the functions of powered operation and cycled operation, the powered manipulator can implement automatic localization motion through the design of PLC program and implement automatic homing motion after the completion of workpiece scanning. In this way, automatic workpiece detection is guaranteed.

The localization should be followed by synchronous workpiece data acquisition. Before acquisition, the relevant scanning parameters such as rotation speed and acquisition step size should be set. Here, by using the P variable in the PMAC card as an intermediate

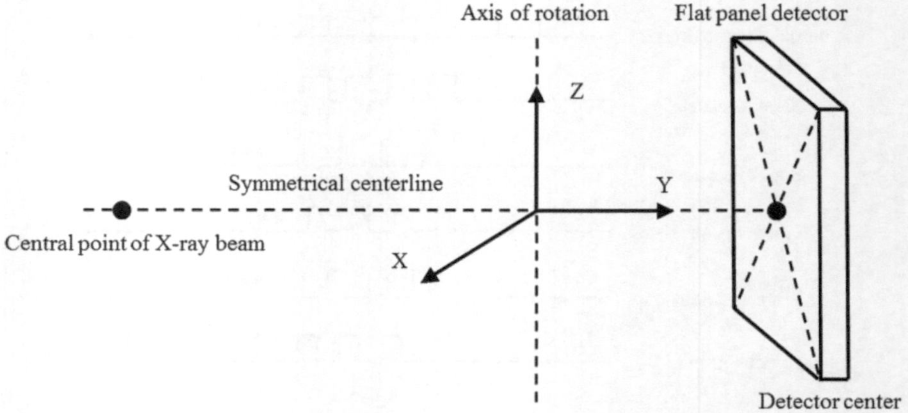

FIGURE 11.24 Diagram of manipulator localization result.

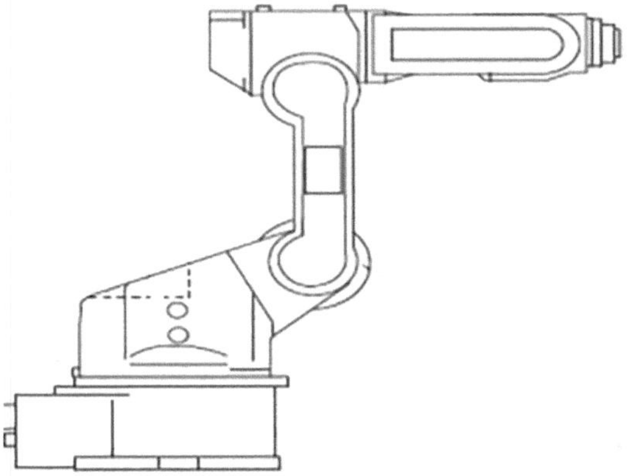

FIGURE 11.25 Initial manipulator position.

variable, the data transmission of upper and lower computers can be realized. The schematic diagram of synchronous scanning motion control process is shown in Figure 11.26.

Here, the control program of synchronous rotational scanning motion is taken as an example to describe the process of motion control program. At first, the upper computer sends the scanning parameters to the PLC program in the lower computer. Then the control variables of scanning motion control program in the PLC are activated to run the scanning motion program. The parameters in the scanning program are controlled or monitored by the upper computer by setting the variables. In the actual scanning motion programming, codes can be written as follows:

```
...........     //P (table of variable macros)
F(SPEED)      //SPEED (a macro of the variable P, representing the
running speed)
WHILE(NUM !> TOTAL_NUM)
   C(ACQ_STEP_LEN)     // ACQ_STEP_LEN (a macro of the variable P,
representing the acquisition step size)
  ........
   M1= TRIH_SIG_HIGH
```

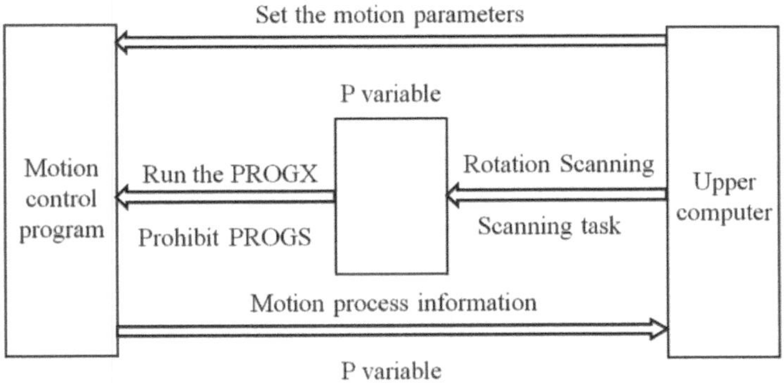

FIGURE 11.26 Schematic diagram of motion program control process.

```
   DWELL PULSE_WITH   // PULSE_WITH (a macro of the variable P,
representing the synchronous trigger pulse width)
   M1= TRIH_SIG_LOW
   DWELL(P19)
......
ENDWHILE
```

According to the above codes, the upper computer can change the synchronous scanning parameters of the system by changing the intermediate variables in the lower computer. The motion information as well as the workpiece image projected to the corresponding position and the workpiece position information can also be obtained from intermediate variables so as to realize synchronous workpiece data acquisition.

11.4.3 Modes of Communication and Control of Upper and Lower Computers

The function of upper-computer application program in cooperative control is mainly to achieve the effective upper control of each control unit in the system, so as to realize cooperative work. For this system, the upper computer program is mainly to effectively control the FPD, the high-energy X-ray source controller PLC and the multi-axis motion control card PMAC of manipulator controller. Therefore, the first problem to be solved is the communication between upper and lower computers and the way to control them. Their basic communication and control modes are shown in Figure 11.27.

For the communication control of the FPD, the projection data should be continuously transmitted at high speed to the upper computer for the purpose of image display. Therefore, in order to meet the demand of high-speed data transmission, the detector is communicated with the upper computer through gigabit Ethernet, that is, its data transmission at the physical layer is guaranteed by Cat6e network cables. In terms of detector control, the interface functions in the application software development kit SIX-HD-SDK provided by Thales are used in this system to control the detector. For example, the SIX::init() function of SIX-HD-SDK is called to initialize the detector; the SIX::Active()

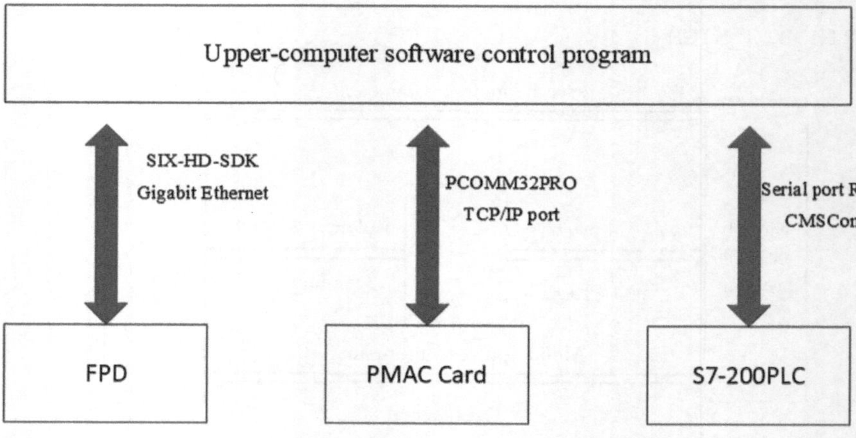

FIGURE 11.27 Schematic diagram of basic communication and control modes.

function is called to activate the acquisition mode loaded in the initialization process and the SIX::SetInputSignal function is called to achieve the function mapping at the external trigger port. The basic control process is shown in Figure 11.28.

The upper control of the manipulator is mainly implemented by its core control unit – PMAC card. The PAMC card can communicate with upper software in many ways, including USB port communication, serial port communication and network port communication. Among them, network port communication is subject to the common TCP\IP communication protocol and has fast control response and data transmission speed. Therefore, network port communication is adopted by this system. Regarding its specific control process, the PMAC software development kit provides a set of interface functions (namely PCOMM32PRO library) for the communication data transmission between upper and lower computers, thus accomplishing the functions such as PMAC card connection, program download, data transmission and intermediate variable monitoring. The main functions we use are shown in the following table (Table 11.1).

With regard to the upper computer control of high-energy X-ray source, the PORT1 serial port of S7–200PLC is used in this system to realize the communication between upper and lower computers. At the same time, the corresponding data transmission message is formulated in the lower program of PLC. Therefore, the CMSComm library in C++ can be used to initialize the serial port of upper IPC and send the control message to the lower computer, so as to control the lower computer. For example, to increase the

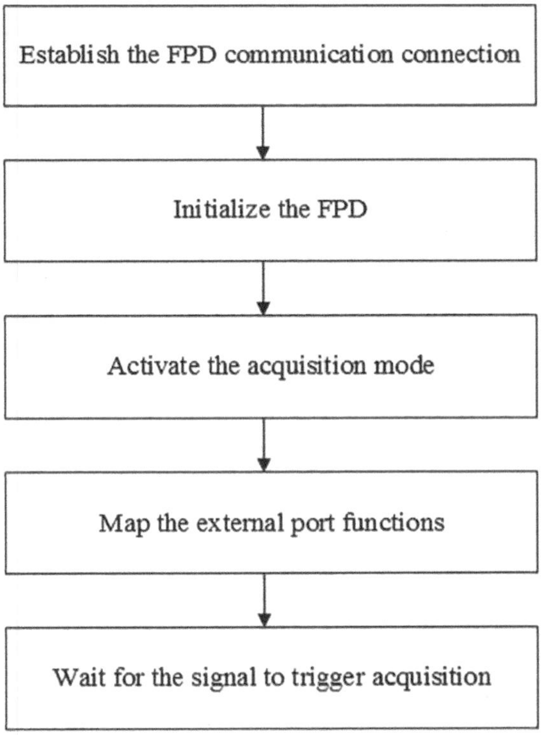

FIGURE 11.28 Control flow diagram of a flat panel detector (FPD).

TABLE 11.1 Common Functions of PMAC Cards

Function	Functionality
Pmac.SelectDevice(0,out dev,out bSuccess)	Select the PMAC
Pmac.Open(m_nDevice,out m_bDeviceOpen);	Switch on the PMAC
Pmac.Close(m_nDevice);	Switch off the PMAC
Pmac.GetnetAcualPosition(int dwDevice, int mtr, double scale, out double pVal)	Acquire the encoder position
Pmac.GetResponseEx(int m_nDevice, string question,bool bAddLF, out string pAnswer, out int pstatus);	Interact with the card and realize the functions such as instruction sending and data reading

exposure voltage of X-rays during exposure, the put_Output() function in the CMSComm library can be used to send the following control message to the downstream computer (Table 11.2):

After receiving the corresponding control message, the lower computer will immediately reply the following response message to the upper computer. In the upper software, the message data can be received through the OnCommMscomm() function in the CMSComm library and then further analyzed to obtain the corresponding control information (Table 11.3).

It should be noted that when setting the current and voltage, the current values are multiplied by 10, rounded off and then sent directly to the upper software for display.

11.4.4 Implementation Method of Cooperative Control Software

In the actual operation of FPD, the X-ray intensity needs to be adjusted by FPD to adapt to different workpieces with different X-ray absorption rates. Therefore, in the upper software, the adaptation of X-ray energy to different workpieces should be achieved through the synchronous control of all the units in the X-ray detection system. Meanwhile, the gray value of the projected image should be verified in order to obtain an image with high quality and contrast.

From the above description, it can be seen that the upper software for synchronous control is mainly to achieve the coordinated control of each control unit in the system and the verification of the acquired image. The so-called image verification is to judge whether the gray value of the acquired image is within the required range, so as to prevent the overexposure or underexposure of the image.

TABLE 11.2 Data Sent by Upper Computer

Start	Control Bit	Placeholder	Voltage	Current	End
FA00	16 bit	16 bit	16 bit	16 bit	00FB
FA00	(0X) 0040	0000	120	2	00FB

TABLE 11.3 Data Received by Upper Computer

Start	Control Bit Return	Defect Character	Voltage	Current	End
FA00	16 bit	16 bit	16 bit	16 bit	00FB
FA00	(0X) 0040	0000	120	2	00FB

In the actual software engineering, the whole control function of the software can be modularized, that is, each control unit can be modularized. Through the effective application of each module, the coordination and synchronization of control units are realized, and the image is verified. The idea of modularization in software design can ensure the data security of modules and the strong logic of the whole software engineering that can't be easily understood or tampered by others. On the other hand, owing to the software modularization, the software code has higher utilization, the whole engineering looks more simple and understandable and the consumption of computer memory is reduced. The modular structure of synchronization control software in this system is shown in Figure 11.29.

As shown above, the synchronization control software can be divided into five submodules. In the actual code implementation, the object-oriented idea of C++ programming language can be used to encapsulate the sub-modules into five categories, namely CManipulatorCtrl, CFPDCtrl, CX-RayCtrl, CImageCheckCtrl and CUserMagament. Each category contains the corresponding member functions and member variables that are used to control each control unit and monitor its running state. In addition, some global intermediate variables or structures are used for the communication between controllers on running state. The basic workflow of this software is shown in Figure 11.30.

In the implementation process of synchronization software, the detection module is called to map the external ports of FPD and initialize their settings. Then the scanning control module of the manipulator is called to set the acquisition parameters for workpiece scanning data. Then the lower program for synchronous scanning acquisition is called by manipulator under the corresponding scanning instruction given by upper computer software, in order to synchronously acquire the projection data of the tested workpiece. If the workpiece is tested for the first time, the acquired projection data (i.e. gray value) can be used for verification. If the gray value is outside the preset range, the member functions in X-ray control module can be called to adjust the current and voltage of X-ray tube as well as the X-ray energy and then to obtain the right image.

After the adjustment of X-ray source intensity, the workpiece can be rotated and scanned. To prevent the image quality degradation caused by the accumulation of charges on FPD, the workpiece data is often collected for three times at the same position and angle, and the data collected in the third projection is kept to ensure the relative stability of the projected image. Then the projection data is collected at the next position and angle. By following this process, the circumferential synchronization data of the tested workpiece can be successfully acquired and saved.

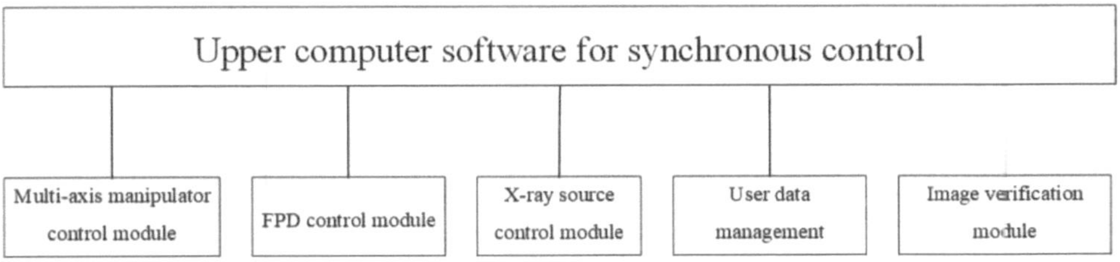

FIGURE 11.29 Framework of synchronization control software.

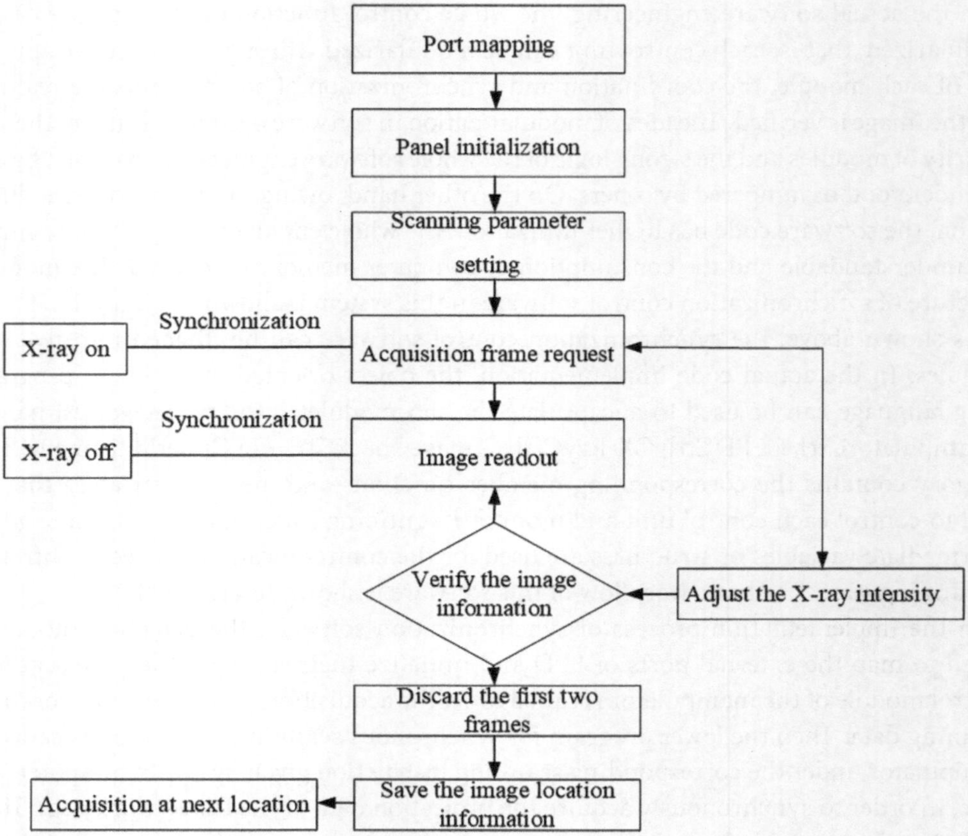

FIGURE 11.30 Basic workflow of synchronize software.

11.5 AN EXAMPLE OF HOLLOW COMPLEX COMPONENT UNDER TEST

In this system, the 3D reconstruction of the tested workpiece is implemented by using the basic calculation principle of X-ray CT 3D image reconstruction algorithm (namely FDK algorithm) and the image reconstruction module in VGStudio software.

First of all, the workpiece to be tested shall be installed on the rotation axis at the end of the manipulator, and then its projection data at different circumferential angles shall be collected. The physical picture of the X-ray CT detection system is shown in Figure 11.31; and the extracted projection data of the three frames at the corresponding angles is shown in Figure 11.32.

In the process of VG software reconstruction, it is necessary to import the projection data of the workpiece at different angles obtained through the collaborative acquisition system. In addition, the orientation of the projected image should be checked and verified through the projection preview function of VGStudio, which can show the 0° and 90° projections.

The gain or bias of the projection data, if not corrected before acquisition, needs to be calibrated by importing the corresponding bright field and dark field images. Then it is required to input the performance parameters of FPD and the distances from the X-ray source to the FPD and the rotation axis of workpiece in space.

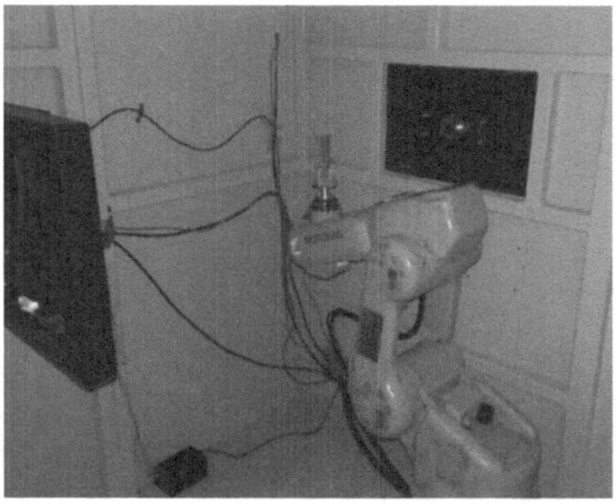

FIGURE 11.31 Physical picture of an X-ray CT scanning system on a manipulator.

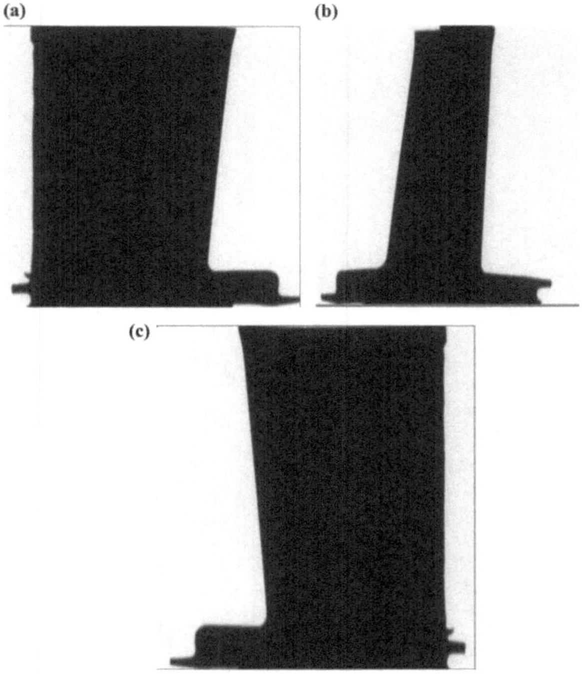

FIGURE 11.32 Projected image of workpiece. (a) Projection imaging of workpiece rotated 0 degrees, (b) projection imaging of workpiece rotated 90 degrees and (c) projection imaging of workpiece rotated 180 degrees.

It is worth noting that the calibration of horizontal and vertical offsets of the rotation axis has been added to the VG software. The measured geometric parameters can be filled into the corresponding dialog boxes so as to put an end to the calibration of the installation geometry parameters in the reconstruction process. Meanwhile, the function of automatic rotation-axis tilt calibration is provided by VG and can automatically calculate the vertical tilt angle of the workpiece's rotation axis.

At this point, the main parameters used for the 3D reconstruction of the projection data have been basically set up, so the projection data of the tested workpiece can be reconstructed in three dimensions. The reconstructed 3D image is shown in Figure 11.33. In the image, the geometric structure inside the turbine blade is basically characterized. By analyzing the internal structural characteristics directly reflected in the image, the defects in the tested workpiece can be accurately evaluated.

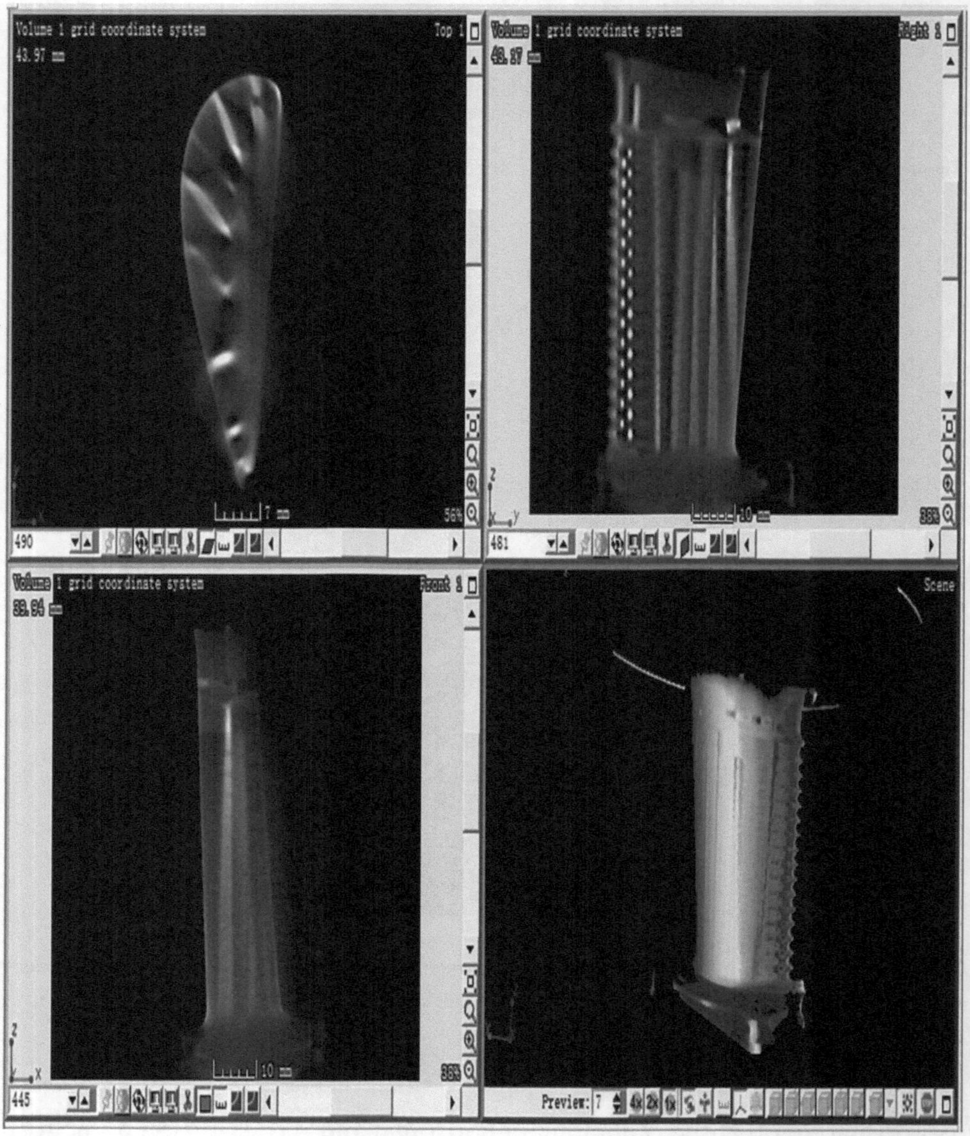

FIGURE 11.33 Schematic diagram of 3D reconstruction.

REFERENCES

1. Persson S., Ostman E., Use of computed tomography in nondestructive testing of polymeric materials [J], *Applied Optics*, 1985, 24(23):4095–4104.
2. Avinash C.K. *Principles of Computerized Tomographic Imaging [M].* New York: IEEE Press, 1988.
3. Zhang C., Guo Z., Zhang P., Technology and principle of industrial CT [J], 2009.
4. Li H. *Research on the Geometric Artifact Correction Technology of Cone-beam CT System [D].* Guangzhou: Southern Medical University, 2015.
5. Toft P.A., Sørensen J.A. *The Radon Transform-theory and Implementation [D].* Technical University of Denmark, Danmarks Tekniske Universitet, Center for Bachelor of Engineering Studies, Center for Diplomingeniøruddannelse, Center for Information Technology and Electronics, Center for Informationsteknologi og Elektronik, 1996.
6. Epstein C.L. *Introduction to the Mathematics of Medical Imaging [M].* Philadelphia: Siam, 2008.
7. Zhao X., Zhang D., Rapid quality test of turbine blades based on cone-beam volume CT [C]. Xi'an: The 9th Young Scientists Forum of Shaanxi Province.
8. Varian Medical System. PaxScan 2520 VIP-9 System & Service Guide, 2000.
9. Wu Z., Image processor in container detection system [J], *Nuclear Electronics & Detection Technology*, 1999, 1:30–33.
10. Ji S., Microcomputer image display system based on memory mapping [J], *Computerworld*, 1995, 5:95–99.
11. Zhou Z. Digital Imaging and Image Calibration of X-ray Flat Panel Detector [J]. Beijing: *Journal of University of Aeronautics and Astronautics*, 2004.
12. Men C., Guan X., Hu M., et al.,Development of open control system of six-DOF robot based on PMAC [J], *Development & Innovation of Machinery*, 2008,04:4–13.

CHAPTER 12

Robotic Electromagnetic Eddy Current Testing Technique

ELECTROMAGNETIC EDDY CURRENT TESTING is based on the principle of electromagnetic induction and is widely used in the nondestructive testing (NDT) of conductive materials [1]. This technique, together with the associated equipment, is widely used in aerospace, metallurgy, machinery, electric power, chemical industry and other fields, thus playing an increasingly important role. Compared with ultrasonic testing (UT), X-ray testing and penetrant testing, this testing technique has its own characteristics. Through the combination with single-manipulator scanning technique, it can achieve satisfactory non-contact automatic scanning. By increasing the eddy current detection frequency, it can detect the defects on or near surface and avoid the need for water coupling (like in NDT) to realize high-frequency UT [2,3]. It is especially suitable for in situ testing.

12.1 BASIC PRINCIPLE OF ELECTROMAGNETIC EDDY CURRENT TESTING

12.1.1 Characteristics of Electromagnetic Eddy Current Testing

The advantages of electromagnetic eddy current testing are as follows:

1. **Fast detection speed**: The eddy current testing does not need any coupling agent and can realize non-contact detection, so its detection efficiency is quite high (for example, for tubes and bars);
2. **High sensitivity**: This technique is very sensitive to the defects on or near work-piece surface, especially the crack defects. Therefore, eddy current testing can be used as one of the testing means in quality management and control;
3. **Easy to realize automation**: Eddy current testing is a process of electromagnetic induction. Its detection signal is processed and identified by circuit and computer, so this testing can be automated without human operation. Since the detection signal

DOI: 10.1201/9781003212232-12

is an electrical signal, the detection result can be digitized and then automatically stored, reproduced and processed;

4. **Suitable for the workpieces with a complex shape**: The probe coil used in eddy current testing can be wound into various shapes, so it can adapt to the cross-section shapes such as triangle, square, rectangle and circle. The coil can also be made into a very small probe to test a variety of workpieces, such as the special workpieces with narrow areas, deep holes and tube walls.

The disadvantages of electromagnetic eddy current testing are as follows. The object of eddy current testing must be a conductive material or a non-metallic material capable of inducing eddy currents. Because this testing is based on the principle of electromagnetic induction, it is mainly used to detect the defects on metal surface and subsurface, rather than inside metal material. When the excitation frequency of electromagnetic eddy current increases, the eddy current density of material surface and the detection sensitivity will increase, but the penetration depth will decrease and vice versa. Therefore, the depth of eddy current detection and the detection sensitivity to surface defects need to be traded off before testing, and the detection scheme and technical parameters need to be determined according to material, surface state, test standard and other factors.

12.1.2 Principle of Electromagnetic Eddy Current Testing

12.1.2.1 Electromagnetism Induction Phenomenon

Electromagnetic induction refers to the mutual induction between electricity and magnetism, including electrically induced magnetism and magnetically induced electricity. The phenomenon that a magnetic field is generated near a live wire is called electrically induced magnetism. When the magnetic flux passing through the area surrounded by a closed conductive loop is changing, a current will be generated in the loop. This phenomenon is known as magnetically induced electricity, as shown in Figure 12.1. The current generated in the loop is called induced current. In addition, when a wire in a closed loop moves in

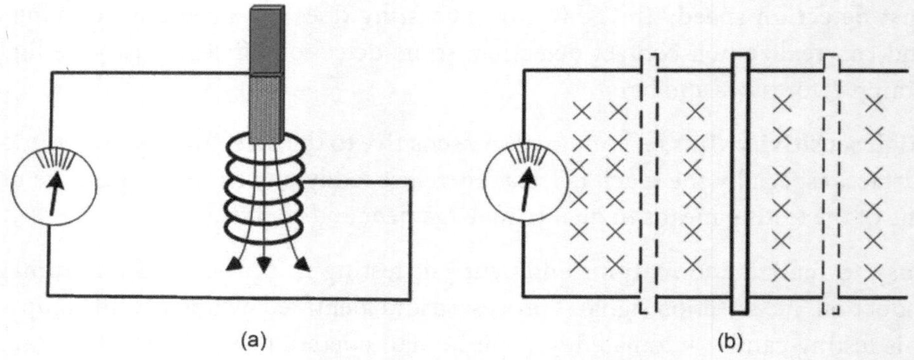

FIGURE 12.1 Electromagnetism induction phenomenon. (a) A magnet passes through a coil and (b) a wire cuts magnetic field lines.

a magnetic field and cuts the magnetic field lines, it will also generate an electric current, which is also magnetically induced electricity.

In the electromagnetic induction, as long as there is a change in the magnetic flux passing through the area surrounded by a closed path, an induced electromotive force (EMF) will be generated. For a non-closed path, as long as the magnetic induction line is cut, an induced EMF will also be generated.

The direction of the induced current can be determined by Lenz's law. The magnetic field generated by the induced current in a closed loop always impedes the change of magnetic flux that generates the induced current. The direction of this current is just the direction of the induced EMF. The direction of the induced EMF generated when a wire cuts magnetic field lines can also be determined using the right-hand rule.

12.1.2.2 Faraday's Law of Electromagnetic Induction

When the magnetic flux passing through the area surrounded by a closed loop is changing, an induced EMF E_i equal to the negative value of the time-dependent change of the magnetic flux Φ in the enclosed area will be generated in the loop:

$$E_i = \frac{d\Phi}{dt} \tag{12.1}$$

where the minus ("–") sign indicates that the magnetic field generated by the induced current in the closed loop always impedes the change of magnetic flux that generates the induced current. This equation is called Faraday's law of electromagnetic induction.

If the above equation is obtained through a tightly wound coil with N turns and the magnetic flux Φ passing through each turn is the same, then the induced EMF of the loop will be:

$$E_i = -N\frac{d\Phi}{dt} = -\frac{d(N\Phi)}{dt} \tag{12.2}$$

When a long wire with a length of l is cutting magnetic field lines in a uniform magnetic field, the induced EMF E_i generated in the wire will be

$$E_i = Blv\sin\alpha \tag{12.3}$$

where B is the magnetic induction intensity, in T; l is the wire length, in m; v is the wire speed, in m/s and α is the angle between the moving direction of the wire and the magnetic field.

12.1.2.3 Self-Inductance

When there is a time-dependent alternating current I in the coil, the alternating magnetic flux generated by it will also generate an induced EMF in the coil. This is just the self-induction phenomenon. The generated EMF is called the self-induced EMF E_L:

$$E_L = -L\frac{dI}{dt} \tag{12.4}$$

where L is the self-induction coefficient (in H), called self-inductance for short.

12.1.2.4 Mutual Inductance

When the two coils with the currents I_1 and I_2 are close to each other, the changing magnetic field caused by the current 1 in coil 1 will generate an induced EMF in coil 2 when passing through the coil 2. Similarly, the changing magnetic field caused by the current in coil 2 will also generate an induced EMF in coil 1 when passing through the coil 1. This phenomenon is called mutual inductance. The generated induced EMF is called mutually induced EMF. When the shape, size, number of turns, mutual position and surrounding magnetic medium of the two coils are fixed, the induced EMF generated by each other will be

$$E_{21} = -M_{21}\frac{dI_1}{d_t} \qquad E_{12} = -M_{12}\frac{dI_2}{d_t} \tag{12.5}$$

where M_{21} and M_{12} are the mutual inductance coefficient of coil 1 to coil 2 and the mutual inductance of coil 2 to coil 1, respectively, referred to as mutual inductance (in H), $M_{21} = M_{21}$.

The first number in each subscript represents the coil inducing the EMF, while the second number represents the coil causing the induction.

Mutual inductance is related not only to the shape, size and surrounding medium of the coils and the magnetic conductivity of the material but also to the mutual position between the coils.

When the above coupling occurs between two coils, the degree of coupling between them can be expressed by the coupling coefficient K, whose magnitude is as follows:

$$K = \frac{M}{\sqrt{L_1 L_2}} \tag{12.6}$$

where L_1 and L_2 are the self-inductance coefficients of coil 1 and coil 2, respectively, and M is the mutual inductance between coil 1 and coil 2.

12.1.3 Eddy Current and Its Skin Effect

Due to electromagnetic induction, a current will be induced in a conductor staying in a changing magnetic field or moving relative to it. This current will form a closed loop in the conductor and flow like an eddy, so it is called eddy current. Eddy current testing is an important application of eddy current effect. When the detection coil with alternating current is close to a conductive specimen, an eddy current will be generated in the specimen due to the effect of a magnetic field created by the excitation coil. Its magnitude, phase and flow form will be influenced by the specimen conductivity. The eddy current will also generate a magnetic field, which, in turn, will change the impedance of the detection coil.

Therefore, by measuring the variation in the impedance of the detection coil, the specimen performance and the existence of a defect can be verified.

When a direct current flows through a wire, the current density across the cross section of the wire is uniform. However, when an alternating current flows through the wire, the changing magnetic field around the wire will generate an induced current in the wire, thus causing uneven distribution of current on the cross section. The current density on the surface is higher. However, as the current moves toward the wire center, its density will attenuate according to the law of negative exponent. Especially at higher frequencies, the current will flow almost in a thin layer near the wire surface. The phenomenon that the current is mainly concentrated near the conductor surface is called skin effect.

The distance at which the eddy current penetrates the conductor is called penetration depth. The penetration depth achieved when the eddy current density attenuates to $1/e$ of its surface value is called standard penetration depth, also known as skin depth. It represents the skin degree of eddy current in the conductor, expressed by the symbol δ and measured in m. From the Maxwell equation for the electromagnetic field in a semi-infinite conductor, the eddy current density at the depth x from the conductor surface can be derived as

$$I_x = I_0 e_X^{\sqrt{xf\mu\sigma}\,x} \tag{12.7}$$

where I_0 is the eddy current density on the semi-infinite conductor surface, in A; f is the frequency of alternating current, in Hz; μ is the material's magnetic conductivity, in H/m and σ is the material's electrical conductivity, in S/m. Then the standard penetration depth is

$$\delta = \frac{1}{\sqrt{\pi f\mu\sigma}} \tag{12.8}$$

It can be seen from Eq. (12.8) that a material with higher frequency or better electrical or magnetic conductivity will demonstrate a more significant skin effect. For a nonferromagnetic material, its magnetic conductivity is $\mu \approx \mu_0 = 4\pi \times 10^{-7}$ H/m. Thus, its standard penetration depth is:

$$\delta = \frac{503}{\sqrt{f\sigma}} \tag{12.9}$$

In practical engineering applications, the standard penetration depth δ is an important data, because the density of eddy current has generally decreased by about 90% at 2.6 penetration depths. In engineering sense, 2.6 standard penetration depths are generally defined as the effective penetration depth of eddy current. This means that 90% of the eddies within 2.6 standard penetration depths have an effective effect on the detection coil, while 10% of the eddies outside 2.6 standard penetration depths have a negligible effect on the coil.

12.1.4 Impedance Analysis Method

The impedance of an ideal coil should include only the inductive impedance [4], and the coil resistance should be zero. However, a coil is actually formed by wound metal wires, which have not only resistance but also inductance. In addition, a capacitance exists between the wire turns. Therefore, a coil can be represented by a circuit where the resistance, inductance and capacitance are connected in series. Generally, the distributed capacitance between the wire turns is ignored. That is to say, a coil is often represented by a series circuit with resistance and inductance, as shown in Figure 12.2. Therefore, the complex impedance of a coil can be expressed by the following equation:

$$Z = R + jX = R + j\omega L \tag{12.10}$$

where R represents resistance; X represents reactance, $X = \omega L$; and ω represents angular frequency, $\omega = 2\pi f$.

In the circuit shown in Figure 12.3a, the two coils are coupled to each other, and the primary coil is infused with an alternating current I_1. According to the above analysis, the circuit in Figure 12.3 can be equivalent to the circuit shown in Figure 12.3b. Due to electromagnetic induction, the secondary coil obtains an induced current, which, in turn, affects the current and voltage of the primary coil. This effect can be represented by the reduced impedance of primary coil circuit reflected by the impedance of secondary coil circuit through mutual inductance. Its equivalent circuit is shown in Figurel2.3c, where the reduced impedance Z is

$$Z_e = R_e + jX_e \quad R_e = \frac{X_m^2}{R_2^2 + X_2^2} R_2 \quad X_e = \frac{X_m^2}{R_2^2 + X_2^2} \tag{12.11}$$

where R_2 is the resistance of the coil; X^2 is the reactance of the coil, $X^2 = \omega L_2$; X_m is the mutual inductive impedance, $X_m = \omega M$; and R_e and X_e are reduced resistance and reactance, respectively.

In addition, the sum of the reduced impedance of secondary coil and the impedance of the primary coil itself is called the current impedance Z:

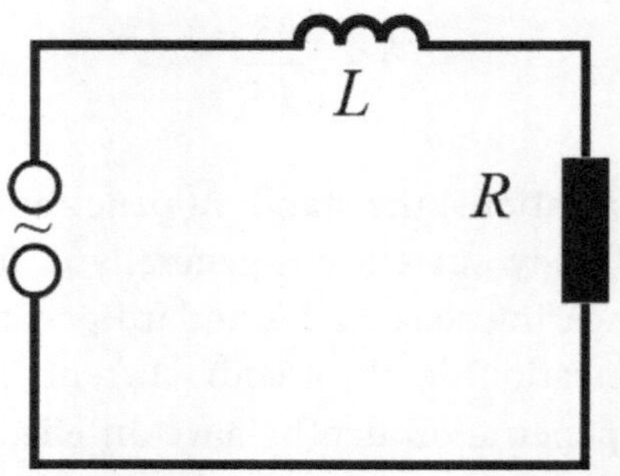

FIGURE 12.2 Equivalent circuit of a single coil.

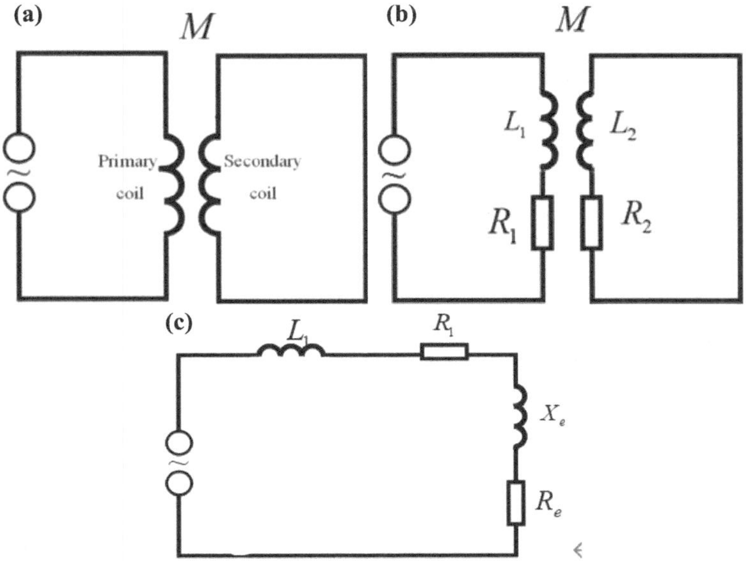

FIGURE 12.3 Equivalent coupled circuit of coils. (a) Coupled circuit of coils, (b) equivalent circuit and (c) equivalent circuit of primary coil converted from secondary coil.

$$Z_s = R_s + X_s \quad R_s = R_1 + R_e \quad X_s = X_1 + X_e \tag{12.12}$$

where R_1 and X_1 are the resistance and reactance of the primary coil, $X_1 = \omega L$; R_s and X_s are apparent resistance and apparent reactance, respectively.

Thus, based on the concept that the apparent impedance of the primary coil is converted from the impedance of the secondary coil, the change of current and voltage in the primary coil can be considered resulting from the change of apparent impedance. From the variation of apparent impedance, the effect of the secondary coil on the primary coil can be known, so that the impedance change of secondary coil circuit can be deduced.

If the secondary coil resistance R_2 gradually decreases from ∞ to 0 or the secondary coil reactance X_2 gradually increases from 0 to ∞, a series of values of the apparent resistance R_s and apparent reactance X_s in the primary circuit can be obtained. Then these values are connected in the coordinate plane with R_s as horizontal axis and X_s as vertical axis to obtain a semicircular curve with a radius of $\frac{K^2 \omega L_1}{2}$, as shown in Figure 2.7. This curve is called coil impedance diagram, where $K = \frac{M}{\sqrt{L_1 L_2}}$ is the coupling coefficient. It can be seen from Figure 2.7 that, as the secondary coil resistance R_2 gradually decreases from ∞ to 0 or the secondary coil reactance X_2 gradually increases from 0 to ∞, the apparent reactance X_s will decrease monotonously from $X_1 = \omega L$ to $\omega L_1 \left(-K^2\right)$, while the apparent resistance R_s will increase from R_1 to the maximum $R_1 + \frac{K^2 \omega L_1}{2}$ and then will gradually fall back to R_1.

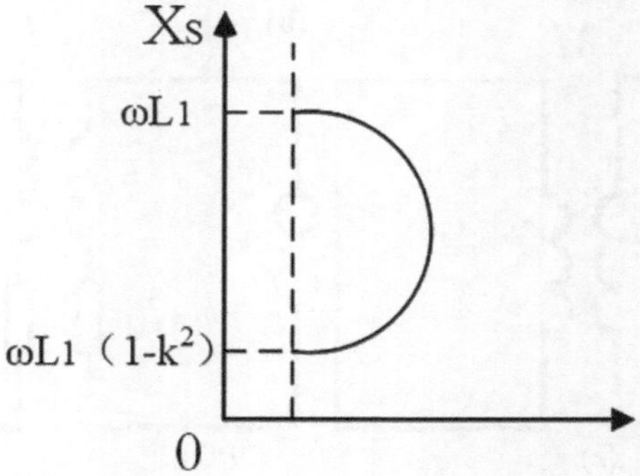

FIGURE 12.4 Apparent impedance of original side coil during the coil coupling.

The impedance diagram shown in Figure 12.4 is intuitive, but the position of the semicircular curve is related to the impedance of the primary coil and the electrical inductance and mutual inductance of the two coils. In addition, the radius of the semicircle is not only affected by the above factors but also varies with frequency. Thus, many semicircular curves with different radii and positions may be obtained when drawing the apparent impedances of the primary coil at different secondary impedances, different frequencies and different coupling coefficient between the two coils. This causes inconvenience to not only the drawing process but also the comparison between the curves obtained under different conditions. Therefore, the impedance normalization method is usually adopted to eliminate the influence of original side coil impedance and excitation frequency on the curve position and to facilitate the comparison between the curves obtained under different conditions.

12.1.4.1 Impedance Normalization

If the curve in Figure 12.4 is shifted to the left by R_1 (i.e. the vertical axis is shifted to the right by R_1) and the new coordinates are divided by X_1 (i.e. the horizontal and vertical coordinates are changed from R_s and X_s into $\frac{R_s R_1}{\omega L_1}$ and $\frac{X_s}{\omega L_1}$), the curve shown in Figure 12.5 can be

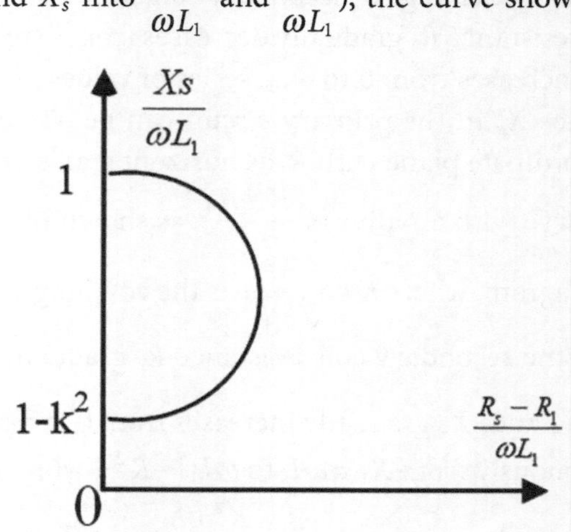

FIGURE 12.5 Graph of normalized impedance.

obtained. As can be seen from the figure, the new trajectory curve is still a semicircle and its diameter coincides with the vertical axis. The coordinates of the upper end of the semicircle are (0,1), the coordinates of its lower end are $(0, 1-K^2)$ and the radius is $K^2/2$. All parameters of the semicircle curve are related only to the coupling coefficient K. Thus, in the new coordinate system, the impedance curve is dependent only on the coupling coefficient K and is independent of the original side coil resistance and excitation frequency. However, the point positions on the curve still vary with R_2 (or X_2). The above method is just the normalization method. The graph of normalized coupled coil impedance is shown in Figure 12.5. As can be seen from the figure, the normalized impedance graph is unified and is only related to the coupling coefficient K, so it is highly comparable. It has the following characteristics:

1) It is a commonly used graph that eliminates the influence of primary coil resistance and inductance;
2) In the impedance curve, a series of factors (such as electrical conductivity and magnetic conductivity) that influence the impedance are taken as parameters;
3) The impedance graph quantitatively shows the magnitude and direction of each effect that influences the impedance, providing a reference basis for selecting the inspection methods and conditions during eddy current testing and for reducing the interference of various effects;
4) Each type of workpiece and detection coil has its own impedance graph.

In actual eddy current testing, the eddy current in metal specimen generated by the electromagnetic induction of a current-carrying excitation coil (primary coil) is like the current flowing through the coils stacked in multiple layers, so the specimen can be treated as a secondary side coil that interlinks with the detection coil. Therefore, from the perspective of circuit, the loop in eddy current testing is similar to coil-coupled circuit, and the coil-coupled impedance analysis mentioned above can be used for the coil impedance analysis in eddy current testing.

12.1.4.2 Effective Magnetic Conductivity and Characteristic Frequency

1. Effective magnetic conductivity

According to the above discussion, the key problem in eddy current testing is the analysis of coil impedance. The key of coil impedance analysis is to analyze the change of a magnetic field where the detection coil has been inserted, so as to determine the change of coil impedance and identify various influencing factors of the workpiece. However, in the actual eddy current testing, the impedances of various coils are very complicated. Therefore, based on the long-term theoretical research and experimental analysis of eddy current testing, Faust proposed the concept of effective magnetic conductivity to greatly simplify the analysis of coil impedance.

A solenoid coil is tightly wound on a long straight cylindrical conductor with the radius a and the relative magnetic conductivity μ. Here, the z-axis coincides with the axis of the solenoid. The edge effect is ignored. An alternating current i is infused into the solenoid. Then, an alternating magnetic field H_z that changes in the radial direction will be generated in the cylindrical conductor inside the solenoid. This magnetic field is the vector superposition of the excitation magnetic field H_0 inside the hollow solenoid coil and the magnetic field generated by eddy currents inside the conductor. Due to skin effect, the distribution of H_z on the cross section of the cylindrical conductor will be non-uniform and will gradually decrease with the increase of its distance from the surface. The magnetic induction intensity in the cylindrical conductor is:

$$\dot{B}_z(r) = \mu_X \mu_X \dot{H}_Z(r) \tag{12.13}$$

where r the distance from any point in the cylindrical conductor to the axis.

A variable with a dot on its top is a complex number. According to the theory of electromagnetic field, $\dot{H}_Z(r)$ can be calculated:

$$\dot{H}_Z(r) = A_1 I_0\left(\sqrt{j}kr\right) + A_2 K_0\left(\sqrt{j}kr\right) \tag{12.14}$$

where $I_0\left(\sqrt{j}kr\right)$ is the first-class zero-order Bessel function of imaginary argument; $K_0\left(\sqrt{j}kr\right)$ is the second-class zero-order Bessel function of imaginary argument; k is electromagnetic propagation constant, $k = \sqrt{\omega\mu\sigma} = \sqrt{\omega\mu_1\mu_0\sigma}$; $A_1 A_2$ is a complex constant and σ is the conductivity of cylindrical conductor.

According to the boundary conditions of magnetic field and the properties of $K_0\left(\sqrt{j}kr\right)$, we can obtain:

$$\dot{H}_z(r) = H_0 \frac{I_0\left(\sqrt{j}kr\right)}{I_0\sqrt{j}ka} \tag{12.15}$$

Therefore, the magnetic flux passing through any cross section of a cylindrical conductor is

$$\dot{\Phi} = \int_s \dot{B}_z ds = \int_0^a 2\pi r \mu_0 \mu_r \dot{H}_z(r) dr = 2\pi\mu_0\mu_r H_0 \frac{a}{\sqrt{-j}k} \cdot \frac{J_1\sqrt{-j}ka}{J_0\sqrt{-j}ka} \tag{12.16}$$

where $J_0\left(\sqrt{j}kr\right)$ and $J_1\sqrt{j}ka$ are zero-order and first-order Bessel functions, respectively.

Regarding the above situation, Faust proposed a hypothetical model to analyze the variation in the apparent impedance of a coil. In his model, the magnetic field on the entire cross section of a cylindrical conductor is constant, while the magnetic conductivity varies in the radial direction of the cross section. The resulting magnetic flux is equal to the magnetic flux generated by the real physical field in a cylindrical conductor. In this way, the actually changing magnetic field H_z and the constant magnetic conductivity μ are replaced by a constant magnetic field H_0 and a changing magnetic conductivity μ_{eff} called effective magnetic conductivity. μ_{eff} is a complex number. For non-ferromagnetic materials, its modulus is less than 1. Thus, the magnetic induction intensity of the hypothetical model can be written as

$$\dot{B} = \mu_0 \mu_r \mu_{eff} H_0 \tag{12.17}$$

The magnetic flux is

$$\dot{\Phi} = \dot{B}S = \mu_0 \mu_r \mu_{eff} H_0 \pi a^2 \tag{12.18}$$

The magnetic flux generated by the real physical field should be equal to that in the hypothetical model, so the effective magnetic conductivity can be calculated:

$$\mu_{eff} = \frac{2}{-\sqrt{-j}ka} \cdot \frac{J_1\sqrt{-j}ka}{J_0\sqrt{-j}ka} \tag{12.19}$$

Thus, it can be seen that the effective magnetic conductivity μ_{eff} is not a constant but a complex variable related to the excitation frequency f as well as the radius r, conductivity σ and magnetic conductivity μ of the conductor.

2. Characteristic frequency

   The imaginary quantity of the Bessel function mentioned above is

$$\sqrt{-j}ka = \sqrt{-j\omega\mu\sigma a^2} = \sqrt{-j2\pi f\mu\sigma a^2} \tag{12.20}$$

Faust defined the frequency corresponding to the modulus 1 of imaginary quantity of Bessel function in the expression of the effective magnetic conductivity μ_{eff} as characteristic frequency (or bounded frequency), which was expressed as f_g. It is an intrinsic property of the workpiece that depends on the electromagnetic characteristics and geometric size of the workpiece itself, that is,

$$\left|\sqrt{-j}ka\right| = \sqrt{2\pi f\mu\sigma a^2} = 1 \tag{12.21}$$

Then $f_s = \dfrac{1}{2\pi\mu\sigma a^2}$

For non-ferromagnetic materials, the magnetic conductivity is $\mu \approx \mu_0 = 4\pi \times 10^{-9}$ H/cm, so the characteristic frequency is

$$f_g = \frac{5066}{\sigma d^2} \tag{12.22}$$

where σ is the electrical conductivity of the material, in MS/m and d is the diameter (cm) of the cylindrical conductor, $d = 2a$.

This is the characteristic frequency of a non-ferromagnetic cylindrical conductive material in the common engineering unit system. In addition, when the electrical conductivity of the material is expressed by the unit of international annealed copper standard (%IACS), the characteristic frequency will be

$$f_g = \frac{8734}{\sigma d^2} \tag{12.23}$$

where σ is the electrical conductivity of the material, in %IACS; d is the diameter of the cylindrical conductor, in cm. It should be noted that, the characteristic frequency for a particular specimen is neither the upper nor lower limit of the test frequency and may not be the optimal test frequency to be used. It is merely a characteristic parameter containing the information about bar size and material performance except for the information about defects.

It is evident that at the general test frequency f, the following relation holds true:

$$ka = \sqrt{\frac{f}{f_g}} \tag{12.24}$$

Therefore, in the analysis of coil impedance, the actual eddy current testing frequency f divided by the characteristic frequency f_g is often used as a reference value and expressed by f/f_g. This frequency ratio can also be used as a variable of the effective magnetic conductivity μ_{eff}. The relation between μ_{eff} and f/f_g is shown in Figure 12.6. It can be seen that, with the increase of f/f_g, the imaginary part of μ_{eff} will increase and then decrease, while its real part will gradually decrease.

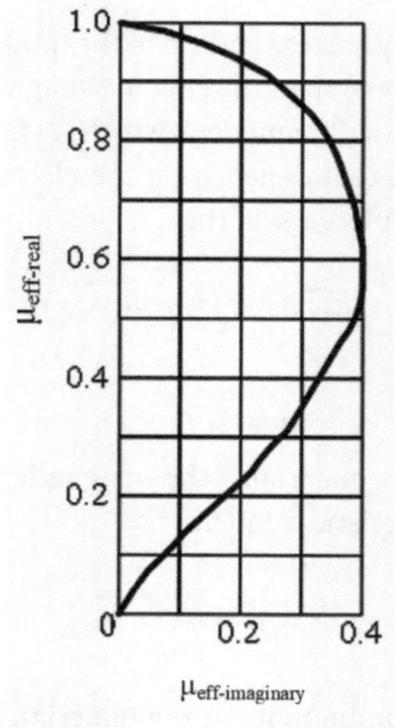

FIGURE 12.6 Relation between μ_{eff} and f/f_g.

3. Similarity law of eddy current test

   The effective magnetic conductivity μ_{eff} completely depends on the frequency ratio f/f_g, while the distribution of eddy current and magnetic field intensity in the specimen depends on μ_{eff}. Therefore, the distribution of eddy current and magnetic field intensity in the specimen changes with f/f_g. Theoretical analysis and derivation can prove that the distribution of eddy current and magnetic field intensity is only a function of f/f_g. Thus, the similarity law of eddy current testing can be obtained as follows: for two different specimens, as long as their frequency ratios $\left(f/f_g\right)$ are the same, the geometric distribution of their effective magnetic conductivities, eddy current densities or magnetic field intensities will also be the same.

12.1.5 Electromagnetic Eddy Current Testing Setup

Eddy current instrument is the core of eddy current testing setup. Depending on the detection object, eddy current instruments can be divided into three types: eddy current flaw detector, eddy current conductometer and eddy current thickness meter. Depending on the application in different detection objects, different types of eddy current devices and even the same type of devices (especially eddy current flaw detectors) will have different testing systems. The testing system generally includes detection coils, a testing instrument and auxiliary devices.

1. Eddy current detection probe

   In the automatic robotic eddy current testing, a differential eddy current probe is often used. Differential probes are divided into differential excitation probes and differential measurement probes. A differential coil is formed by two identical windings very close to each other and in reverse connection. Affected by the differential effect of the two windings, the coils are less sensitive to the slowly changing features (such as specimen material, electrical conductivity and temperature) but more sensitive to the suddenly changing features (such as defect cracks). Meanwhile, another advantage of the differential probe is that the probe jitter, lift-off clearance and other factors have a smaller influence on defect detection.

   A differential excitation eddy current probe is composed of three parallel coils with symmetric spaces. The coils on both sides are two excitation coils in differential connection, and the coil in the middle is the detection coil, as shown in Figure 12.7. When a controllable sinusoidal voltage is applied to the excitation coils, the tested component will generate an eddy current under the action of the magnetic field generated by excitation coils. Then the eddy current will generate alternating magnetic field in the detection coil, which, in turn, will output a detection signal. The two excitation coils are in differential connection, so the eddy currents generated by excitation coils will cancel each other out. When the differential excitation probe passes through the detected defect, the output signal of the detection coil will experience two abrupt changes in opposite directions.

   The differential measurement eddy current probe is similar to the differential excitation probe, except for two detection probes in reverse connection on two internal

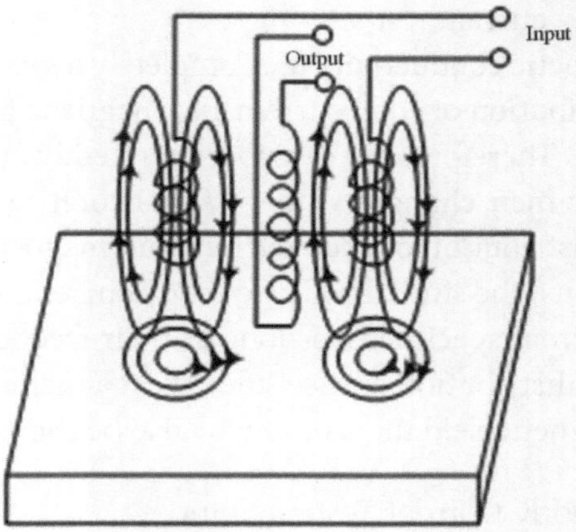

FIGURE 12.7 Structural diagram of a differential excitation probe.

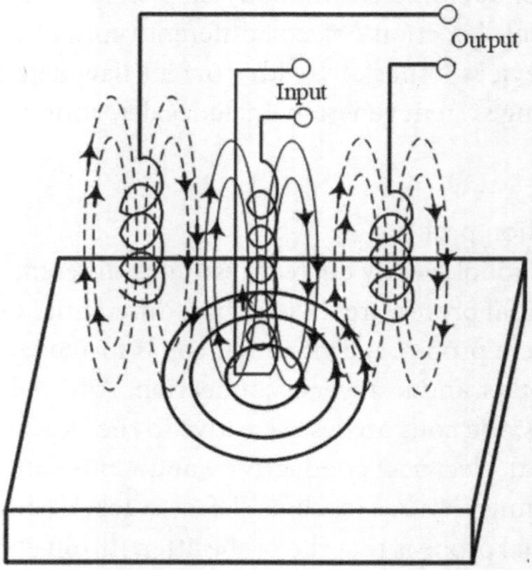

FIGURE 12.8 Structural diagram of a differential measurement probe.

sides and an excitation coil in the middle. This probe is to apply a controllable sinusoidal voltage to the excitation coil and generate eddy currents in the tested component. The two detection coils will receive the detection signals containing the characteristic information on their positions. When the differential measurement probe passes through the detected defect, the output signal of the detection coils will experience two abrupt changes in opposite directions, as shown in Figure 12.8.

2. Eddy current instrument

In the testing process, the impedance of an eddy current probe coil is affected by various factors, so the output AC signal will change. It is necessary to process the signals according to their difference, so as to suppress the interference caused

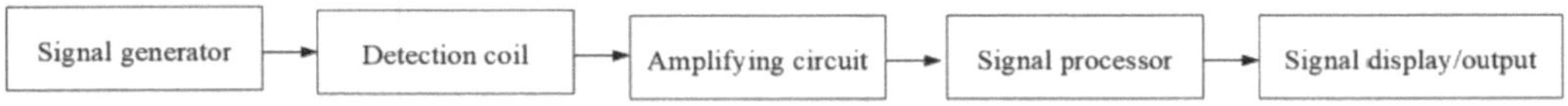

FIGURE 12.9 Basic composition of eddy current instrument.

by unnecessary factors and detect the defect signal. The eddy current instrument is mainly used to analyze the signals generated by the eddy current probe and detect the defect in the specimen.

The eddy current instruments can be classified into many types with different circuit forms but basically the same working principle. Their working principle is as follows. The signal-producing circuit generates alternating current and supplies it to the detection coil. The alternating magnetic field of the coil induces eddy currents in the workpiece, which, in turn, cause the coil impedance to change due to the influence of specimen material or defects. The amplifying circuit amplifies the weak detection signal obtained by the detection coil. Then, the signal processing circuit eliminates the interference factors causing the impedance change, in order to identify the defect effect. Finally, the results are displayed. The working principle is shown in Figure 12.9.

12.2 COMPOSITION OF A ROBOTIC ELECTROMAGNETIC EDDY CURRENT TESTING SYSTEM

12.2.1 Hardware Composition

The structure of the robotic electromagnetic eddy current nondestructive testing (NDT) system is shown in Figure 12.10. The hardware of this system mainly includes a scanning motion module, an electromagnetic eddy current detection module, a computer control module and other mechanical support structures. The scanning motion module is the core unit of the whole system, including an articulated manipulator, a manipulator controller, a detection worktable and a quick-change fixture module. It plays a vital role in the testing of curved workpieces. This module is a manipulator with high precision and six degrees of freedom (DOFs), whose flexibility is leveraged to track the profile of a curved workpiece along a planned trajectory. The electromagnetic eddy current detection module is mainly to transmit, receive and process the eddy current signals. The computer control module includes an industrial computer integrating the hardware such as data acquisition card and manipulator position/attitude acquisition card, as well as a software system special for the equipment. Similar to the single-manipulator UT system, the robotic electromagnetic eddy current testing system can achieve the detection purpose in two modes. In the first mode, the manipulator holds an eddy current probe to detect a fixed workpiece, which is usually rotating or has a large complex curved surface. In the second mode, the workpiece under test, which is usually small, is held by a manipulator and detected by a fixed eddy current probe. The second mode is often used for the rapid automatic testing of small components.

1. Manipulator and controller

   In this system, a high-precision six-DOF manipulator is used for scanning. Its end-effector has high localization accuracy and a good development prospect.

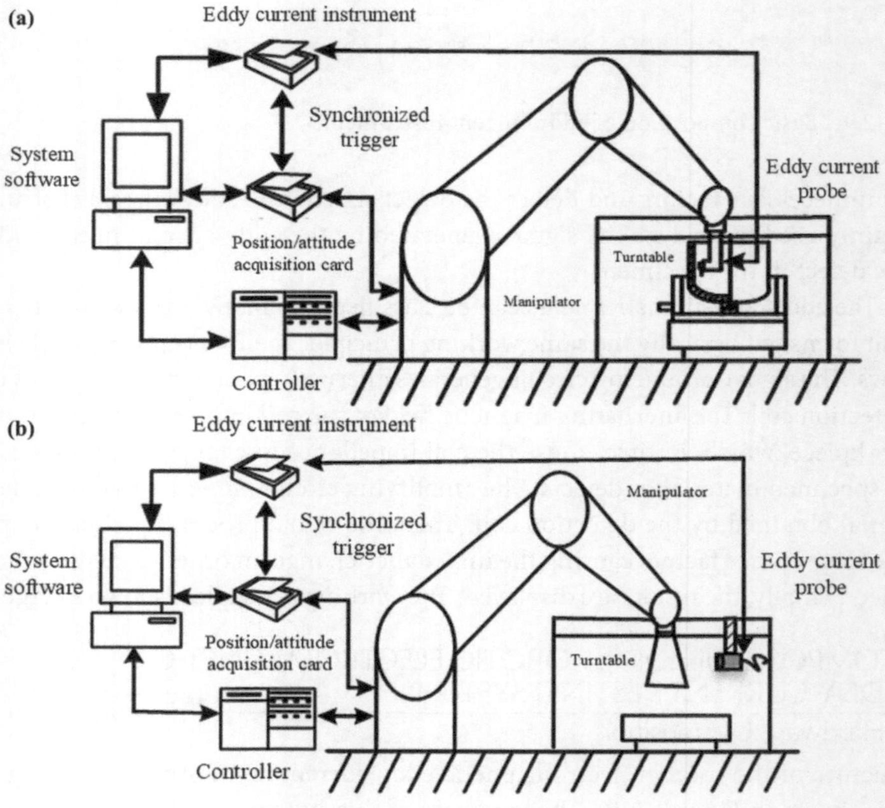

FIGURE 12.10 Composition of robotic eddy current testing equipment. (a) Testing mode with a manipulator holding the probe and (b) testing mode with a manipulator holding the workpiece.

The advantages of this manipulator are high localization accuracy, flexible attitude, high response sensitivity and fast moving speed. Compared with a traditional scanning rack, the manipulator is very suitable for the automatic measurement of complex contours. In the system, the flexible manipulator clamps an eddy current probe or a curved workpiece to scan along the established trajectory in order to complete the detection task. The model of this manipulator and the distribution of its six joints are shown in Figure 12.11.

The manipulator controller is a CS8C controller, including communication I/O interface, motion control card, processor and network interface. It is also equipped with a teaching box that serves as the human-computer interaction window. The CS8C controller is composed of digital power amplifier, manipulator power supply, manipulator auxiliary power supply, manipulator' safety integrated board, CPU, main power switch and power module. This controller has a programmable function module, which is connected to the industrial personal computer (IPC) through RS232/422 serial communication protocol and Ethernet Modbus server.

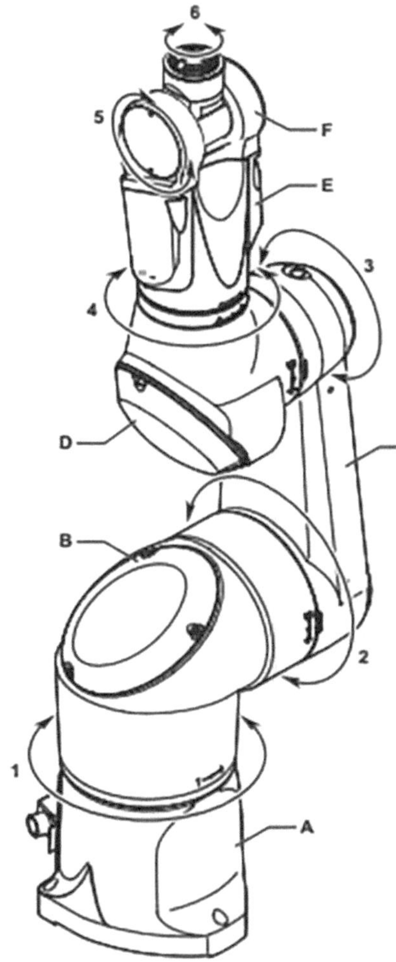

FIGURE 12.11 Manipulator model.

2. Other necessary hardware

   Like the single/dual-manipulator ultrasonic NDT systems, the robotic electromagnetic eddy current NDT system also needs a manipulator-position acquisition card and a data acquisition card. The main function of manipulator-position acquisition card is to obtain the scanning position information, collect and calculate the manipulator position/attitude information in real time and transfer the manipulator position/attitude to the workpiece coordinate system for imaging display. At the same time, the position acquisition card has two trigger modes: position trigger and time trigger. It can output a pulse signal to trigger the data acquisition card and save the new position data into memory.

   The data acquisition card has two trigger modes: external trigger and software trigger. Software trigger is just time trigger. External trigger is mainly to receive a trigger signal from the position acquisition card to realize synchronous trigger.

The data acquisition card is divided into several levels within ±50 mV to ±10 V. A level can be selected according to the amplitude of the input signal. If the level is too high, the actual accuracy will be too low. If the selected level is too low, the signal will be clipped and then distorted. The sampling frequency range of data acquisition card is 10 kHz–10 MHz. After selecting the sampling frequency, the total sample size and the number of samples collected before trigger should be set. The difference between them is just the number of samples to be collected after trigger.

12.2.2 Software Composition

Like the control system in the robotic ultrasonic NDT system, the control system in the robotic electromagnetic eddy current NDT system also includes two parts: system software and lower-computer motion control. Through software, it achieves the functions like detection subsystem control, trajectory planning, motion control, electromagnetic eddy current test data acquisition, defect characterization, verification and system management. This is a real-time multi-task operation and control system with complex calculation. While controlling the ultrasonic signal excitation receiver to obtain ultrasonic detection signals, this system needs to complete the coordinated action of a multi-DOF high-precision motion control system, as well as other work (such as signal processing, defect feature extraction and image rendering).

The software system runs on the upper computer and is mainly used to realize the functions such as detection system control, trajectory planning, motion control, eddy current detection data acquisition and display, defect characterization analysis, image display, verification, test report output and system management. The software system mainly includes four modules, namely manipulator motion control module, data acquisition module, signal processing module and automatic detection data management. The functional block diagram is shown in Figure 12.12.

When the receiving device acquires the eddy current detection signals, the software system needs to complete the coordinated action of the multi-DOF high-precision motion control system, as well as other work (such as signal processing, defect feature extraction and image rendering). At the same time, as a type of engineering detection software, the software system should show good robustness and man-machine interaction while achieving the predetermined functions. Therefore, the above factors must be fully considered when designing the software framework, data structure and data relation.

The software can display the impedance graphs. It can also display the eddy current C-scan images in real time according to the *XYZ* coordinates at the probe-scanning positions and the characteristics of eddy current signals at the corresponding positions. The C-scan supports the imaging of flaw-echo amplitude and defect depth. From the C-scan images, the amplitudes and depths of all the scanning points in the C-scan image area can be directly read. On the C-scan images, the defect area and the distance between defects can be directly mapped, the defects can be evaluated and abnormal areas can be marked. In addition, the A-scan waveforms at any positions can be displayed simultaneously.

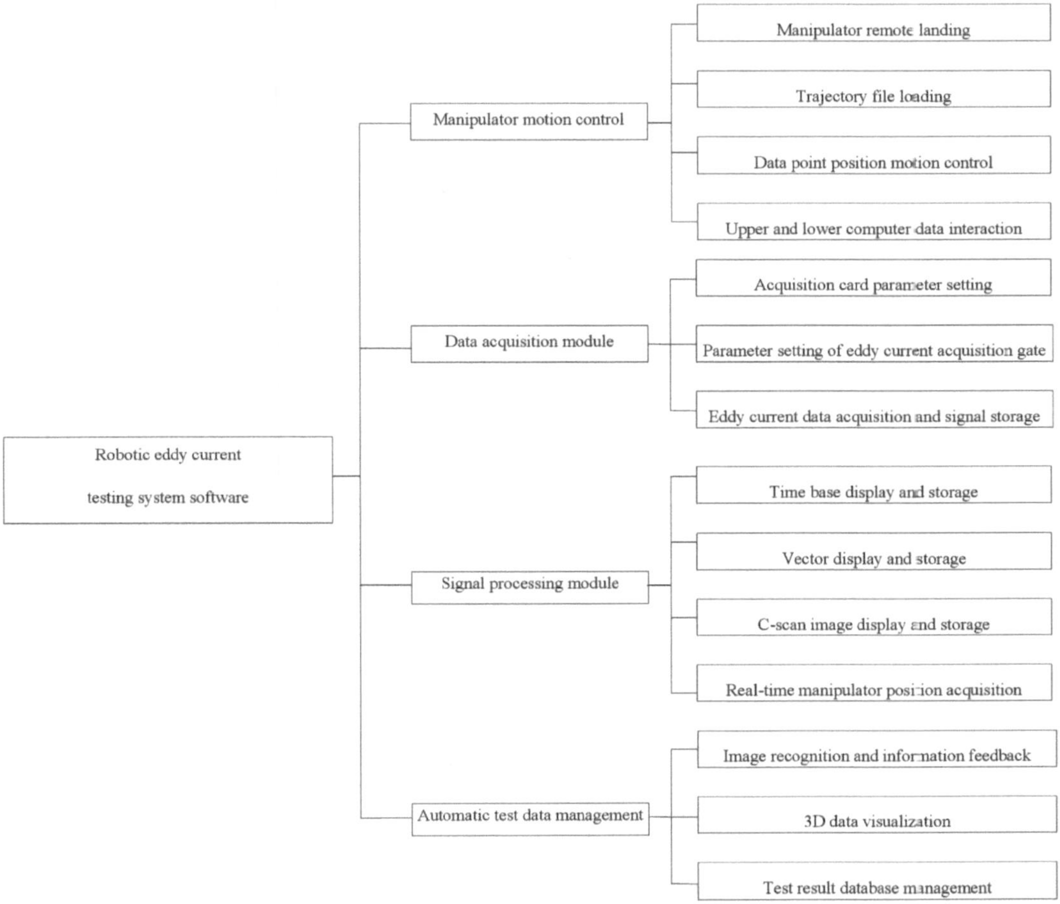

FIGURE 12.12 Functional block diagram of system software.

The scanning result diagrams of various windows can also be displayed simultaneously. All the data files of the whole scanning process can be recorded to check the scanning data in real time.

The motion control software is mainly to realize the two-level manipulator control. The IPC is installed as upper computer in the control cabinet of this automatic UT system, while the manipulator motion controller is installed as lower computer in the cabinet of manipulator controller. They are connected through an Ethernet interface. The multi-axis motion trajectory of the manipulator is controlled by the lower computer software. The switching between manual control and automatic control, the transmission of scanning trajectory data, the acquisition of scanning position data and the display of human-computer interaction interface for automatic motion control and other functions are implemented by the upper computer software. The motion control of each manipulator joint can be implemented in manual control mode, while automatic surface tracking can be implemented in automatic control mode.

12.3 METHOD OF ELECTROMAGNETIC EDDY CURRENT DETECTION IMAGING

Flaw detection is the main application of eddy current testing technique. It can be used to detect the defects on the surface or subsurface of conductive materials [5–9]. The operating principle of eddy current testing is electromagnetic induction. Therefore, when the material, shape, size and other factors of the specimen are changing, the induced magnetic field and the eddy current distribution will change to generate the corresponding eddy current signals. The frequency range commonly used in eddy current flaw detection is about tens of Hz to 10 MHz. Before detection, an appropriate eddy current probe should be selected according to the component material, the defect location and type, the detection sensitivity and other conditions. While meeting the detection requirements, a higher detection frequency should be considered to achieve better detection sensitivity. For example, a probe up to several megahertz can be selected to detect the surface cracks. Electrical conductivity and magnetic conductivity are two important factors affecting the penetration depth and distribution density of eddy currents. They can be analyzed by the following formula:

$$\delta = 1/\sqrt{\pi f \mu_0 \mu_r \sigma}$$

where δ is standard penetration depth, in m;

f is the operating frequency, in Hz;

μ_0 is vacuum permeability, $\mu_0 = 4\times10^{-7}$, in H/m;

μ_r is the relative permeability and is a dimensionless constant;

σ is electrical conductivity, in S/m.

From this formula, it can be inferred that the conductivity and permeability are inversely proportional to the penetration depth of eddy current.

In the process of eddy current scanning and flaw detection, the edge effect will also affect the detection result. When the probe is close to the edge of a component or its structural discontinuities during scanning, the flow path of eddy current will be distorted, as shown in Figure 12.13.

The eddy current signal generated by edge effect is often stronger than the defect signal to be obtained. If the edge effect is not eliminated, the defects on or near the edge of the tested component (for example, on the intake and exhaust edges of an aero-engine blade)

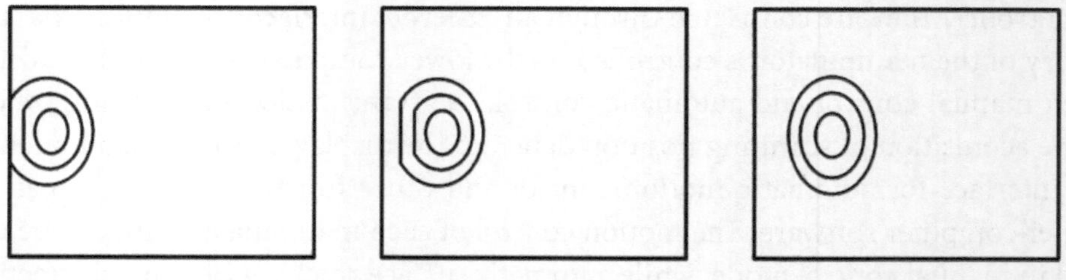

FIGURE 12.13 Edge effect of electromagnetic eddy current.

cannot be detected. The range of edge effect is related to the electrical and magnetic conductivities of the tested material as well as the size and structure of the probe coil.

The lift-off effect of the eddy current probe should also be taken into account when designing the eddy current scanning path of the manipulator. Lift-off effect refers to the phenomenon that the induced eddy current reaction will change with the distance between the departing eddy current probe and the tested component surface.

12.3.1 Display Method of Eddy Current Signals

The impedance graph shows the change of impedance caused by the influencing factors such as the electrical conductivity, magnetic conductivity, thickness, position, coating and defects of a material. The phase uniqueness of an impedance graph allows the operator to separate signals from noise, thus reducing the errors. Eddy current signals are often displayed in two ways: vector display and time-based display.

1. Vector display:

   Vector display is also known as Lissajous figure. When an AC voltage is applied to the eddy current detection coil, the current will flow through the equivalent reactance X_L and the coil resistance R. Denote the voltage on the reactance as E_1 and the voltage on the resistance as E_2. The voltage on the detection coil is the vector sum of E_1 and E_2, and the phase difference between E_1 and E_2 is 90°. When R and X_L are changing, the voltage drop of the detection coil will also change. The total voltage E is the vector sum of E_1 and E_2:

$$Z = jX_L + R \tag{12.25}$$

   The reactance and the resistance, which are perpendicular to each other, are combined into the impedance. The phase angle α is the angle between the voltage phase and the current phase behind it. In the vector diagram, the x-axis is the resistance R and the y-axis is the reactance X_L, so the impedance Z can be represented by the point P determined by the two orthogonal components X_L and R. When the probe is not on the specimen, the point P_0 can be represented by X_{L0} and R_0. If the eddy current field of the probe coil is changing in the detection process under the influence of the tested component, the point P_1 can be represented by X_{L1} and R_1. The magnitude and direction of impedance change depend on the component properties and the instrument characteristics. The component properties mainly include electrical conductivity, component size, magnetic permeability and material discontinuity. The instrument characteristics include the frequency, size and shape of the probe as well as its distance from the specimen.

2. Time-based display:

   The full name of time-based display is linear time-based display. In the time-based coordinate system, the x-axis is the number of acquisition points of time or signal, and the y-axis is the signal amplitude. The variation trend of signal amplitude is the

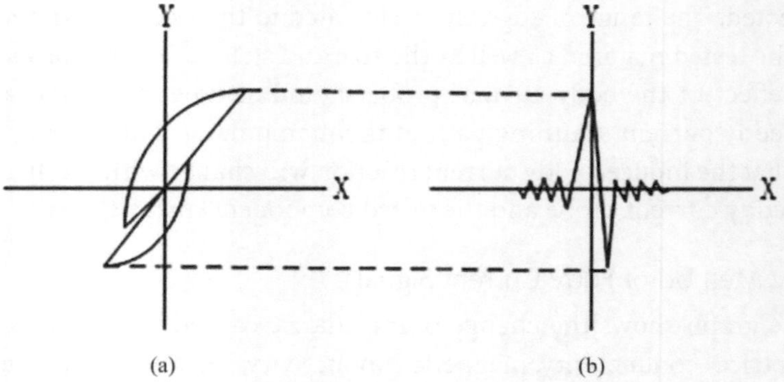

FIGURE 12.14 (a) Vector display and (b) time-based display.

main basis to judge whether there is a defect in the tested object. If the curve of signal amplitude is relatively flat with small fluctuation, it means that there is no defect in the area scanned by the probe. However, if the signal amplitude in a certain segment of the curve increases sharply and the resultant peak value is much higher than the average amplitude, it indicates that there is a defect in the area scanned by the eddy current probe. Compared with vector display, the time-based display is more user-friendly and intuitive. Because the *x*-axis is time or a time-related quantity, the signal changes during the whole scanning process can be recorded continuously in time-based display. The height of pulse signal obtained by time-based display is equal to the projection of the signal obtained by vector display on the *y*-axis, as shown in Figure 12.14.

12.3.2 Method of Eddy Current C-Scan Imaging

The main working mode of the robotic electromagnetic eddy current testing system is C-scan. Through the C-scan results, the location, shape and size of defects can be determined. The imaging quality has a direct influence on the detection accuracy of specimen defects.

The basic principle of scanning imaging in the robotic electromagnetic eddy current system is that its impedance will change with the properties of the tested component (electrical conductivity, component size, magnetic permeability, material discontinuity, etc.). When the characteristics of eddy current signal (such as peak and phase) corresponding to each point on the tested sample are displayed in different colors or with different gray values, the defects on the surface or subsurface of the sample can be identified. C-scan imaging is to display the pulse wave intensities of the defect signals in different colors or with different gray values, so as to draw the specimen defect diagram or conductivity distribution diagram.

Robotic electromagnetic eddy current C-scan imaging is generally to display the features of time-based signals or vector signals in colors or with gray values. The quality of the imaging algorithm has a direct influence on the imaging accuracy and identifiability. The imaging methods include peak imaging method and phase imaging method. The principle and implementation methods of the commonly used peak imaging technique are mainly presented below.

Due to the existence of defects on the material surface or subsurface, the impedance of the eddy current probe will change, directly resulting in the change of its pulse signal intensity shown in time-based display. The existence of defects will lead to the change of material continuity, that is, the change of impedance distribution. The distribution of defects will also lead to the change in the energy of reflected echo (namely, the intensity of echo), so the distribution of echo intensity will also reflect the continuity of material surface or subsurface. Therefore, echo peak value can be used for eddy current C-scan imaging.

The method of peak imaging can be briefly described as follows:

1. Adjust the relative distance between the eddy current probe and the sample, observe the change of impedance and pulse wave in time-based display, and focus the transducer on the depth of the detection area. Then, set the tracking gate for the echo of the upper surface, and set the data gate for the area to be detected.
2. In the scanning process, the manipulator moves relative to the workpiece/probe while clamping the probe/workpiece. When the probe arrives at the preset position, the eddy current instrument and the data acquisition card will be triggered automatically and the impedance signal will be received.
3. Take out the peak value P from the gate range and convert it into the corresponding gray value or color value (pseudo-color mode). Next, the 256-level (8-bit, binary) gray value is taken as an example to illustrate its calculation method. The gray value G can be obtained from the following equation:

$$G = \frac{P}{V_{max}} \times 255 \tag{12.26}$$

   where V_{max} is the full-range positive voltage value of the current pulse echo. For example, if the peak value P of the gate at a certain point is 0.32 V and the full-range positive voltage value is set as 0.5 V, then the gray value at this point will be $G = P/V_{max} \times 255 = 0.32/0.5 \times 255 = 163$.
4. Repeat Steps 2–3. Take out the peak value from the gate of each scanning point one by one, convert it into the corresponding gray value or color value and fill it into the defined bitmap, so as to construct a 2D peak image.

In addition, it can be seen from Figure 12.15 that the echoes generated by eddy currents at the interfaces with different impedances contain both positive and negative peaks. Therefore, different peaks are optional for imaging. Generally speaking, several ways are available for the selection of peak value:

1. **Positive peak**: Only the positive peak value is selected for calculation;
2. **Negative peak**: Only the negative peak value is selected for calculation;

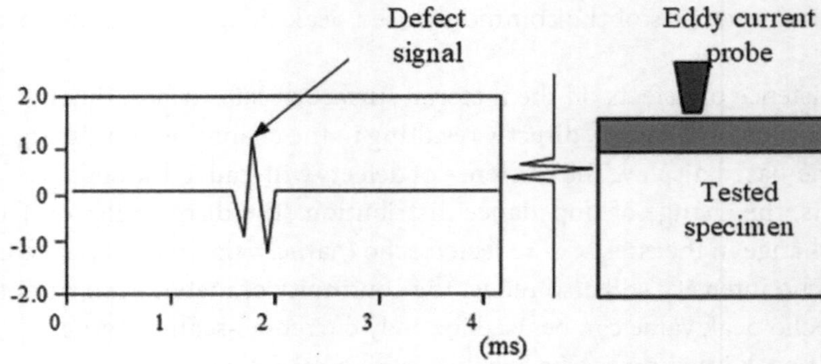

FIGURE 12.15 Schematic diagram of a robotic electromagnetic eddy current C-scan.

3. **Peak-to-peak imaging**: Both positive and negative peaks are considered, and the average of the absolute values of positive and negative peaks is used for calculation. This method can give consideration to both positive and negative peaks.
4. **Maximum absolute peak value**: Both positive and negative peaks are considered, but only the peak with the maximum absolute value in the gate is selected for calculation.

Take the imaging of an aero-engine blade and an aluminum alloy block as an example. The electromagnetic eddy current C-scan images obtained by peak imaging are shown in Figures 12.16 and 12.17. It can be seen that the peak imaging can represent the results of defect detection.

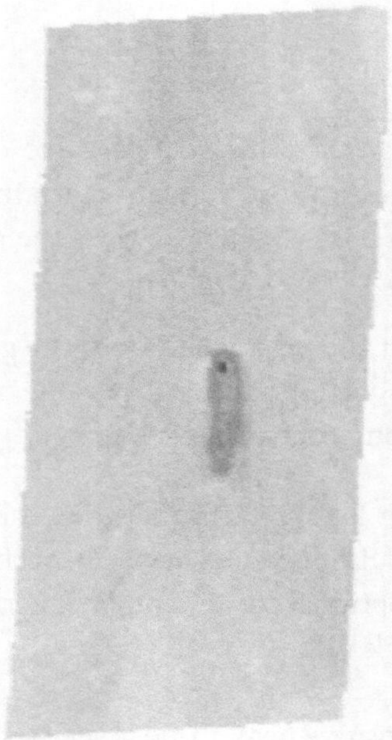

FIGURE 12.16 Eddy current C-scan image of an aero-engine blade.

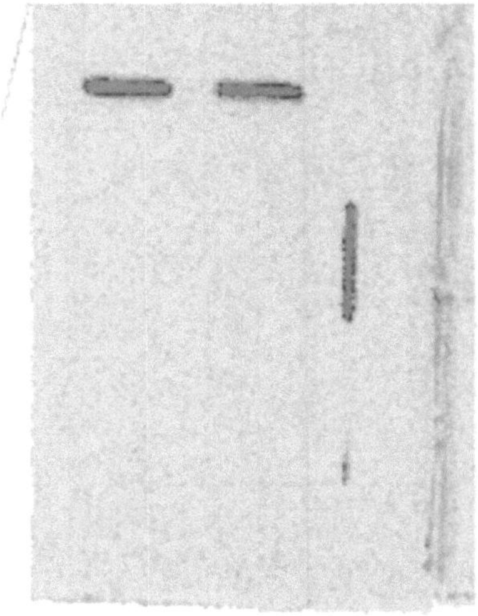

FIGURE 12.17 C-scan image of an aluminum alloy block.

Peak imaging is the main traditional time-domain imaging method. It is simple to implement and is capable of real-time imaging with high quality. It is generally used as the default imaging method of the robotic NDT system.

REFERENCES

1. Li J., Jimao C. *Nondestructive Testing [M]*. Beijing: China Machine Press, 2004:391–412.
2. Kebei X., Zhou J. *Eddy Current Testing [M]*. Beijing: China Machine Press, 2004:1–27.
3. Fan H. *Eddy Current Testing of Metal Materials [M]*. Beijing: China Science and Technology Press, 2006:65–98.
4. Liu K., Analysis of eddy current impedance diagram [J], *Nondestructive Testing*, 1996, 18(2):50–56.
5. Feng Q. *Research on the Eddy Current Scanning Detection of Defects in Aviation Turbine Blades [D]*. Beijing: Beijing Institute of Technology, 2016.
6. Sheng H., Cao B., Zhang X., Fan M., Research on the defect quantification and detection based on eddy current C-scan imaging [J], *China Science Paper*, 2018, 13(2):121–125.
7. Liu B., Luo F., Hou L., Research on the C-scan imaging technique with eddy current array detection and correction [J], *Chinese Journal of Sensors and Actuators*, 2011, 24(8):1172–1177.
8. Wang N., Xiucheng L., Bin Z., Eddy current O-scan and C-scan imaging techniques for macro-crack detection in silicon solar cells [J], *Nondestructive Testing and Evaluation*, 2019(34):389–400.
9. Song K., Wang C., Zhang L., et al., Design and experimental study of automatic profiling eddy current detection system for cracks on aero-engine turbine blades [J], *Aeronautical Manufacturing Technology*, 2018, 61(19):45–49.

CHAPTER 13

Manipulator Measurement Method for the Liquid Sound Field of an Ultrasonic Transducer

An ultrasonic transduction system is an important carrier for analyzing and quantitatively evaluating the detection characteristics. The modeling method has long been used by the researchers at home and abroad to quantitatively study the ultrasonic transduction system. In this chapter, the method to study the overall model of the ultrasonic transduction system is the focus. The process of ultrasonic pulse excitation and receiving is studied based on the equivalent circuit model and lossy transmission model of the ultrasonic transducer. Then, a modeling method based on an equivalent circuit is given. By using this method, the pulse excitation and echo prediction of the ultrasonic system are studied. Moreover, the characteristics of ultrasonic pulse in a steel block, such as surface reflection echo and bottom echo, leap time and wave attenuation, are obtained by simulation calculation. The comparison shows that the modeling result and the actual measurement result are in good agreement.

13.1 MODEL OF AN ULTRASONIC TRANSDUCTION SYSTEM

The operation process of any acoustoelectric conversion system can be divided into the following steps: electric pulse excitation, cable transmission, acoustic radiation, propagation in media, acoustic signal reception and so on. In the ultrasonic transduction system, the performance of key components will change or deteriorate over time, or the performance of the system composed of a different type of transducer and a different type of testing instrument may not match with that of the original system. All these cases will lead to the failure to give full play to the detection ability correctly. However, by establishing

DOI: 10.1201/9781003212232-13

the sub-model for each step and organically combining those sub-models into the overall model of an ultrasonic transduction system, the entire system can be simulated and analyzed. This new analysis method can save the time and cost of system development and better grasp the direction of detection system improvement [1].

Pulse-reflection ultrasonic nondestructive testing (NDT) system is widely used in industrial NDT and medical testing. The characteristics of each component and the global characteristics of all components will have an influence on the detection accuracy of ultrasonic echo wave. To explore the influence degree and correctly predict the defect echo, the researchers proposed a lot of equivalent models for ultrasonic transduction steps. However, the accurate simulation of the ultrasonic system is very difficult, because it involves the acoustic theory, material theory and electronic circuit knowledge and covers complicated physical phenomena such as electromechanical integration and acoustic-mechanical coupling. The equivalent model of transducers has always been the research focus in this field. In 1948, M. P. Mason [2] deduced the mathematical relationship between the mechanical and electrical quantities on the end face of a piezoelectric vibrator from the piezoelectric equation, wave equation and mechanical boundary conditions and then obtained a six-terminal equivalent network to describe the mechanical and electrical characteristics of piezoelectric vibrator. This is the famous Mason electromechanical equivalent model. Later, Redwood [3] and Krimholtz et al. [4] modified Mason's equivalent circuit in 1961 and 1970, respectively, for example, by using the lumped parameter to describe the electrical characteristics of a piezoelectric wafer and the transmission line to describe its mechanical characteristics so as to obtain the KLM equivalent circuit. However, a negative capacitance introduced to Mason's equivalent circuit has no definite physical significance, and the primary/secondary ratio of transformer in KLM equivalent circuit varies with frequency. This makes it difficult to simulate the piezoelectric wafer on a computer. In 1994, Leach [5] proposed an equivalent circuit model based on a controlled source, thus avoiding the main shortcomings of the above equivalent model. The main feature of his model is that the control equations of the piezoelectric wafer are equivalent to its transmission line equations, and that a controlled source is used to ensure the energy conversion between the acoustic end and the electric end. In 1997, Alf Puttmer et al. [6] proposed a lossy equivalent model of an ultrasonic piezoelectric transducer based on the Leach circuit model and simulated it on the computer. In their model, the lossy transmission line model parameter was introduced to represent the loss of the piezoelectric wafer during transmission. In 2001, J. Johansson and P. Martinsson et al. proposed an equivalent model of sound field diffraction based on transmission line theory [7]. In 2009, Noureddine Aouzale et al. verified the sound-field diffraction effect in the transmission line model through experiments [8].

In this chapter, the equivalent circuit model of the Leach transducer and the lossy transmission line model are equivalent to the propagation medium. They, together with the introduced ultrasonic pulse excitation model, constitute a complete ultrasonic transduction system model. The system model can be divided into three parts: the ultrasonic transducer transceiving model, pulse excitation receiving process and ultrasonic propagation process. Each sub-model has accurate mathematical parameters and circuit parameters, which are obtained through simple measurement. Finally, the process of ultrasonic excitation and pulse echo is simulated, and the model results are verified by experiments.

13.1.1 Equivalent Circuit Model of an Ultrasonic Transducer

Generally, the vibration modes of piezoelectric ceramic wafers include stretching vibration, shear vibration and bending vibration [9]. In this paper, the direct-incident piezoelectric transducers are studied. Usually, the main vibration mode of their piezoelectric wafers is thickness extension vibration, as illustrated in Figure 13.1 [10].

The polarization direction of a piezoelectric wafer with extensional thickness is parallel to the direction of thickness, that is, along the Z-axis direction. The electrode surface covers the main surface, freely stretching and vibrating along the Z-axis direction. Suppose that the radial dimension of a thin piezoelectric wafer is much larger than its thickness δ and the wavelength λ of sound wave propagating in the wafer. In this case, the strain components inside the wafer can be approximately considered as, $S_3 \neq 0$, $S_1 = S_2 = S_4 = S_5 = S_6 = 0$. This is equivalent to the lateral clamping state. Meanwhile, the components of electric field intensity are, $E_1 = E_2 = 0$ $E_3 \neq 0$. Under this condition, the piezoelectric equation is [11]:

$$T_3 = c_{33}^D S_3 - h_{33} D_3 \tag{13.1}$$

$$E_3 = -h_{33} S_3 + \beta_{33}^S D_3 \tag{13.2}$$

where T_3 is the normal stress of thin piezoelectric wafer, c_{33}^D is the elastic constant, h_{33} is the piezoelectric constant, D_3 is the electrical displacement and β_{33}^S is the dielectric constant.

Eq. (13.2) can be converted into

$$D_3 = \frac{E_3}{\beta_{33}^S} + \frac{h_{33}}{\beta_{33}^S} S_3 \tag{13.3}$$

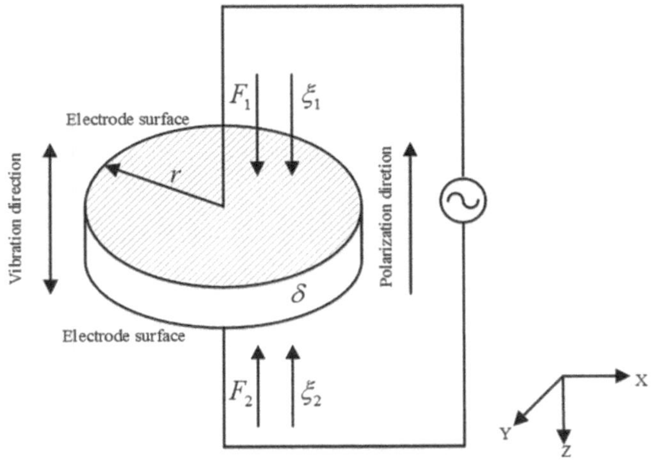

FIGURE 13.1 Thickness extension vibration mode of a piezoelectric wafer.

Considering that there is no free charge in the wafer, the partial derivative of Eq. (13.3) will be

$$\frac{\partial D_3}{\partial z}=0 \tag{13.4}$$

Because

$$S_3=\frac{\partial \zeta}{\partial z} \tag{13.5}$$

the substitution of Eqs. (13.3) and (13.5) into Eq. (13.4) can yield

$$\frac{\partial E_3}{\partial z}=-h_{33}\frac{\partial^2 \zeta}{\partial z^2} \tag{13.6}$$

The integral of Eq. (13.6) is

$$E_3=-h_{33}\frac{\partial \zeta}{\partial z}+a \tag{13.7}$$

where a is an integral constant determined by the voltage V applied on the wafer. Because

$$V=\int_0^{\delta} E_3\,dz \tag{13.8}$$

can be calculated:

$$a=\frac{V}{\delta}-\frac{h_{33}}{\delta}(\xi_1+\xi_2) \tag{13.9}$$

where ξ_1 and ξ_2 are the normal displacements of end face at $x=0$ and $x=\delta$, respectively. Thus, Eqs. (13.7) and (13.9) can be combined into:

$$E_3=-h_{33}\frac{\partial \xi}{\partial Z}+\frac{V}{\delta}-\frac{h_{33}}{\delta}(\xi_1+\xi_2) \tag{13.10}$$

The substitution of Eqs. (13.5) and (13.10) into Eq. (13.3) can yield

$$D_3=\frac{V}{\beta_{33}^S\delta}-\frac{h_{33}}{\beta_{33}^S}(\xi_1+\xi_2) \tag{13.11}$$

According to Eq. (13.11), the current in the wafer is

$$I=i\omega AL_zD_3=i\omega C_0V-n(v_1+v_2) \tag{13.12}$$

where $v_1 = \dot{\xi}_1, v_2 = \dot{\xi}_2$ are the normal velocities of end face at $x=0$ and $x=\delta$, respectively, $C_0 = \frac{\pi r^2}{\delta \beta_{33}^S}$ is the wafer cutoff capacitance, $n = C_0 h_{33} = \frac{\pi r^2 h_{33}}{\delta \beta_{33}^S}$ is the electromechanical conversion coefficient and $A_z = \pi r^2$ is the end-face area of the wafer.

Suppose that the piezoelectric wafer is in a three-dimensional coordinate system, where the longitudinal plane wave propagates only in the Z direction and its displacement in the Z direction is ξ. According to Newton's second law, its equation of motion is:

$$\rho_0 \frac{\partial^2 \xi}{\partial t^2} = \frac{\partial T_3}{\partial z} \tag{13.13}$$

where ρ_0 is the density of thin piezoelectric wafer. The substitution of the piezoelectric Eq. (13.1) into Eq. (13.13) can yield:

$$\rho \frac{\partial^2 \xi}{\partial t^2} = c_{33}^D \frac{\partial S_3}{\partial z} - h_{33} \frac{\partial D_3}{\partial z} \tag{13.14}$$

The substitution of Eqs. (13.4) and (13.5) into Eq. (13.14) can yield

$$\frac{\partial^2 \xi}{\partial t^2} = v^2 \frac{\partial^2 \xi}{\partial z^2} \tag{13.15}$$

where $v = \sqrt{\frac{c_{33}^D}{\rho_0}}$ is the longitudinal wave velocity of thickness extension vibration.

For a simple harmonic excitation, Eq. (13.15) is

$$\frac{\partial^2 \xi}{\partial z^2} + \left(\frac{\omega}{v}\right)^2 \xi = 0 \tag{13.16}$$

Let $k = \frac{\omega}{v}$. Then the solution of Eq. (13.16) is

$$\xi = A \sin kz + B \cos kz \tag{13.17}$$

where A and B are the constants to be determined according to the boundary conditions. The displacements of both ends of the wafer are:

$$\xi\big|_{z=0} = \xi_1, \xi\big|_{z=\delta} = -\xi_2 \tag{13.18}$$

By substituting Eq. (13.18) into Eq. (13.17), the constants A and B can be determined:

$$A = -\frac{\xi_2 + \xi_1 \cos k\delta}{\sin k\delta}, B = \xi_1 \tag{13.19}$$

The substitution of Eq. (13.19) into Eq. (13.17) can yield

$$\xi = \frac{\xi_1 \sin k(r-z) - \xi_2 \sin kz}{\sin k\delta} \tag{13.20}$$

At the two ends of the wafer, the stress should be balanced with the external force, so

$$\begin{cases} -F_1 = c_{33}^D \left(\dfrac{\partial \xi}{\partial z} \right)_{z=0} A_z - h_{33} D_3 A_z \\ -F_2 = c_{33}^D \left(\dfrac{\partial \xi}{\partial z} \right)_{z=\delta} A_z - h_{33} D_3 A_z \end{cases} \tag{13.21}$$

where F_1 and F_2 are the external forces on the wafer end faces at $z = 0$ and $z = \delta$, respectively. Eq. (13.21) is combined with Eq. (13.11) to obtain

$$h_{33} D_3 A_z = nV + i \frac{n}{\omega C_0} (v_1 + v_2) \tag{13.22}$$

By substituting Eqs. (13.20) and (13.22) into Eq. (13.21), the wafer vibration equation under simple harmonic excitation can be obtained as follows:

$$\begin{cases} F_1 = \left(i \dfrac{n^2}{\omega C_0} - i \dfrac{Z_0^a}{\sin k\delta} \right) (v_1 + v_2) + i z_0^a \tan \dfrac{k\delta}{2} v_1 + nV \\ F_2 = \left(i \dfrac{n^2}{\omega C_0} - i \dfrac{Z_0^a}{\sin k\delta} \right) (v_1 + v_2) + i z_0^a \tan \dfrac{k\delta}{2} v_2 + nV \end{cases} \tag{13.23}$$

where $Z_0^a = \rho_0 v A_z$ is the radiation impedance of thin piezoelectric wafer.

From the circuit state Eq. (13.12) and the mechanical vibration Eq. (13.23), the electromechanical equivalent model of a thin piezoelectric wafer in thickness extension vibration can be derived. The wafer can be viewed as a reciprocal three-terminal system, including one electrical terminal and two acoustic terminals. The electrical terminal is represented by (V, I) and the two acoustic terminals are represented by (F_1, v_1) and (F_2, v_2), respectively. Then the electromechanical conversion relation in the wafer can be obtained as follows [2]:

$$\begin{Bmatrix} F_1 \\ F_2 \\ V \end{Bmatrix} = i \begin{bmatrix} Z_0^a \cot(k\delta) & \dfrac{Z_0^a}{\sin(k\delta)} & \dfrac{h_{33}}{\omega} \\ \dfrac{Z_0^a}{\sin(k\delta)} & Z_0^a \cot(k\delta) & \dfrac{h_{33}}{\omega} \\ \dfrac{h_{33}}{\omega} & \dfrac{h_{33}}{\omega} & \dfrac{1}{\omega C_0} \end{bmatrix} \begin{Bmatrix} v_1 \\ v_2 \\ I \end{Bmatrix} \tag{13.24}$$

Equivalent simulation model of an ultrasonic transducer

In this section, the equivalent circuit model of the piezoelectric transducer is established on the basis of the above equivalent theory, and the calculation method of model parameters is given. According to Eqs. (13.1) and (13.13), the following equivalent transformation is made for some of the parameters [12]:

$$T_3 = F/A_z \, , S_3 = d\zeta/dz \, , u = \zeta/\omega \tag{13.25}$$

where F is the wafer surface pressure and μ is the vibration velocity. So,

$$\begin{cases} \dfrac{dF}{dz} = -\rho A_z \omega u \\ c_{33}^{D} \dfrac{d\zeta}{dz} = h_{33} D_3 - \dfrac{F}{S} \end{cases} \tag{13.26}$$

At the same time, the piezoelectric Eq. (2.2) turns into:

$$E_3 = -h_{33} \frac{d\zeta}{dz} + \beta_{33}^{S} D_3 \tag{13.27}$$

If I is the external current of the electrode, then D_3 is $D_3 = q/A_z$ where q is the quantity of wafer electrode charges, $q = I/\omega$. So,

$$D_3 = \frac{I}{\omega A_z} \tag{13.28}$$

According to Eqs. (13.4), (13.25) and (13.28), we can obtain:

$$\frac{d\left(h_{33} I/\omega\right)}{dz} = 0 \tag{13.29}$$

By subtracting Eq. (13.29) from Eq. (13.26) and replacing the variables, Eq. (13.26) turns into:

$$\begin{cases} \dfrac{d}{dz}\left(F - \dfrac{h_{33} I}{\omega}\right) = -\rho A_z \omega u \\ \dfrac{du}{dz} = -\dfrac{\omega}{A_z c_{33}^{D}}\left(F - \dfrac{h_{33} I}{\omega}\right) \end{cases} \tag{13.30}$$

By calculating the integrals of both sides of Eq. (13.27) at the same time according to Eq. (13.8), the voltage V between electrodes can be obtained. The integration range of z is the distance between piezoelectric wafers, i.e. $z = 0, z = \delta$. So,

$$V = \frac{h_{33}}{\omega}\left(u_1 - u_2\right) + \frac{I}{C_0 \omega} \tag{13.31}$$

where C_0 is as shown in Eq. (13.12); $u_1 = u(z=0), u_2 = u(z=\delta)$.

Eq. (13.30) is compared with the voltage-current relationship in the transmission line equation (also known as telegraph equation) of transmission line theory [13]:

$$\begin{cases} \dfrac{dv}{dz} = -L\omega I \\ \dfrac{dI}{dz} = -C\omega V \end{cases} \tag{13.32}$$

It can be seen that when an additional term of zero (Eq. 13.29) is added to the control Eq. (2.26) of the transducer wafer, the control equation and the telegraph equation of transmission line model will have the same form, where:

$$\begin{cases} V \Leftrightarrow \left(F - \dfrac{h_{33} I}{\omega}\right) \\ I \Leftrightarrow u \end{cases} \tag{13.33}$$

This indicates that there is a certain equivalent transformation relationship between the two equations. According to the commonness of their forms, Leach proposed the equivalent circuit model of the piezoelectric transducer based on the transmission line model and controlled source model [5]. The equivalent circuit in his model is shown in Figure 13.2, where the port E represents the electrical terminal of the piezoelectric transducer and is characterized by two parameters, namely voltage and current (V, I). The ports F and B

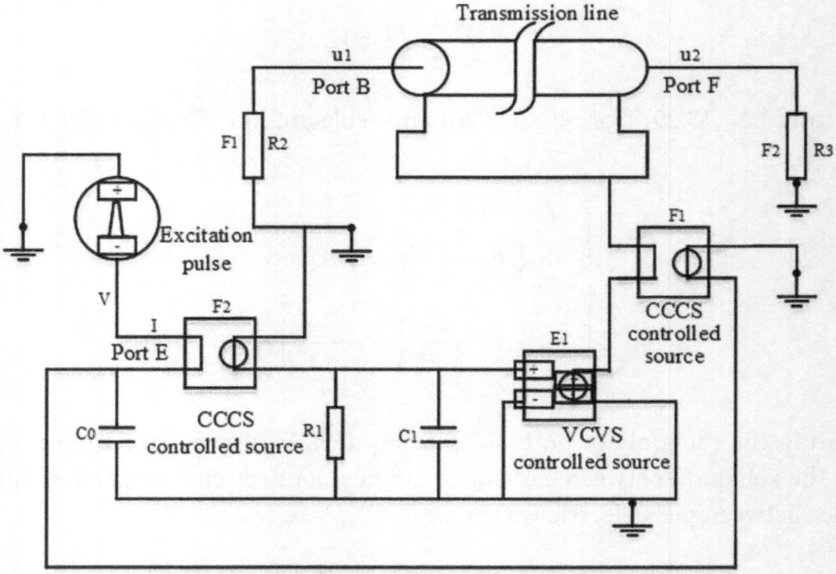

FIGURE 13.2 Equivalent circuit model of an ultrasonic transducer.

are the two acoustic terminals of the transducer. The port F is the transmitting terminal, characterized by pressure and particle vibration velocity (F_2, u_2). The port B is the backing terminal (R_2), characterized by backside pressure and particle vibration velocity (F_1, u_1).

The model parameters are calculated as follows. The static capacitance between the two electrodes of the piezoelectric wafer is:

$$C_0 = \frac{\beta_{33}^S A_z}{\delta} \tag{13.34}$$

According to the telegraph equation in Eq. (13.32), the basic parameters for characterizing the equivalent transmission line model are the transmission line length Len and the resistance R, inductance L, capacitance C and conductance G per unit length. Among them, the resistance R represents the energy loss caused by absorption, and the conductance G represents the energy loss caused by diffraction or acoustic beam diffusion. Suppose that there is no diffraction loss in the piezoelectric wafer, i.e. $G = 0$. So,

$$\begin{cases} \text{Len} = \dfrac{V_c}{2f_a} \\ L = \rho A_z \\ C = \dfrac{1}{V_c^2 \rho A_z} \\ R = \dfrac{Z_0}{(2hC_0)^2} \end{cases} \tag{13.35}$$

where V_c is the sound velocity in the piezoelectric material; f_a is the inverse resonance frequency of piezoelectric wafer; ρ is the density of the wafer material; Z_0 and h represent the acoustic impedance and piezoelectric field coefficient (N/C) of the wafer material, respectively and are determined by the following equations [14]:

$$V_c = 1/(LC)^{1/2} = \left(c_{33}^D/\rho\right)^{1/2} \tag{13.36}$$

$$Z_0 = (L/C)^{1/2} = A_z\left(c_{33}^D \rho\right)^{1/2} = \rho A_z V_c \tag{13.37}$$

$$h = \frac{h_{33}}{\beta_{33}^S} \tag{13.38}$$

13.1.2 Ultrasonic Excitation and Propagation Medium

1. Model of pulse excitation source

   Ultrasonic pulse excitation source (generator) is a complex voltage driving circuit. Its output characteristics mainly include output waveform, voltage amplitude and characteristic impedance. In this paper, the pulse generator is simplified as a linear equivalent circuit. In fact, the pulse generator is not globally linear under different settings due to the existence of diode protection circuit. However, if the external load of the circuit remains unchanged and various parameters (such as attenuation rate and energy level) are determined, the model can accurately simulate the pulse generator.

Usually, the electric excitation signal of the input transducer in ultrasonic pulse detection is a negative spike pulse or pulse train. A typical pulse excitation device has three main parameters: pulse repetition frequency (PRF), excitation energy (voltage) and damping. The excitation pulses with different characteristics can be obtained by adjusting different parameter combinations.

The pulse excitation circuit model established in this section is shown in Figure 13.3, mainly including pulse modulation, energy regulation, waveform regulation, protection circuit and other functions. The main components and their functions are: *Vc*: high-voltage source (50~400 V); *Vp*: pulse source; diode network: protection circuit; field effect tube (M1): pulse modulation switch; combination of *Rc* and *Cd*: pulse energy regulation; *Cd*: main energy storage element and *Rp* and *Lp*: spike pulse waveform adjustment.

The excitation simulation waveforms of negative spike pulse and square wave pulse with different parameter combinations are shown in Figures 13.4 and 13.5, respectively. These simulation waveforms were obtained from the output port without load (piezoelectric transducer wafer). The excitation waveform with load is slightly different mainly in the rear rising edge. The simulation parameters of the waveform in Figure 13.4a are $Vc = 400$ V; $Rc = 15$ KΩ; $Rp = 100$ Ω; $Lp = 56$ μH; $Cd = 1.37$ nF and Figure 13.4b $Vc = 100$ V; $Rc = 14$ KΩ; $Rp = 450$ Ω; $Lp = 22$ μH; $Cd = 1.87$ nF. The simulation parameters of the waveform in Figure 13.5a are $Vc = 400$ V; $Rc = 18$ KΩ; $Rp = 210$ Ω; $Lp = 132$ μH; $Cd = 23.3$ nF and Figure 13.5b $Vc = 100$ V; $Rc = 15$ KΩ; $Rp = 220$ Ω; $Lp = 214$ μH; $Cd = 51.8$ nF.

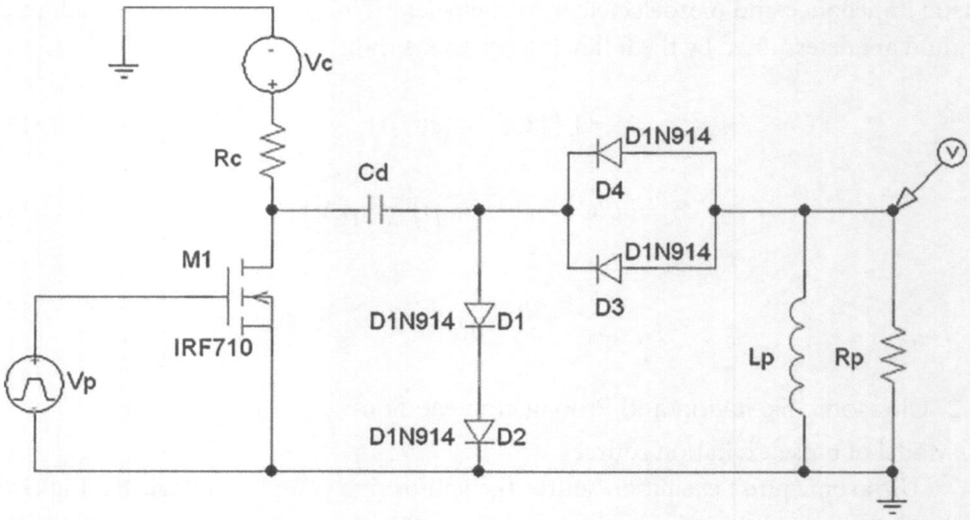

FIGURE 13.3 Circuit model of an ultrasonic pulse excitation source.

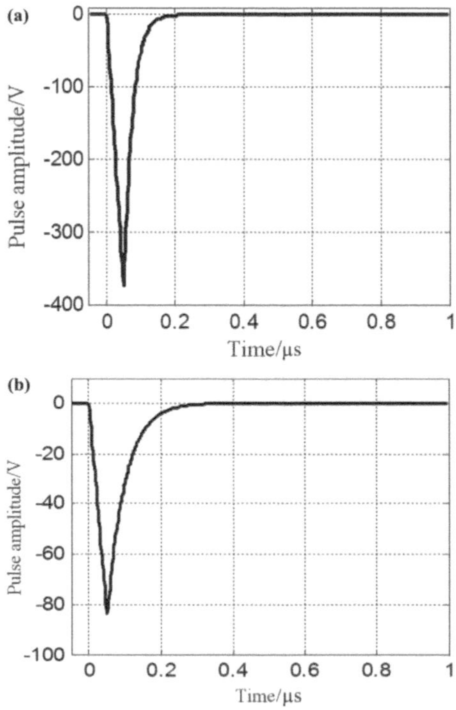

FIGURE 13.4 Simulation waveform of an excitation negative spike pulse (without load). (a) The excitation simulation waveform of the negative spike and (b) the excitation simulation waveform of the square wave pulse.

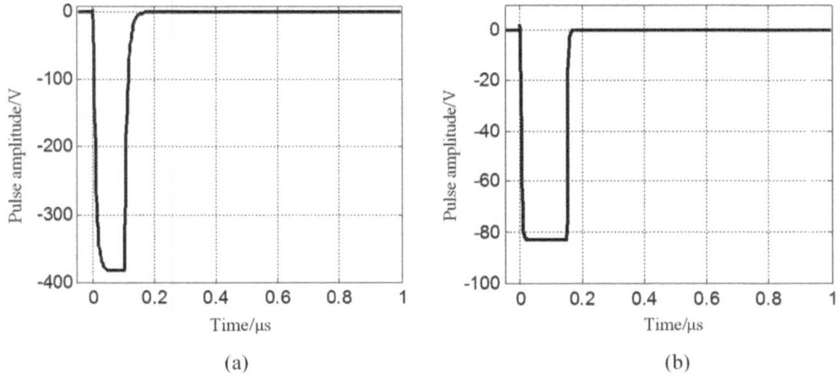

FIGURE 13.5 Simulation waveform of an excitation square wave pulse (without load). (a) The excitation simulation waveform of the negative spike and (b) the excitation simulation waveform of the square wave pulse.

2. Propagation medium model

Propagation medium is the carrier of sound wave. When an ultrasonic wave propagates in the medium, the following phenomena will usually appear. First, its sound pressure intensity will gradually decrease with the increase of the propagation distance. This is called attenuation phenomenon. There are many reasons for the attenuation of sound intensity, mainly including absorption attenuation, scattering attenuation and diffusion attenuation. Second, when traveling from one medium into another medium, the ultrasonic wave will be reflected, refracted and transmitted. This is a process of wave mode conversion, accompanied by energy redistribution. Third, when passing through the interfaces with different shapes, the ultrasonic wave will be focused or defocused by certain rules. In this section, the phenomena like absorption attenuation, diffusion attenuation and interface reflection are specially considered in the modeling of propagation medium. As the phenomena of attenuation and interface reflection also exist in the transmission line, the transmission line model can be used to simulate these characteristics.

The transmission line equivalent model for the ultrasonic transmission medium also has five parameters: $\text{Len}_m, R_m, L_m, C_m$ and G_m [15]. Len_m represents the actual propagation distance of sound wave in the medium. L_m and C_m are used to control the sound velocity in the medium, as shown in Eq. (13.40). R_m represents the energy loss caused by medium absorption:

$$v_m = \frac{1}{\sqrt{L_m C_m}} \tag{13.39}$$

$$R_m = 2 v_m \rho_m A_z \alpha_m \tag{13.40}$$

where v_m is the sound velocity in the propagation medium, ρ_m is the density of the propagation medium and α_m is the absorption attenuation factor of sound wave in the propagation medium.

The diffraction loss is realized through the conductance parameter G_m [16], which is calculated in accordance with the following equation:

$$G_m = \frac{-2}{Z_0 A_z \text{Len}_m} \text{In}\left(A_{\text{dif}}(z)\right) \tag{13.41}$$

where Z_0 is the acoustic impedance of the propagation medium; $A_{\text{dif}}(z)$ is the diffraction loss factor and can be expressed as the ratio between the average sound pressure at the axial distance z, denoted as $P(z)$, and the average sound pressure on the transducer surface, denoted as $P(0)$:

$$A_{\text{dif}}(z) = \frac{P(z)}{P(0)} \tag{13.42}$$

For a circular plane transducer, its $P(z)$ and $P(0)$ can be estimated with a numerical method such as Gaussian beam superposition algorithm or space impulse response method, or measured by experiment, as detailed in the relevant literature. In this paper, the piezoelectric liquid-immersed focusing transducer is studied. Since the sound beam diffusion loss of the focusing transducer in the focal area is very small, it is ignored here to simplify the calculation:

$$G_m = 0 \tag{13.43}$$

3. Simulation of the ultrasonic transduction system

   The main sub-models of the ultrasonic transduction system discussed above can be combined together and arranged along the ultrasonic propagation path to form a complete simulation model of this system. The connection diagram of equivalent circuit model of a pulsed ultrasonic transduction system is shown in Figure 13.6. For the convenience of representation, the Leach transducer model is replaced by a three-port network, whose pins are marked as shown in Figure 13.2.

   The phenomena in ultrasonic pulsed reflection echo system include the ultrasonic pulse excitation, the propagation of pulsed acoustic wave in water, the reflection echo at the water-steel interface and the reflection echo from the bottom surface of steel block. In the simulation process, the ultrasonic excitation source is used as negative spike pulse source, and the collection point of ultrasonic excitation pulse and echo voltage is set as the port E. The terminal B (R_2) represents the backing terminal. In order to reduce multiple reflections on the backside and improve the axial resolution, a larger resistance value is generally taken. The terminal F is connected to the propagation medium model. The parameters of water medium and steel block in the circuit simulation model, as well as the characteristic parameters of other components used in calculation and simulation, are shown in Table 13.1.

According to the simulation model and parameters, an experimental verification system was established in this section, as shown in Figure 13.7. During the measurement, the center frequency of the transducer was 5 MHz and the wafer diameter was 12.7 mm. The thickness of steel test block in the direction of normal ultrasound incidence was 30 mm.

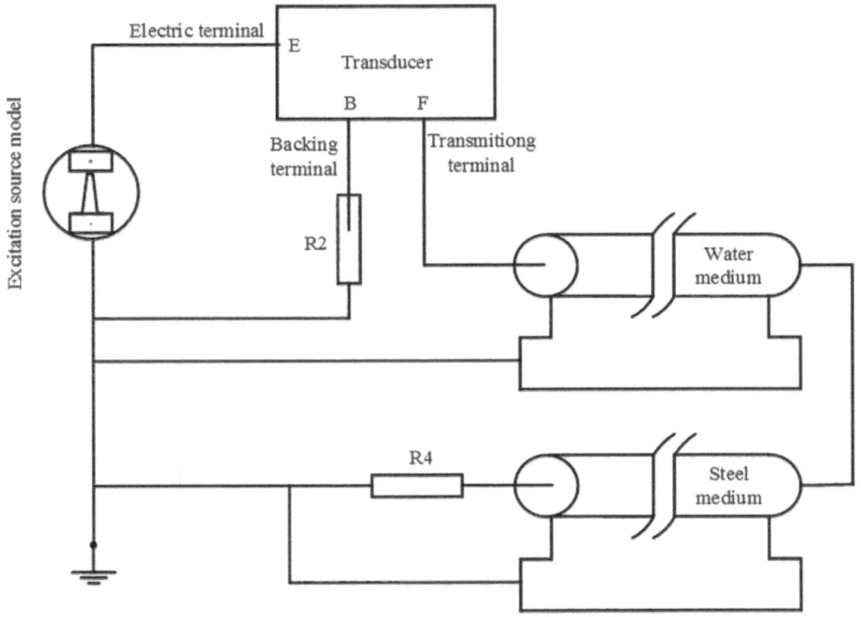

FIGURE 13.6 Connection diagram of sub-models of an ultrasonic transduction system.

TABLE 13.1 Parameters of the Transducer Model and Transmission Medium Model

Parameter Name	Symbol	Value
Wafer radius	A_z	9.43×10^{-6} m²
Wafer density	ρ	7.55×10^{3} kg/m³
Wafer thickness	δ	455μm
Piezoelectric constant	h_{33}	16C/m²
Sound velocity in wafer	V_c	4350 m/s
Relative dielectric constant	β_{33}^{s}	386
Sound velocity in water	v_w	1480 m/s
Sound velocity in steel	v_s	5900*m/s*
Water density	ρ_w	1000 kg/m³
Steel density	ρ_s	8000 kg/m³
Loss factor in water	α_w	0.13 Np/m
Loss factor in steel	α_s	0.11 Np/m

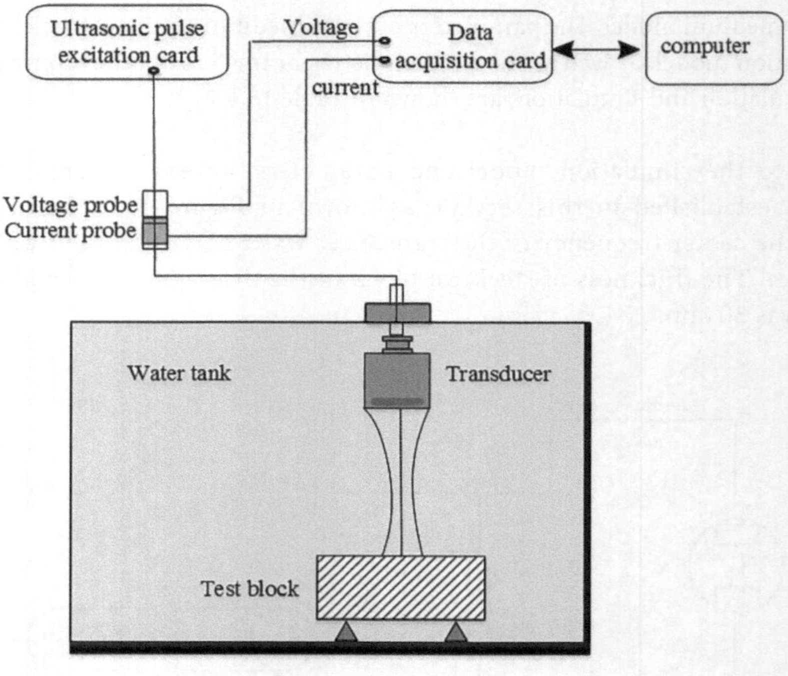

FIGURE 13.7 Verification experiment setup.

The simulation results and experimental measurement results of ultrasonic pulse excitation voltage and current under load (transducer) are shown in Figures 13.8–13.10 (for −50, −150 and −300 V, respectively). It can be seen that the simulation waveforms are in good agreement with the experimental waveforms. Compared with the no-load condition (Figure 13.5), the simulation waveforms show some fluctuation under load at the rear rising edge. This is because in the transceiving case, the vibration on the front and rear surfaces of the piezoelectric wafer is also received by the wafer during the excitation pulse phase and overlapped with the excitation pulse.

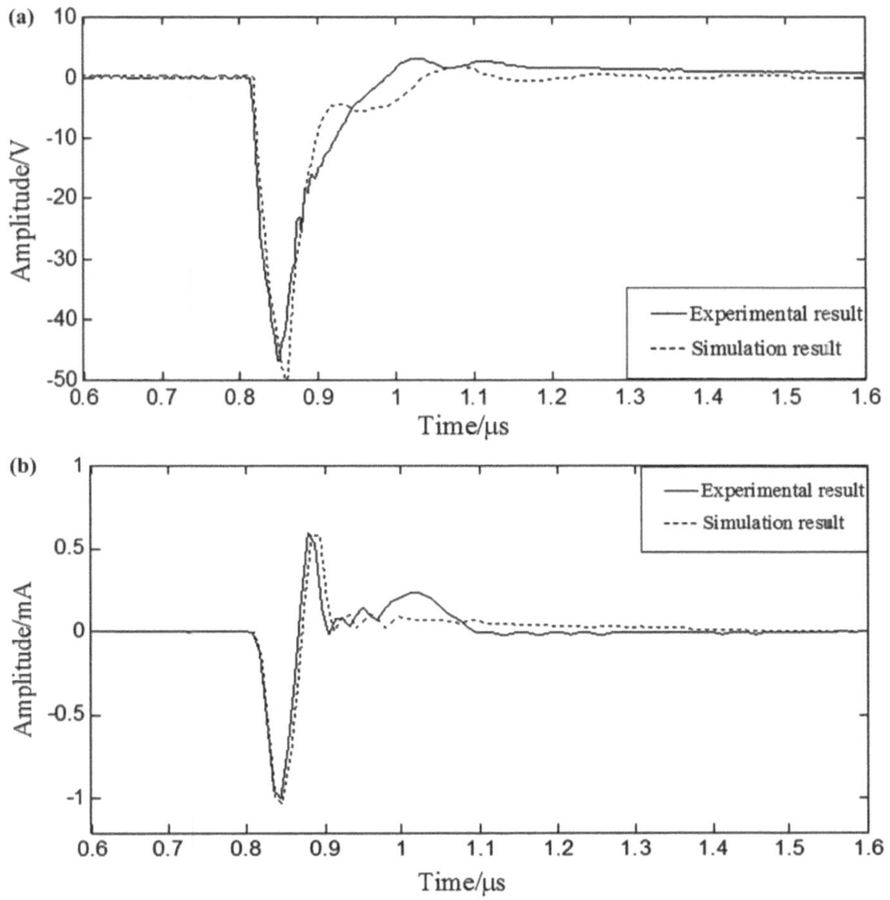

FIGURE 13.8 Negative spike pulse excitation voltage and current under load (−50 V). (a) The negative spike pulse excitation voltage and (b) the negative spike pulse excitation current

The comparison between the simulation results and experimental results of upper surface reflection echo and bottom echo is shown in Figure 13.11. For the main peak waveforms of surface echo, the experimental results and the simulation results were in good agreement. The durations of echo pulse in experiment and simulation were basically the same, except for some differences in the late oscillation. It can be seen in Figure 13.11b that the sound path lengths of surface echo and bottom echo in simulation are very consistent with the experimental measurement results. Given the sound velocity of 5,900 m/s in steel, the theoretical result of sound path length is 10.17 μs, the simulation result is 10.16 μs and the experimental measurement result is about 10.18 μs. The energy attenuation in the simulated propagation medium is also basically consistent with that in the experiment. Through the normalized comparison of echo amplitudes, it can be seen that the amplitude ratio between the bottom echo and the surface echo in the simulation is basically consistent with that in the experiment.

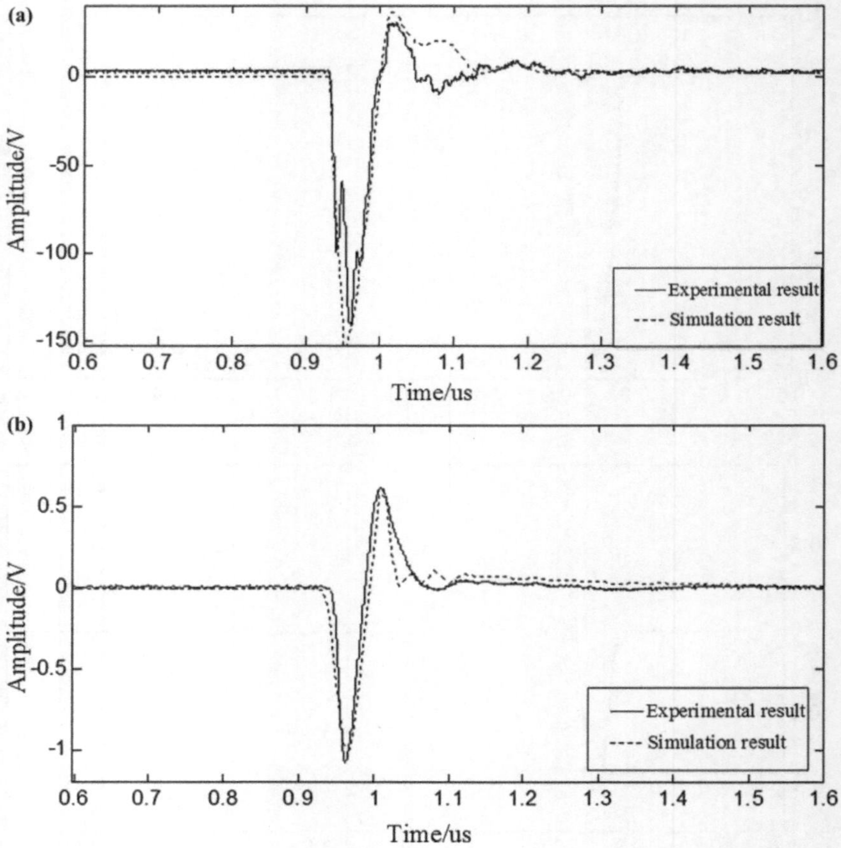

FIGURE 13.9 Negative spike pulse excitation voltage and current under load (–150 V). (a) The negative spike pulse excitation voltage and (b) the negative spike pulse excitation current.

13.2 SOUND FIELD MODEL OF AN ULTRASONIC TRANSDUCER BASED ON SPATIAL PULSE RESPONSE

In this chapter, the sound field model of a round planar/focused transducer is calculated based on the spatial response of this transducer to an excitation pulse. The radiation characteristics of sound field in the transducer are studied. For common transducers, the calculation results of sound field distribution and sound propagation law are given.

13.2.1 Theory of Sound Field in an Ultrasonic Transducer

1. Sound-field radiation model of the transducer

   Based on the theory of piston bore, a planar vibration aperture on the plane $z = 0$ with an area of S is assumed to be embedded in an infinite hard baffle, as shown in Figure 13.12. Except for vibration aperture, the vertical vibration velocities of other parts of the transducer are 0. In addition, the sound wave is transmitted without loss. Then the sound pressure field that the vibration aperture generates in the medium with an density of ρ_0 can be obtained from the following equation [9,17]:

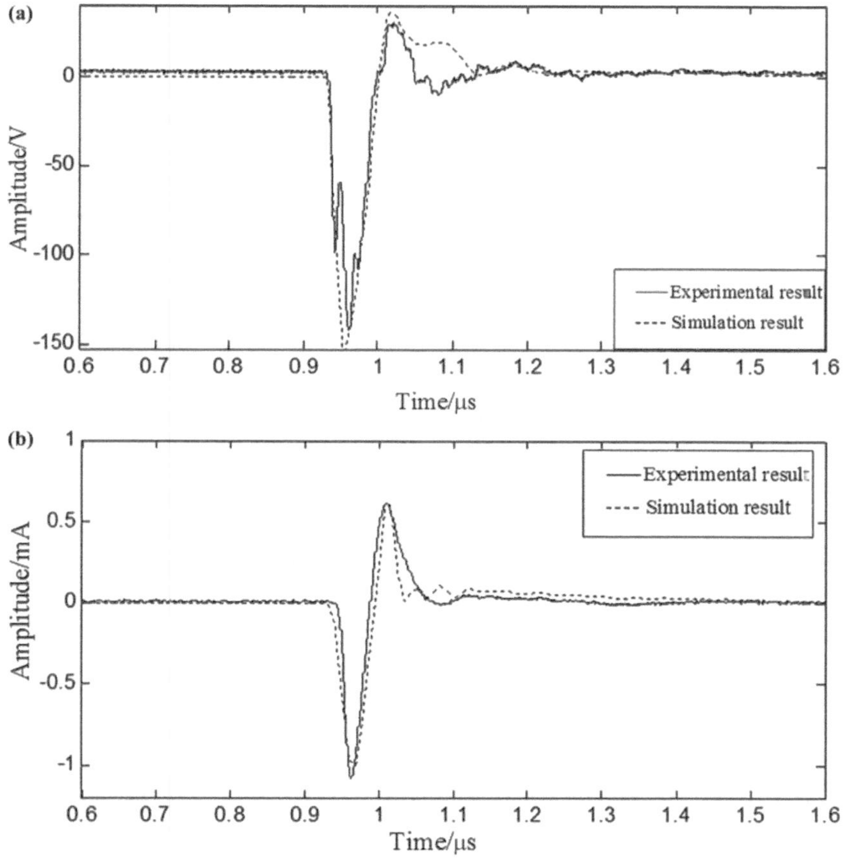

FIGURE 13.10 Negative spike pulse excitation voltage and current under load (−300 V). (a) The negative spike pulse excitation voltage and (b) the negative spike pulse excitation current.

$$p(\vec{r}_1,t)=\frac{\rho_0}{2\pi}\int_S \frac{\dfrac{\partial v_n\left(\vec{r}_2,t-\dfrac{|\vec{r}_1-\vec{r}_2|}{c}\right)}{\partial t}}{|\vec{r}_1-\vec{r}_2|}dS \tag{13.44}$$

where v_n is the vibration velocity of the aperture surface of the vertical transducer, $\vec{r}_1$ represents the coordinates of a field point and $\vec{r}_2$ represents the coordinates of an aperture point.

After the integral transformation, Eq. (13.44) can be rewritten as:

$$p(\vec{r}_1,t)=\frac{\rho_0}{2\pi}\frac{\partial\displaystyle\int_s \frac{\dfrac{\partial v_n\left(\vec{r}_2,t-\dfrac{|\vec{r}_1-\vec{r}_2|}{c}\right)}{\partial t}}{|\vec{r}_1-\vec{r}_2|}dS}{\partial t} \tag{13.45}$$

For the convenience of representation, the velocity potential $\Psi(\vec{r},t)$ is introduced while satisfying:

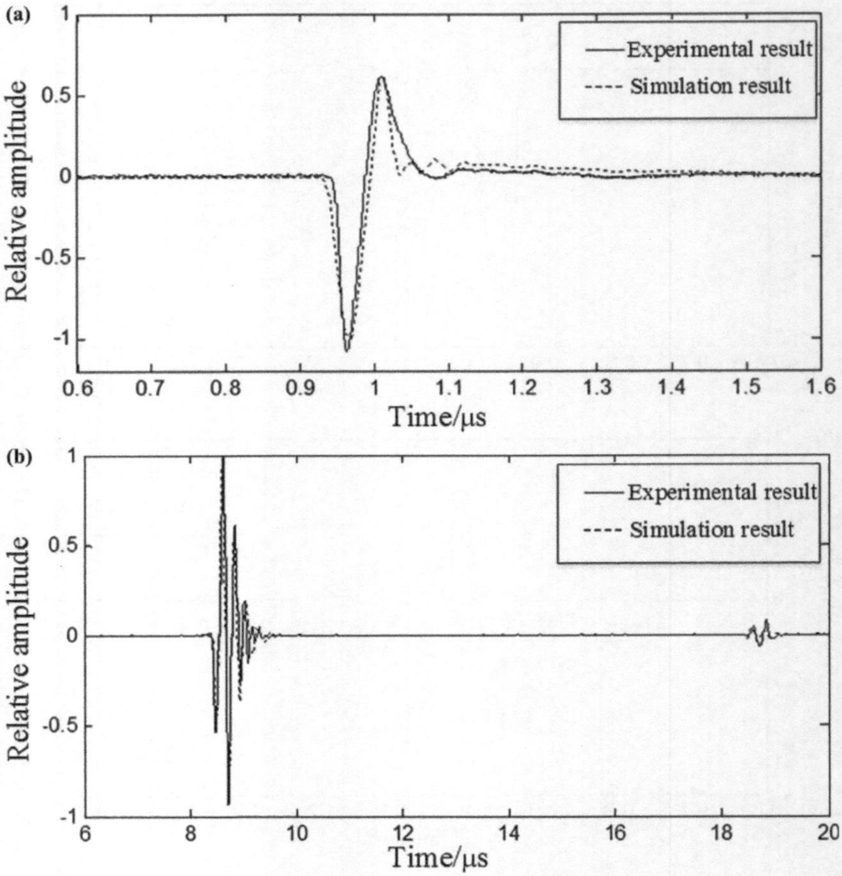

FIGURE 13.11 Experimental and simulation results of pulse echoes from the upper and lower surfaces of a 20# steel block. (a) Surface echo and (b) surface echo and bottom echo.

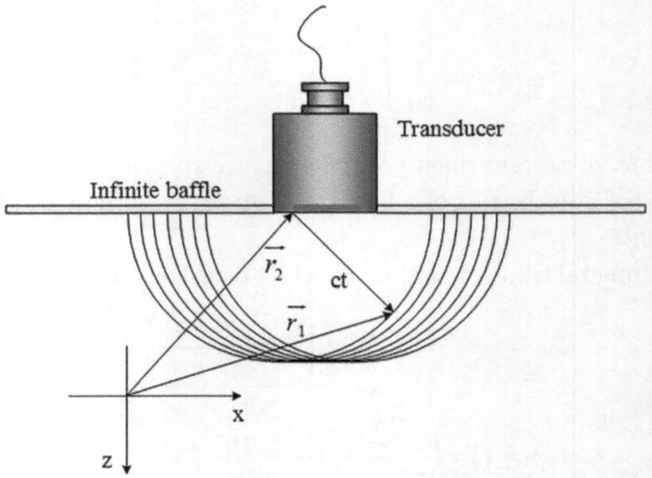

FIGURE 13.12 Principle of sound field radiation of a transducer aperture.

$$v_n(\vec{r},t) = -\nabla\Psi(\vec{r},t) \tag{13.46}$$

$$p(\vec{r},t) = \rho_0 \frac{\partial\psi(\vec{r},t)}{\partial t} \tag{13.47}$$

From Eqs. (13.45) and (12.47), the following equation can be derived:

$$\Psi(\vec{r}_1,t) = \int_s \frac{v_n\left(\vec{r}_1, t\frac{|\vec{r}_1 - \vec{r}_2|}{c}\right)}{2\pi|\vec{r}_1 - \vec{r}_2|} dS \tag{13.48}$$

The pulse excitation component is separately extracted and its time-domain convolution with the function δ is introduced. Then Eq. (13.48) can be rewritten as:

$$\Psi(\vec{r}_1,t) = v_n(t) \otimes \int_s \frac{\delta\left(t - \frac{|\vec{r}_1 - \vec{r}_2|}{c}\right)}{2\pi|\vec{r}_1 - \vec{r}_2|} dS \tag{13.49}$$

where $\otimes$ represents the time-domain convolution. The integral part of the above equation represents the characteristics of spatial pulse response of the transducer-aperture radiation field, which can be expressed as:

$$h(\vec{r}_1,t) = \int_s \frac{\delta\left(t - \frac{|\vec{r}_1 - \vec{r}_2|}{c}\right)}{2\pi|\vec{r}_1 - \vec{r}_2|} dS \tag{13.50}$$

By using the spatial pulse response function, $h(\vec{r}_1,t)$, the sound pressure in the radiated sound field of the transducer can be expressed as:

$$p(\vec{r}_1,t) = \rho_0 \frac{\partial v_n(t)}{\partial t} \otimes h(\vec{r}_1,t) \tag{13.51}$$

2. Calculation method of spatial pulse response

   As known from Eq. (13.51), if the spatial pulse response of the ultrasonic transducer is known, the sound pressure value at any field point in the radiation space and the spatial distribution characteristics of sound field can be deduced from the aperture vibration function $v_n(t)$ [18]. The principle of spatial pulse response and its rapid calculation method are the focus of this section. The spatial pulse response of the transducer is based on the linear system theory. When the transducer input is an ideal pulse excitation, the pulsed sound pressure response of a spatial field point can be determined. According to the rule of acoustic reciprocity, the pulse response generated by the excitation aperture at a certain field point can be obtained from the intersection between the spherical wave emitted by that field point and the array element [19]. Therefore, to determine the spatial pulse response, the field point should be projected onto the plane where the array element is located, and then the intersection between the projected spherical wave and the excited array element should be calculated.

Based on the dimensions and coordinate parameters in Figure 13.13 and Eq. (13.50), the spatial pulse response $h(\vec{r}_1,t)$ of the transducer can be expressed as:

$$h(\vec{r}_1,t)=\frac{1}{2\pi}\int_s \frac{\delta(R/c)}{R}dS \tag{13.52}$$

where R is the distance between the field point and the radiation source point, $R = ct$. The integral of Eq. (13.52) can also be written as the following polar coordinates:

$$h(\vec{r},t)=\frac{1}{2\pi}\int_{\Theta 1}^{\Theta 2}\int_{d_1}^{d_2}\frac{\delta(t-R/c)}{R}r\,dr\,d\Theta \tag{13.53}$$

where r is the radius of the projected circle; R is given by the formula $R^2=r^2+z_p^2$, where z_p is the height of the field point above the x–y plane of the aperture; the projection distances d_1 and d_2 are the minimum and maximum distances from the aperture, respectively, and depend on the aperture and Θ_1 and Θ_2 are the angles of two end points of the aperture arc intersected by the projected circle with radius r at a given time. By using the relation $2RdR=2rdr$, Eq. (13.53) can be transformed into:

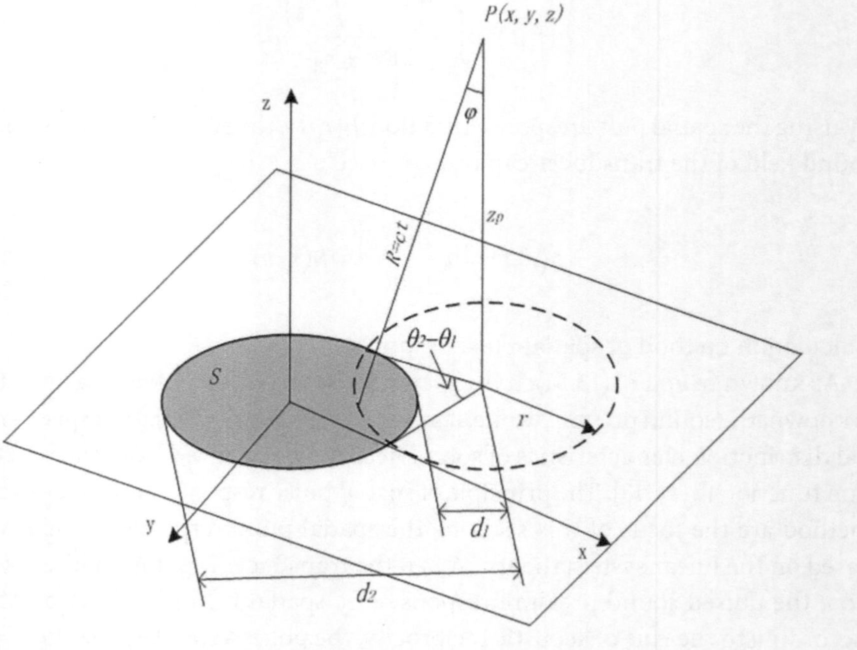

FIGURE 13.13 Coordinate system for calculating the spatial pulse response.

$$h(\vec{r},t)=\frac{1}{2\pi}\int_{\Theta_1}^{\Theta_2}\int_{d_1}^{d_2}\delta\left(t-\frac{R}{c}\right)dR\,d\Theta \tag{13.54}$$

Because $R=ct'$, Eq. (13.54) can be written into the following form by using $dR=cdt'$:

$$h(\vec{r},t)=\frac{c}{2\pi}\int_{\Theta_1}^{\Theta_2}\int_{t_1}^{t_2}\delta(t-t')dt'\,d\Theta \tag{13.55}$$

For a given time value, the integral can be written as the following relation, namely the simplified calculation formula of spatial pulse response:

$$h(\vec{r},t)=\frac{\Theta_2-\Theta_1}{2\pi}c \tag{13.56}$$

The above formula shows that at the time t, the real contribution to the sound field in the aperture is made by the aperture arc intersected by the sphere with P as the center and ct as the radius. For the convenience of numerical calculation, Eq. (13.56) can be changed into the following discrete form:

$$h(\vec{r}_1,t)=\frac{c}{2\pi}\sum_{i=1}^{N}\left[\Theta_2^{(i)}(t)-\Theta_1^{(i)}(t)\right] \tag{13.57}$$

where N is the number of aperture arcs intersected by the sphere at the time t and $\Theta_1^{(t)}(t)$ and $\Theta_2^{(t)}(t)$ are the angles of the intersection points on the corresponding arcs. It can be seen from the above equation that the calculation of spatial pulse response is related to the shape of the transducer aperture, that is, the calculation method varies with the aperture shape. The following sections will mainly analyze the numerical calculation methods of spatial pulse response of several kinds of transducer apertures studied in this paper and the characteristics of their spatial sound field distribution.

13.2.2 Sound Field of a Planar Transducer

1. Spatial pulse response of a planar transducer

   For a circular monocrystal planar transducer, the calculation of its spatial pulse response is shown in Figure 13.14. O_1 is the center of the circular aperture with radius r_a, and O_2 is the center of spherical-wave projection circle on the aperture plane at the field point P at a certain time. The calculation of spatial pulse response of this transducer should consider two situations: the spherical wave projection is outside the aperture circle, and the spherical wave projection is inside the aperture circle.

   The parameters in Figure 13.14 are defined as follows: $r_b(t)$ is the radius of the projected circle at the time t, a is the distance between the centers of the two circles and $h_a(t)$ is the distance between the focal points of the two circles and the x-axis, as specifically defined as follows [20]:

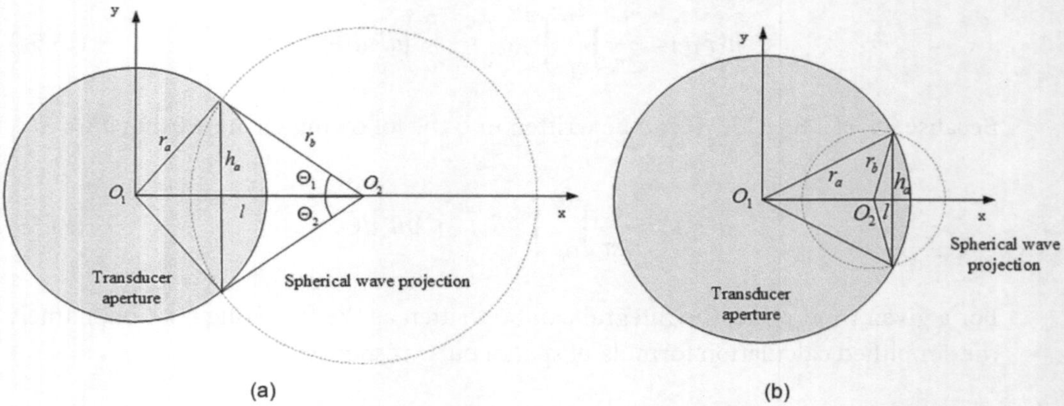

FIGURE 13.14 Calculation of spatial pulse response of a circular planar aperture. (a) Spherical-wave projection center is outside the aperture and (b) spherical-wave projection center is inside the aperture.

$$\begin{aligned} r_b(t) &= \sqrt{(ct)^2 + z_p^2} \\ a &= \|O_1 - O_2\| \\ p(t) &= \frac{a + r_a + r_b(t)}{2} \\ I &= \sqrt{r_b^2(t) - h_a^2(t)} \\ h_a(t) &= \frac{2\sqrt{p(t)\big(p(t)-a\big)\big(p(t)-r_a\big)\big(p(t)-r_b(t)\big)}}{a} \end{aligned} \tag{13.58}$$

When the spherical wave at the calculated field point is projected outside the transducer aperture (see Figure 13.14a), the calculation Eq. (13.56) of spatial pulse response can be changed into the following equation because of $\Theta_2 = \arctan\left(\frac{h_a(t)}{l}\right) = -\Theta_1$

$$h(r_1,t) = \frac{c}{2\pi}\|\Theta_1 - \Theta_2\| = \frac{c}{\pi}\arctan\left(\frac{h_a(t)}{l}\right) \tag{13.59}$$

$$\begin{aligned} h(r_1,t) &= \frac{c}{2\pi}\arcsin\left(\frac{2\sqrt{p(t)\big(p(t)-a\big)\big(p(t)-r_a\big)\big(p(t)-r_b(t)\big)}}{r_b^2(t)}\right) \\ &= \frac{c}{2\pi}\arcsin\left(\frac{ah_a(t)}{r_b^2(t)}\right) \end{aligned} \tag{13.60}$$

The conditions for starting and ending the calculation period are, respectively, the start and end of the intersection of the two circles, namely $r_a + r_b(t) = \|O_1 - O_2\|$ and $r_b(t) = r_a + \|O_1 - O_2\|$.

$$t_s = \frac{\sqrt{r_b^2(t)+z_p^2}}{c} = \frac{\sqrt{\left(\|O_1 - O_2\| - r_a\right)^2 + z_p^2}}{c} \tag{13.61}$$

$$t_e = \frac{\sqrt{r_b^2(t)+z_p^2}}{c} = \frac{\sqrt{\left(\|O_1 - O_2\| - r_a\right)^2 + z_p^2}}{c} \tag{13.62}$$

When the spherical wave at the calculated field point is projected inside the transducer aperture (see Figure 13.14b), the calculation Eq. (13.56) of spatial pulse response can be changed into Eq. (13.63).

At the time of $\frac{z_p}{c} \le t \le \frac{\sqrt{\left(r_a - \|O_1 - O_2\|\right)^2 + z_p^2}}{c}$, the two circles do not intersect, namely $h(r_1,t)=c$. After that, the intersection arc is located outside the aperture circle, so:

$$h(r_1,t)=\frac{c}{2\pi}\left(2\pi - \|\Theta_1 - \Theta_2\|\right) \tag{13.63}$$

The condition for ending the calculation period is the end of the intersection of the two circles (that is, $r_b(t)=r_1+\|O_1-O_2\|$):

$$t_e = \frac{\sqrt{\left(\|O_1 - O_2\| + r_a\right)^2 + z_p^2}}{c} \tag{13.64}$$

By using the above calculation process, the spatial pulse response of a planar transducer with a 12.7 mm aperture was calculated as shown in Figure 13.15. The starting point of calculation time is the time when the plane wave arrives at the corresponding distance on the axis. The radial calculation step is 0.5 mm, and the time step is 0.1 μS. It can be seen from the calculation results that with the increase of the axial distance, the response delay of the off-axis field point will gradually decrease. This also verifies the hypothesis that the far-field wave of the transducer can be generally regarded as a plane wave in the sound field calculation.

2. Sound field of a planar transducer

According to Eq. (3.8), the sound pressure in the pulsed sound field can be determined by the convolution of the acceleration of the transducer surface and the spatial pulse response. That is, the sound pressure $p(\vec{r},t)$ at any point $\vec{r}$ in space is equal to the convolution of the derivative of the normal velocity $v_n(t)$ of the transducer surface and the spatial pulse response of the transmitting aperture. If the normal velocity $v_n(t)$ of any micro-element on the transducer surface is the same, the vibration velocity of the surface of a transducer array element can be approximately calculated by the following equation in the numerical calculation [21]:

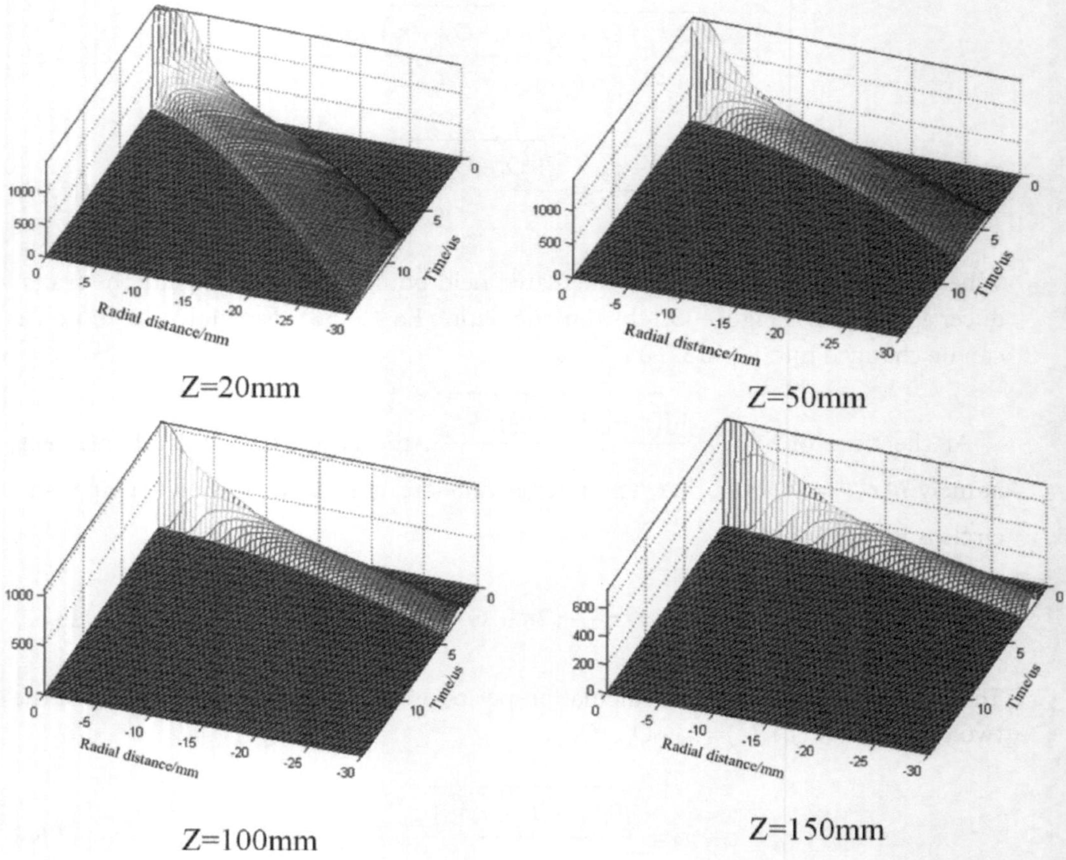

FIGURE 13.15 Spatial pulse response of a planar transducer.

$$v_n(t) = t_3 e^{-kf_0 t} \cos(2\pi f_0 t) \tag{13.65}$$

For example, when the center frequency of the transducer is $f_0 = 5\,\text{MHz}$ and the broadband pulse coefficient is $k = 3.8$, the waveform and frequency spectrum of array element surface velocity are as shown in Figure 13.16a and b, respectively.

Take the circular liquid-immersed piezoelectric transducer as an example. The central frequency of the transducer is set as 2 MHz, the size of the piezoelectric wafer is 10 mm, the ultrasound velocity in water is $c = 1480\,\text{m/s}$ and the ultrasound wavelength is 0.74 mm. According to Eq. (3.8), the three-dimensional spatial distribution of the sound field emitted by the transducer can be determined. However, for a circular symmetric transducer, several of its main sections can fully represent the characteristics of sound field. The numerical calculation results of the transducer axis section are shown in Figure 13.17. It can be seen from the figure that its near-field distance (natural focus) is about $N = 35\,\text{mm}$.

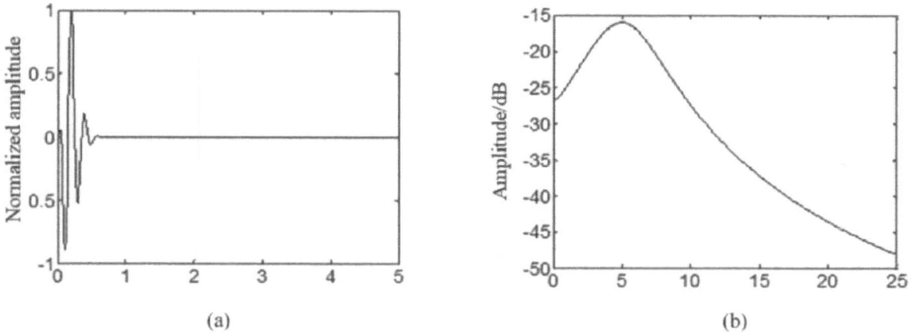

FIGURE 13.16 (a) Waveform and (b) frequency spectrum of array element surface velocity.

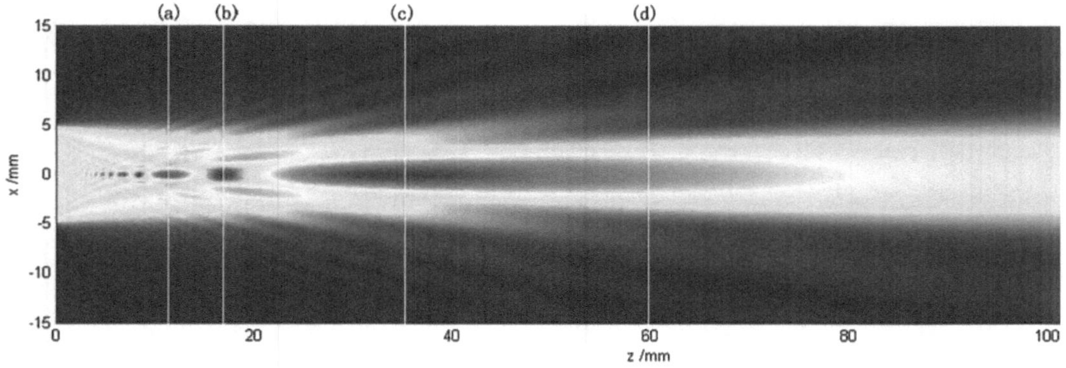

FIGURE 13.17 Numerical calculation result of transducer-radiated sound field distribution.

The sound pressure distribution on four sound beam sections (see Figure 13.17) is shown in Figure 13.18. In the figure, (a) and (b) are the sound pressure distributions at the maximal point ($z = 12$ mm) and minimal point ($z = 17$ mm) on the near-field axis; (c) is the sound pressure distribution at the interface between near and far fields (natural focus: $z = 35$ mm); and (d) is the far-field sound pressure distribution ($z = 60$ mm).

13.2.3 Sound Field of a Focusing Transducer

1. Spatial pulse response of a focusing transducer

The calculation principle of spatial pulse response of a spherical focusing transducer is the same as that of a planar transducer. However, because the transmitting surface of the focusing transducer is concave, the projection of a spatial field point on the transmitting aperture surface will not be a single plane. For a spherical reflection aperture transducer, its radiation sound field space can be divided into two areas according to their geometric relationship in order to simplify the calculation. As shown in Figure 13.19, the geometrical focal point is taken as the origin of coordinates, and the beam emission direction is the z-axis direction. The conical

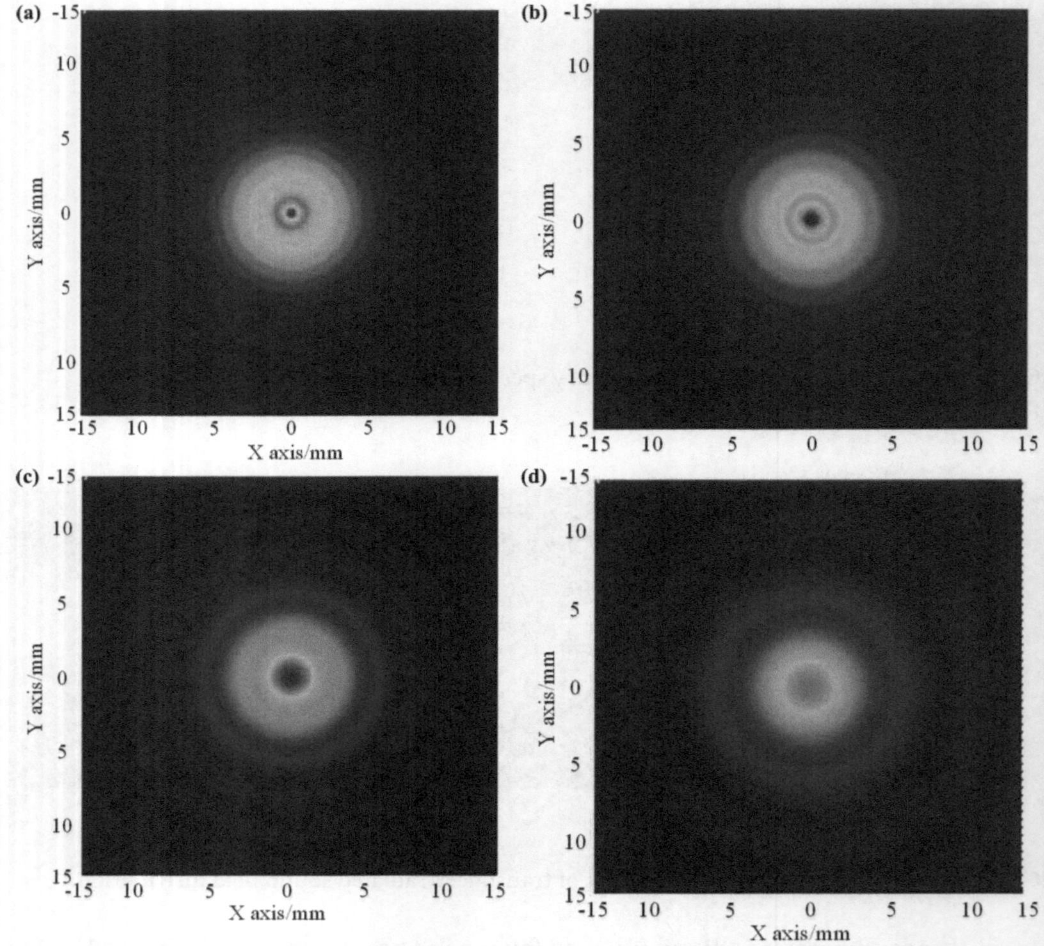

FIGURE 13.18 Calculation results of transverse sound pressure distribution of sound beam at different axial distances. (a) $z = 12$ mm, (b) $z = 17$ mm, (c) $z = 35$ mm and (d) $z = 60$ mm.

area composed of the origin and the transducer reflection sphere, together with the infinite conical area symmetrical to it about the X-axis, is the first zone, while the rest is the second zone.

When the calculated field point is in the first zone of sound field space, the spatial pulse response of the transducer can be calculated as follows [22]:

$$h(\vec{r},t)=\begin{cases} 0 & z<0 & z>0 \\ cR/r & ct<r_0 & ct>r_0 \\ \dfrac{cR}{r}\dfrac{1}{\pi}\arccos\left(\dfrac{\eta(t)}{\sigma(t)}\right) & r_0<ct<r_1 & r_0>ct>r_1 \\ & r_1<ct<r_2 & r_1>ct>r_2 \\ 0 & r_2<ct & r_2>ct \end{cases} \tag{13.66}$$

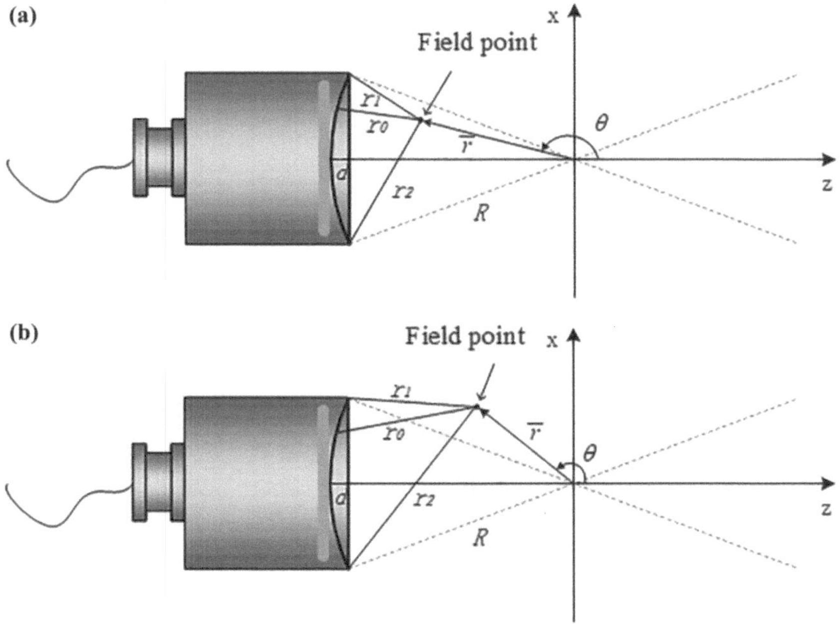

FIGURE 13.19 Calculation of spatial pulse response of a spherical focusing transducer. (a) Field point in the first zone and (b) field point in the second zone.

The geometric parameters are marked in Figure 13.19a and b. $\eta(t)$ and $\sigma(t)$ are represented by the following equations:

$$\eta(t)=R\left\{\frac{1-d/R}{\sin\theta}+\frac{1}{\tan\theta}\left(\frac{R^2+r^2-c^2t^2}{2rR}\right)\right\} \tag{13.67}$$

$$\sigma(t)=R\sqrt{1-\left(\frac{R^2+r^2-c^2t^2}{2rR}\right)^2} \tag{13.68}$$

When the calculated field point is in the second zone of sound field space, the spatial pulse response of the transducer can be calculated as follows:

$$h(\vec{r},t)=\begin{cases} 0 & ct<r_1 \\ cR/r & - \\ \dfrac{cR}{r}\dfrac{1}{\pi}\arccos\left(\dfrac{\eta(t)}{\sigma(t)}\right) & r_1<ct<r_2 \\ 0 & r_2<ct \end{cases} \tag{13.69}$$

For the transducers with different apertures, their spatial pulse responses on the radial lines of the focusing plane are shown in Figures 13.20 and 13.21, respectively. It can be seen from the figures that with the increase of the off-axis distance, the

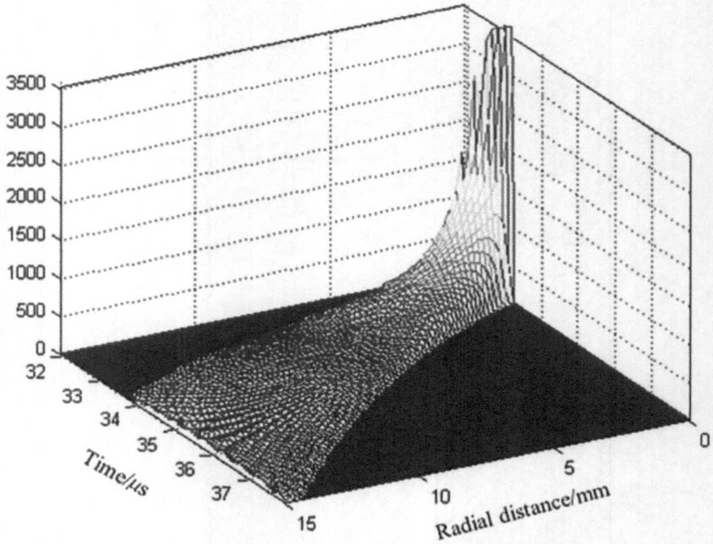

FIGURE 13.20 Spatial pulse response of a spherical focusing transducer (12 mm aperture, $z = 50$ mm).

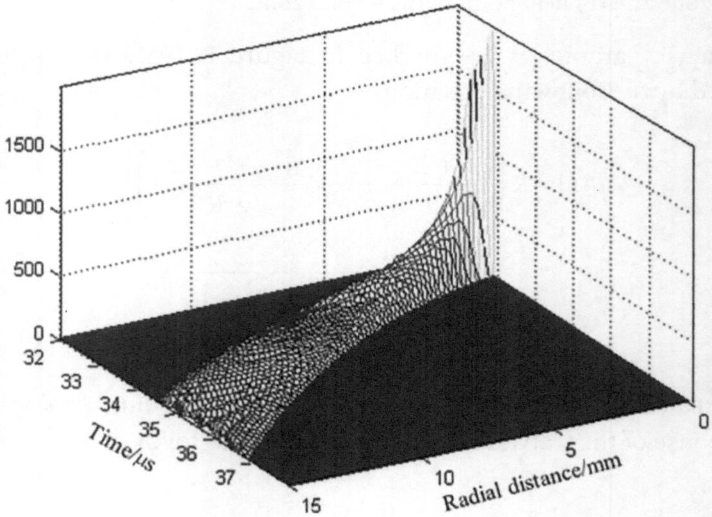

FIGURE 13.21 Spatial pulse response of a spherical focusing transducer (6 mm aperture, $z = 50$ mm).

spatial pulse response will be gradually widened on the time axis. Moreover, with the increase of the transducer aperture diameter, the spatial pulse response will also be widened on the time axis. This is because the time difference between the two edge waves in radial direction arriving at the off-axis field point will increase with the diameter of the transducer aperture.

2. Sound field of a focusing transducer

According to Eq. (3.8), the sound pressure $p(\vec{r},t)$ at any point $\vec{r}$ in space is equal to the convolution of the derivative of the normal velocity $v_n(t)$ of the transducer surface and the spatial pulse response $h(\vec{r},t)$ of the transmitting aperture. Similarly, it is assumed that the normal velocity of any micro-element on the transducer surface is the same. In the numerical calculation, the vibration velocity of the transducer array surface is approximately calculated by Eq. (3.22). Take a spherical focusing piezoelectric transducer as an example. The central frequency of the transducer is set as 5 MHz, the size of the piezoelectric wafer is 10 mm and the focal distance is 30 mm. The ultrasound velocity in water is $c = 1480$ m/s, and the ultrasound wavelength is about 0.3 mm. According to Eq. (3.8), the three-dimensional spatial distribution of the sound field emitted by the transducer can be determined. The numerical calculation result of the transducer axis section is shown in Figure 13.22. From the figure, it can be seen that the focal distance is consistent with the preset focal distance. The sound pressure distributions on the beam sections (a) and (b) (see Figure 13.22) are shown in Figure 13.23, where (a) is located on the focal plane ($z = 30$ mm) and (b) is located at the rear focal area ($z = 40$ mm).

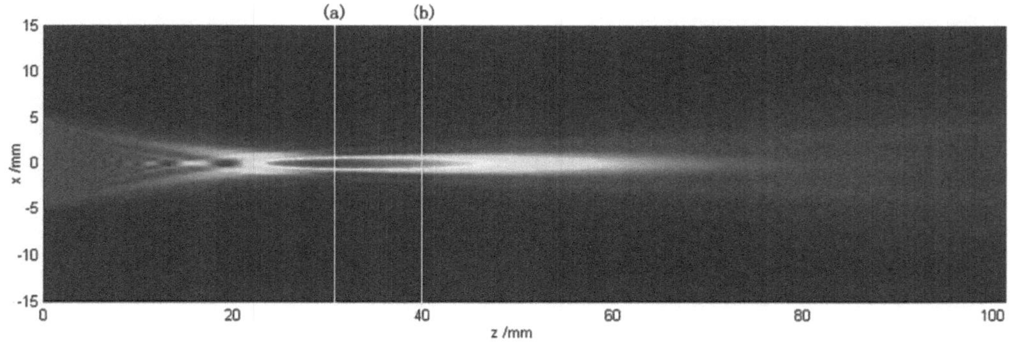

FIGURE 13.22 Numerical calculation result of axial section of the sound field radiated by a focusing transducer.

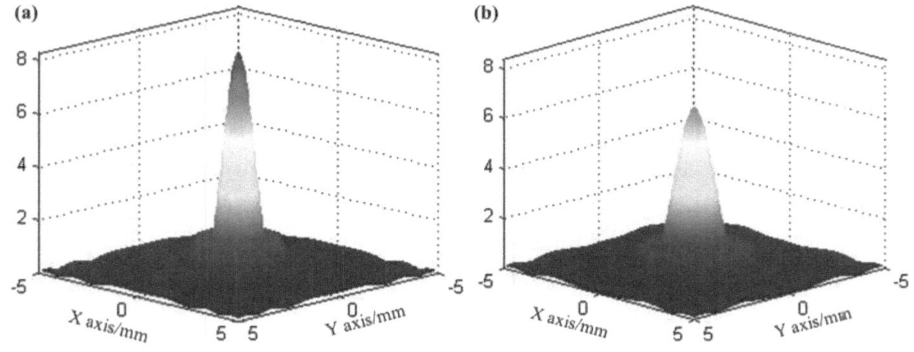

FIGURE 13.23 Numerical calculation result of horizontal distribution of the sound beam radiated by a focusing transducer. (a) Focal plane: $z = 30$ mm and (b) rear focal area: $z = 40$ mm.

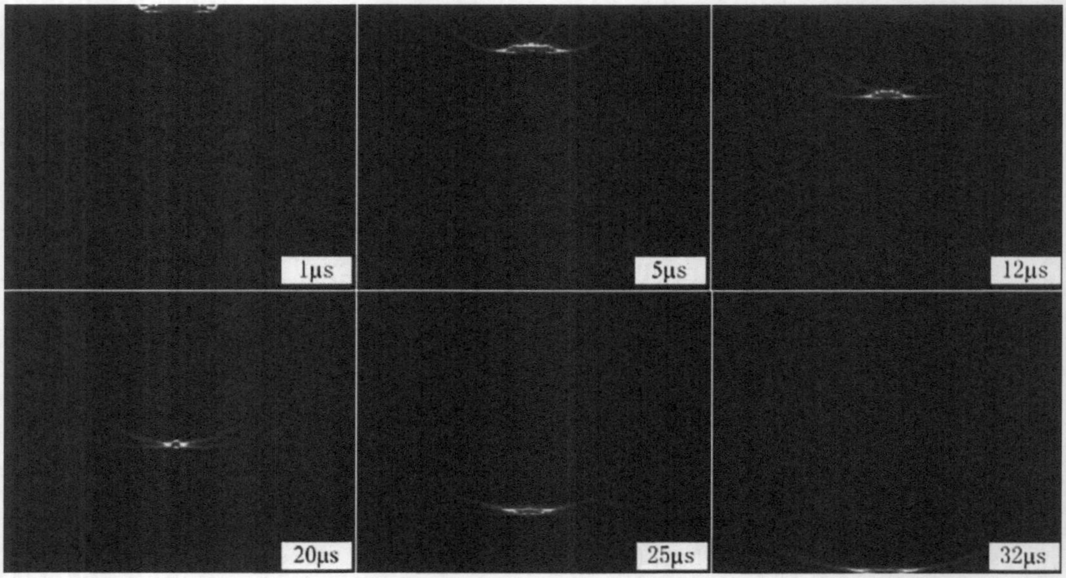

FIGURE 13.24 Sound beam propagation simulation result of a focusing transducer with beam-focusing characteristics.

The numerical simulation result of radiated sound field calculated by using the spatial pulse response of the transducer is shown in Figure 13.24. It can be seen that the sound beam arrives at the focal point at 20 μs, when the beam is the narrowest. Later the beam begins to diffuse.

13.3 MEASUREMENT MODEL AND METHOD OF SOUND FIELD OF AN ULTRASONIC TRANSDUCER

To improve the precision of defect detection, higher requirements are put forward for the measurement method of the sound-field distribution of the ultrasonic transducer. As described in Chapter 1, the researchers have proposed many sound-field measurement methods over the decades, such as hydrophone method, ball reflection method, sonoluminescence method, schlieren method and dynamic photoelastic method. However, each measurement method has its own disadvantages, such as complex and expensive system, limited object, inconvenient operation and poor measurement accuracy. According to the project requirements and several problems in sound field measurement, this chapter studies the measurement model based on ball reflection, the measurement method based on straight side reflection, the automatic collimation method of hydrophone, the correction model of average effect of hydrophone aperture, the simple measurement method of the sound-field distribution of an air-coupled ultrasonic transducer and other methods.

13.3.1 Ball Measurement Method of Sound Field of an Ultrasonic Transducer

Considering the measurement timeliness, result stability, repeatability and operation difficulty, ball reflection method is one of the most commonly used sound-field measurement methods at home and abroad. In this method, a small ball with appropriate diameter and material is selected and then spatially scanned by the sound beam of the transducer under test, and finally, the amplitude of the reflected pulse echo is extracted to obtain the information on ultrasonic beam distribution. During the measurement, only the characteristics of the transducer under test and of the reflecting ball can receive the system response. The biggest advantage of ball reflector is that the sound beam in any position of ultrasonic field can be aligned with the normal direction of the ball, so the maximum reflection echo can be obtained. Moreover, the ball method has a simple device and low cost. However, this method has certain requirements for the diameter and surface finish of the reflecting ball. Too small diameter is likely to cause the generation of a creeping wave, which will interfere with the surface echo signal. The surface roughness will reduce the signal-to-noise ratio (SNR) and change the echo spectrum.

1. Model of ball-scattered sound field

   In actual sound field measurement, a rigid ball is often used as the reflection target for sound pressure measurement. The sound wave is scattered complicatedly on the ball. The characteristics of scattered wave have a lot to do with the wavelength of sound wave and the ball size. The scattered wave overlaps with the incident wave field to generate interference, resulting in a complex interfering sound field and an error between the measurement result and the theoretical truth value [23]. Therefore, it is very important to analyze the sound field scattered by the small ball when studying the ball-reflection measurement technique of the transducer sound field.

   The characteristics of sound field of a water-immersed ultrasonic transducer depend on both the radiation characteristics of the transducer and the nature of propagation medium. Since water can be used as a uniform propagation space, the characteristics of the sound field radiated by the transducer mainly depend on the radiation frequency, radiation aperture and radiation mode of the transducer. In this section, the ball-reflection measurement model of the transducer sound field will be built upon the ball-scattered sound field and the transducer receiving model.

   As shown in Figure 13.25, the incident wave propagating in the z direction is scattered by a rigid ball with radius a, which is located at the origin. The spherical coordinate system $r\theta\phi$ is introduced, whose z-axis is $\theta=0$. The whole system is symmetric with respect to the z-axis and has nothing to do with ϕ. In the spherical coordinate system, the sound pressure of incident wave can be expressed as

$$P_i(x,t)=P_0e^{i(\omega t-kx)}=P_0e^{i\omega t}e^{-ikr\cos\theta} \tag{13.70}$$

For a rigid ball with radius a, the sound pressures of radiation wave and scattered wave should both meet the wave equation and the boundary condition of zero normal displacement [24]:

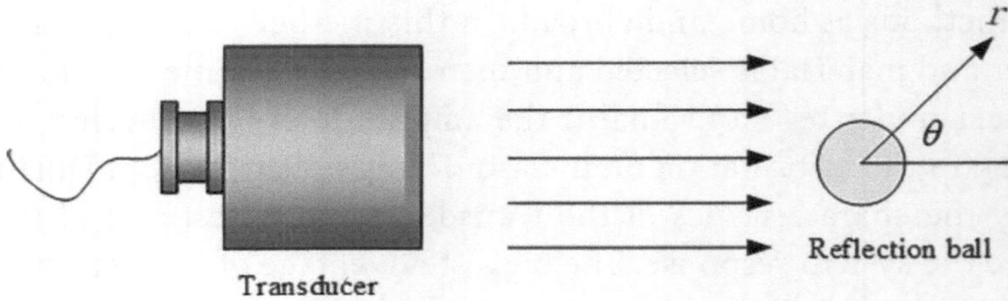

FIGURE 13.25 Schematic diagram of plane wave scattered by ball.

$$U_r\big|_{r=r_0}=U_{ri}\big|_{r=r_0}+U_{rr}\big|_{r=r_0}=0 \tag{13.71}$$

where U_{ri} and U_{rr} are the normal velocities of incident wave and scattered wave, respectively. After deducing U_{rr} from the velocity U_{ri} of incident wave at the interface, this problem can be transformed into a radiation problem. Since the solution of wave equation can be expressed as the superposition of ball functions, the incident plane wave can also be expanded into the series of ball functions. So, the plane wave is expanded into the series of Legendre functions:

$$\exp(ikr\cos\theta)=\sum_{l=0}^{\infty}C_lP_l(\cos\theta) \tag{13.72}$$

By using the orthogonality of Legendre functions

$$\int_0^{\pi}P_l(\cos\theta)P_n(\cos\theta)\sin\theta\, d\theta=\int_{-1}^{1}P_n(x)P_l(x)dx=\begin{cases}0 & n\neq l\\ \dfrac{2}{2l+1} & n\neq l\end{cases} \tag{13.73}$$

we can obtain

$$C_l=\frac{2l+1}{2}\int_{-1}^{1}P_l(\cos\theta)\exp(ikr\cos\theta)d(\cos\theta)=(2l+1)i^l j_l(kr) \tag{13.74}$$

where the integral can be verified by Taylor series expansion. Thus, the equation of plane wave represented by spherical harmonics is obtained:

$$P_i=P_0\exp(ikz)=P_0\sum_{l=0}^{\infty}(2l+1)i^lP_l(\cos\theta)j_l(kr) \tag{13.75}$$

It is known from the above equation that, plane wave decomposition is to change a single-frequency wave with equal amplitude on an infinitely great plane into the superposition of infinite spherical sound wave components of the same frequency with different orders and amplitudes. The normal velocity of incident wave at the interface is:

$$U_{ri}\Big|_{r=r_0} = \frac{P_0}{i\rho_0 c}\sum_{l=0}^{\infty}(2l+1)i^l P_l(\cos\theta)\left[j_l(ka)\right] \tag{13.76}$$

The scattered waves can be expressed as the superposition of ball functions:

$$P_r = P_0\sum_{l=0}^{\infty} A_l h_l^{(2)}(kr)P_l(\cos\theta) \tag{13.77}$$

The velocity normal to the sphere is:

$$U_{rr}\Big|_{r=r_0} = \frac{P_0}{i\rho_0 c}\sum_{l=0}^{\infty} A_l P_l(\cos\theta)\left[h_l^{(2)}(ka)\right]' \tag{13.78}$$

Eqs. (13.75) and (13.77) cancel each other out. Because they both are the series of Legendre functions, they must be cancelled out term by term, that is:

$$A_l = -\frac{(2l+1)i^l\left[j_l(ka)\right]'}{\left[h_l^{(2)}(ka)\right]'} \tag{13.79}$$

By substituting Eq. (13.80) into Eq. (13.76), the scattering field is obtained:

$$P_r = -P_0\sum_{l=0}^{\infty}\frac{(2l+1)i^l\left[j_l(ka)\right]'}{\left[h_l^{(2)}(ka)\right]'}h_l^{(2)}(kr)P_l(\cos\theta) \tag{13.80}$$

The scattered wave is the superposition of various orders of axisymmetric spherical waves, so it demonstrates an axisymmetric directivity. The sound pressure amplitude of the scattered wave is very complicated and is proportional to that of the incident wave. Different orders of scattered wave components have different amplitudes. Because $h_l^{(2)}(ka)$ is a complex function, the initial phases of different components are also different.

By using the far field formula of radiated sound field, the far field solution $(kr \geq 1)$ is obtained:

$$P_r \to -P_0 a\frac{\exp(ikr)}{r}R(\theta) \tag{13.81}$$

where $R(\theta)$ is the directivity function of the scattered sound field,

$$R(\theta)=\frac{1}{ka}\sum_{l=0}^{\infty}\frac{(2l+1)(-i)^l\left[j_l(ka)\right]'}{\left[h_l^{(2)}(ka)\right]'}\exp\left(\frac{l+1}{2}\pi\right)P_l(\cos\theta) \tag{13.82}$$

If, $b_l=\dfrac{(2l+1)(-i)^l\left[j_l(ka)\right]'}{\left[h_l^{(2)}(ka)\right]'}$ then

$$R(\theta)=\frac{1}{ka}\sum_{l=0}^{\infty}b_l\exp\left(\frac{l+1}{2}\pi\right)P_l(\cos\theta) \tag{13.83}$$

Based on the principle of linear superposition of sound fields, the transient scattered sound field can be calculated through the Fourier transform of steady scattered sound field. It is assumed that the pulsed sound field is excited by the transducer in Figure 13.25. By using the Fourier transform, the pulsed sound field of the incident wave can be expressed as:

$$P_i(z,t)=\frac{1}{2\pi}\int_{-\infty}^{\infty}A(z,\omega)e^{j\omega t}e^{-jkr\cos\theta}\,d\omega \tag{13.84}$$

where $A(z,\omega)=F\left(p_i(z,t)\right)$ is the frequency spectrum of the incident sound field. The pulsed scattered sound field is the superposition of the corresponding single-frequency continuous scattered sound fields:

$$P_s(r,\theta,t)=\frac{1}{2\pi}\int_{-\infty}^{\infty}A(z,\omega)D(\omega,\theta)e^{j\omega t}e^{-jkr\cos\theta}\,d\omega \tag{13.85}$$

where $D(\omega,\theta)$ is the directivity function of the single-frequency continuous scattered sound field at the frequency ω. Similarly, according to the definition of directivity function, the directivity function of the pulsed sound field scattered by the ball is [25,26]:

$$D_p(\theta)=\frac{\int_{-\infty}^{\infty}A(z,\omega)D(\omega,\theta)e^{j\omega t}e^{-jkr\cos\theta}\,d\omega}{\int_{-\infty}^{\infty}A(z,\omega)e^{j\omega t}e^{-jkr\cos\theta}\,d\omega}\approx\frac{\sqrt{\sum_{\omega=0}^{\infty}\left(A(z,\omega)D(\omega,\theta)\right)^2}}{\sqrt{\sum_{\omega=0}^{\infty}A(z,\omega)^2}} \tag{13.86}$$

The directivities of the effective sound pressures calculated at different *ka* values are shown in Figure 13.26, where the curves are normalized by the maximum value. It can be seen that, when the visible frequency is relatively low and the ball is relatively small, the scattering phenomenon will mainly appear on the ball side facing the sound source. With the increase of frequency and ball size, the directivity will

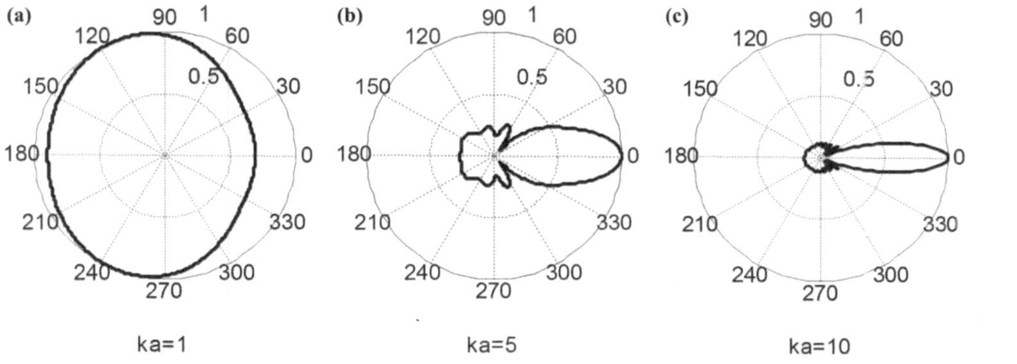

FIGURE 13.26 Directivity of far-field effective sound pressure scattered by a rigid ball.

become very complicated. When the frequency is very high and the ball is very large, a strong scattered wave will be generated on the ball backside and offset by the direct wave to form a shadow zone, and uniform scattering will be seen in other directions.

The calculated results of planar-pulse-wave sound fields scattered by different balls are shown in Figure 13.27a and b, respectively, representing the pulsed scattered sound fields with different *ka* values (5 and 20). The vertical stripe shown in the figure is a planar pulse wave with multiple Gaussian envelopes propagating from left to right. The left side of the rigid ball is the irradiated area and the right side is the shadow area. It can be seen from the figure that the sound field of pulse wave scattered by the rigid ball is approximate to that of a spherical wave, and the sound pressure decreases gradually with the diffusion. When the *ka* value is relatively large, the scattered sound field in the shadow area on the ball backside is sharp with large sound pressure amplitude. This agrees with the analysis conclusion of the scattered sound field of a small ball, that is, with the increase of ball radius, the acting surface of the ball and the spherical wave energy scattered by the ball will also increase.

When the sound wave scattered by the rigid ball propagates to the radiant surface of the transducer, the transducer can receive the sound pressure of the scattered wave and convert it into the output voltage. The measurement model of the transducer sound field in the ball method is shown in Figure 13.28. If the received total sound pressure is proportional to the average sound pressure of the transducer receiving surface, then:

$$P_{\text{ave}}(\omega)=\frac{P(r,\omega)}{S_r}=\frac{1}{S_r}\int_{S_r} P_s(s,\omega)dS_r \tag{13.87}$$

where S_r is the area of the receiving surface of the transducer, $P(r,\omega)$ is the total sound pressure received by the transducer and $P_s(s,\omega)$ is the sound pressure of the scattered sound field on the ball surface. The relationship between the average velocity field of the particles on the transducer radiant surface, denoted as $v_C(\omega)$ and the frequency spectrum component of the excitation pulse voltage, denoted as $V_i(\omega)$ can be expressed as [27]:

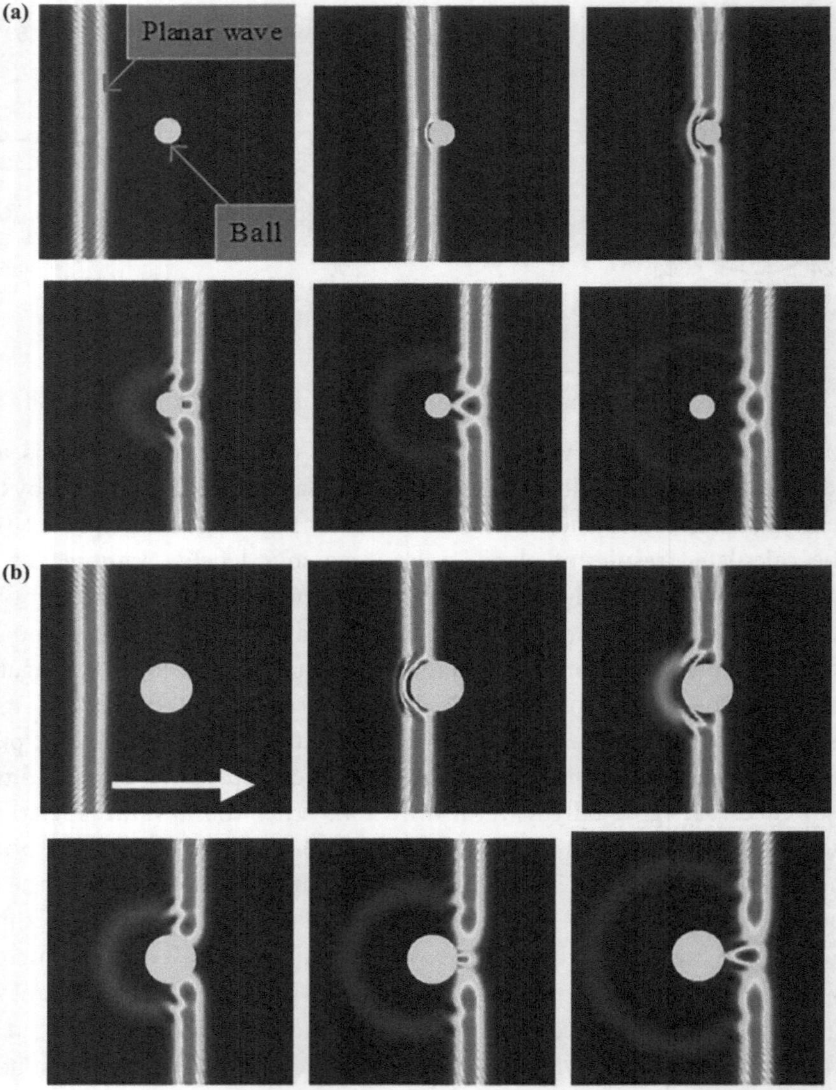

FIGURE 13.27 Sound field of planar pulse wave scattered by a rigid ball. (a) $ka = 5$ and (b) $ka = 20$.

$$v_0(\omega) = \frac{\beta_i(\omega) V_i(\omega)}{\rho_0 c} \tag{13.88}$$

where $\beta_i(\omega)$ is the input scaling factor. When the transducer receives the sound pressure, the following equation can be derived according to the reciprocity theorem:

$$V_R(\omega) = \beta_r(\omega) P_{\text{ave}}(\omega) \tag{13.89}$$

where $\beta_r(\omega)$ is the output scale factor. By multiplying Eqs. (4.19) and (4.20), we can obtain

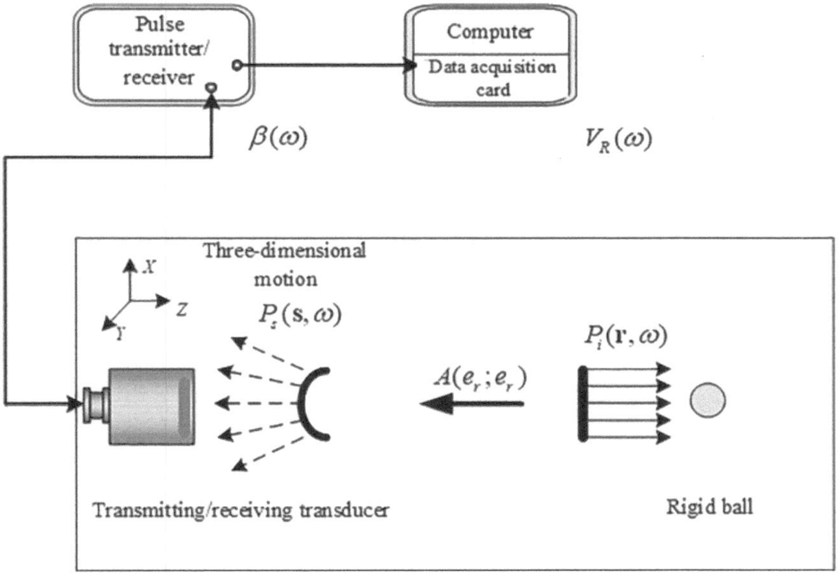

FIGURE 13.28 Measurement model of transducer sound field in the ball method.

$$V_R(\omega)v_0(\omega) = \frac{\beta_r(\omega)P_{\text{ave}}(\omega)\beta_i(\omega)V_i(\omega)}{\rho_0 c} \tag{13.90}$$

$$V_R(\omega) = \beta(\omega)\frac{P_{\text{ave}}(\omega)}{\rho_0 c v_0(\omega)} \tag{13.91}$$

where $\beta(\omega)$ is the system influence factor, $\beta(\omega) = \beta_i(\omega)\beta_r(\omega)V_i(\omega)$. It describes the influence of the whole ultrasonic testing system (including pulse transmitter/receiver, amplifier circuit, cable, as well as the energy conversion caused by the piezoelectric effect of a transducer) on ultrasonic echo signal [27]. The influence factor can be obtained from the deconvolution of Eq. (13.91).

The physical meaning of Eq. (13.91) is that the echo response received by the measurement system is directly proportional to the average sound pressure of the ball-scattered sound field received by the transducer surface. This scaling factor is just the system influence factor $\beta(\omega)$. $\beta(\omega)$ can be determined experimentally by a planar reflector, and $V_R(\omega)$ can be measured experimentally. The mean sound pressure term $P_{\text{ave}}(\omega)$ here can be calculated by using the corresponding reference model. For example, in the experiment of planar reflector, Rogers and Van Buren concluded that the mean sound pressure term was approximately [28]:

$$P_{\text{ave}}(\omega) = \rho_0 c v_0 R_{12}\exp(2ikD)\left\{1 - \exp\left(\frac{ika^2}{2D}\right)\cdot\left[J_0\left(\frac{ka^2}{2D}\right) - iJ_1\left(\frac{ka^2}{2D}\right)\right]\right\} \tag{13.92}$$

The experimentally measured ultrasonic echo voltage signal of a 5 MHz ultrasonic transducer with a central frequency of 5 MHz is shown in Figure 13.29a. The system

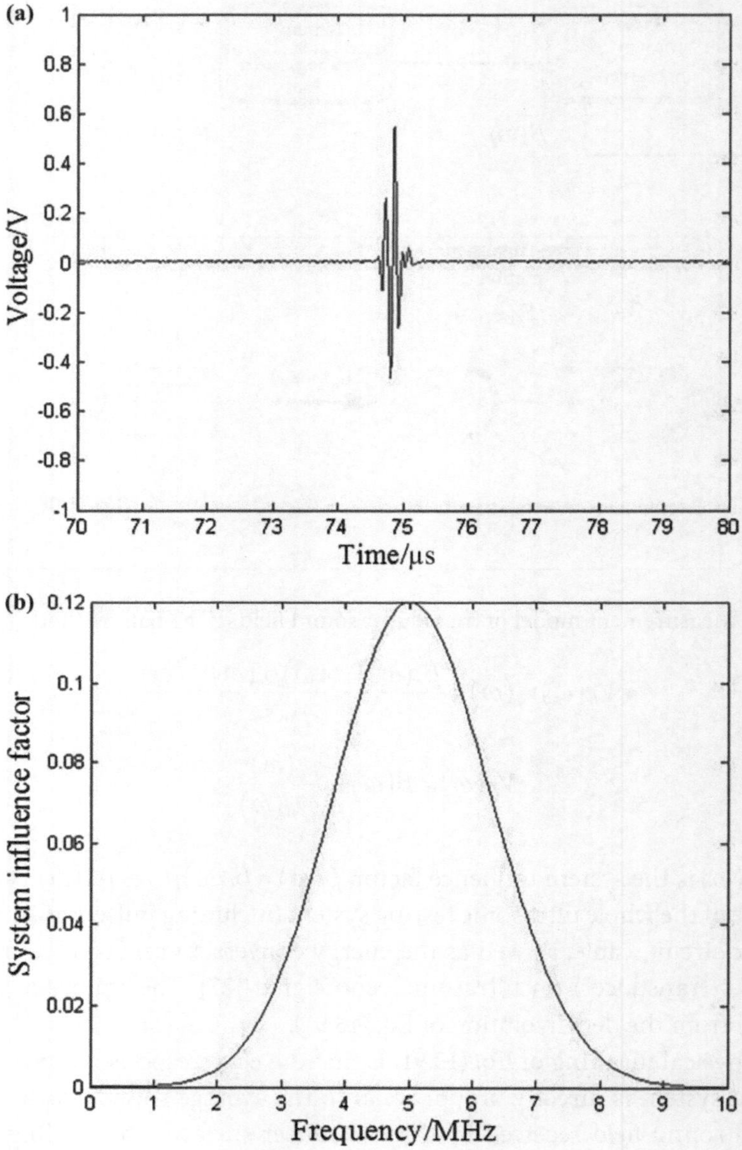

FIGURE 13.29 Time domain waveform and system influence factor of a transducer. (a) Time domain waveform and (b) system influence factor.

influence factor obtained after the deconvolution of frequency domain is shown in Figure 13.29b.

By combining Eqs. (13.87) and (13.91), the voltage response $V_R(\omega)$ of the ball-scattered sound field received by computer in the model can be obtained (Figure 13.30):

$$V_R(\omega)=\beta(\omega)\frac{P_{\text{ave}}(\omega)}{\rho_0 c v_0(\omega) S_r}=\frac{\beta(\omega)}{\rho_0 c v_0(\omega) S_r}\int_{S_r} P_s(s,\omega)dS_r \tag{13.93}$$

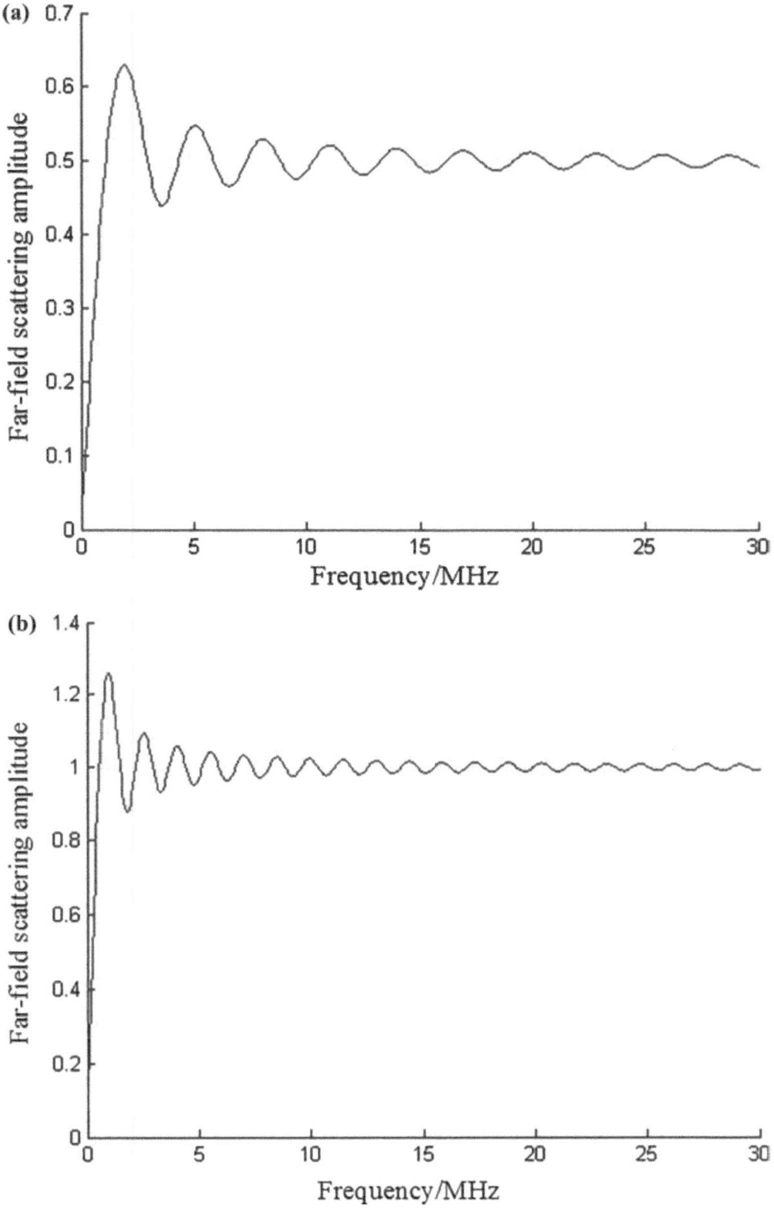

FIGURE 13.30 Far-field scattering amplitude of steel balls (1 and 2 mm) in water. (a) $d = 1$ mm and (b) $d = 2$ mm.

For the convenience of integral calculation, the sound pressure $P_s(\mathbf{r},\omega)$ of the scattered sound field is expressed as:

$$P_s(\mathbf{s},\omega) = P_0 A(e_r;e_r)\frac{\exp(ikr)}{r} = \frac{-i\omega\rho_0 v_0(\omega)}{2\pi}\int_{S_t}\frac{\exp(ikr)}{r}dS_t \cdot A(e_r;e_r)\frac{\exp(ikr)}{r} \tag{13.94}$$

where S_t is the radiant surface area of the transmitting transducer, $S_t = S_r$. In this model, $A(e_r;e_r) = a\exp(-2ika)/2$ is the far-field scattering amplitude of small ball [27], and a is the ball radius. The far-field scattering amplitudes of steel balls (1 and 2 mm) in water are shown in Figure 13.30.

According to the above theoretical derivation, the following conclusions can be drawn:

The intensity of the scattered wave varies with time. In other words, the scattering capacity of the ball varies with the frequency of incident wave. On the other hand, in the sound field with sound waves at a certain frequency, the sound waves scattered by the balls with different diameters will have different intensities and directivities. The scattered power depends entirely on ka (i.e. a/λ), in addition to the intensity of the incident wave. The scattered power will decrease with ka. The intensity of the scattered wave is proportional to that of the incident wave and diffuses as a spherical wave in the far field. Therefore, the intensity of the scattered wave is inversely proportional to the square of the distance r.

The intensity of the scattered wave is not uniform in space but varies in different directions. Its directional characteristics are expressed by the directivity function. At the same time, the value of b_l varies with ka, that is, the amplitude and energy distribution of each order of scattered spherical wave component vary with ka. Thus, the directional characteristics of the scattered wave also change with ka.

When the ball reflection method is used to measure the sound field, ka should be kept as small as possible in order to improve the resolution of sound beam distribution measurement and reduce the fuzziness caused by sound pressure integral in the reflection area of the ball. However, a smaller ka will lead to the reduction of the scattered power and SNR. In the actual measurement, the ball radius should be determined according to the requirement for measurement precision, the frequency response characteristic of the transducer under test as well as the performance of instrument.

2. Model of a ball measurement system

   The sound-field measurement system in the ball reflection method includes a multi-axis scanning and motion control system, an ultrasonic pulse transceiving card, a data acquisition system, a water tank and an inclined reflection base. The parameters and performance of these instruments and equipment will be described in detail in the next section. This section will mainly present several problems involved in the scanning and measurement of sound field with the ball reflection method:

   1. Ball selection

      It can be seen from the calculation of scattered sound field in the above section that a smaller ball should be used in the sound field measurement in order to improve the resolution of sound beam distribution measurement. However, when the ball radius is small, part of the sound wave incident on the front ball surface will become a creeping wave to propagate around the ball. Then part of the creeping wave will turn into a longitudinal wave and return to the transducer, so as to affect the measurement accuracy of sound field. On the contrary, for a larger ball, its reflection area cannot be approximated as a point reflection, thus reducing the resolution of sound beam measurement and bringing about the fuzziness. Therefore, the selection of ball size has a great impact on the measurement accuracy. At present, no consensus has been reached on the ball size among the domestic and foreign sound-field measurement standards. In China, both the mechanical industry standard JB/T10062-1999 and the national standard GB/T 18694-2002 directly stipulate that the ball diameter is 4 mm. The standard E1065–2008 issued by the American Society for Testing Materials (ASTM) stipulates that the ball diameter is ten times the radiation wavelength of the tested transducer in water. The European standard BS EN12668-2:2001 gives a simple formula for estimating the ball size. It can be seen from the above three standards that the European standard gives the criteria for selecting the ball size at different transducer center frequencies. The ASTM criteria are unreasonable for the low-frequency transducers tested by big balls. The Chinese standards give only a general rule on the selection of ball size.

According to the theoretical analysis of the pulsed sound field scattered by a small ball as described in the above section, the balls with different diameters are used in this paper for sound-field measurement experiment. Based on the measurement results, the selection rule for ball radius is discussed. The ball selection rule obtained by experimental and theoretical analysis in this paper is presented here in advance:

When $0.5\ \text{MHz} \le f \le 3.0\ \text{MHz}$ $3\ \text{mm} \le d \le 5\ \text{mm}$

When $3.0\ \text{MHz} \le f \le 10\ \text{MHz}$ $2\ \text{mm} \le d \le 3\ \text{mm}$

When $10\ \text{MHz} < f \le 15\ \text{MHz}$ $d = 1\ \text{mm}$

When $f > 15\ \text{MHz}$, the ball reflection method is no longer applicable.

2. Path planning

To obtain the sound-field characteristic parameters of the transducer, the scanning motion path needs to be planned. Generally speaking, the sound beam of a circular transducer is approximately axisymmetric, so the characteristic parameters can be calculated by measuring the distribution of sound field on several characteristic planes. Through the one-dimensional scanning along the transducer's acoustic axis, the sound pressure distribution on the acoustic axis as well as the characteristic parameters (focal distance and focal area length) can be obtained. Through the two-dimensional scanning vertical to the transducer's radiation surface, the distribution of sound field and pressure on the transducer's acoustic axis as well as the characteristic parameters (focal area widths and beam diffusion angles in two directions) can be obtained.

To ensure the high efficiency of scanning and the complete coverage of scanning scope, the scanning path should be planned before scanning. Generally, there are two patterns of planar scanning: one-way scanning and two-way scanning, as shown in Figure 13.31. One-way (also known as "*W*" type) scanning

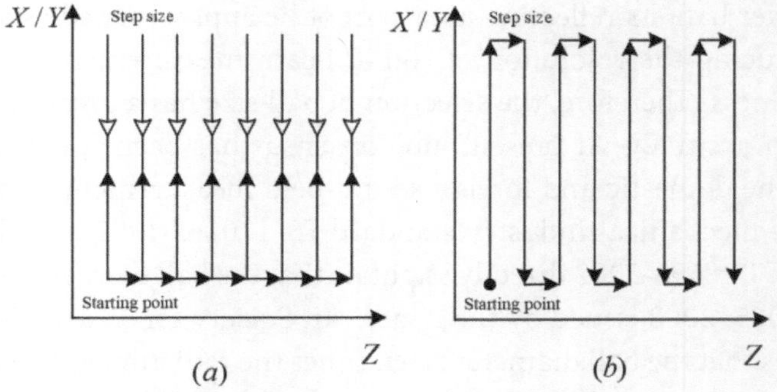

FIGURE 13.31 Two-dimensional scanning modes. (a) One-way (also known as "W" type) scanning and (b) two-way (also known as "M" type) scanning

is shown in (a), where the solid arrow indicates moving while reading the data and the hollow arrow indicates only moving without reading the data. Two-way (also known as "*M*" type) scanning is shown in (b), where the data is read in the round-trip motion. The differences between the two scanning modes are as follows. Due to the lack of an idle stroke, the efficiency of two-way scanning is much higher than that of one-way scanning. However, due to the positioning error and clearance in the motion control system, two-way scanning will cause the dislocation of the data obtained between reciprocations during the position mapping, thus resulting in the formation of a sawtooth edge and the generation of a large error in precise measurement. This phenomenon will not occur in one-way scanning.

The scanning area and planned path of the proposed sound field measurement system are shown in Figure 13.32. To ensure the measurement accuracy, two-way sampling and scanning is adopted for low-frequency transducers and one-way sampling and scanning is used for high-frequency transducers.

Usually, the ball is in a fixed position and the transducer is driven by a moving mechanism for scanning purpose. Before the measurement, the position/attitude regulator (AB axis) of the transducer is adjusted until the transducer and the ball are coaxial, that is, the sound beam axis of the transducer is adjusted to the *Z*-axis (the vertical normal line of the ball).

3. Result representation

In the scanning process with sound pressure distribution on axial section or line, as the distance between the transducer and the ball gradually increases (or decreases, but usually increases in order to prevent collision), the reflection echo signal will fall outside the observation interval and cannot be found in the waveform displayed in real time. There are two ways to solve this problem: (1) software delay following and (2) setting the echo-sampling interval as the maximum. Because the second method results in the formation of too large analysis interval, the consumption of a lot of computer resources and the loss of real-time sound-field imaging data, the first method (delay following) is adopted here to ensure that

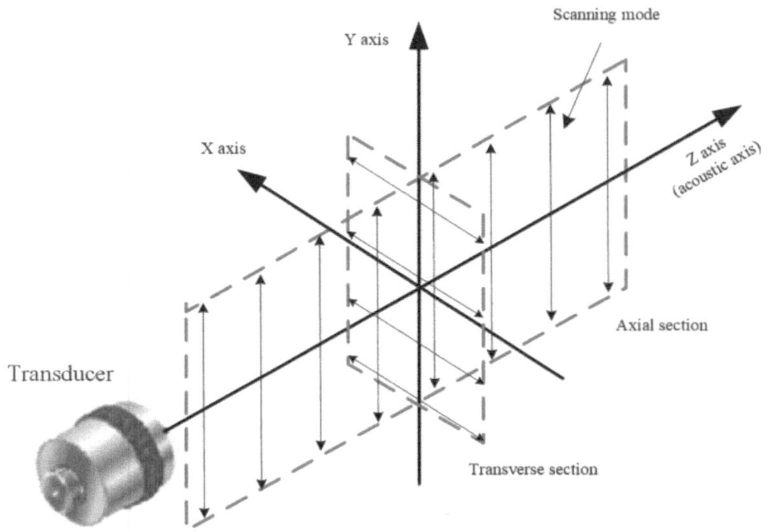

FIGURE 13.32 Planned scanning path of sound field of a circular transducer.

the primary reflection echo signal always falls within an effective analysis interval during the transducer scanning. Usually, the peak-to-peak value of primary surface echo signal of the ball is taken as the echo acoustic pressure at that point according to the characteristics of sound field distribution, so the amplitudes at all sampling points in the effective analysis interval must be peak-detected. In this system, the pressure amplitude distribution of sound field is displayed with the color method. The sound pressure amplitudes are zoned, the colors of different zones are set in the palette and the sound pressure amplitudes distributed in each zone are colored in the corresponding zone color during imaging. The color mapping of peak point is used as the imaging data to describe the characteristics of sound field distribution. Then, according to the scanning range and the preset sound field accuracy (scanning space), the point location image data corresponding to the sound field distribution is created and the pixel points are drawn. It can be seen that the sound pressure distribution measured by the ball reflection method is a relative sound pressure field, where the point values are dimensionless and only reflect the distribution of sound field intensities. Finally, the sound-field distribution data in the point location image file is searched and the characteristic parameters of sound field are calculated.

In the measurement of sound pressure distribution on focal plane, the distance between the transducer and the ball is set as focal distance or near-field length so that the reflected sound pressure at that location is the maximum. It should be pointed out that the nominal focal distance of a transducer is usually not the real focal distance, so the focal plane must be determined during measurement through the sound pressure scanning on axial section or line.

13.3.2 Hydrophone Measurement Method of Sound Field of an Ultrasonic Transducer

1. Autocollimation method of hydrophone

When a hydrophone is used to measure the sound field of the transducer under test, the hydrophone shall be placed in front of the transducer and its receiving directivity needs to be aligned with the axial direction of the transducer sound beam. In almost all the standards, this process is briefly summarized as "aligning the hydrophone axis with the transducer axis to the same line". Actually, this adjustment process is often carried out manually in line with the same collimation principle that the signal received by hydrophone shall be the strongest. This process often wastes time and energy and is hard to guarantee accuracy. For a wide sound beam, when the hydrophone deviates from the acoustic axis by a small distance, its waveform amplitude at that location will be not significantly different from that on the acoustic axis and its signal can't be easily identified as the "strongest" through vision. In this section, the method of hydrophone collimation based on the peak value of edge wave in the near field of the transducer will be studied. The hydrophone collimation algorithm and criterion embedded in the control system module of the software can automatically collimate the hydrophone.

The radiated pulse sound fields of circular transducers can be divided into plane waves and edge waves according to the radiation principle. Plane wave is the sound wave generated by an infinite radiant surface of the ultrasonic transducer in theory, while edge wave is generated by a finite radiant surface of the ultrasonic transducer in real world. The two waves are opposite in sign and are separated in time domain by a time interval determined by the position of the measuring point. In the cylindrical area opposite to the radiant surface, the plane wave arrives at the measuring point first and then the edge wave arrives. Outside the cylindrical area, only the edge wave exists. In the far field of the transducer, the two waves are superimposed on each other in the time domain and cannot be distinguished. However, in the near field of the transducer, they can be completely separated. Figure 13.33 shows the time-domain waveforms of radiated sound pressures at different distances along the transducer

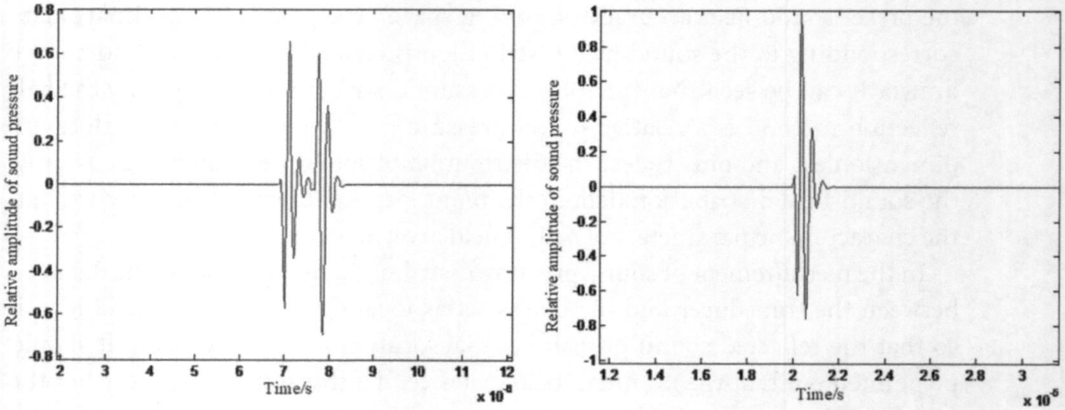

FIGURE 13.33 Theoretically calculated time-domain waveforms at different axial distances.

beam axis, which were calculated by using the sound field model. Only the edge wave on the acoustic beam axis of the transducer can obtain the maximum amplitude. Moreover, the edge wave is very sensitive to off-axis position. Figure 13.34 shows the calculated time-domain waveforms of sound pressures at different off-axis positions of a planar transducer with an aperture of 6.3 mm. The axial distance of these points from the emitting surface of the transducer is 5 mm. The waveform on the left of each diagram is a planar direct wave signal followed by an edge wave. It can be seen that when the field point is located on the axis of the acoustic beam ($x = 0$ mm), there is little difference between the amplitude of the edge wave and that of the planar direct wave. However, with the continuous increase of off-axis distance, the edge wave will rapidly attenuate. When the off-axis distance is 0.3 mm, the edge wave will attenuate to about 10% of that on the axis. However, the amplitude of the planar direct wave will not change much. In other words, compared with plane wave, the edge wave is more sensitive to off-axis position.

To verify the feasibility of this method, this section presents the experimental measurement results of the hydrophone, as shown in Figure 13.35.

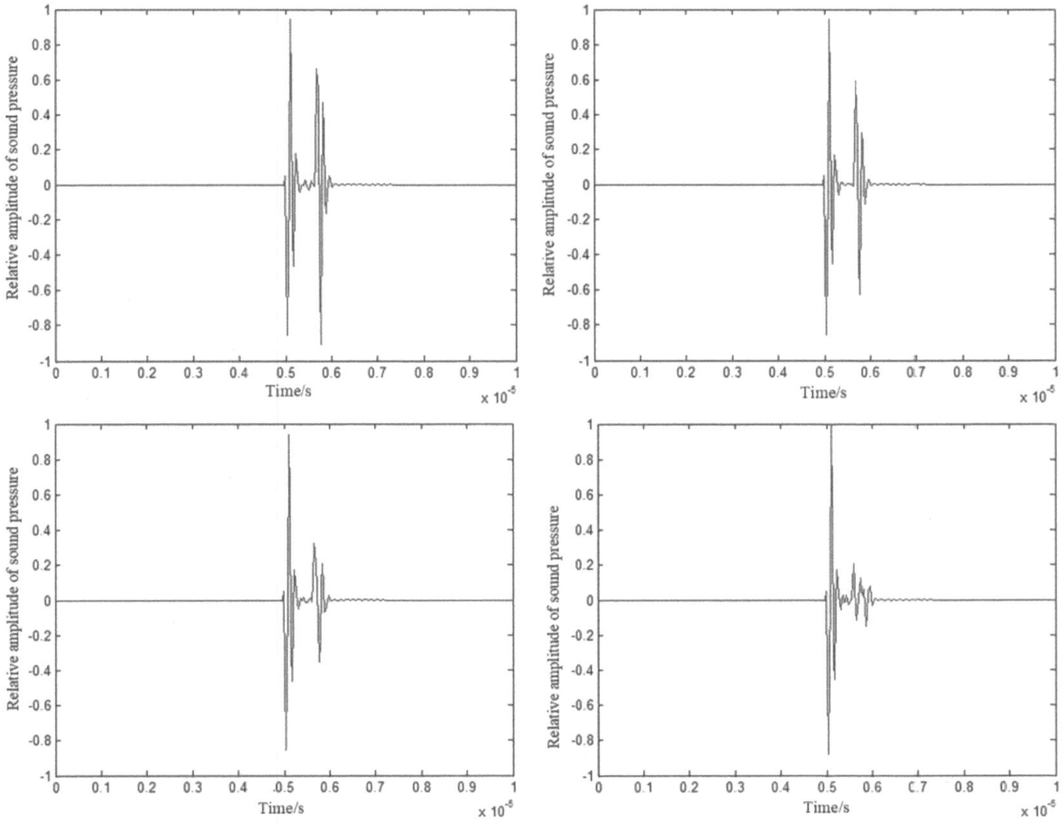

FIGURE 13.34 Theoretically calculated time-domain waveforms at different off-axis distances ($z = 5$ mm).

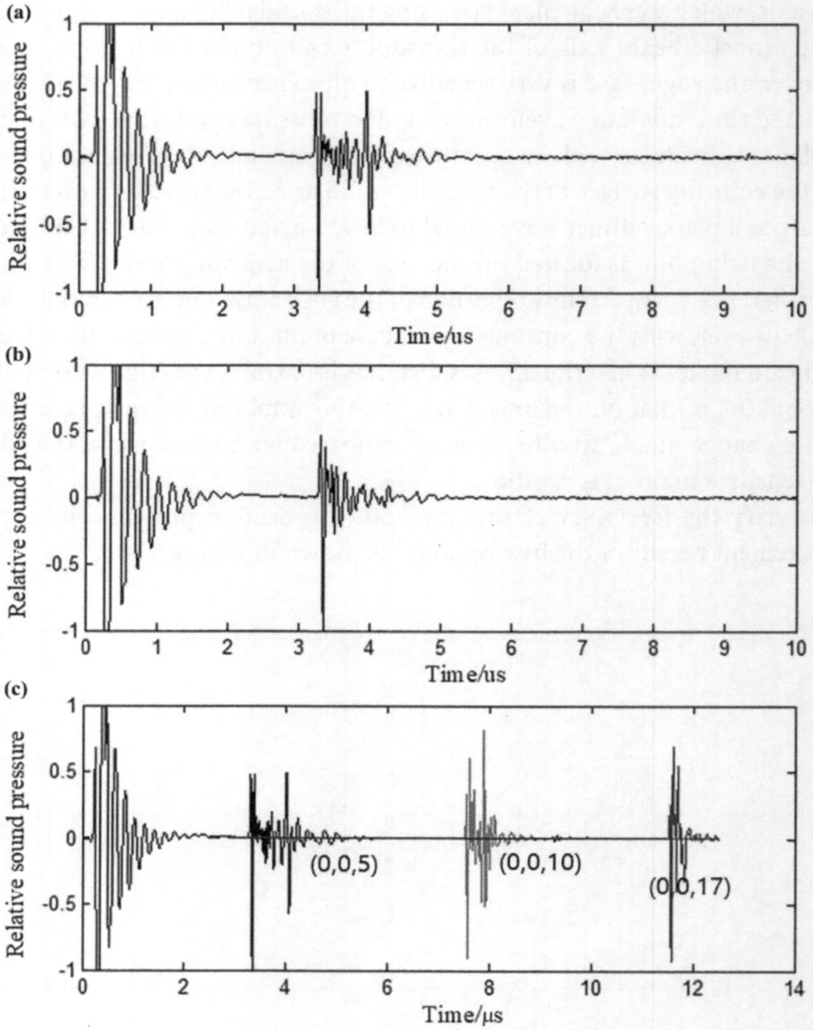

FIGURE 13.35 Experimental measurement results of hydrophone. (a) Waveforms measured on acoustic axis $(x, y, z) = (0, 0, 5)$ mm, (b) waveforms measured off acoustic axis $(x, y, z) = (0.3, 0, 5)$ mm and (c) waveforms measured at different axial distances, $x = y = 0$ mm, $z = 5, 10, 17$ mm.

The experimental results are completely consistent with the results of theoretical analysis. The edge wave received by hydrophone near the transducer surface is very sensitive to radial off-axis position. By using the amplitude of the edge wave, the acoustic axis can be accurately aligned. Based on this principle, the hydrophone can be automatically collimated. Of course, the settings of hardware and software are also needed. The motion control algorithm can be implemented by a software program module. The control flow of the collimation process, as shown in Figure 13.36, is divided into two steps: rough scanning and fine scanning. The purpose of rough scanning is to quickly locate the general position of the acoustic axis and then set a small step size for fine scanning. In the process of fine scanning, two-dimensional

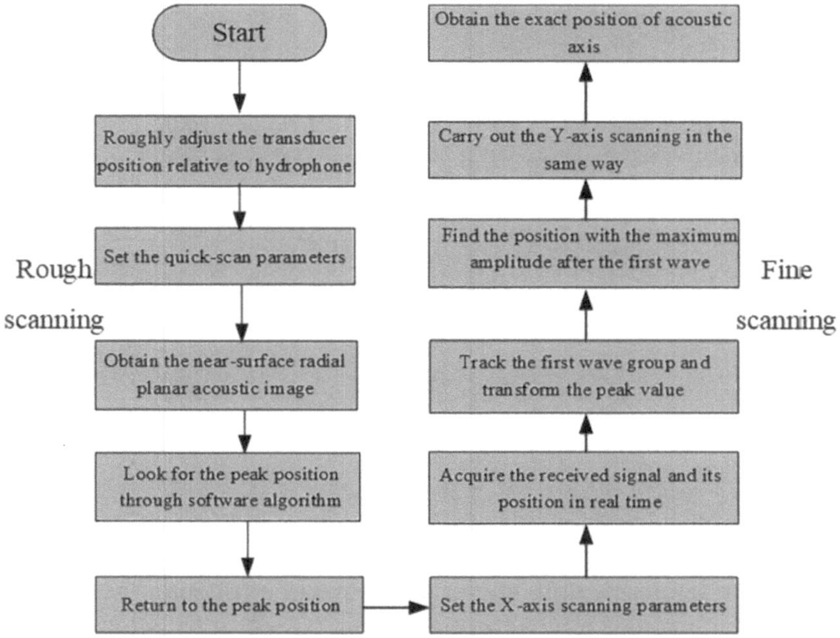

FIGURE 13.36 Control flow of automatic collimation process of hydrophone.

scanning is not required, and only the scanning on two radial vertical axes (such as *X*-axis and *Y*-axis) of the transducer is required. The accurate scanning process needs to calculate the waveform eigenvalues in real time, track the amplitude transformation of the second wave group (edge wave) and find the peak position of edge wave on the two scanning axes, namely the real center of the acoustic axis. It is found in the experiment that, the center of the acoustic axis does not coincide exactly with – but slightly deviates from – the geometric center of the transducer. The deviation value varies from transducer to transducer.

2. Aperture averaging effect

In theory, the effective radius of hydrophone sensor is less than one-fourth wavelength in sound field measurement, so the change of phase and amplitude has no significant influence on the measurement uncertainty. However, to meet the actual requirements for SNR or other factors, the size of hydrophone component is much larger than the size recommended by theory. In this way, when the sound field of an ultrasonic transducer is measured in liquid, the averaging effect of field-point sound pressure will be inevitably introduced. In other words, the sound pressure signal output by hydrophone is not the field-point sound pressure at that location but the sound pressure integral on the surface of hydrophone sensor. In particular, the aperture averaging effect is quite serious when the sound field of a high-frequency ultrasonic transducer (whose frequency is higher than 20 MHz) is measured with a hydrophone. The direct result of aperture averaging effect is that the measured transverse sound beam becomes wider and brings a large error to the measurement results. Strictly speaking, the hydrophone with an aperture diameter of 0.5 mm can measure a transducer no more than 3 MHz, and the 75 μm hydrophone can measure a transducer no more than 20 MHz. In the existing sound-field measurement standards, the effective receiving aperture of a hydrophone can be up to 0.6 mm. However, aperture averaging effect is generally ignored or slightly mentioned in many measurement standards.

To reduce the averaging effect of the receiving aperture in the sound field measurement, a simple empirical formula for selecting the effective receiving aperture of a hydrophone is presented as follows [29]:

$$\begin{cases} d_e \le \dfrac{z\lambda}{2d_s}, \dfrac{z}{d_s} \ge 1 \\ d_e < \dfrac{\lambda}{2}, \dfrac{z}{d_s} < 1 \end{cases} \tag{13.95}$$

where d_e is the effective receiving aperture diameter of the hydrophone, z is the distance between the hydrophone and the transducer surface, λ is the wavelength of sound wave in liquid medium and d_s is the maximum size of the emitting surface of the transducer under test. According to Eq. (12.93), such conditions may not be completely satisfied for the high-frequency ultrasonic transducers used in NDT. For example, when the radiation wave frequency of a transducer is greater than 20 MHz and the apertures of the existing hydrophones are greater than one wavelength, the aperture averaging effect will be inevitable.

Currently, the aperture averaging effect of hydrophones can be corrected by many methods, most of which are applicable to the correction of focal plane and far field. The estimation accuracy of beam size depends on the calculation accuracy of the corrected model of aperture averaging effect, which, in turn, heavily depends on the signal SNR, signal bandwidth and axial distance. For example, a simple correction coefficient method was proposed by Harris et al. [30]:

$$c = (3 - \beta)/2 \tag{13.96}$$

where β is the ratio of sound pressure at the field point with a hydrophone-radius distance from the acoustic axis to the sound pressure on the corresponding axis.

Zeqiri et al. [31] proposed a beam-width correction model to remove the "fuzziness" of acoustic beam. The model is expressed as:

$$P_{\text{avg}}(r) = \frac{\iint_s rp(r,\varphi)\,dr\,d\varphi}{\iint_s r\,dr\,d\varphi} \tag{13.97}$$

where $P_{\text{avg}}(r)$ is the integral average pulsed sound pressure on the hydrophone surface. When the spatial sound pressure $p(\vec{r},t)$ of the transducer is symmetric about the acoustic axis and the radius of hydrophone aperture is b, Eq. (13.97) can be simplified as:

$$P_{avg}(r) = \frac{1}{b}\int_{r-b/2}^{r+b/2} \frac{p(r)}{p(0)}dr \tag{13.98}$$

where $p(r)$ and $p(0)$ are, respectively, the sound pressure at the field point with a hydrophone-radius distance from the acoustic axis and the sound pressure on the acoustic axis.

In this paper, a two-dimensional Wiener spatial deconvolution filter is constructed to correct the aperture averaging effect of the hydrophone, so as to improve the resolution of hydrophone aperture detection. A circular piston transducer is considered. On the transverse cross section (x, y, z_0) of its radiated sound field, z_0 is the axial distance from the transducer surface. Its sound field is measured by a hydrophone. Then the measured value of its radiated sound pressure can be expressed as [32]:

$$V(x,y,z_0) = H(x,y) \otimes_{xy} P(x,y,z_0) + N(x,y,z_0) \tag{13.99}$$

where $H(x,y)$ is the impulse response function of the hydrophone, $p(x,y,z_0)$ is the sound pressure value of the sound field radiated by the transducer, $N(x,y,z_0)$ is the signal-independent noise component and $\otimes_{xy}$ is the convolution operator in two-dimensional space.

According to Eq. (13.99), the aperture averaging of a hydrophone with circular receiving aperture can be simulated. The calculated results of a liquid-immersed planar transducer with the center frequency of 10 MHz and the transmitting aperture of 6 mm as well as the result of hydrophone aperture averaging are shown in Figure 13.37. It is assumed that the pulse response of hydrophone aperture is a two-dimensional Gaussian distribution and the aperture diameter is 0.5 mm (wavelength: 0.15 mm).

In order to obtain the optimal estimation $\hat{p}(x,y,z_0)$ of the real sound pressure amplitude $P(x,y,z_0)$ from the measured signal affected by hydrophone aperture averaging and noise, the relation can be expressed as:

$$\hat{P}(x,y,z_0) = H_F(x,y) \otimes P(x,y,z_0) \tag{13.100}$$

where $H_F(x,y)$ is the response transfer function, including the aperture averaging effect and noise component of the hydrophone. Based on the two-dimensional linear system theory, a spatial filter can be constructed for deconvolution operation and the optimal estimation of real sound pressure can be obtained from the measured results. In this paper, a two-dimensional Wiener spatial deconvolution filter is adopted to achieve this goal. The two-dimensional Wiener filter is based on the minimum mean square error and can achieve the optimal estimation [33,34]:

$$E\left\{\hat{P}(x,y,z_0)(x,y,z_0) - P(x,y,z_0)\right\} \Rightarrow \min \tag{13.101}$$

where E represents the mathematical expectation. The transfer function of the deconvolution filter can be expressed as:

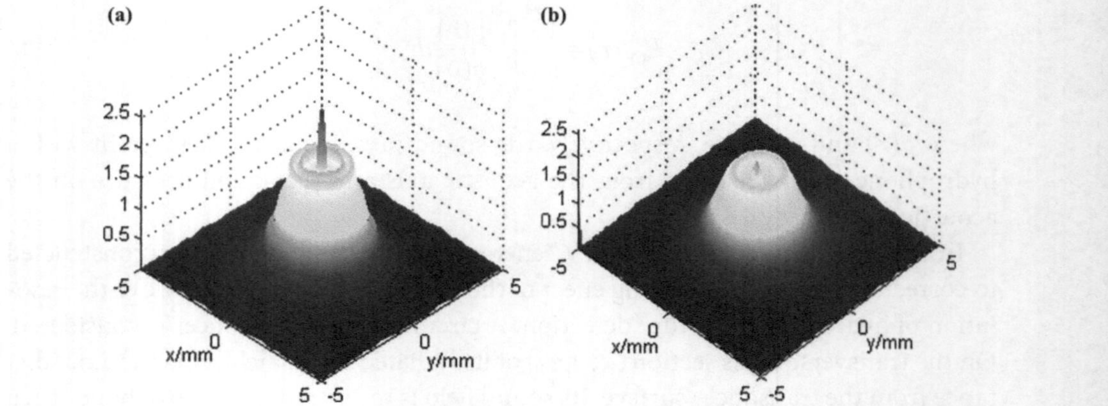

FIGURE 13.37 Sound pressure distribution on radial transducer section, and result of hydrophone aperture averaging ($z_0 = 15\,\text{mm}$). (a) Original distribution and (b) result obtained after aperture averaging.

$$H_{F\omega}\left(k_x,k_y\right)=\frac{H_\omega^*\left(k_x,k_y\right)\Phi_{p\omega}\left(k_x,k_y,z_0\right)}{\left|H_\omega\left(k_x,k_y\right)^2\Phi_{p\omega}\left(k_x,k_y,z_0\right)+\Phi_{n\omega}\left(k_x,k_y,z_0\right)\right|} \tag{13.102}$$

where $H_\omega\left(k_x,k_y\right)$ is the pulse response in frequency domain of the hydrophone, $H_\omega^*\left(k_x,k_y\right)$ is its conjugate function and $\Phi_{p\omega}\left(k_x,k_y,z_0\right)$ and $\Phi_{n\omega}\left(k_x,k_y,z_0\right)$ are the power spectral densities of sound pressure field and noise component, respectively. As the power spectrum of noise, denoted as $\Phi_{n\omega}\left(k_x,k_y,z_0\right)$, is hard to estimate, it is usually considered as the constant A [35]. The value of A varies with the condition of system noise. Since the acquired real-time waveforms were averaged for many times in this experiment and most of the Gaussian white noise was eliminated, the value of A could be very small. Here the value of A is $A = 0.05$.

The impulse response function images $H(x,y)$ of the hydrophones with different apertures based on Gaussian estimation are shown in Figure 13.38, and their two-dimensional frequency spectrum is just the impulse response $H_\omega\left(k_x,k_y\right)$ in the frequency domain. The power spectral density function image $\Phi_{p\omega}\left(k_x,k_y,z_0\right)$ of sound pressure distribution in the transverse section of the transducer (see Figure 13.37a) is illustrated in Figure 13.39.

The corrected result of transverse acoustic beam obtained after the deconvolution and reconstruction of aperture averaging result of the transducer with transverse sound field distribution (see Figure 13.37) is shown in Figure 13.40, where the hydrophone aperture averaging is 0.5 mm. It can be seen from the result comparison that the corrected result is quite consistent with the original result. In spite of a minor difference between them, the resolution of hydrophone aperture has been greatly improved, thus demonstrating the effectiveness of two-dimensional Wiener deconvolution correction algorithm. In the measurement of sound field of the high-frequency transducer described in Chapter 6, the corrected results of more experimental measurements will be given.

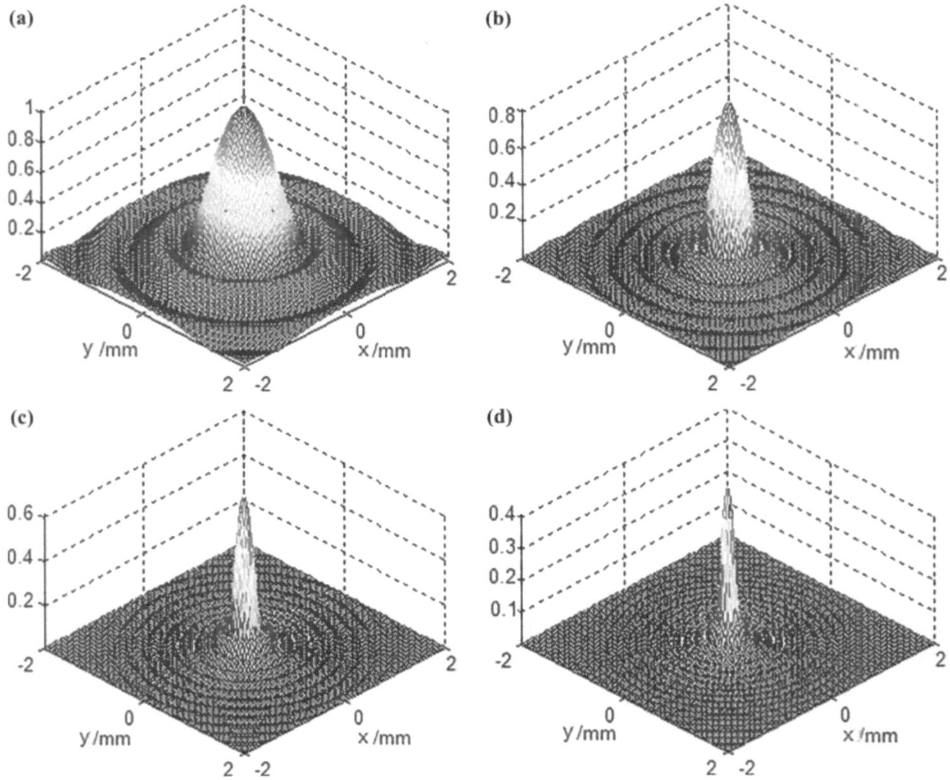

FIGURE 13.38 Impulse response characteristics $H(x,y)$ of hydrophones with different receiving apertures. (a) 1 mm hydrophone, (b) 0.5 mm hydrophone, (c) 0.2 mm hydrophone and (d) 0.075 mm hydrophone.

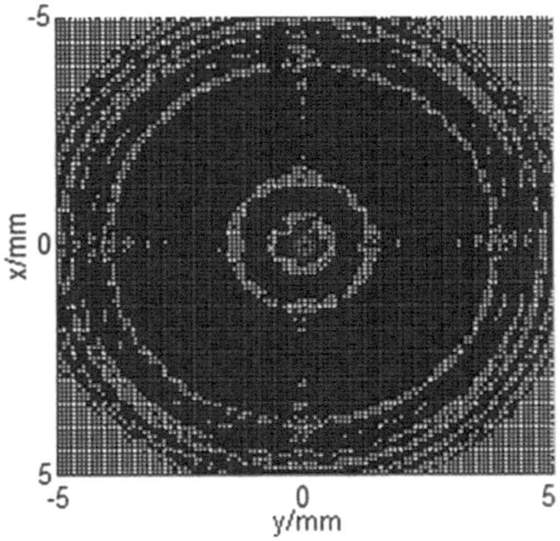

FIGURE 13.39 Two-dimensional power spectral density $\Phi_{p\omega}(k_x,k_y,z_0)$ of sound-field distribution image.

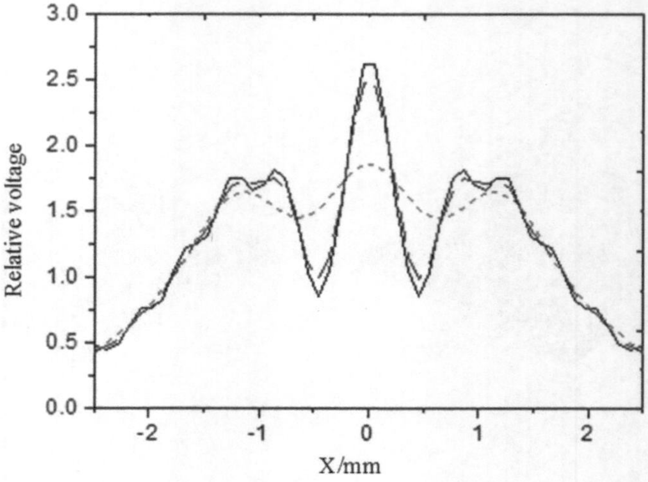

FIGURE 13.40 Deconvolution aperture correction result (hydrophone with 0.5 mm aperture).

13.4 SOUND-FIELD MEASUREMENT SYSTEM OF ROBOTIC ULTRASONIC TRANSDUCER

The study of this section is based on the transducer sound-field measurement system of a manipulator. The main function of this system is to measure the sound field of water-immersed ultrasonic transducer by taking advantage of the flexible spatial motion and high positioning accuracy of the manipulator. According to the principle of transducer sound field measurement and the design requirements of automatic measurement and imaging display, the system is expected to implement the automatic scanning, the transmission, acquisition and processing of ultrasonic signals, the measurement of transducer sound field, the imaging analysis of results, the parameter calculation, the generation of test report and other functions, so as to realize the automatic measurement and image output of transducer sound field characteristics.

According to the requirements for measurement process and equipment of sound field characteristics, the target functions of the system are set as follows:

1. **Reliable scanning capability**: A six-axis manipulator with the maximum scanning range of 200×200×300 mm and the maximum scanning speed of 200 mm/s is used to control the movement. The minimum spacing between scanning trajectories is up to 0.15 mm, and the minimum spacing between scanning points on a trajectory is also 0.15 mm. At the same time, the system provides the stroke protection for six axes of the manipulator and the feedback control for the encoder. The manipulator's repeated positioning accuracy is up to ±0.02 mm.

2. **The measuring system can clamp an ultrasonic transducer reliably and is easy to operate**: Moreover, the clamping structure is required to have sufficient rigidity and flexibility to adapt to the transducers in different sizes. Meanwhile, it has the adjustment function to facilitate the transducer alignment during debugging.

3. **High-quality ultrasonic signal excitation, acquisition and processing**: This is the core function of the whole system and has a direct influence on the system performance. The following requirements have been raised for the system performance:

   a. Strong ability to excite the transducer pulse and to control and adjust the excitation voltage, transmission impedance matching and PRF. In particular, the excitation voltage is 100–400 V adjustable step by step or continuously, the PRF is 0.5–5 KHz adjustable step by step or continuously and the frequency acceptance band is 0.5–15 MHz.

   b. Adaptable to the conditioning of a wide range of ultrasonic signals. Since the transducer signal received by hydrophone has a weak amplitude, the total amplification/attenuation gain of the received signal is required to be >100 dB. The gain can be adjusted to guarantee the SNR after the signal amplification.

   c. Reliable ultrasonic signal processing. The highest sampling frequency is 200 MHz, the sampling accuracy is 12 bit and the onboard memory is 64 MB.

4. **High-quality sound field imaging and image display**: In addition to A-scan, B-scan and C-scan imaging, the display of transducer sound field image is also required. The measurement image should be displayed in a friendly and intuitive manner with adjustable resolution.

5. **Capable of post-processing and analysis**: This capability includes drawing the sound field images, measuring the characteristic parameters of sound field and generating the measurement reports. The measurement results of a sound beam can be displayed with various graphs, including: axial sound pressure distribution curve, radial sound pressure distribution curve, axial cross-sectional sound pressure distribution diagram and radial cross-sectional sound pressure distribution diagram.

6. **Capable of calculating the characteristic parameters of sound field**: The parameters to be calculated include focal distance, focal area width, focal area length and sound beam diffusion angle. In terms of the measurement ranges and uncertainties of various parameters, the measurement uncertainties of focal distance, focal area width and focal area length are less than 5% ($K = 2$) and the measurement uncertainty of sound beam diffusion angle are less than 3% ($K = 2$).

7. **Capable of data management**: The system can store and export the preset parameters, waveform data, graphic data and other measurement data. The measurement parameters are input into an interface to directly generate the measurement report. The contents in the generated report include the transducer information, system settings, measurement conditions and measurement results.

13.4.1 Composition of a Hardware System

The sound-field measurement system of ultrasonic transducer studied in this paper is shown in Figure 13.41. During measurement, the manipulator grips a hydrophone and moves relative to the fixed transducer under test. By leveraging the manipulator's spatial flexibility and localization accuracy, the location information is obtained. Through a high-speed data acquisition card, the sound pressure information is acquired. Then the information from the two sources is processed by computer to obtain a sound field image, based on which the characteristic parameters of transducer sound field can be measured.

The overall block diagram of the system is shown in Figure 13.41. The measurement system consists of an upper computer system, a software system, a scanning motion system, a signal acquisition system and an auxiliary system. During the sound field measurement, the manipulator grips a hydrophone and moves relative to the fixed transducer under test, in order to measure the sound field of water-immersed ultrasonic probe. The upper computer system and the software system provide a perfect human-computer interaction platform, which can control the measurement process of transducer sound field and visualize and analyze the measurement results. The scanning motion system consists of a six-degree of freedom (DOF) manipulator, a manipulator controller, a teaching box and other devices, and controls the manipulator to move with hydrophone along the preset spatial path. The signal acquisition system is composed of an ultrasonic transceiver, a high-speed acquisition card, a transducer, a hydrophone and other devices, capable of exciting, conditioning, collecting and digitally processing ultrasonic signals. The auxiliary system provides a stable environment and reliable support for sound field measurement, thus ensuring the quality of sound field measurement.

The sound-field measurement system uses the hydrophone method to measure and visualize the sound field of ultrasonic transducer. The whole measurement process is controlled by a software system. Before the measurement, the system should be connected and debugged. Meanwhile, the transducer should be collimated so that the transducer axis and the hydrophone axis are on the same line. Then the hydrophone is centered and

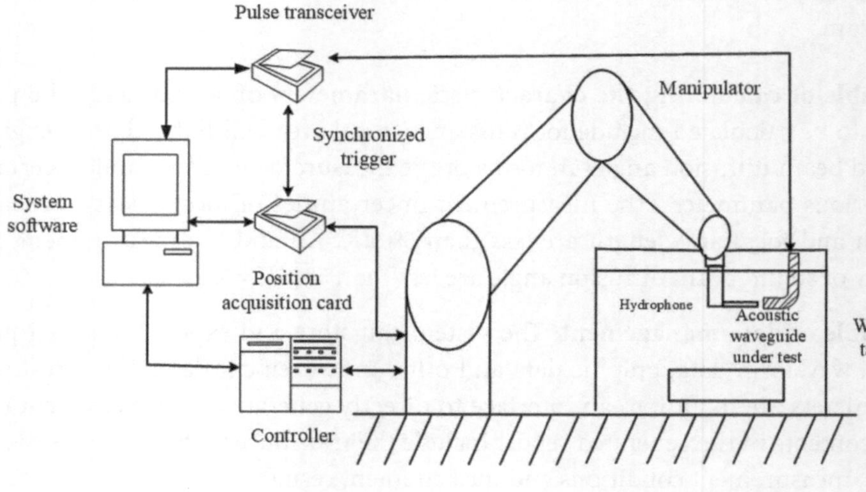

FIGURE 13.41 General system layout.

the data acquisition parameters are set. After the centering observation, the coordinate value of the manipulator is noted down as a relative zero. Next, the manipulator is controlled, logged in and powered. The coordinate value of the relative zero and the scanning range are based on for calculation. Then the starting point of scanning and the scanning parameters are set. After returning to the starting point, the manipulator can start scanning. The measurement results of sound field are displayed on the software interface. Finally, the sound pressure distribution of sound field can be visually represented in the form of color image, and the results can be analyzed and stored.

13.4.2 Composition of a Software System

The main functions of the software include ultrasonic transceiver parameter setting, data acquisition, real-time waveform display and storage, motion control, sound-field scanning setting and automatic report generation. In addition, the software can deduce the characteristic parameters from the collected original data and obtain the image results. The results include the distribution of radial sound pressure in the sound field of the transducer, the distribution of axial sound pressure in the sound field, the distribution of sound pressure in the radial section of sound field, the distribution of sound pressure in the axial section of sound field and the three-dimensional sound pressure distribution in the sound field; the focal distance, focal distance, focal width and beam diffusion angle of the focusing transducer; and the deflection angle and focusing characteristic of phased array transducer.

The measurement software of transducer sound field characteristics is a digital-signal processing system with complex algorithm and high real-time requirement. The measurement of sound field characteristics should not only control the scanning mechanism to scan the sound field along the predetermined trajectory but also collect the sound signals in real time, analyze the signal waveform, image the sound field and store the data. Therefore, the measurement system requires high data-handling capacity and real-time performance of the software. At the same time, as a type of engineering measurement software, the software system must demonstrate good robustness and man-machine interaction while performing the predetermined functions. Therefore, the above factors must be fully considered when designing the software framework and data structure [9]. In this paper, the Windows XP operating system is selected as the software running platform, and the object-oriented Visual C# is adopted as the software development tool.

The functional block diagram of transducer calibration software is shown in Figure 13.42. This software mainly includes the modules for ultrasonic signal transmission/reception and acquisition, time-frequency domain measurement, sound field measurement and system management.

The following is a brief introduction to the functions and settings of main interfaces, including the main interface, the setting of sound field scanning parameters, the scanning display interface for each sound field and the calculation module of sound field parameters.

(1) **Main interface**: This application is a multi-document interface. It consists of a menu, a toolbar, a display area and a parameter setting panel. The parameter setting panel is mainly used to adjust the excitation parameters, display parameters and motion parameters. Among them, the excitation parameters include excitation voltage, pulse energy and

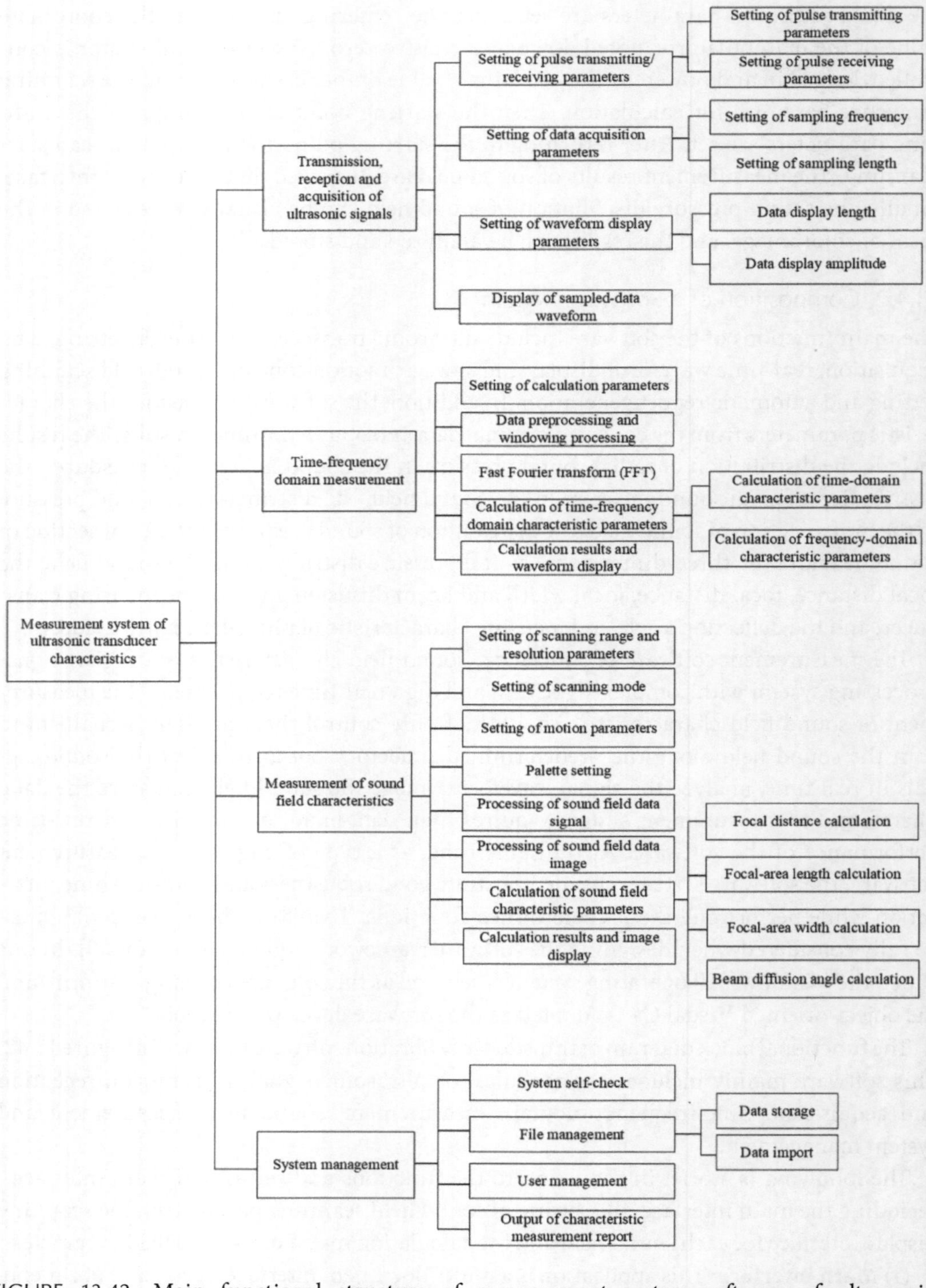

FIGURE 13.42 Main functional structure of measurement system software of ultrasonic transducer characteristics.

width. The display parameters include waveform delay, display width, sampling frequency, filtering parameters and gain. The motion parameters include motion speed, motor acceleration and mechanism parameters.

This calibration software is used to analyze the time-domain and frequency-domain characteristics of the transducer. It can measure the initial wave width, echo width and peak-to-peak value of the transducer time-domain signal, while measuring and analyzing the center frequency, peak frequency and bandwidth of the transducer in frequency domain.

13.5 MEASUREMENT VERIFICATION OF SOUND FIELD OF MANIPULATOR TRANSDUCER

13.5.1 Measurement of Sound Field of a Planar Transducer

Figure 13.43 shows the theoretical calculation results of the planar transducer V312, as well as the actual axial sound-pressure distribution curves obtained by 2 mm ball reflection and 0.2 mm hydrophone. It can be seen that, the theoretical calculation result of sound pressure distribution in the near field region (before the peak) presents a sharp oscillation, while the sound pressure curves obtained by hydrophone method and ball reflection method are relatively smooth with a small oscillation in the near field. This is because the receiving aperture (reflecting surface) of either a hydrophone or a ball has a geometric size and is equivalent to a spatial smoothing filter. As the distance from the transducer surface increases, the sound pressure will increase sharply to the axial peak (the natural focus of planar transducer). It can be seen that the distances of the three peaks are basically the same, but those of the hydrophone method and ball method are slightly smaller. The sound pressure in the far field will decrease gradually with the increase of distance. The sound absorption in water medium tends to attenuate at a certain rate during the sound propagation, but the absorption attenuation is not considered in the theoretical calculation. Therefore, with the increase of distance, the actually measured sound pressure amplitude in the far field will attenuate faster than the theoretical calculation result.

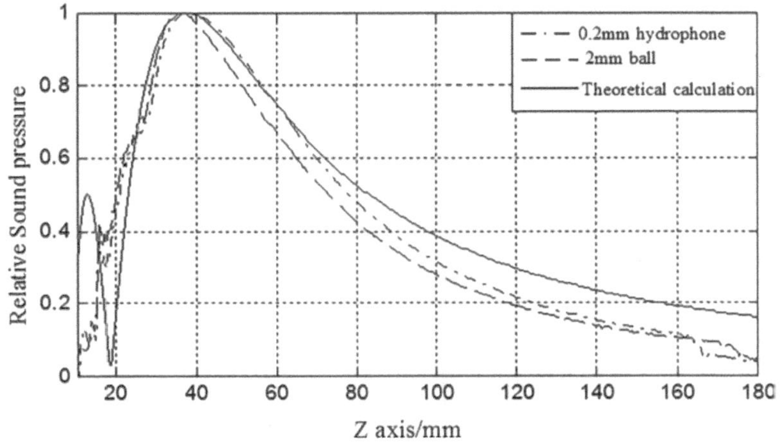

FIGURE 13.43 Calculation and measurement results of sound pressure distribution of a planar transducer (V312).

Figures 13.44 and 13.45 show the sound pressure distributions on the axial and radial (focal plane) sections of sound field of a planar transducer, which were, respectively, obtained by the numerical calculation with spatial pulse response method and by the measurement with the balls of different diameters and the hydrophones of different specifications. The comparison results of transducer sound-beam widths at the focus obtained by different methods are shown in Figure 13.46. The values of sound field characteristic parameters calculated from experimental data are listed in Table 13.2.

13.5.2 Measurement of Sound Field of Focusing Transducer

Figure 13.47 shows the theoretical calculation result of the focusing transducer C309, as well as the actual axial sound-pressure distribution curves obtained by 2 mm ball reflection and 0.2 mm hydrophone. Similar to the calculation result of a planar transducer, the calculation result of a focusing transducer shows violent oscillation in the near field of the transducer. Due to the smoothing filtering effect of an aperture, the sound pressure curves in the near field obtained by hydrophone method and ball reflection method are relatively smooth. The attenuation characteristics of sound pressure in the far field area are similar to those of a planar transducer. When the absorption attenuation in water medium is ignored, the measured sound pressure amplitude in the far field will attenuate faster than the theoretical result.

Figures 13.48 and 13.49 show the sound pressure distributions on the axial and radial (focal plane) sections of sound field of a focusing transducer, which were, respectively, obtained by the numerical calculation with spatial pulse response method and by the measurement with the balls of different diameters and the hydrophones of different specifications. The beam widths measured at the focus by using the hydrophones with different receiving apertures are shown in Figure 13.50. The beam widths measured at the focus by using the reflecting balls with different diameters are shown in Figure 13.51. The values of sound field characteristic parameters calculated from experimental data are listed in Table 13.2.

The hydrophones with a smaller aperture (75 μm) and a larger aperture (1 mm) than when measuring the planar transducer V312 were used to measure the sound field distribution of this focusing transducer. Through comparison, the influence of hydrophone aperture on the measurement result of sound pressure distribution in the focal plane of the transducer was studied. For a focusing transducer with the nominal frequency of 5 MHz, its center frequency measurement is about 4.7 MHz and its corresponding wavelength in water is 0.32 mm. Theoretically, in order to avoid the measurement error introduced by hydrophone aperture, the effective receiving aperture of the hydrophone should be less than one-fourth of the wavelength, that is, the hydrophone aperture should be less than 80 μm. The 75 μm hydrophone used in this experiment can accurately measure the sound field of the transducer. Based on this measurement "standard", other measurement results will be analyzed.

Table 13.2 shows the scanned axial sound pressure distribution, focal-plane sound pressure distribution and axial cross-sectional sound pressure distribution of this transducer obtained with various experimental measurement methods, as well as the sound-field characteristic parameters obtained by the sound-field parameter calculation module in the calibration software system of this transducer.

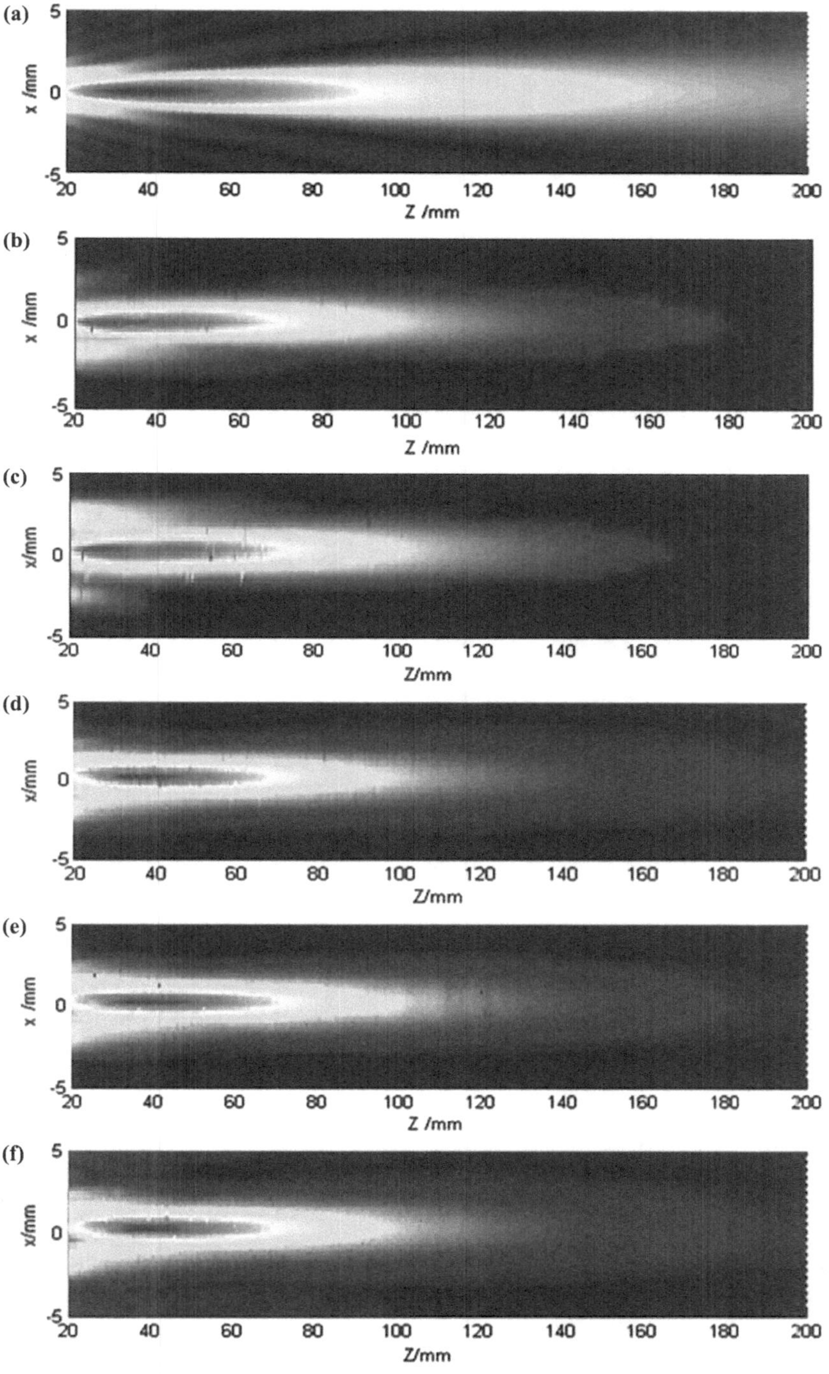

FIGURE 13.44 Calculation and measurement results of sound pressure distribution on axial beam section of planar transducer (V312). (a) Theoretical calculation result, (b) measurement result with PAL-0.2 mm hydrophone, (c) measurement result with Onda-0.5 mm hydrophone, (d) measurement result with 2 mm ball, (e) measurement result with 3 mm ball and (f) measurement result with 4 mm ball.

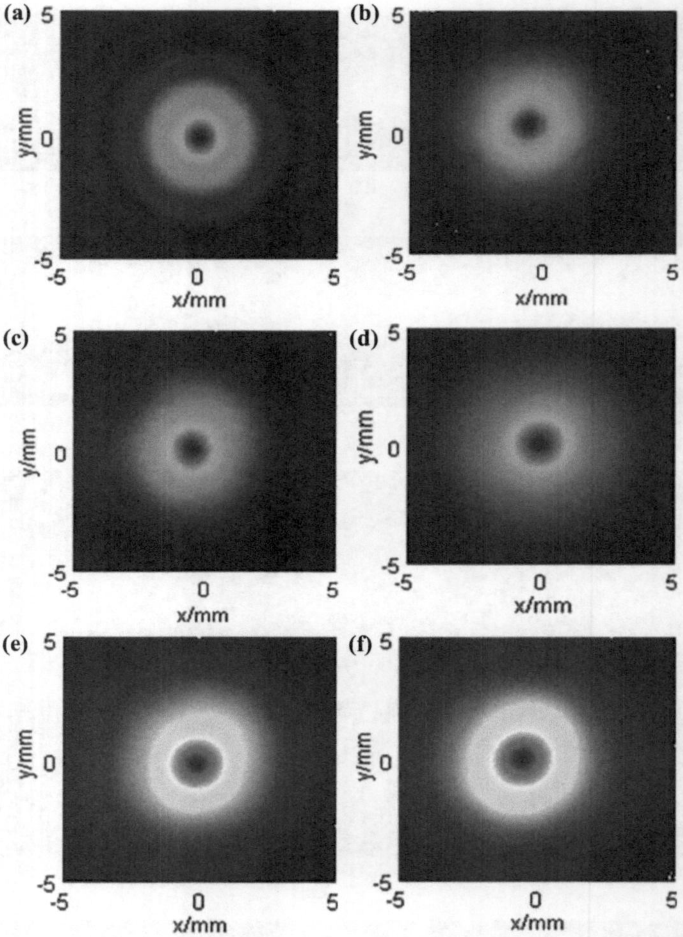

FIGURE 13.45 Calculation and measurement results of sound field distribution on transverse beam section of planar transducer (V312) ($z = 38$ mm). (a) Theoretical result, (b) PAL-0.2 mm hydrophone, (c) Onda-0.5 mm hydrophone, (d) 2 mm ball, (e) 3 mm ball and (f) 4 mm ball.

By comparing Figures 13.43–13.51 and the results of sound-field characteristic parameters in Table 13.2, the following conclusions can be drawn:

1. **Focal distance/near-field distance**: For either a focusing transducer or a non-focusing transducer, the actually measured focal distance/near-field distance has a certain error from the theoretical value. However, this error is much larger for non-focusing transducer. This is because the energy emitted by the non-focusing transducer does not converge at the focus as the focusing transducer does, so that the peak of sound pressure on the axis may cover a wider range and stand in a position very sensitive to external disturbance.

2. **Focal area size**: The dimensions of focal area can be divided into focal area width and focal area length. It can be seen from Figures 13.45 and 13.49 that, except for

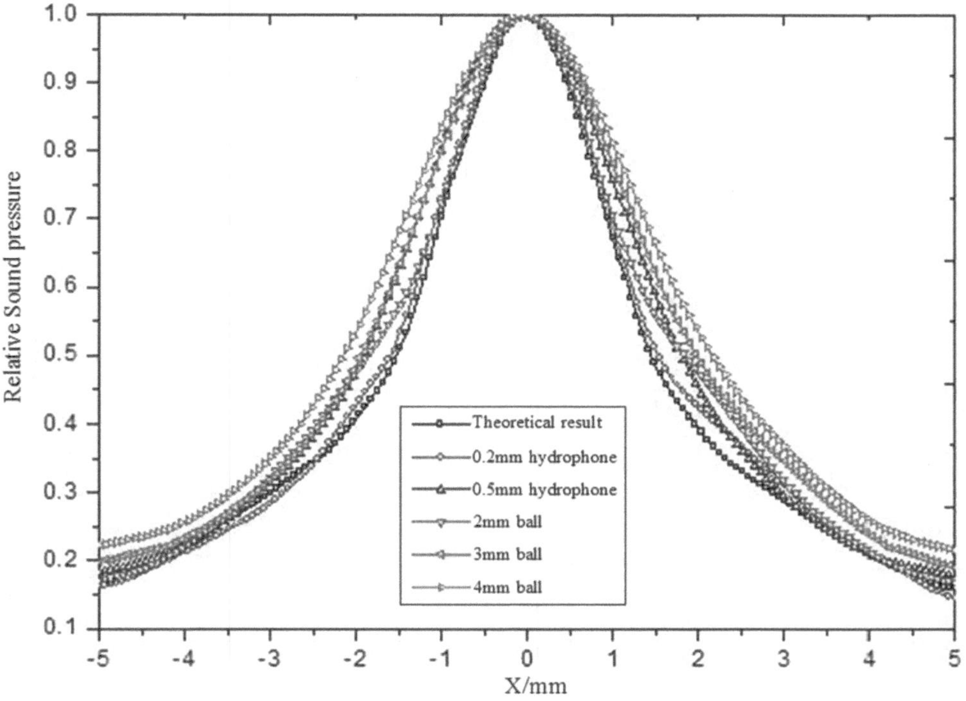

FIGURE 13.46 Calculation and measurement results of radial beam width of a planar transducer (V312) ($z = 38$ mm).

TABLE 13.2 Measurement Results of Sound-Field Characteristic Parameters of a Liquid-Immersed Transducer (−3 dB)

Transducer	Measurement Method	Focal Distance (mm)	Focal Area Length (mm)	Beam Width (mm)	Beam Diffusion Angle (°)
V312 (10 MHz/6.4 mm)	2 mm ball	38.3	28.6	2.7	4.8
	3 mm ball	40.1	29.7	2.9	5.1
	4 mm ball	41.0	30.5	3.1	5.6
	0.2 mm hydrophone	38.5	28.4	2.5	4.7
	0.5 mm hydrophone	39.7	31.3	2.8	4.9
	Theoretical calculation result	43.7	33.9	2.0	4.6
C309 (5 MHz/12.7 mm)	2 mm ball	76.5	38.8	1.8	3.7
	3 mm ball	77.5	40.1	2.0	3.4
	4 mm ball	78.9	41.2	2.4	3.1
	75 µm hydrophone	74.7	38.5	1.6	3.6
	0.2 mm hydrophone	76.1	39.1	1.7	3.8
	0.5 mm hydrophone	78.1	40.4	1.8	4.6
	1 mm hydrophone	84.3	52.4	2.8	5.2
	Theoretical calculation result	75.0	39.2	1.7	4.1

Note: For a planar transducer, the focal distance refers to its near field length (or "natural focus").

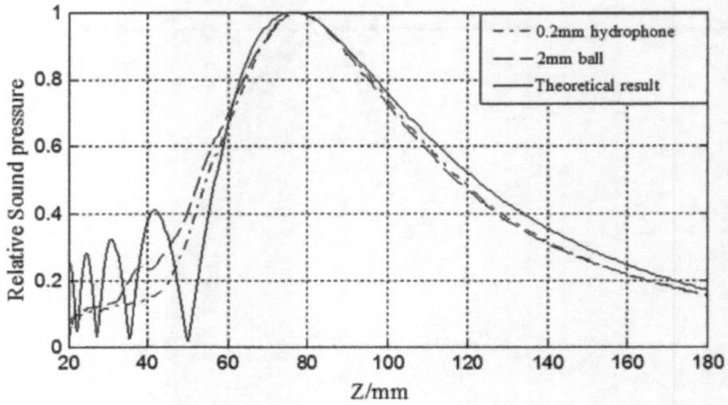

FIGURE 13.47 Calculation and measurement results of axial sound pressure distribution of a focusing transducer (C309).

theoretical calculation result, the measured sound pressure distribution on the focal plane is not a standard circle. Therefore, the focal area sizes measured in different directions are different to some extent. The focal area width of the transducer listed in Table 6.3 refers to the minimum measured value.

3. **Beam diffusion angle**: The sound pressure distribution in the focal plane of the transducer is irregular, so is the sound field distribution on the axial section in the sound field space. Moreover, this irregularity tends to increase with the axial distance. The calculation of beam diffusion angle involves the sound pressure distribution on the axial section. This means that the beam diffusion angles obtained from the axial sections at different radial deflection angles are different. The higher the beam irregularity is, the greater the deviation of beam diffusion angle will be. Therefore, to accurately characterize the beam diffusion angle of the transducer, it should be measured for several times in different directions and be related to the deflection angle of axial transducer section. This aspect is not clarified in various standards and is also one of the research directions of transducer beam performance characterization.

By comparing the measurement methods, the following conclusions can be drawn:

1. In this experiment, the hydrophone with an aperture of 75 μm meets the requirements for aperture in the accurate measurement of sound field and can be used as a reference "standard". It can be seen that with the increase of hydrophone aperture size, the errors in the measured characteristic parameters of each sound field tend to increase to varying degrees. For example, for a C309 transducer with the nominal frequency of 5 MHz (whose center frequency is measured to be 4.7 MHz and whose wavelength in water is 0.32 mm), the errors in focal distance measurements are 1.8% for 75 μm hydrophone and 0.2 mm hydrophone, 3.2% for 0.5 mm hydrophone and 12.9% for 1 mm hydrophone. The measurement results of other parameters also have a similar deviation trend. It can be seen from the measurement results of the C309 focusing transducer that the sound pressure distribution measured with a hydrophone with 1 mm receiving aperture is significantly different from those measured with other hydrophones.

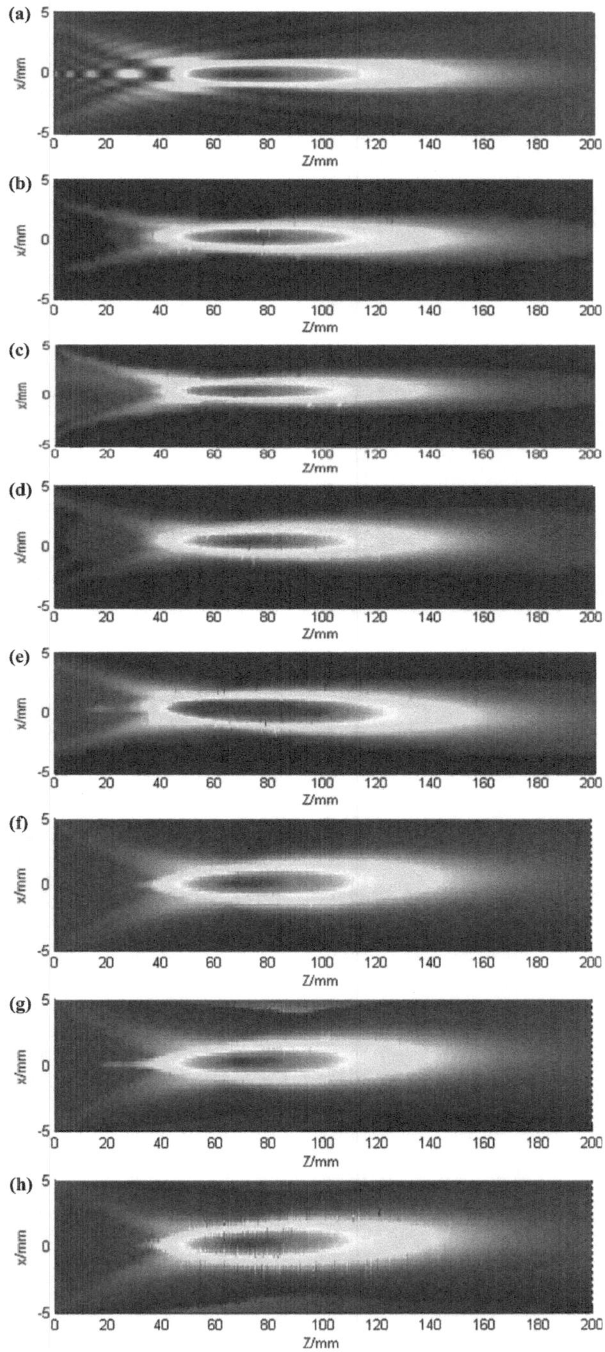

FIGURE 13.48 Calculation and measurement results of sound field distribution on axial beam section of focusing transducer (C309) obtained with different measurement methods. (a) Theoretical calculation result, (b) measurement result with PAL-75 μm hydrophone, (c) measurement result with PAL-0.2 mm hydrophone, (d) measurement result with Onda-0.5 mm hydrophone, (e) measurement result with NCS-1 mm hydrophone, (f) measurement result with 2 mm ball, (g) measurement result with 3 mm ball and (h) measurement result with 4 mm ball.

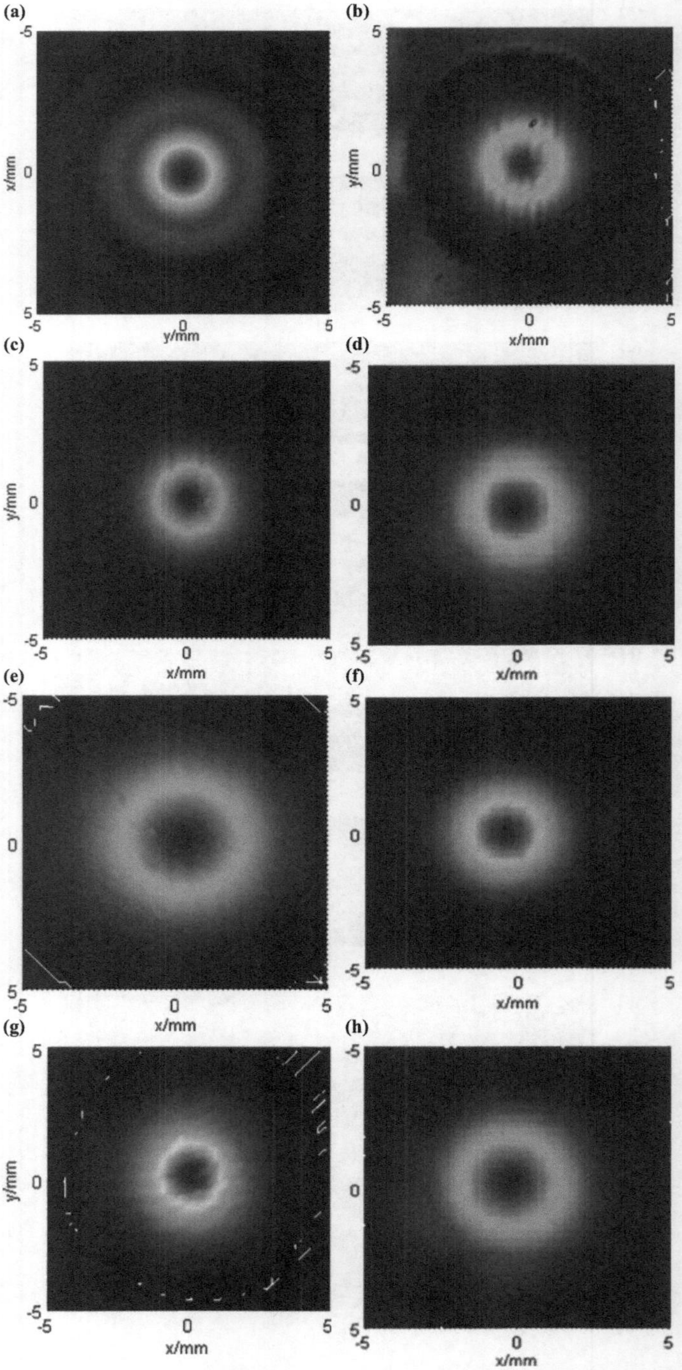

FIGURE 13.49 Calculation and measurement results of sound field distribution on transverse section (focal plane) of a focusing transducer (C309) obtained with different measurement methods. (a) Theoretical result, (b) PAL-75 μm hydrophone, (c) PAL-0.2 mm hydrophone, (d) Onda-0.5 mm hydrophone, (e) NCS-1 mm hydrophone, (f) 2 mm ball, (g) 3 mm ball and (h) 4 mm ball.

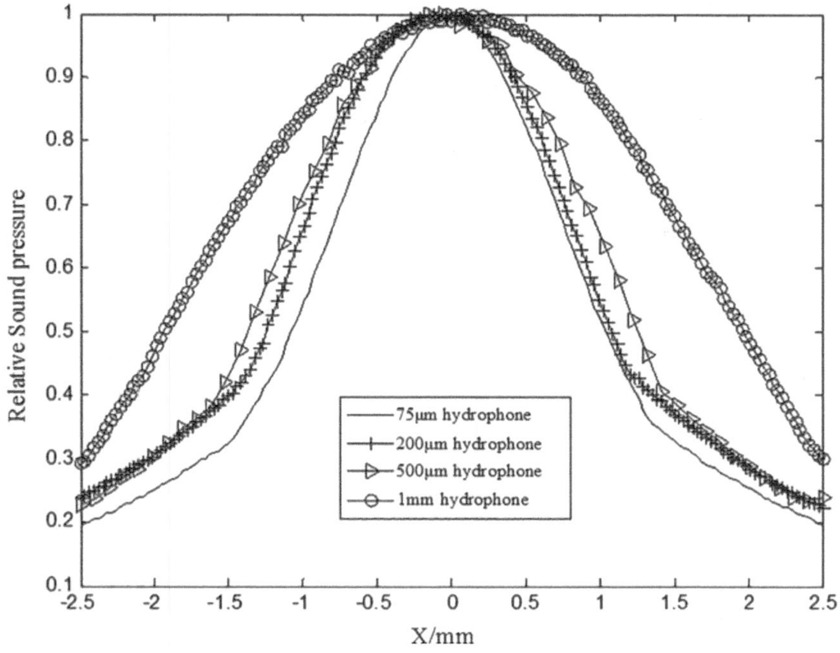

FIGURE 13.50 Measurement results of transverse width of a focusing transducer (C309) obtained by the hydrophones with different apertures.

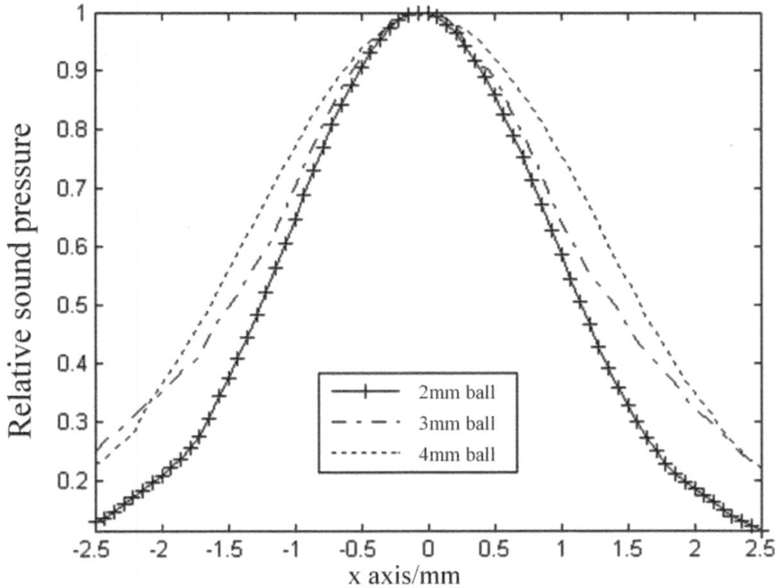

FIGURE 13.51 Measurement results of transverse width of a focusing transducer (C309) obtained by the balls with different diameters.

The measurement deviations of 0.2 and 0.5 mm hydrophones are within the acceptable range, because the hydrophone apertures less than 1/4 wavelength can accurately measure the sound field parameters and those comparable to the wavelength can correctly measure the sound field parameters. The 0.2 and 0.5 mm apertures can basically or barely meet the requirements of correct measurement. The 1 mm aperture is much larger than a wavelength and thus is no longer suitable for the correct measurement of transducer sound field at this frequency. The ultrasonic reception by this aperture can no longer be approximated as point reception but surface reception.

2. By observing the sound-field distribution images and sound field parameters measured by different balls, the same trend with the hydrophone measurement results can be found. The measurement results obtained with the balls in some sizes didn't deviate greatly from those obtained with micro-aperture hydrophones. This demonstrates the feasibility of measuring the focal-area parameters with the ball method.
3. By comparing the measurement results obtained by the balls with different diameters and those obtained by the hydrophones with different apertures, it can be found that the deviation in the measured sound field of the non-focusing transducer V312 used in this experiment will increase with the ball diameter. This is because a larger ball diameter indicates a larger reflection area, which leads to a greater spatial averaging effect. For the non-focusing transducer V312 and focusing transducer C309, the measurement results of sound field obtained by a ball with 2 mm diameter are close to those obtained by a hydrophone with 0.2 mm aperture. The measurement results of sound field obtained by a ball with 3 mm diameter are close to those obtained by a hydrophone with 0.5 mm aperture. There is an approximately equivalence relation between the ball radius and the hydrophone diameter. When the transducer frequency is higher than 10 MHz (the corresponding wavelength in water is less than 0.15 mm), both the hydrophone with an aperture greater than 0.2 mm and the ball with a diameter greater than 2 mm will induce large measurement errors. The above sound field measurement is also made for the transducers at other frequencies. After comparison, the principle of adapting the ball diameter to the transducers at different frequencies is obtained as follows:

   When $0.5\ \text{MHz} \leq f \leq 3.0\ \text{MHz}$ $\quad$ $3\ \text{mm} \leq d \leq 5\ \text{mm}$

   When $3.0\ \text{MHz} \leq f \leq 10\ \text{MHz}$ $\quad$ $2\ \text{mm} \leq d \leq 3\ \text{mm}$

   When $10\ \text{MHz} < f \leq 15\ \text{MHz}$ $\quad$ $d = 1\ \text{mm}$

   When $f > 15\ \text{MHz}$, the ball reflection method is no longer applicable.

Compared with other relevant standards (such as GB/T 18694-2002, BS EN12668-2:2001 and E1065–2008), the ball selection rule proposed for sound field measurement in this paper is more detailed. For example, for the measurement of sound field characteristics of high-frequency transducers (10–15 MHz), the proposed ball selection rule is consistent with that in the ASTM standard E1065–2008. However, for low-frequency transducers, the two rules are quite different. For the transducers with a center frequency of 0.5 MHz, the ball diameter calculated by the recommended formula in E1065–2008 is 30 mm, which is too large to meet the approximation conditions in the sound field measurement with ball method.

By comparing the European standard BS EN12668-2:2001, the American standard E1065–2008 and the Chinese standard GB/T 18694-2002 and analyzing a large amount of experimental measurement data, we can find that the measurement result of transducer sound field obtained by using the ball specified in BS EN12668-2:2001 is the most accurate. When using the ball specified in E1065–2008 to measure the sound field of a transducer with the frequency higher than 5 MHz, a moderately accurate result can be obtained. In contrast, the Chinese standard GB/T 18694-2002 is more general and only applies to the measurement within a certain frequency range. Although there is a certain deviation between the length and width of focal area obtained by the ball method and those obtained by the micro-aperture hydrophone, this error can be minimized by the selection of an appropriate ball. In the ultrasonic industrial detection, this error has no impact on the transducer evaluation result.

REFERENCES

1. Zhang W., Que P., Song S., Modeling of ultrasonic testing system [J], *Computer Simulation*, 2006(23):78–80.
2. Mason W.P. *Electromechanical Transducers and Wave Filters [M].* New York: Van Nostrand, 1942.
3. Redwood M., Transient performance of a piezoelectric transducer [J], *Journal of the Acoustical Society of America*, 1961(33): 527–536.
4. Krimholtz R., Leedom D.A., Matthaei G.L., New equivalent circuits for elementary piezoelectric transducers [J], *Electronics Letters*, 1970, 16(13): 398–399.
5. Leach W.M., Controlled-source analogous circuits and SPICE models for piezoelectric transducers [J], *IEEE Transactions on Ultrasonics, Ferroelectrics, and Frequency Control*, 1994, 41(1): 60–66.
6. Püttmer A., Hauptmann P., Lucklum R., Krause O., Henning B., SPICE model for lossy piezoceramic transducer [J], *IEEE Transactions on Ultrasonics, Ferroelectrics, and Frequency Control*, 1997, 44(1): 60–66.
7. Johansson P.M., Incorporation of diffraction effects in simulations of ultrasonic systems using PSpice models [C], *IEEE Ultrasonics Symposium*, 2001:405–410.
8. Aouzale N., Chitnalah A., Jakjoud H., Experimental validation of SPICE modeling diffraction effects in a pulse-echo ultrasonic system [C], *IEEE Transactions on Circuits and Systems II: Express Briefs*, 2009, 56(12): 911–915.
9. Lin S. *Principle and Design of Ultrasonic Transducer [M].* Beijing: Science Press, 2004:1–49.
10. Luan G., Zhang J., Wang R. *Piezoelectric Transducers and Transducer Arrays [M].* Beijing: Peking University Press, 2005:297–324.
11. Chongfu Y. *Ultrasonics [M].* Beijing: Science Press, 1990:1–10.
12. Johansson J., Martinsson P., Incorporation of diffraction effects in simulations of ultrasonic systems using SPICE models [J], *Proceedings of the IEEE Ultrasonics Symposium*, 2001(1): 405–410.
13. Ramo S., Whinnery J.R., Van Duzer T. *Fields and Waves in Communication Electronics, 2nd edition[M].* New York: John Wiley, 1984.

14. Aouzale N., Chitnalah A., Jakjoud H., Kourtiche D., SPICE modeling of an ultrasonic setup for materials characterization [J], *Ferroelectrics*, 2008, 372(1): 107–114.
15. Johansson J., Optimization of a piezoelectric crystal driver stage using system simulations [J], *IEEE Ultrasonics Symposium*, 2000(2): 1049–1054.
16. Kaufman J.J., Xu W., Chiabrera A.E., Siffert R.S., Diffraction effects in insertion mode estimation of ultrasonic group velocity [J], *IEEE Transactions on Ultrasonics, Ferroelectrics, and Frequency Control*, 1995, 42(2): 232–242.
17. Zhang H. *Theoretical Acoustics [M]*. Beijing: Higher Education Press, 2007:174–324.
18. Xu X., Ge J., Liu Z., Application of convolution-impulse response method in linear phased array [J], *Journal of Tongji University (Natural Science Edition)*, 2005, 33(12):1700–1703.
19. Wu P., Stepinski T., Spatial impulse response method for predicting pulse-echo fields from a linear array with cylindrically concave surface [J], *IEEE Transactions on Ultrasonics, Ferroelectrics, and Frequency Control*, 1999, 46(5): 1283–1297.
20. Stepanishen P.R., Wide bandwidth near and far field transients from baffled pistons [J], *Proceedings of the IEEE Ultrasonics Symposium*, 1977: 113–118.
21. Li S. *Research on Key Theory of Ultrasonic Phased Array [D]*. Beijing: Beijing Institute of Technology, 2009.
22. Arditi M., Forster F.S., Hunt J., Transient fields of concave annular arrays [J], *Ultrasonic Imaging*, 1981(3): 37–61.
23. Xu Y., Xu C., Xiao D., et al., Research on the directivity of ball-scattered ultrasonic pulse sound field [J], *Technical Acoustics*, 2009, 28(5):615–619.
24. Cheeke J.D. *Fundamentals and Applications of Ultrasonic Waves [M]*. New York: CRC press, 2002: 21–38.
25. Ma D. *Theoretical Basis of Modern Acoustics [M]*. Beijing: Science Press, 2004:74–124.
26. Hailan Z. *Theoretical Acoustics [M]*. Beijing: Higher Education Press, 2007:174–192.
27. Dang C., Schmerr L.W., Sedov A., Ultrasonic transducer sensitivity and model-based transducer characterization [J], *Research Nondestructive Evaluation*, 2002, 14: 203–228.
28. Zhao X., Nonparaxial multi-Gaussian beam models and measurement models for phased array transducers [J], *Ultrasonics*, 2009, 49: 126–130.
29. Schafer E. *Techniques of Hydrophone Calibration in Ultrasonic Exposimetry [M]*. M.C. Ziskin and P.A. Lewin, Eds. Boca Raton, FL: CRC Press, 1993: 216–255.
30. Harris G.R., Medical ultrasound exposure measurements: update on devices, methods, and problems [J], *Proceedings of the IEEE Ultrasonics Symposium*, 1999(2): 1341–1352.
31. Zeqiri B., Bond A.D., The influence of waveform distortion on hydrophone spatial-averaging correction—theory and measurement [J], *Journal of the Acoustical Society of America*, 1992, 92(4): 1809–1821.
32. Huttunen, T., Kaipio, J.P., Hynynen K., Modeling of aNomalies due to hydrophones in continuous-wave ultrasound fields [J], *IEEE Transactions on Ultrasonics, Ferroelectrics, and Frequency Control*, 2002, 50(11): 1486–1500.
33. Zhao F., Lan C., A time-domain deconVolution algorithm used to improve resolution for ultrasonic reflection mode computerizes tomography [J], *ACTA Acoustica*, 1995, 20(6): 413–416.
34. Guo J., Lin S., A modified Wiener inverse filter for deconVolution in ultrasonic detection [J], *Applied Acoustics*, 2005(24): 97–102.
35. Kong T., Xu C., Xiao D., Experimental ESD method for restoration of blurry image in ultrasonic C-scan [C]. 2010 IEEE International conference on mechanic automation and control engineering, 2010: 2632–2635.

CHAPTER 14

Robotic Laser Measurement Technique for Solid Sound Field Intensity

With good directionality, rich information, no damage to sample, strong penetration and other excellent characteristics, ultrasonic wave is playing an important role in various fields such as materials, biology, national defense, energy, electronics, industry, agriculture and environmental protection. The working environments of ultrasonic wave can be classified into liquid, solid, gas and organism. In terms of working power, ultrasonic waves can be divided into detection ultrasound and power ultrasound. The detection ultrasound is mainly used in ultrasonic nondestructive testing (NDT) devices and various ultrasonic sensors. The power ultrasound is widely used in ultrasonic welding, ultrasonic cleaning, ultrasonic stress reduction and other applications [1].

14.1 SOLID SOUND FIELD AND ITS MEASUREMENT METHOD

14.1.1 Definition, Role and Measurement Significance of Solid Sound Field

In recent years, power ultrasonics has become an important technique in the fields of advanced machining and material processing, such as metal processing, welding of dissimilar materials, metallurgical processing and residual stress regulation and control. High-intensity ultrasound is to gather the low-intensity ultrasonic waves outside the structure into the material body by some means to form a high-intensity sound field. It is a new metal processing technique that uses the thermal effect, mechanical effect and kinetic energy effect of power ultrasound to change an intergranular structure in an instant (0.5–2 s) without damaging or destroying the nearby normal structures. In national economic construction, high-intensity ultrasound has a unique potential in improving the product quality, reducing the production cost, preventing environmental pollution and improving the production efficiency [2,3]. The study on the basic theory of power ultrasound will not

DOI: 10.1201/9781003212232-14

only make a great breakthrough in the basic research but also have a great impact on the industrial manufacturing of power ultrasound equipment.

The propagation law in media is very important for the application of ultrasound. Just like the case of the ultrasonic method used for quality testing, the propagation law in a medium is very important for the industrial application effect of power ultrasound, that is, the sound field distribution in the medium has a guiding significance for the application of power ultrasound. However, the sound field distribution law of power ultrasound in the solid medium is still unknown at present. This makes it impossible to obtain an ideal application effect and may even affect the performance of industrial products. For example, in ultrasonic welding, too high ultrasonic power will cause large plastic deformation above the maximum limit, a thinner contact area and a shorter product life. On the contrary, too low ultrasonic power and incomplete welding will also shorten the service life and possibly trigger an accidental danger. Therefore, the magnitude of ultrasonic power is particularly important for studying the propagation law and sound field distribution of power ultrasound in a solid.

The research on the sound field distribution law of power ultrasound in a solid can produce a huge impact on power ultrasound applications, provide an important reference basis for ultrasonic processing and other ultrasonic applications and avoid the formation of many defects that affect the normal use of the system, such as the performance degradation, safety degradation and loss of life caused by improper power selection. It is very important to study the measurement technique of power ultrasound field in a solid. As a measurement technique of power ultrasound, solid power ultrasound measurement is not only beneficial to the development of power ultrasound industry but also necessary for improving the quality of civil and national defense manufacturing. It is of great significance to the technical progress, cost reduction and quality improvement in high-end manufacturing.

14.1.2 Current Domestic and Overseas Measurement Methods and Their Problems

Power ultrasound is widely used in various measurement environments, which can be generally divided into liquid environment and solid environment. The main detection principles of liquid power ultrasound field include mechanical effect, thermal effect, chemical effect and optical effect [4]. However, the measurement principles and methods of solid power ultrasound field are very few and need to be further explored.

The measurement methods of mechanical effect include the radiation force method, hydrophone method and photodetector method. As one of the basic measurement methods of ultrasound field power in liquid, the radiation force method is widely used in the field of medical ultrasound to measure the radiation field power of medical devices. This method is mainly used to measure the sound power of plane waves in water or liquid. In 1998, Shouwende et al. successfully used the radial force derived from geometric acoustics to measure the sound power of a high-intensity focusing ultrasonic transducer with a deviation of less than 3% [5]. Xu Jian et al. studied the water column method based on the radiant force method, converted acoustic radiant force into the potential energy of water and used the height of water column to represent the radiant force value at that point.

The upper limit of power measurement in the water column method was 2 KW/cm^2. The hydrophone method [6,7] is used to convert a pressure signal into an electrical signal by using the inverse piezoelectric effect of piezoelectric materials. After being amplified, the electrical signal can be displayed directly and conveniently on the acquisition and display equipment. With the development of piezoelectric materials, the performance of hydrophones has been constantly improved with the measurement bandwidth up to 150 MHz and the measurement range of sound pressure over 100 Mpa. Moreover, due to a small size, the hydrophones can achieve a spatial resolution up to 40 μm and a sensitivity up to 20 nV/Pa [8]. Based on the hydrophone method, the optical detection method has the potential to withstand high sound field intensity and to provide higher accuracy (resolution: 0.1 mm, bandwidth: 3 GHz, sensitivity: 4 nV/Pa).

The measurement methods based on thermal effect include the thermosensitive probe method and crystal color development method [9]. The thermosensitive probe is cheap and easy to operate, with a resolution of 10 μm. However, the sensor is susceptible to cavitation erosion in the high-intensity sound field and to possible reactions such as flowing and shielding. Thus, it cannot measure the high-intensity sound field. The multielement thermal probes emerging later can only measure the sound field at the frequency of 20 kHz. The method of crystal color development is intuitive but complex in operation. Moreover, it is susceptible to interference and has a large error, so it cannot achieve high-precision measurement [10].

The measurement methods based on chemical effect include the film corrosion method, dyeing method and iodine release method. The film corrosion method is simple and convenient but has a large measurement error. The dyeing method can only measure the low-frequency sound field. Like the iodine release method, the dyeing method is also not technologically mature and can't achieve the quantification level of other detection methods. Moreover, the chemical methods cannot be used to measure the sound fields in solids. Although the measurement method based on optical effect can visually describe the shape of sound field and the sound intensity distribution, it can only be used for the measurement of low-frequency sound field in a liquid.

In the 1970s, Hall first reported the dynamic photoelastic imaging results of ultrasonic pulse internationally [11]. The basic principle of the photoelastic method is to study the problem of elastic mechanics with the optical method and visualize the propagation of elastic wave with the aid of transparent elastic material and high-speed camera, so as to intuitively explore the propagation law of sound wave in the elastic body. It is physically based on the temporary birefringence phenomenon and mathematically described by the stress optical law. The dynamic photoelastic method has made important contributions to the research on the stress wave propagation and dynamic stress concentration under an impact load as well as the crack propagation process and law in structures.

In the early 1980s, the academicians Ying Chongfu, Zhang Shouyu and Shen Jianzhong from the Institute of Acoustics, Chinese Academy of Sciences, built a set of dynamic photoelastic laser imaging equipment in order to study in depth the propagation law and scattering phenomenon of ultrasonic waves in solids. They did a lot of experiments on the system, achieved a series of research results and realized the visualization of ultrasonic field. As a result, they made great contributions to the research of ultrasonic field and received attention from international counterparts. In 2014, Wang Xiaomin et al. used the dynamic photoelastic method to measure the directivity of radiated sound field in a cylindrical cavity for the first time. By summarizing the propagation law of creeping wave on the cavity wall, they proposed the use of transverse creeping wave to detect the cracks on the cavity wall. The relevant results show that the dynamic photoelastic method can provide not only an intuitive image of ultrasonic scattering field but also the result of quantitative analysis. It is an effective experimental means for studying ultrasonic propagation and scattering laws [12]. In 2017, Yu Qun et al. proposed a sound intensity evaluation method based on near field measurement in their study on the measurement method of sound intensity and sound power of focusing transducers. By measuring the sound pressure and sound intensity on two planes in the focusing region, they obtained the sound intensity distribution of focusing transducers [13].

In 1964, Mihailov I.G. and Shutilov V.A. realized the absolute measurement of ultrasonic field in a solid by using the knowledge on electrodynamics. At first, a small light metal strap was attached as a receiver to the fixed surface and was positioned opposite to the radiator. Then, the fixed surface was placed in a magnetic field with the magnetic induction intensity B. When the metal strap is forced to vibrate in the magnetic field, a voltage related to vibration velocity will be generated. From the voltage and magnetic field strength, the vibration velocity can be deduced. This method is relatively simple in principle, but the precision of the magnetic field strength will directly affect the measurement precision of the ultrasonic field [14]. In 2009, Wu Jing studied the solid sound field of an ultrasonic transducer and its measurement method, explored the sound field characteristics of the impulse wave excited by the ultrasonic transducer with the aid of double probes and measured the sound field directivity and the sound pressure on acoustic axis, which proved to be consistent with the theoretical calculation results. However, no effective method was proposed to measure the sound field distribution of continuous acoustic waves in a solid [15]. The innovation of the above study is to turn vibration into an electric signal and mainly to measure the vibration. Currently, the most popular vibration sensor is laser vibrometer [16]. It has the advantages of non-contact and high precision, thus eliminating the influence of magnetic field precision. One of the experimental tools used in this book is laser interference vibrometer, which can pick up the surface vibration of a vibrating object. With this tool, the sound field on transducer surface and the intensity of radiated sound field were studied.

14.2 SOUND SOURCE CHARACTERISTICS OF SOLID SOUND FIELD AND ITS CHARACTERIZATION PARAMETERS

14.2.1 Structure and Characteristics of Exciter Sound Source

With the rapid development of power ultrasound technique, this technique has been widely used in ultrasonic welding, ultrasonic cleaning, ultrasonic stress relief, ultrasonic emulsification and other fields. The composition of the power ultrasound system depends on its application field. The simplest system consists of a signal generator, a power amplifier and a power transducer. Its working principle is shown in Figure 14.1.

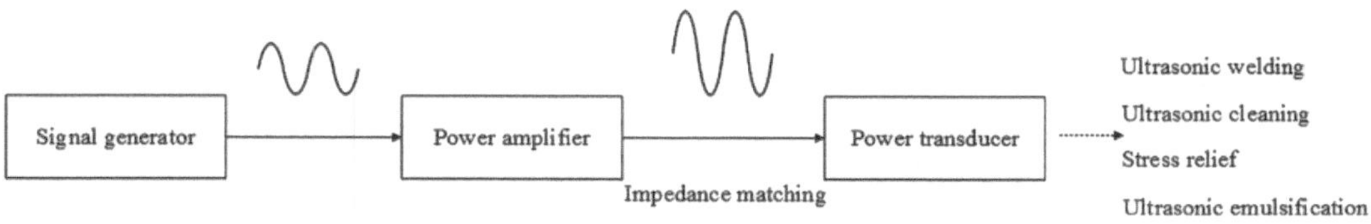

FIGURE 14.1 Schematic diagram of a power ultrasound system.

The role of the power ultrasound transducer in the power ultrasound application system is to convert high-voltage high-frequency electrical signals into high-frequency mechanical vibration, through which the transducer will disturb the liquid, solid or gas and achieve the application purpose. Since the power transducer realizes the conversion from electrical energy to mechanical energy, it serves as the transduction component in the power ultrasound system. Its performance will have a direct impact on the system efficiency.

Power ultrasound transducers are mainly divided into two types: magnetostrictive transducer and piezoelectric transducer. The piezoelectric transducer is most widely used. Generally, a piezoelectric transducer is composed of a piezoelectric vibrator, electrode plates, pre-tightening bolts, an insulation tube, front and rear cover plates, an insulation sleeve and locking nuts. It is also known as the sandwiched transducer, as shown in Figure 14.2.

The working principle of the piezoelectric transducer is to vibrate continuously along the axis under continuous excitation by leveraging the piezoelectric effect of a piezoelectric vibrator. Therefore, the piezoelectric vibrator is the most critical transduction component. Piezoelectric vibrators can be made of a variety of materials. Among them, piezoelectric ceramic is most widely used owing to its high performance and is mostly based on PZT. The applications and outstanding characteristics of common components of piezoelectric ceramics are shown in Table 14.1. PZT-8 is used as piezoelectric material for most of the power ultrasound transducers.

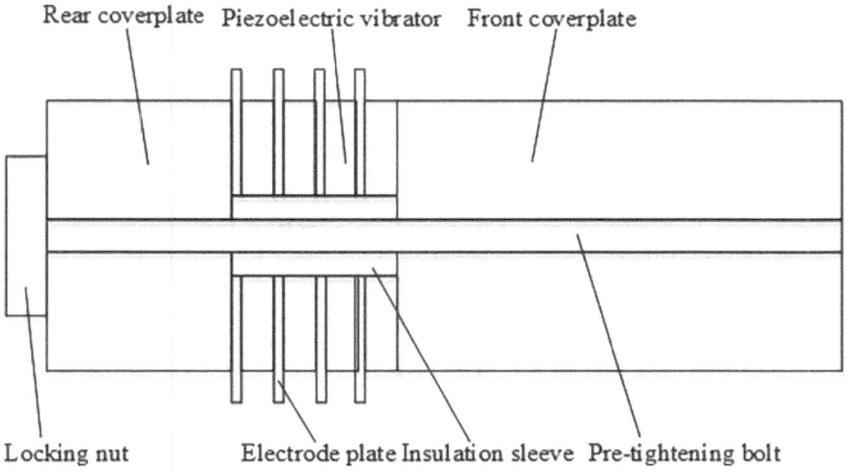

FIGURE 14.2 Schematic diagram of a piezoelectric power ultrasound transducer.

TABLE 14.1 Common Components of Piezoelectric Ceramics [17]

Material	Outstanding Features	Common Applications
PZT-4	High coupling coefficient and high excitation performance	Sonar radiator, high-voltage generator
PZT-5A，PZT-5H	High dielectric constant and high coupling coefficient	Hydrophone, rheomicrophone
PZT-7A	Small aging coefficient and small dielectric constant	Shear modulus transducer, delay line transducer
PZT-8	Outstanding excitation performance	Power ultrasound apparatus

Theoretically, an elastomer may have an infinite number of vibration modes corresponding to different frequencies, while a piezoelectric vibrator used as a transducer has a limited number of vibration modes. The vibration mode of piezoelectric ceramics is mainly determined by the size, shape, polarization direction and excitation direction of the vibrator [18]. Therefore, the vibration mode that meets the requirements can be obtained only by selecting an appropriate material and shape and designing an appropriate size and excitation mode. The vibration modes of piezoelectric ceramic vibrators are diversified but usually divided into longitudinal and transverse effect vibrations. In this paper, thin disc-type piezoelectric vibrators with thickness vibration, namely longitudinal effect vibration, are the focus [19,20].

In piezoelectric effects, the conversion relationship between the electrical quantity derived from the electric field E and the electrical displacement D and the mechanical quantity derived from the stress T and the strain S is known as electromechanical coupling [21,22]. The mechanical boundaries of piezoelectric vibrators are divided into free condition and clamped condition, and their electrical boundaries are divided into short circuit and open circuit. By combining two mechanical boundary conditions and two electrical boundary conditions, four different boundary conditions [23] can be obtained, as shown in Table 14.2.

TABLE 14.2 Four Boundary Conditions of Piezoelectric Vibrator

Type	Boundary Condition	Physical Quantity	Piezoelectric Equation
Type 1	Mechanical clamping; electrical open circuit	$S=0, T\neq 0$ $D=0, E\neq 0$	$\text{Type h:}\begin{cases} T_h = c_{hk}^{D} S_k - h_{jh} D_j & h,k=1,2,\ldots,6 \\ E_i = -h_{ik} S_k + \beta_{ij}^{S} D_j & i,j=1,2,3 \end{cases}$
Type 2	Mechanical free condition; electrical open circuit	$T=0, S\neq 0$ $D=0, E\neq 0$	$\text{Type e:}\begin{cases} T_h = c_{hk}^{E} S_k - e_{jh} E_j & h,k=1,2,\ldots,6 \\ D_i = e_{ik} S_k + \varepsilon_{ij}^{T} E_j & i,j=1,2,3 \end{cases}$
Type 3	Mechanical clamping; electrical short circuit	$S=0, T\neq 0$ $E=0, D\neq 0$	$\text{Type g:}\begin{cases} S_h = S_{hk}^{D} T_k + g_{jh} D_j & h,k=1,2,\ldots,6 \\ E_i = -g_{ik} T_k + \beta_{ij}^{T} D_j & i,j=1,2,3 \end{cases}$
Type 4	Mechanical free condition; electrical short circuit	$T=0, S\neq 0$ $E=0, D\neq 0$	$\text{Type h:}\begin{cases} T_h = c_{hk}^{D} S_k - h_{jh} D_j & h,k=1,2,\ldots,6 \\ E_i = -h_{ik} S_k + \beta_{ij}^{S} D_j & i,j=1,2,3 \end{cases}$

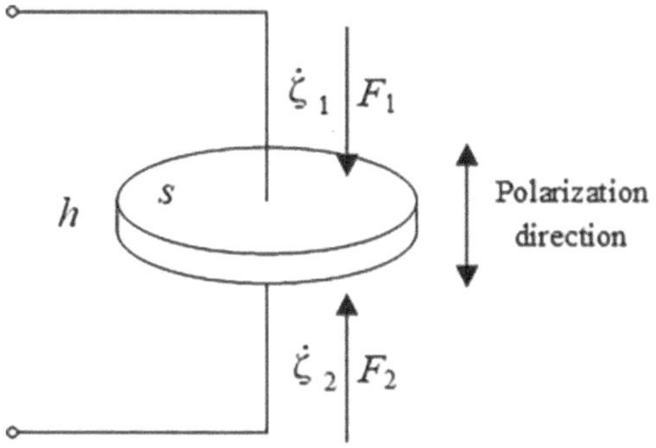

FIGURE 14.3 Piezoelectric vibrator with a thickness vibration mode.

The thin disc-type piezoelectric vibrator with thickness vibration is shown in Figure 14.3. It is polarized in the thickness direction. It has a thickness of h and a round electrode surface with the area s. Suppose that its transverse dimension is much larger than the wavelength. So, the wafer can be considered to be transversely static [24].

Each part of the transducer in this system is a cylinder with the same cross section. Here we start our discussion with the application of variable cross sections – from an ordinary section to a special section. The axial section of a variable-section body is illustrated in Figure 14.4.

Suppose that the bar with variable cross sections is made of a homogeneous isotropic material with the same stress distribution on various sections and that its mechanical loss is ignored. Note: these assumptions are true when the cross-section dimensions are much smaller than the wavelength [25].

While meeting the engineering requirements without affecting the transducer performance, the following assumptions need to be made:

1. The motion of each cross section of the power transducer is the same, that is, plane waves are generated;
2. The excitation received by each ceramic slice in the piezoelectric ceramic stack is in the same phase, that is, the whole piezoelectric ceramic stack can be regarded as a piezoelectric ceramic slice.

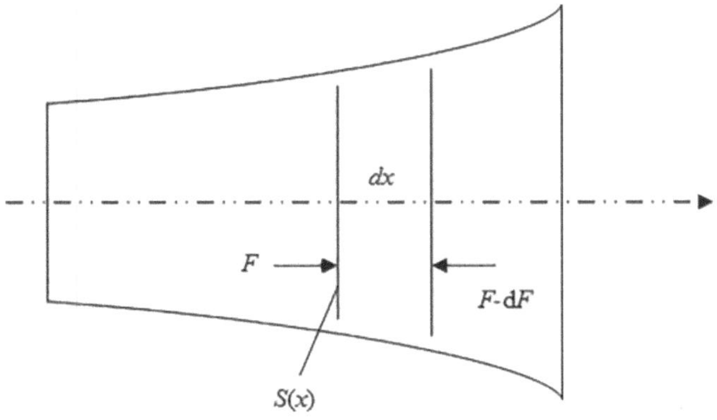

FIGURE 14.4 Schematic diagram of variable sections.

In addition, other conditions for designing a sandwiched piezoelectric transducer can be considered basically satisfied, for example:

1. The length of the transducer is larger than its diameter;
2. The total length of the piezoelectric ceramic stack is close to 1/3 of the total length of the transducer.

14.2.2 Characterization Method of Solid Sound Field

Ultrasonic transducers are widely used in the field of ultrasonic NDT. The sound field generated by the transducer (especially piston probe) in an infinitely large medium is a basic issue in the field of acoustics. In this section, the sound fields radiated by a detection ultrasound transducer and a power ultrasound transducer are analyzed and calculated with the analytical method, semi-analytical method and numerical method, respectively. The detection ultrasound transducer has a center frequency of 5 MHz and a wafer radius of 5 mm. The power ultrasound transducer has a center frequency of 15 MHz, a wafer radius of 30 mm and a wafer vibration velocity of 0.1 m/s. The propagation media are all aluminum.

14.2.2.1 Analytical Method

Among various analytical methods, the superposition model of multiple Gaussian beams is selected as the analysis method in this paper. Although the normal velocity distribution of particles on the transducer surface is not Gaussian distribution, it can be obtained by the superposition of several Gaussian functions. Wen and Breazeale used the superposition of ten Gaussian beams to simulate the normal velocity of piston probe surface, which was expressed as [26]

$$v_z\left(x_0, y_0, z_0 = 0\right) = A_n e^{-\frac{B_n\left(x_0^2 + y_0^2\right)}{r^2}} \tag{14.1}$$

where A_n, B_n is the Gaussian complex coefficient. The A_n, B_n values corresponding to ten Gaussian functions are listed in Ref. [27]. r is the radius of the piston source. According to the above equation, we can get the diagram of the vibration velocity distribution on piston source surface, as shown in Figure 14.5. It can be seen that the vibration velocity field of the particles near the surface is basically identical with that of actual piston sound source.

According to the Gaussian superposition function on the probe surface, the calculation equation of the superposed sound field of multiple Gaussian beams is obtained, namely

$$p(x, y, z) = \sum_{n=1}^{10} \frac{\rho c v_0 A_n}{1 + iB_n z/x_r} e^{ikz} e^{-\frac{B_n\left(x^2 + y^2\right)/r^2}{1 + iB_n z/x_r}} \tag{14.2}$$

where k is the wave number; x_r is the Rayleigh distance, $x_r = 0.5kr^2$.

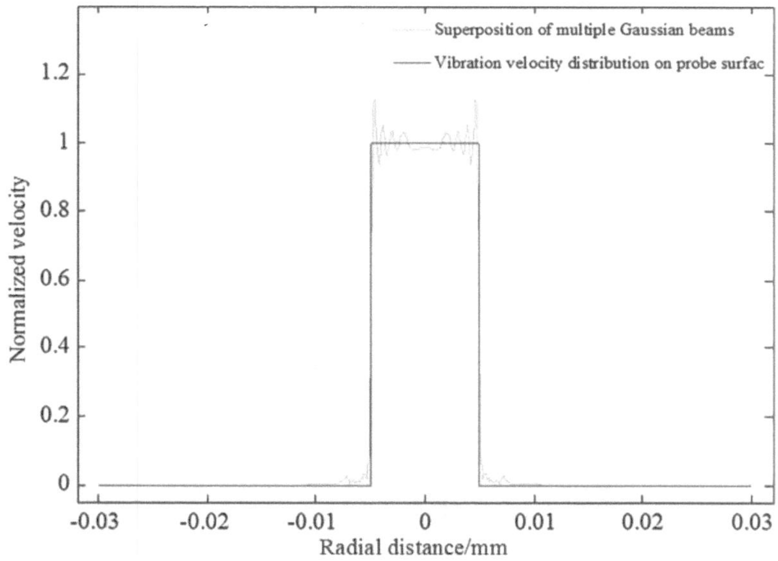

FIGURE 14.5 Vibration velocity distribution on a piston source surface.

The superimposed radiated sound field of multiple Gaussian sound beams calculated by Eq. (14.2) is shown in Figure 14.6.

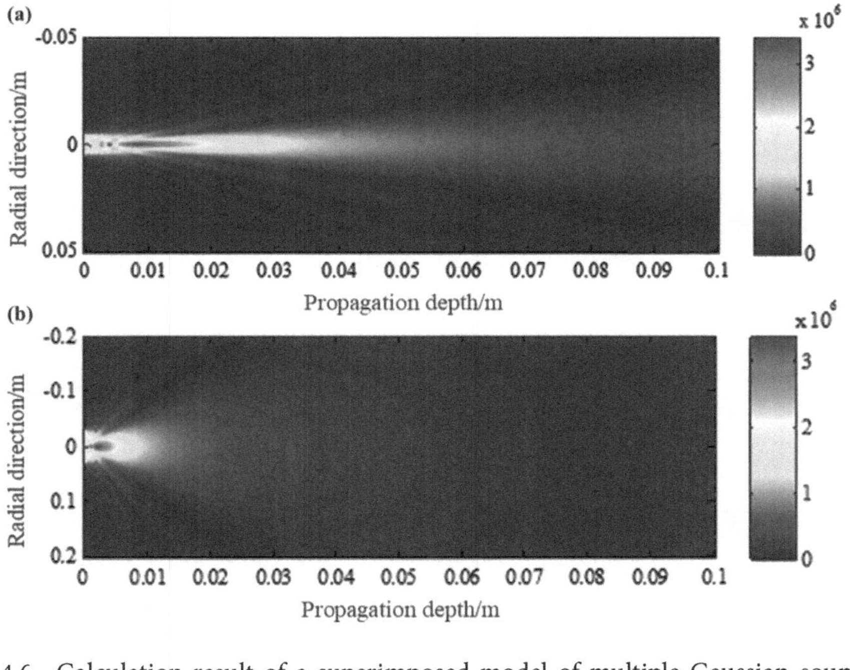

FIGURE 14.6 Calculation result of a superimposed model of multiple Gaussian sound beams. (a) Radiated sound field of a detection ultrasound transducer and (b) radiated sound field of a power ultrasound transducer.

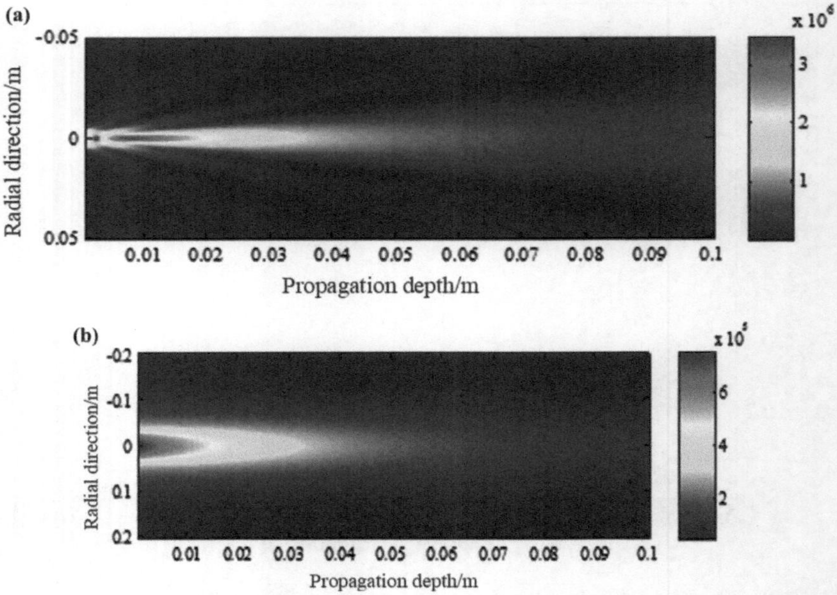

FIGURE 14.7 Calculation result of the Rayleigh integral model. (a) Radiated sound field of a detection ultrasound transducer and (b) radiated sound field of a power ultrasound transducer.

14.2.2.2 Semi-Analytical Method

Among various semi-analytical methods, Rayleigh integral model is chosen as the analysis method in this paper. Rayleigh integral is a sound field calculation method based on Huygens principle. The calculation equation of the transducer sound field is [28]:

$$p(R)=\frac{i\omega\rho}{2\pi}\iint_{s} v_z\left(x_0,y_0,z_0=0\right)\frac{e^{-ikR}}{R}ds \tag{14.3}$$

where ρ is the density of the propagation medium, and $R=\sqrt{z^2+(x-x_0)^2+(y-y_0)^2}$ is the distance between any point (x, y, z) in the medium and any point $\left(x_0,y_0,z_0=0\right)$ on the transducer surface. The Rayleigh integral can be considered as the radiated sound field of a sound source with area s represented by the superposition of sound field responses of the point sources (ds).

The radiated sound field in Rayleigh integral calculated by using Eq. (14.3) is shown in Figure 14.7.

14.2.2.3 Numerical Method

Among various numerical methods, finite element method is chosen as the analysis method in this paper. COMSOL Multiphysics is selected as the platform for finite element analysis. COMSOL Multiphysics is comprehensive advanced numerical simulation software widely applied to the engineering calculation and scientific research in various fields, as well as to various physical processes in engineering and simulation sciences. The current COMSOL version includes an acoustic module, which provides tools to simulate various acoustic scenarios, such as acoustic absorber, NDT, acoustic radiation and transducer. In this paper, the pressure-acoustics physical field in the module is used to analyze the frequency domain radiation field of transducers [29].

To simplify the calculation process, a two-dimensional axisymmetric model was selected based on the fact that both the transducer and the radiated sound field were axisymmetric. The modeling and gridding processes were the same as other finite element software, as shown in Figure 14.8. The transducer acted on the boundary ①. Because both multi-beam Gaussian model and Rayleigh integral model analyzed the sound field distribution of transducers in a semi-infinite area, here the radiated field in a limited area was simulated and the same impedance as the medium material was provided at all boundaries of the medium. The analysis results of sound field distribution on surface before and after the impedance matching are shown in Figure 14.9. It can be seen that the impedance matching has effectively avoided the boundary effect. The results of finite element analysis are shown in Figure 14.10.

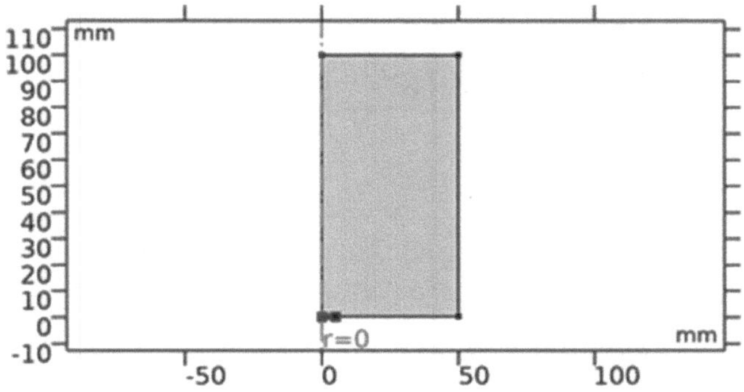

FIGURE 14.8 COMSOL sound field simulation model.

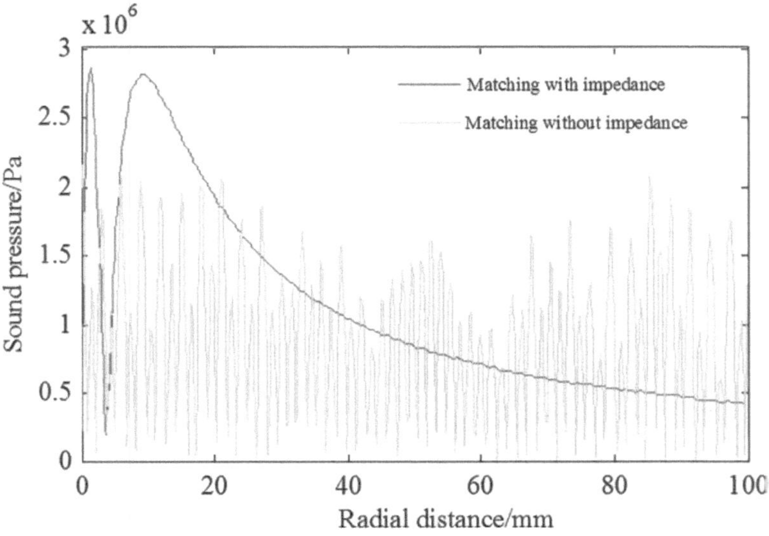

FIGURE 14.9 Comparison of boundary effects.

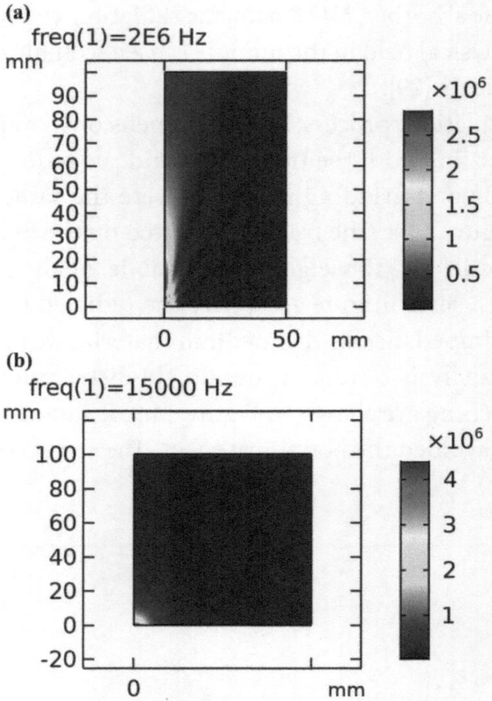

FIGURE 14.10 Results of finite-element sound field analysis. (a) Radiated sound field of a transducer and (b) radiated sound field of a detection ultrasound power ultrasound transducer.

14.2.2.4 Measurement Method of Ultrasonic Intensity in Solids

To investigate the radiation intensity distribution of power transducers in solids, a method of layered measurement was proposed and then the measurement results were fitted to obtain the radiation intensity distribution.

1. Layered measurement

   Since the sound field radiated by the transducer is axisymmetric, our research is focused on the sound field distribution on the axis. In other words, the propagation depth is a variable and the medium with a typical depth is selected as the research object. By designing the medium with a typical depth, the sound field distribution on the medium surface is scanned.

   In this program, aluminum was selected as the propagation medium. Then, according to the curve of sound pressure distribution on theoretical acoustic axis and the experimental conditions, typical propagation depths (medium thicknesses) were determined to be 10, 14, 19, 25 and 35 mm. The medium was cylindrical with a diameter of 200 mm, as shown in Figure 14.11.

2. Fitting calculation

   The media with different thicknesses were coupled with the transducer. Then, the vibration velocities or displacements of end faces of the media were picked up

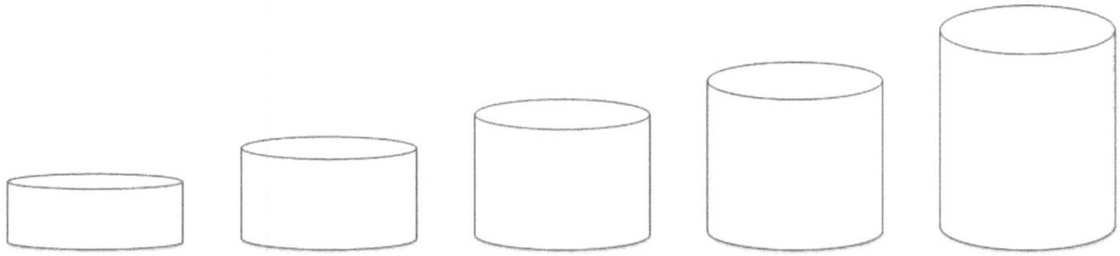

FIGURE 14.11 Schematic diagram of medium blocks with different thicknesses.

by a sensor, and the sound intensities were calculated. Here the distribution of sound field is expressed in a cylindrical coordinate system. The value measured in the medium is assumed to be $I(r,z)$. Due to the boundary effect of a bounded medium (that is, the complex superposition of sound beams emitted by ultrasonic waves in the medium), the measured value is not equal to the sound intensity at the original position in an infinite medium, namely $I'(r,z)$. The relationship between the two values is

$$I'(r,z) = f(r,z)I(r,z) \tag{14.4}$$

where $f(r,z)$ is the conversion coefficient between finite and infinite media and is related to the position of the measurement point. Therefore, after the conversion coefficient is determined, the sound intensity value at that point in the infinite medium can be inversely calculated.

Each group of $I'(r,z_i)i = 1,2,\ldots,5$ values obtained by inverse calculation was fitted by means of fitting algorithm. According to the fitting results, the sound field distribution was obtained by interpolation. In this paper, the calculation data of multi-beam Gaussian model is taken as an example to compare two types of algorithms, namely polynomial fitting and fractional linear fitting.

The data set of depth and sound intensity at the same radial position point in five layers of the medium is assumed to be $(x_i, y_i)(i = 1,2,\ldots,5)$.

1. The fitting principle of the least square polynomial is as follows:
   The target is a m-fitting polynomial $(m < 5)$, i.e.

$$P_m(x) = a_0 + a_1x + \cdots + a_mx^m \tag{14.5}$$

Then the coefficient $a_k\,(k = 0,1,\ldots,m)$ of the least-square fitting polynomial should satisfy the following equation set

$$
\begin{aligned}
& na_0 + a_1 \sum_{i=1}^{5} x_i + a_2 \sum_{i=1}^{5} x_i^{\,2} + \cdots + a_m \sum_{i=1}^{5} x_i^{\,m} = \sum_{i=1}^{5} y_i \\
& a_0 \sum_{i=1}^{5} x_i + a_1 \sum_{i=1}^{5} x_i^{\,2} + a_2 \sum_{i=1}^{5} x_i^{\,3} + \cdots + a_m \sum_{i=1}^{5} x_i^{\,m+1} = \sum_{i=1}^{5} y_i x_i \\
& a_0 \sum_{i=1}^{5} x_i^{\,m} + a_1 \sum_{i=1}^{5} x_i^{\,m+1} + a_2 \sum_{i=1}^{5} x_i^{\,m+2} + \cdots + a_m \sum_{i=1}^{5} x_i^{\,2m} = \sum_{i=1}^{5} y_i x_i^{\,m}
\end{aligned}
\tag{14.6}
$$

If the above equation has a unique solution, the polynomial corresponding to its solution will be the least-square m-fitting polynomial corresponding to the data set. The results of piecewise linear fitting, cubic fitting and quartic fitting based on the above algorithm are shown in Figure 14.12. According to the observation, the fitting results within the measurement depth range (10–35 mm) are basically useable. However, when the measurement depth is greater than 35 mm, no reference data will be available for linear fitting and no linear fitting result can be obtained. In addition, the results of other types of polynomial fitting (such as cubic fitting and quartic fitting) are greatly different from the variation in sound field shown in Figure 14.12. Therefore, polynomial fitting is not applicable to the sound field outside the measured range.

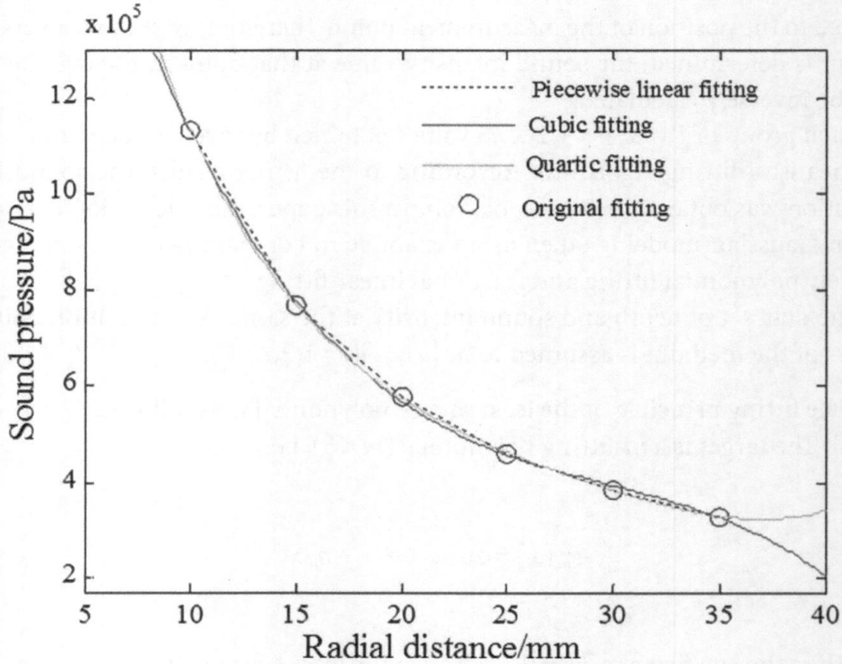

FIGURE 14.12 Polynomial fitting results.

2. The principle of fractional linear fitting is as follows:
   Suppose that the objective fractional linear function is

$$y = \frac{1}{ax+b} \tag{14.7}$$

At first, transform Eq. (14.7) into

$$\tilde{y} = \frac{1}{y} = ax+b \tag{14.8}$$

$$\left(x_i, \frac{1}{y_i}\right)(i=1,2,\ldots,5)$$

Linearize the fitting problem and obtain the least-square linear fitting polynomial based on the data set.

$$\tilde{y} = ax+b \tag{14.9}$$

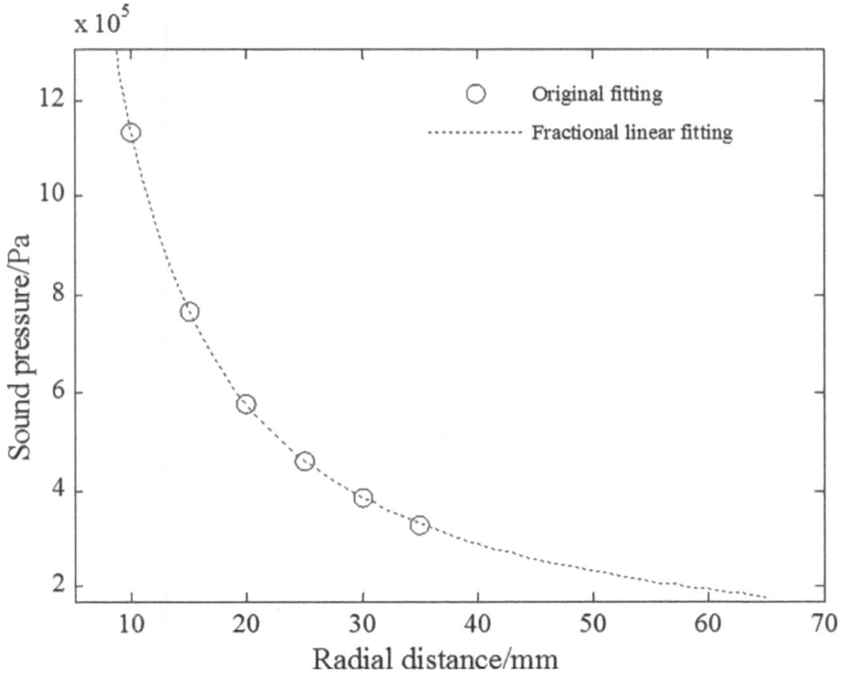

FIGURE 14.13 Result of a fractional linear fitting.

Then, calculate the reciprocal of the linear polynomial and obtain the least-square fractional linear fitting function of the data set $(x_i, y_i)(i=1,2,\ldots,5)$.

The fitting result of the original data in ① is shown in Figure 14.13. It can be seen that the fitting effect is very good, and that the variation trend of the fitting result is consistent with that of the theoretical calculation result.

Compared with polynomial fitting, fractional linear fitting is more suitable for sound field fitting. After the points in all the positions are fit according to the above process, the corresponding fractional linear equation will be obtained. Then, according to the equation and coordinates, the distribution of the entire sound field can be calculated and obtained.

14.3 COMPOSITION OF A ROBOTIC MEASUREMENT SYSTEM FOR SOUND FIELD INTENSITY

Based on the discussion of the measurement system, a manipulator-based power ultrasound intensity measurement system is proposed in this book. In this system, a six-degree of freedom (DOF) manipulator holds a laser vibrometer and moves along the preset trajectory to scan the surface contour. The position acquisition card and the data acquisition card collect the manipulator position coordinates and the surface vibration data, respectively. The position acquisition card triggers the data acquisition card to realize synchronous acquisition and transfers the data to the upper computer software via PCI bus. The upper computer software processes the data in real time, displays the vibration information through A-scan, extracts its eigenvalues for one-to-one matching with the position coordinates and displays the sound field distribution through C-scan. In the sound field fitting, the B-scan image of sound field is obtained through fitting and interpolation by using the fitting algorithm and the sound field distribution on the existing cross section. The structure of the power ultrasound intensity measurement system is shown in Figure 14.14.

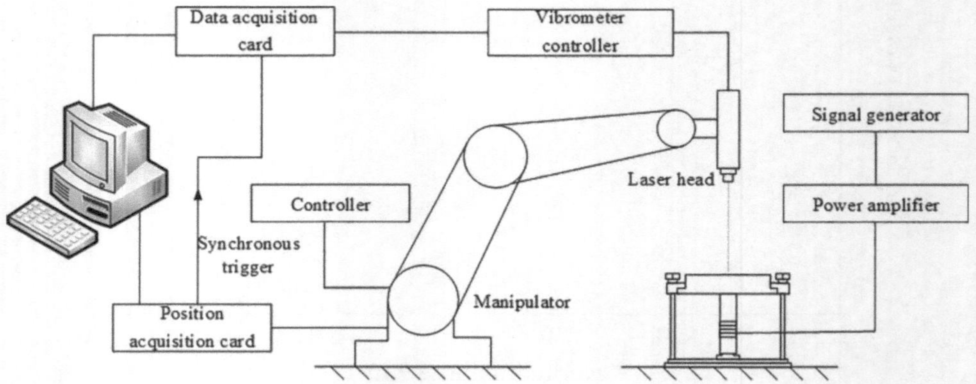

FIGURE 14.14 Structure of a power ultrasound intensity measurement system.

14.3.1 Hardware Composition

The hardware in the power ultrasound intensity measurement system includes the manipulator and its controller, laser vibrometer, industrial personal computer (IPC), data acquisition card, position acquisition card, pressure acquisition module, power ultrasound system and other mechanical support structures. As a motion control unit, the manipulator scans the surface contour according to the trajectory point packet. As a non-contact vibration sensor, the laser vibrometer picks up the vibration information of particles on the tested component surface and converts mechanical vibration signals into voltage signals. The data acquisition card is to convert the analog signals output by laser virometer into digital signals. The position acquisition card is to monitor the position of manipulator end-effector in real time and output the pulsed trigger signals to realize the acquisition synchronization with data acquisition card, that is, to ensure the accurate matching between position and data. The power ultrasound system includes a signal generator, a power amplifier and a power transducer. In the sound field fitting module, the pressure acquisition module is responsible for ensuring uniform distribution of the coupling forces between the power transducer and solid medium [30].

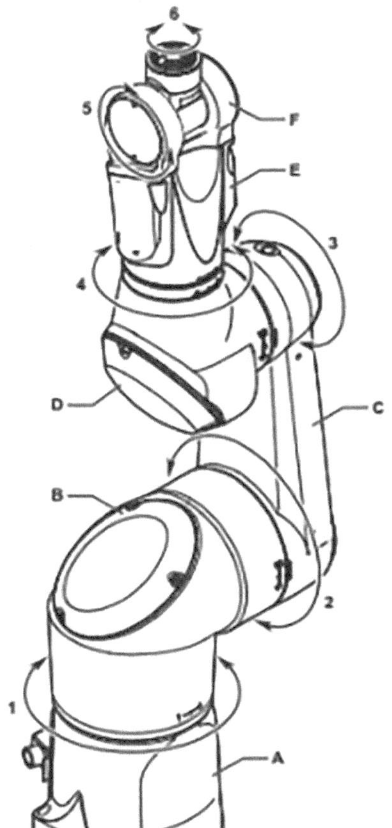

FIGURE 14.15 Manipulator model. A, base; B, shoulder; C, arm; D, elbow; E, forearm; F, wrist.

1. Manipulator and its controller

   At present, the six-DOF articulated manipulator has been widely used in various manufacturing fields, such as automobile assembly line, painting, welding, handling and loading/unloading. The measurement system studied in this paper mainly takes advantage of the flexibility of a manipulator. The manipulator holds a laser vibrometer and scans along the given trajectory in order to measure the sound field on the component surface. The advantages of six-DOF manipulator are high localization accuracy, flexible attitude, high response sensitivity and fast movement speed. Compared with a traditional scanning rack, the manipulator is very suitable for the automatic measurement of complex contours [31]. In this paper, a manipulator with six-DOFs is adopted. Its end-effector has high localization accuracy and a load capacity equal to the laser head of laser vibrometer. The model of this manipulator and the distribution of its six joints are shown in Figure 14.15.

   The manipulator controller is a CS8C controller, including communication I/O interface, motion control card, processor and network interface. It is also equipped with a teaching box that serves as the human-computer interaction window. The CS8C controller is composed of digital power amplifier, manipulator power supply, manipulator auxiliary power supply, manipulator' safety integrated board, CPU, main power switch and power module. This controller has a programmable function module, which is connected to the IPC through RS232/422 serial communication protocol and Ethernet Modbus server.

2. Laser vibrometer

   The Doppler laser vibrometer is widely used in mechanical engineering, acoustics and other engineering disciplines. Its non-contact, nondestructive, accurate and fast measurement performance has further expanded its use in the vibration measurement market. In industrial research and development, laser vibrometer is used to study the dynamic and acoustic characteristics of the objects of various sizes – from the whole vehicle body, aircraft parts, engines and buildings to MEMS systems and data hard-disk components [32].

   Polytec's single-point laser interference vibrometer, including an OFV-5000 controller and a OFV-503 laser head, is used in this measurement system. Laser vibrometer is a precision instrument for surface vibration measurement based on optical heterodyne interference. The principle of optical heterodyne interference is that the interference between two coherent beams with a very small frequency difference results in the change of interference fringes in the interference field, then the optical signal in the interference field is converted into an electrical signal by photoelectric detector and finally the phase difference of interference field is calculated of by computer [i,ii]. The laser beam shoots vertically at the vibrating object and is then reflected vertically back into the optical head. According to the Doppler frequency shift effect, the vibration of the object surface will modulate the frequency or phase of the laser, and then the vibration information including vibration velocity and vibration displacement will be demodulated through signal processing and outputted [33]. The calculation formula of vibration velocity is as follows:

$$|v| = \frac{\lambda f_D}{2} \tag{14.10}$$

where v is the vibration velocity, λ is the laser wavelength and f_D is the calculated frequency shift difference.

3. Position acquisition card

To achieve the surface sound field scanning with high accuracy and the one-to-one matching between vibration information and position information, a position acquisition card is needed to acquire and calculate the position of the sensor held by manipulator relative to the workpiece. When the position change is equal to the pre-set step size, the position acquisition card will output a pulse signal to trigger the data acquisition card and save the new position data into its memory. During the sound field scanning, the scanning points are generally required to be as dense as possible to improve the scanning accuracy, so the position acquisition card is required to output high-frequency trigger pulses.

The acquisition card of manipulator position/attitude data simultaneously collects and processes six signals from an EnData2.2 protocol encoder through FPGA and transmits the data to an IPC via a PCI interface that can meet the requirement for data transmission speed. The acquisition card has two working modes: time trigger and position trigger. The corresponding trigger step lengths can be set in the upper computer, with the accuracy and range satisfying the practical application. The highest synchronous pulse frequency of position acquisition card is 2.5 kHz, which meets the demand of high-frequency pulse output.

Data acquisition card

Data acquisition card is the core part of the measurement system. Its function is to sample the analog signal of vibration velocity outputted by laser virometer and convert it into a digital signal that can be used by IPC. The main indicators of signal acquisition include sampling frequency and maximum sampling voltage. To ensure the signal integrity, the sampling frequency is usually more than five times the input signal frequency. In this system, the maximum input signal frequency is 50 kHz, while the maximum sampling voltage is determined by the output signal of laser virometer, with the maximum peak-to-peak value of 5 V. Meanwhile, in order to guarantee enough measurement accuracy, the A/D conversion digits of data acquisition card mustn't be too few. To improve the scanning speed, the card needs to transmit data via DMA (Direct Memory Access).

The data acquisition card supports two trigger modes: external trigger and software trigger. Software trigger, namely time trigger, is generally used for the debugging signals, while external trigger is generally used for the synchronization signals. In this system, external trigger is selected, and a position signal is used to trigger the data acquisition card to achieve the accurate matching between position and data. The data acquisition card is divided into several levels within ±50 mV to ±10 V. A level can be selected according to the amplitude of the input signal. If the level is too high, the actual accuracy will be too low. If the selected level is too low, the signal will be clipped and then distorted. The sampling frequency range of data acquisition card is 10 kHz–10 MHz. After selecting the sampling frequency, the total sample size and the number of samples collected before trigger should be set. The difference between them is just the number of samples to be collected after trigger. The data acquisition card has no function of internal data processing or eigenvalue processing. After the digital signal is transmitted to IPC, it must be processed by upper computer program to obtain the required signal eigenvalues, such as peak-to-peak value and frequency.

4. Arbitrary waveform generator and power linear amplifier.

   Arbitrary waveform generators are generally divided into two types. One type is signal generator, a stand-alone device that sets the output waveform by use of buttons. The other type is board card, which sets the desired output waveform with the help of PC. The stand-alone type can be independently used in different occasions, while the board card type can be integrated into a software system and is highly programmable. When providing signal excitation to the power transducer, the signal source should send a sinusoidal signal corresponding to the natural frequency of the transducer. The function of power amplifier is to receive a small excitation signal from the signal source, process it through linear amplification and impedance matching and then send it to the power transducer for excitation use.

   The power amplifier used in this system is a broadband power amplifier with large output current and high voltage. It can drive the resistive, capacitive and inductive loads. Meanwhile, it has a perfect protection circuit that provides high reliability. Its maximum continuous power output is 1000 VA, its operating frequency range is 200 Hz–60 kHz and its maximum input signal amplitude is 1 V_{rms}. Through the selection switch on the panel, its output impedance can be matched with the impedances of different loads. In addition, the power amplifier is provided with the monitoring ports that can directly monitor the voltage/current signals output by this device.

5. Pressure sensor and acquisition module

   When studying the solid sound field of the power transducer, the experimental medium is pressed tightly on the transducer by six bolts. The pressure sensor can monitor the pre-tightening force of those bolts. The pressure sensor module includes a pressure sensor and a digital acquisition module. The performance parameters of pressure sensor are shown in Table 14.3.

TABLE 14.3 Performance Parameters of Pressure Sensor

Range	Output Sensitivity	Null Balance	Combined Error	Output Impedance
1000 N	1.100 mV/V	±1% F.S	0.2% F.S	350 Ω

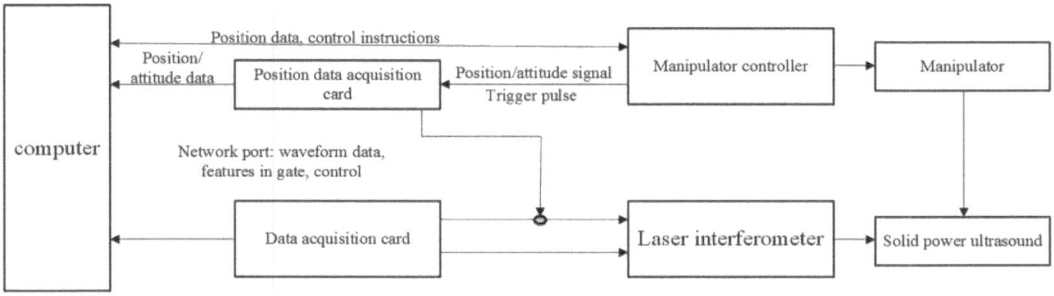

FIGURE 14.16 Data flow of a power ultrasound intensity measurement system.

14.3.2 Software Function

After building the hardware subsystem of the power ultrasound intensity measurement system, a set of software measurement system should be designed according to the data flow of the system as shown in Figure 14.16.

This measurement system mainly includes a data acquisition module, a position acquisition module, a sound field scanning module, a sound field fitting module, a pressure acquisition module and a data storage and graphic redrawing module.

1. Data acquisition module

   The main function of data acquisition module, namely A-scan module, is to collect data through the data acquisition card, transmit it to the memory of upper computer through DMA, process the collected waveform through the data processing thread, analyze the waveform eigenvalues and finally display the collected data in the form of graph. It is the combination of data acquisition, display adjustment, eigenvalue extraction, waveform display and other functions. The data acquisition module can display each triggered A-scan waveform in real time and provide an eigenvalue data interface for sound field scanning module.

   Before the switch-on of data acquisition card, the sampling parameters need to be set. The sampling parameters include sampling frequency, sampling voltage and the number of input channels, sampling depth and triggering mode. In addition, the A-scan display parameters, including excitation frequency and the number of display cycles, should be set. The voltage range of A-scan display is consistent with that of data acquisition card.

2. Position acquisition module

   The function of position acquisition module is to provide the configuration of position acquisition card and acquire the position of a laser vibrator relative to the workpiece origin. After being acquired, the number of triggers and the coordinates can be displayed on the interface in real time for verification and observation and provide a position data interface for sound field scanning module.

3. Sound-field scanning module

Sound-field scanning module is the core of the whole measurement system. It can realize the one-to-one matching between vibration information and position information and reflect the distribution of sound field with the color method. It is very important for the study of sound field.

The image display of sound field scanning is shown in Figure 14.17. The displayed information in the module also includes the projection plane, imaging range, cylindrical axis and radius and sound-field imaging distance.

In the process of sound field scanning, the functions such as vibration data acquisition, position acquisition, data calculation and graph drawing need to be performed by CPU. Therefore, in order to smooth the programmed motion and display the waveforms and graphs in real time, four sub-threads have been developed in this software system to implement the data acquisition and data processing. The main thread is only to redraw the waveforms and graphs with the aid of a timer and perform other message response functions. The data flow of the software is shown in Figure 14.18.

FIGURE 14.17 C-scan display interface.

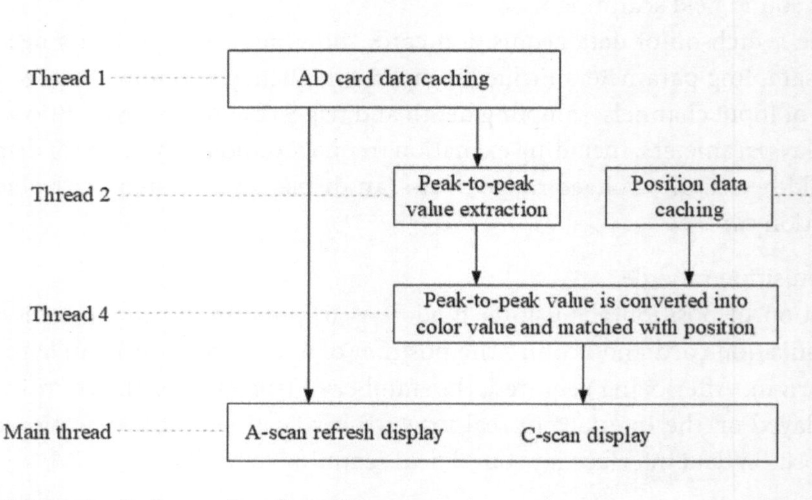

FIGURE 14.18 Software data flow diagram.

The main thread is used to draw and refresh the A-scan signal and C-scan image by means of a timer. The four sub-threads are, respectively, the vibration signal data acquisition thread, the vibration signal eigenvalue extraction thread, the position acquisition thread and the data position matching thread.

The vibration signal data acquisition thread is mainly to read the data from the acquisition card buffer into the computer memory. A 128 MB buffer is defined in the memory of the acquisition card. The data collected by each trigger is written into the buffer in sequence. According to the requirements of DMA communication mode, a block region is defined to accommodate the data acquired in several triggers. Then the Block state is continuously accessed. When the Block is full, a Block of data will be read into the computer memory via DMA. Then, the memory in the acquisition card is emptied. In other words, the acquired data is circularly stored in the buffer of the acquisition card, and then read out segment by segment. In this way, the acquisition speed can be effectively improved when the memory space of the acquisition card is limited.

The vibration signal eigenvalue extraction thread is mainly used to extract the peak-to-peak value of the data collected by each trigger and provide a complete frame of data for the A-scan signal. Since the data read each time in the acquisition thread contains the data collected in several triggers and the acquisition card is not able to extract the peak-to-peak values, each datum collected in each trigger should be traversed one by one in the upper computer thread to obtain the peak-to-peak value for each trigger. To shorten the traversal time, a single cycle can be selected for traversal. Based on overall consideration, the data processing capacity can be improved by treating the step of peak-to-peak value extraction as a separate sub-thread.

The position acquisition thread is mainly used to read the position data from the position acquisition card into the computer memory. Since each trigger corresponds to only one set of data, the reading mode of position acquisition thread is different from that of data acquisition card, that is, the position acquisition thread continuously reads the buffer in the position acquisition card. Any new data in the buffer will be immediately read into the memory. This reading strategy will not affect the normal operation of both the software and the acquisition card.

4. Pressure acquisition module

The pressure acquisition module is used to ensure the consistency of the coupling force between the power transducer and solid medium when measuring the solid sound field for many times. This module has two working modes: manual acquisition and automatic acquisition. The time interval of automatic acquisition can be set arbitrarily. The eight-channel acquisition module supports the simultaneous acquisition of six pressure sensors. After receiving the acquisition command, the acquisition module returns the pressure information to each sensor. After the data processing, the pressure value of each sensor is displayed in the interface. In addition, this module has the function of zero calibration.

14.4 PRINCIPLE OF LASER MEASUREMENT FOR SOUND FIELD INTENSITY DISTRIBUTION

14.4.1 Measurement Principle of Laser Displacement Interferometer

The biggest advantage of the vibration measurement based on optical instrument is non-contact measurement. Laser has high brightness and good monochromaticity, so it is often used as a measuring light source. However, the measurement accuracy varies with the optical measurement principle. Among various optical measurement techniques, the vibration measurement technique based on the Doppler Effect has high measurement accuracy with a bandwidth up to tens of megahertz and thus is often used in the high-precision measurement of the vibration system. Therefore, in view of the measurement requirements of this experiment, the Doppler laser vibrator is selected as the vibration measurement equipment to pick up the vibration information of the transducer end face.

The laser virometer converts the detected vibration velocity change into an electrical signal. The change of vibration velocity with time, namely the time-domain signal of vibration velocity, can be observed on the display. The purpose of our experiment is to obtain the natural frequency of the transducer, so the frequency spectrum of vibration velocity signal needs to be analyzed. Frequency spectrum analysis is to transfer a signal from the time domain to the frequency domain. Each frequency in the obtained frequency spectrum corresponds to a periodic harmonic component of the signal. The most commonly used transformation method is fast Fourier transform (FFT).

The experimental measuring instrument is shown in Figure 14.19. This experiment is to measure the vibration velocity response of the output transducer end under the pulsed excitation signal, collect the vibration velocity signal in its free attenuation process and analyze the frequency component of free transducer vibration.

The principle diagram of a laser vibration measurement system is shown in Figure 14.20. The basic principle of this measurement system is as follows. An arbitrary waveform generator is used to generate the pulse signals, from which the sinusoidal signal of any frequency in a period can be selected, as shown in Figure 14.21. After being amplified by the power amplifier, the signal drives the power transducer to vibrate with free attenuation. Then the vibration velocity of the transducer end face is measured by a laser virometer, and the

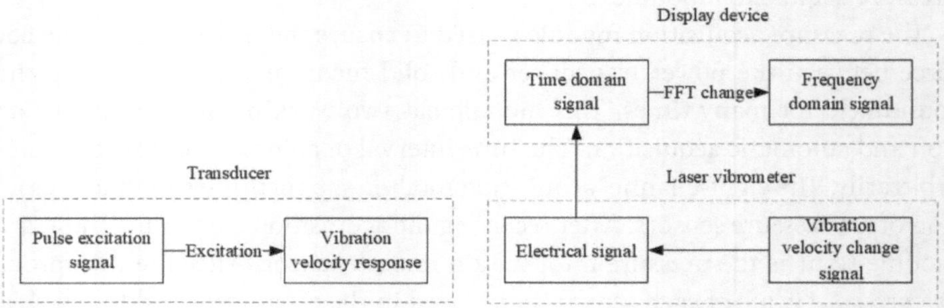

FIGURE 14.19 Experimental measuring instrument.

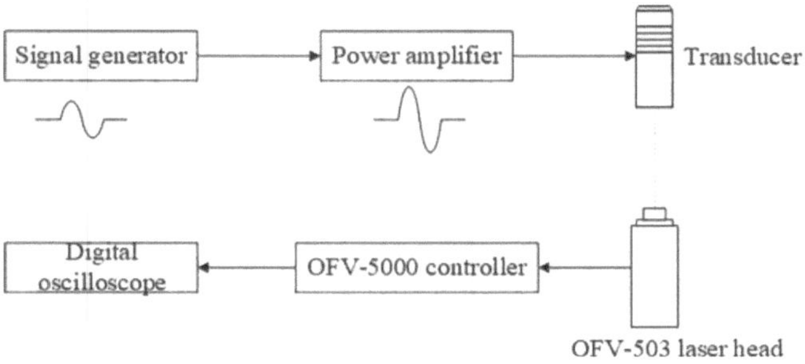

FIGURE 14.20 Principle diagram of a laser vibration measurement system.

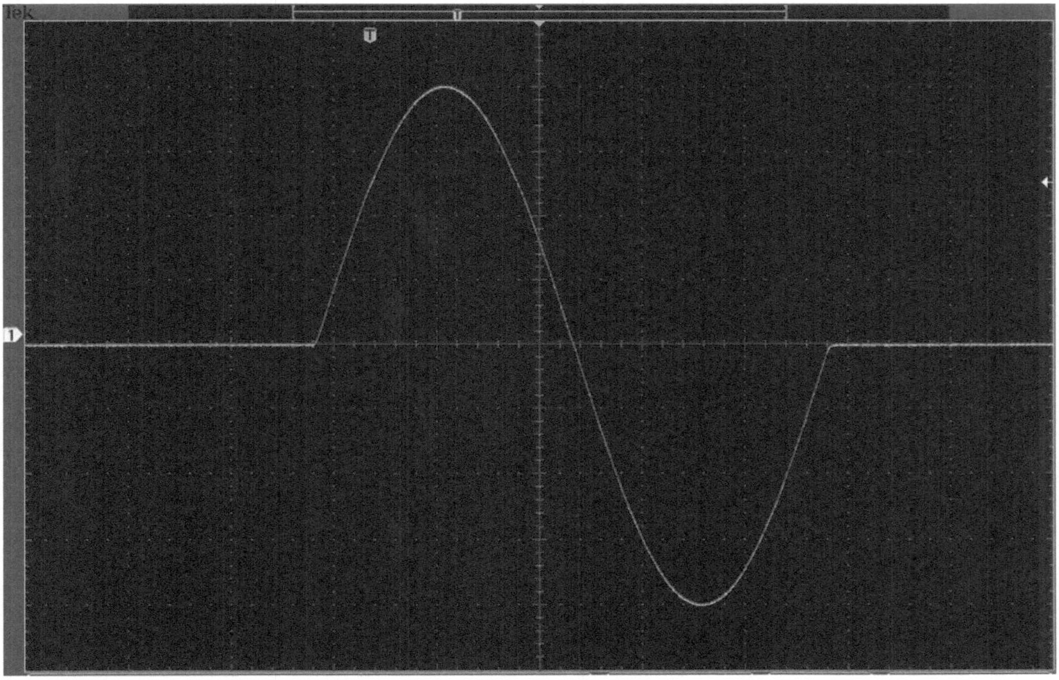

FIGURE 14.21 Sinusoidal excitation signal.

converted voltage signal is displayed on a digital oscilloscope through the cable. Finally, the frequency spectrum is analyzed to obtain the frequency of free attenuation vibration.

The light path of the laser head of the OFV-505 vibrator is shown in Figure 14.22.

In the above figure, BS1, BS2 and BS3 are polarization beam splitters. A helium-neon laser is used as light source to provide the polarized light. After passing through the beam splitter BS1, the beam is divided into two beams: one for signal light or signal carrier and the other for reference light. The signal light passes through the beam splitter BS2 and a quarter-wave plate. Then it is focused by a prism onto the surface of the vibrating object and is reflected by the surface. The reflected light passes through the quarter-wave plate again, where its polarization direction is changed. After passing through the beam splitters BS2 and BS3, the polarized light is received by a photoelectric detector. The reference light

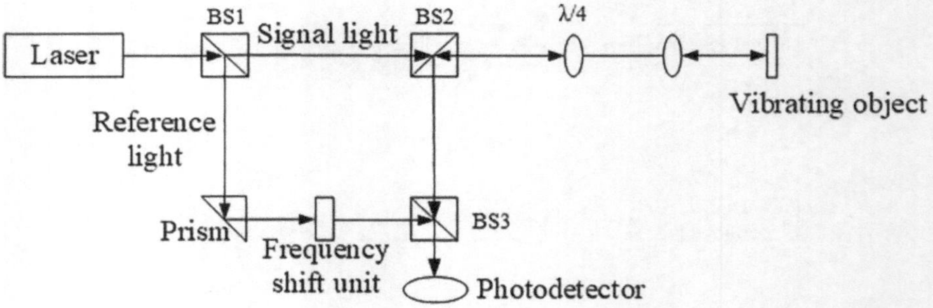

FIGURE 14.22 Light path of a laser head.

is also received by the photodetector after passing through the prism, frequency shift unit and BS3. When the two coherent beams interfere, a beat signal can be obtained. From the interference signal, the information on surface vibration can be extracted. In this system, a Bragg element is introduced to cause a slight frequency shift of the reference light. The frequency shift here is 40 MHz. It can improve the signal-to-noise ratio (SNR) and distinguish the vibration directions.

14.4.2 Measurement Principle of Normal Displacement of Sound Wave

The sound fields radiated by a power transducer in liquid and gas can be scanned by an interventional sensor. However, solids can't be intervened, so the sound field distribution inside a solid can only be derived from its surface sound field, that is, the radiated sound field of a transducer in an infinite medium can be inversely derived from that in a finite medium. Before analyzing the measurement method, sound intensity is determined as the physical quantity representing the sound field intensity.

The eigenvalues of ultrasonic field are mainly sound pressure, sound intensity and sound impedance. In this paper, sound intensity is determined as the physical quantity representing the sound field intensity. The derivation process of sound intensity is as follows [34]:

Take the propagation of a planar cosine longitudinal wave in a pencil rod as an example, as shown in Figure 14.23. When the sound wave propagates to the volume element

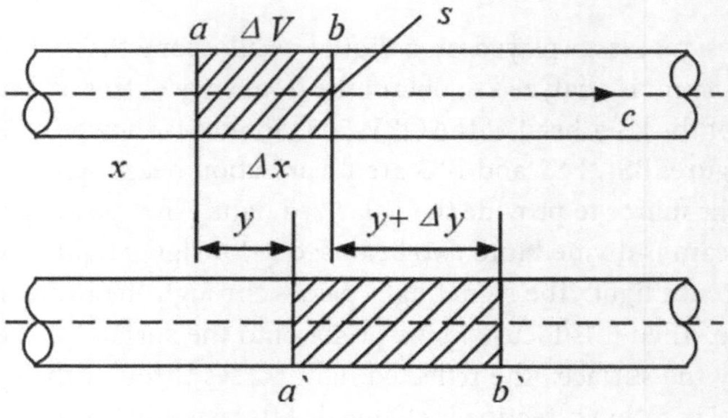

FIGURE 14.23 Propagation of longitudinal wave in a pencil rod.

$\Delta V = s\Delta x$, the vibration will cause deformation, namely the displacement y and the deformation quantity Δy.

The vibration kinetic energy W_k caused by sound wave propagation to the volume element is

$$W_k = \frac{1}{2} m v^2 \tag{14.11}$$

where

$$m = \rho \Delta V \tag{14.12}$$

$$v = \frac{dy}{dt} = -\omega A \sin\omega\left(t - \frac{x}{c}\right) \tag{14.13}$$

So

$$W_k = \frac{1}{2} \rho \Delta V A^2 \omega^2 \sin^2\omega\left(t - \frac{x}{c}\right) \tag{14.14}$$

The elastic-deformation potential energy W_p caused by sound wave propagation to the volume element is

$$W_p = \frac{1}{2} K (\Delta y)^2 \tag{14.15}$$

According to Hooke's law: $F = \frac{E_s}{\Delta x} \Delta y$

$$K = \frac{E_s}{\Delta x} \tag{14.16}$$

According to the wave equation: $y = A\cos\omega\left(t - \frac{x}{c}\right)$

$$\frac{\partial y}{\partial x} = A \frac{\omega}{c} \sin\omega\left(t - \frac{x}{c}\right) \tag{14.17}$$

The wave velocity in the rod is $c = \sqrt{E/\rho}$, that is,

$$E = c^2 \rho \tag{14.18}$$

so

$$W_p = \frac{1}{2} K (\Delta y)^2 = \frac{1}{2} \rho \Delta V A^2 \omega^2 \sin^2\omega\left(t - \frac{x}{c}\right) \tag{14.19}$$

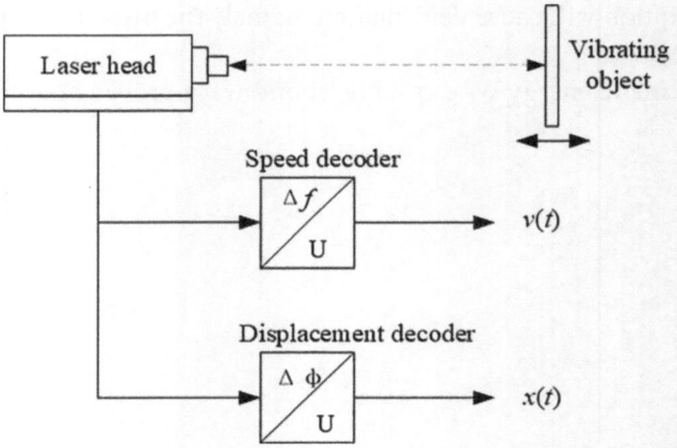

FIGURE 14.24 Schematic diagram of vibration measurement by a laser vibrometer.

Therefore, the total energy of the volume element is

$$W = W_k + W_p = \rho \Delta V A^2 \omega^2 \sin^2 \omega \left(t - \frac{x}{c} \right) \tag{14.20}$$

Then the average energy of the volume element is

$$W = \frac{1}{2} \rho \Delta V A^2 \omega^2 \tag{14.21}$$

Its average sound intensity is

$$I = \frac{W}{s\Delta t} = \frac{1}{2} \rho c v^2 \tag{14.22}$$

According to the calculation formula of sound intensity, the sound intensity I depends not only on the density ρ and sound velocity c of the medium material but also on the vibration velocity v at that point. After determining the type of medium material, only the physical quantity v needs to be measured. Therefore, a velocity sensor is needed. To achieve high sound field accuracy, the position resolution and measurement accuracy of the sensor are required to be high, so laser vibrometer is the best choice. The schematic diagram of vibration measurement is shown in Figure 14.24.

14.5 MEASUREMENT METHOD FOR TRANSVERSE WAVE AND LONGITUDINAL WAVE BY A DUAL-LASER VIBROMETER

In the laser vibration measurement, the laser beam is focused on the measured structure, whose moment of force causes the Doppler Effect in laser reflection. If the object can reflect the beam correctly, its velocity and displacement can be calculated. Because of the high frequency of the laser, direct demodulation is not possible. Instead, the scattered beam is coherently mixed with the reference beam through an interferometer. The schematic diagram of laser interferometry is shown in Figure 14.22. The laser source emits a spatially and temporally coherent beam (in which all the photons have the same frequency, direction and phase), which is divided into a reference beam and an objective beam. The scattered beam and the reference beam are recombined and received by a photodetector that measures the intensity of the mixed light. The intensity varies with the phase difference $\Delta\Phi$ between the two beams according to the following formula [35,36]:

$$I(\Delta\phi)=\frac{I_{\max}}{2}(1+\cos\Delta f) \tag{14.23}$$

The phase difference $\Delta\Phi$ is a function of the path difference ΔL between the two beams and depends on:

$$\Delta\phi=2\pi\cdot\frac{\Delta L}{\lambda} \tag{14.24}$$

λ is laser wavelength. The optical path difference (OPD) becomes a function of time $(\Delta L=\Delta L(t))$ when the object is moving at a constant speed v. The interference fringe patterns move on the detector. The object displacement can be determined by counting the passing fringe patterns. The detector intensity shows a sinusoidal variation. The frequency generated by velocity, called the Doppler shift f_D, is a function of the velocity component in the direction of the object beam.

$$f_D=2\cdot\frac{|V|}{\lambda} \tag{14.25}$$

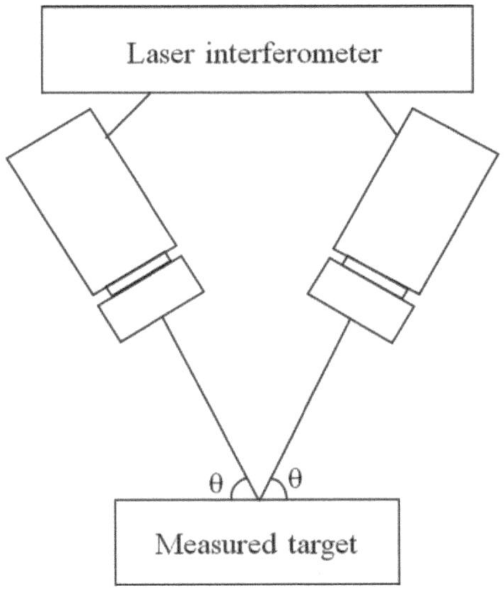

FIGURE 14.25 Schematic diagram of vibration measurement of transverse and longitudinal waves.

The vibration measurement in and out of the plane using a laser interferometer is shown in Eq. (14.24). If the laser beams are perpendicular to the object surface, the out-of-plane vibration can be derived from the above theory. To measure the in-plane vibration, the two laser beams are arranged at an angle to the object surface. The in-plane displacement U_x can be expressed as $U_x = 2U\cos\theta, U_y = U\sin\theta$, where U is the displacement measured by one of the beams. If θ is set as 60° (see the Figure 14.25), then the in-plane displacement will be $U_x = U$.

14.6 APPLICATION OF A SOUND FIELD INTENSITY MEASUREMENT METHOD

The sandwiched transducer can take advantage of the vibration mode of piezoelectric ceramics along the direction of thickness to realize its axial vibration. In this transducer, mechanical energy is transferred to the contact medium through the end face. Ideally, piezoelectric ceramics is in uniform vibration. However, because the transducer is pressed onto the medium by the pre-tightening bolts, the radial vibration amplitudes and phases of the particles on the end face will be different and the application of the power transducer system will be affected. This is similar to the case of bending vibration. For example, as the transducer bends and vibrates in ultrasonic welding, the amplitude transformer will also bend and vibrate to cause the degradation of welding quality. Therefore, it is necessary to explore the sound field distribution on the end face of a sandwiched transducer.

Because a sandwiched transducer is a vibration system, its vibration mode is related to excitation frequency. The function of the sandwiched transducer is to transfer energy by means of longitudinal vibration, so the frequency corresponding to longitudinal vibration shall be determined. In this paper, by analyzing the freely attenuated vibration of two transducers, we obtained their first-order and second-order natural frequencies, as shown in Figure 14.26. It is empirically proved that the voltage and frequency of pulse excitation

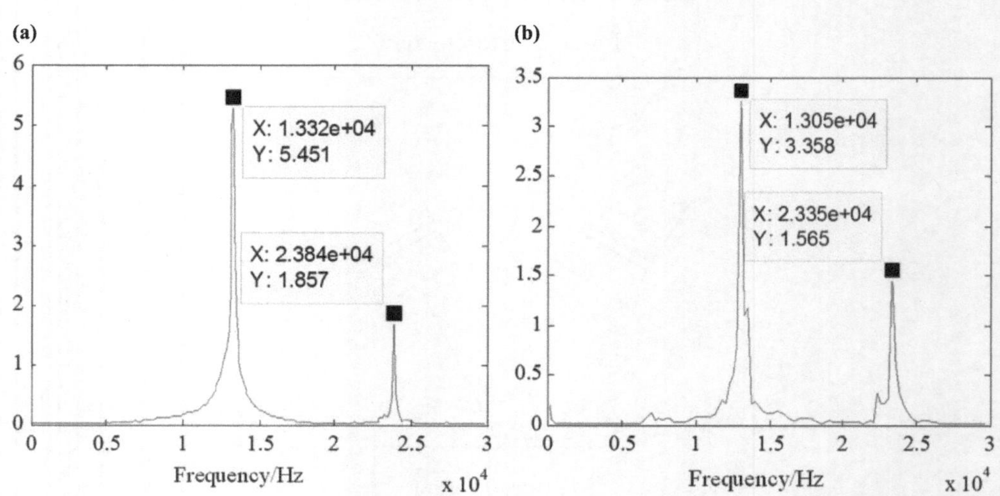

FIGURE 14.26 Analysis of natural frequencies of transducers. (a) Natural frequency of 1# transducer and (b) natural frequency of 2# transducer.

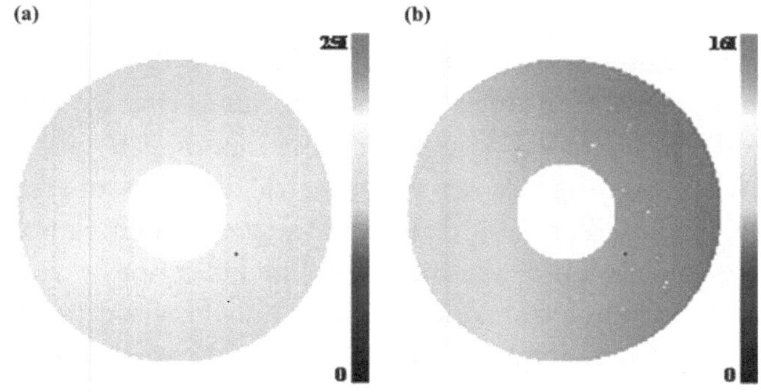

FIGURE 14.27 Radiation intensity distribution on the end face of 1# transducer. (a) 13.32 kHz radiation intensity and (b) 23.84 kHz radiation intensity.

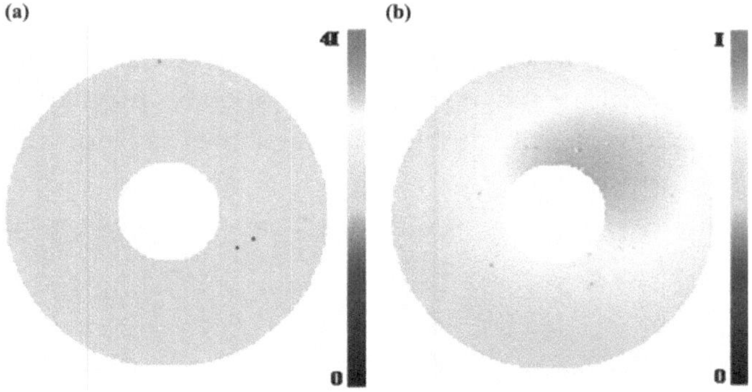

FIGURE 14.28 Radiation intensity distribution on the end face of 2# transducer. (a) 13.05 kHz radiation intensity and (b) 23.35 kHz radiation intensity.

have no influence on the analysis results of natural frequencies. Then, continuous waves were applied to excite the two transducers one by one and the results of radiation intensity distribution were obtained, as shown in Figures 14.27 and 14.28. Because the pre-tightening bolt runs through the whole transducer, the working area on the end face is annular.

In the figure, I is one unit of sound intensity, $I = 85{,}050\ \mathrm{W/m^2}$. Its corresponding vibration velocity in aluminum medium is 0.1 m/s. It can be clearly seen from the comparison of first-order and second-order radiation intensity distributions that the second-order radiation intensity on the transducer end face is reduced to about 1/4 of the first-order radiation intensity. According to the calculation formula of sound intensity, the vibration velocity of particles on end face is reduced to about 1/2. The first-order frequency component is much higher than the second-order frequency, that is, its energy share is bigger. This also proves that a sandwiched transducer usually operates only at the first-order natural frequency.

As observed in Figures 14.27a and 14.28a, the first-order radiation intensity distribution is relatively uniform, indicating that the vibration amplitudes of first-order particles on end face are basically the same. This also means that basically no other vibration modes than longitudinal vibration exist or other vibration modes account for a very small proportion. The second-order sound field distribution, as shown in Figures 14.27b and 14.28b, is not uniform, especially for 2# transducer. We can first judge that the vibration amplitudes are not constant, and that the radiation intensity distribution is not concentric. This indicates that the structural coupling between piezoelectric ceramics and front and rear cover plates and between front and rear cover plates and the pre-tightening bolts is not uniform. As a result, the overall vibration of the transducer includes not only axial vibration but also bending vibration. Ideally, a transducer vibrates only longitudinally, that is, it doesn't vibrate radially. In this case, its vibration does not include the small lateral vibration caused by material deformation. However, when the bending vibration exists, the vibration in the normal direction of the transducer cylinder will be produced and superposed to longitudinal vibration. Therefore, the radiation intensities were scanned for the cylinders of the two transducers. The scanning area was the 90° side area corresponding to front cover plate. The scanning results are shown in Figures 14.29 and 14.30.

Compared with the radiation intensity of end face, the radiation intensity of side face is very small, indicating that the radial vibration amplitude is relatively small. The results of first-order side radiation intensity distribution are shown in Figures 14.29a and 14.30a.

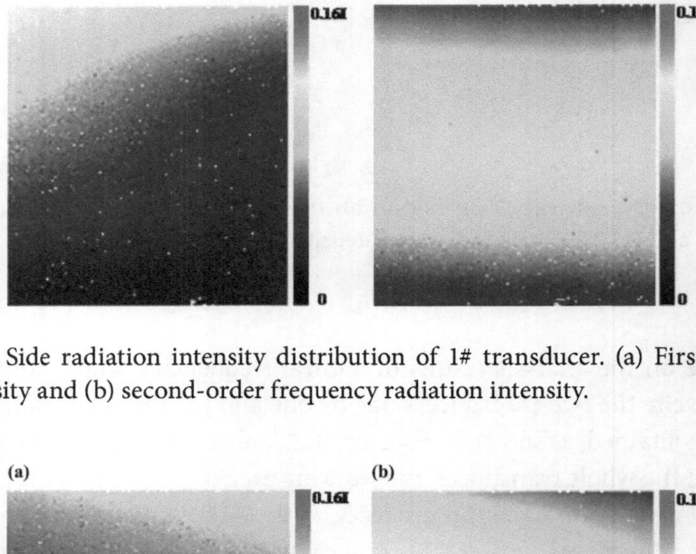

FIGURE 14.29 Side radiation intensity distribution of 1# transducer. (a) First-order frequency radiation intensity and (b) second-order frequency radiation intensity.

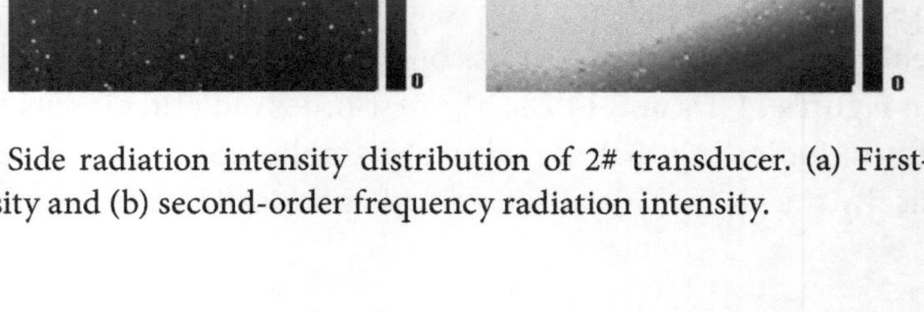

FIGURE 14.30 Side radiation intensity distribution of 2# transducer. (a) First-order frequency radiation intensity and (b) second-order frequency radiation intensity.

Although the circumferential distribution is not uniform, its value is very low. Thus, the influence of bending vibration on the radiation intensity distribution on end face can be ignored. This is consistent with the measurement results in Figures 14.29a and 14.30a.

REFERENCES

1. Li J. *Influence of Middle Bolt and Prestress on the Performance Parameters of Piezoelectric Transducer [D]*. Xian: Shaanxi Normal University, 2008.
2. Chen S., Recent developments and application progress of power ultrasound technology in China [J], *Technical Acoustics*, 2002, 21(1):46–49.
3. Wang Y., Research status and application progress of power ultrasound technology [J], *Mechanical Research & Application*, 2006, 19(4):41–43.
4. Zhou Z. *Research on the Automatic Robotic Technology of Sound Field Characteristics of Ultrasonic Transducer [D]*. Zhejiang University, 2016.
5. Shi Y. *Design and Implementation of Performance Testing System of Ultrasonic Transducer [D]*. Dalian University of Technology, 2008.
6. Al-Bataineh O.M., Meyer R.J., Newnharn R.E., et al. Utilization of the high-frequency piezoelectric ceramic hollow spheres for exposimetry and tissue ablation [C]//2002 IEEE Ultrasonics Symposium, 2002. Proceedings. IEEE, 2002, 2:1473–1476.
7. Hodnett M., Zeqiri B., A strategy for the development and standardisation of measurement methods for high power/cavitating ultrasonic fields: review of high power field measurement techniques [J], *Ultrasonics Sonochemistry*, 1997, 4(4):273.
8. He Z., He Y., Shang D., Error analysis and calibration for underwater sound intensity measuring system [J], *Chinese Journal of Acoustics*, 2000, 19(3):193–206.
9. Romdhane M., Gourdon C., Casamatta G., Development of a thermoelectric sensor for ultrasonic intensity measurement [J], *Ultrasonics*, 1995, 33(2):139–146.
10. Cook B.D., Werchan R.E., Mapping ultrasonic fields with cholesteric liquid crystals [J], *Ultrasonics*, 1971, 9(2):88–94.
11. Jin S., An Z., Lian G-x., et al., Photoelastic measurement of sound field of ultrasonic transducer [J], *Applied Acoustics*, 2014, 33(02):107–111.
12. Wang X. Research on the dynamic photoelastic imaging of ultrasonic field in solids [A]. Acoustical Society of China. Proceedings of the 2014 National Acoustical Academic Conference of the Acoustical Society of China [C]. Acoustical Society of China: Editorial Department of Technical Acoustics, 2014:4.
13. Yu Q., Wang Y., Cao W., et al., Research on the measurement method of sound intensity and sound power of focusing transducer [J], *China Measurement & Test*, 2017, 43(01):27–32.
14. Mao L.C., Absolute measurement of ultrasonic field in solids [J], *Journal of Acoustics: Chinese Edition*, 1964(2):45–46. (in Chinese).
15. Wu J. *Research on the Calculation and Measurement Method of Solid Sound Field of Ultrasonic Transducer [D]*. Beijing Institute of Technology, 2009:61–66.
16. Selbach A., Laser-interferormeter zur Posidions-undSchwignungsmes sung. F&M, 1988, 96(l/2), S. 33–36.
17. Pang X. *Research on the Bending-Bending-Composite Piezoelectric Motor Driven by Separated Sinusoidal/Resonant Synthesized Square Wave [D]*. Hefei University of Technology, 2018:40–55.
18. Cheng S. *Research on the Key Technology of Ultrasonic Cleaner [D]*. Northeastern University, 2006:20–45.
19. Duan Y. *Performance Analysis and Simulation Software Development of Piezoelectric Composites of Types 1-3 [D]*. Xidian University, 2014:50–65.
20. Kong F. *Research on the Measurement Technology of Electrical Characteristics of Ultrasonic Transducers [D]*. Beijing Institute of Technology, 2011:50–70.

21. Gonnard P., Schmitt P.M., Brissaud M., New equivalent lumped electrical circuit for piezoelectric transformers [J], *Transactions of Ultrasonics, Ferroelectrics, and Frequency Control*, 2006, 53(4):802–809.
22. Ohki M., Toda K., Comparison between Mason's equivalent circuit and complex series dynamics from energetic point of view [J], *National Defense Academy, Yokosuka*, 2002:45–48.
23. Cheng W. *Test and Research of Valveless Piezoelectric Pump at the Bottom of Asymmetric Slope Cavity [D].* Nanjing University of Aeronautics and Astronautics, 2010.
24. Zhang Z. *Design and Modal Frequency Optimization of Lead-welding High-frequency Ultrasonic Transducer [D].* China Jiliang University, 2013.
25. Feng D., Zhao F., Xu Z., Modal analysis of ultrasonic amplitude transformer using inventor software [J], *Applied Acoustics*, 2010, 29(01):69–73.
26. Zhao X. *Research on the Nondestructive Evaluation of Austenitic Stainless Steel Welds based on Ultrasonic Model [D].* Harbin Institute of Technology, 2008:35–40.
27. Wen J.J., Breazeal M.A. A diffraction beam field expressed as the superposition of Gaussian beams.
28. Ding H. *Computational Ultrasonics [M].* Science Press, 2010:50–66.
29. Zhao X., Qi Q., Guo C., Implementation method of COMSOL and MATLAB® joint simulation [J], *China New Technologies and Products*, 2014(24):17–19.
30. Xiao Z. *Robotic Ultrasonic Testing Technology [D].* Beijing Institute of Technology, 2017:55–80.
31. Zongxing L. *Research on Key Problems in the Dual-manipulator Scanning Technology of Ultrasonic Nondestructive Testing [D].* Beijing Institute of Technology, 2016:30–56.
32. Chen V.C., Micro-Doppler effect of micro-motion dynamics: a review [J]. *Proceedings of SPIE*, 2003(5102):240–249.
33. Wang S. *Research on the Simulation and Measurement Method of Ultrasonic Surface Wave Sound Field [D].* Beijing Institute of Technology, 2014:20–51.
34. Zheng H., Lin S. *Ultrasonic Testing [M].* China Human Resources & Social Security Publishing Group Co., Ltd., 2008:29–35.
35. Yu X., Manogharan P., Fan Z., et al., Shear horizontal feature guided ultrasonic waves in plate structures with 90° transverse bends [J], *Ultrasonics*, 2015(65):370–379.
36. Fan Z., Applications of guided wave propagation on waveguides with irregular cross-section (Ph.D. Thesis), Imperial College London, 2010.

CHAPTER 15

Typical Applications of Single-Manipulator NDT Technique

CONSIDERING THE CHARACTERISTICS OF complex curved metal components, their ultrasonic reflection testing can be realized by using a water spray-coupled focusing transducer or a water immersion-coupled ultrasonic transducer. At first, the position/attitude of a sound beam in space is automatically controlled by using the flexible multi-dimensional motion of a multi-joint manipulator. Then the internal defects are automatically detected by controlling the vertically incident ultrasonic beam to scan the component along its contour. Finally, the detection results are displayed by using a special data processing algorithm.

Based on the characteristics of six-degree of freedom (DOF) articulated manipulator (including flexible motion in three-dimensional space, high localization accuracy and fast speed), the position/attitude of an ultrasonic beam in space can be precisely controlled in order to adapt to the automatic ultrasonic testing (UT) of complex curved metal components to the maximum extent. The established reflection-type robotic ultrasonic automatic nondestructive testing (NDT) system has two working modes. One mode is the establishment of a turntable-manipulator system to test complex curved rotary components. In this mode, the manipulator clamping an ultrasonic transducer moves along the component generatrix, and the component under test rotates along with the turntable. This can further expand the motion space and DOFs of beam position/attitude and can provide fast high-precision full-coverage scanning for both small complex curved components and curved rotary components (mainly including turbine disks, vehicle hubs and shafts). The other mode is the establishment of a manipulator-transducer system where the manipulator clamping a workpiece rotates in space around a fixed transducer in order to rapidly scan and image the workpiece. This method is mainly to achieve the rapid automatic UT of small light-weight complex curved components such as aero-engine blades.

DOI: 10.1201/9781003212232-15

15.1 CONFIGURATION OF A SINGLE-MANIPULATOR NDT SYSTEM

To meet the requirements of automatic testing of curved components (especially curved rotary components), a robotic UT system is proposed in this section. The component under test is placed on a turntable and fixed with a three-jaw chuck. The manipulator clamping an ultrasonic transducer tracks and scans the component contour. For a curved rotary component, when the manipulator moves along the component generatrix, the turntable will rotate to realize spiral scanning and further improve the detection efficiency. The robotic UT system with an additional turntable further expands the scope of detection objects to cover not only complex curved components (such as boxes) but also rotary components (such as turbine disks, vehicle hubs and shafts).

To meet the requirements of automatic detection of small complex curved components (such as aero-engine blades), a robotic NDT system is proposed in this section. In this system, the transducer is fixed, and the manipulator clamping the component under test moves relative to the transducer for surface contour scanning. For small complex curved components, a pneumatic quick-change fixture can be selected for the manipulator to automatically replace the components so as to effectively improve the detection efficiency.

The structure of a robotic ultrasonic NDT system is shown in Figure 15.1. The hardware in this system mainly includes a scanning motion module, a UT module, a computer control module and other mechanical support structures. The scanning motion module is the core unit of the whole system and plays a crucial role in testing the curved components. It is a high-precision six-DOF manipulator composed of an articulated manipulator, a manipulator controller, a special water immersion tank and a quick-change fixture module. The manipulator takes advantage of its flexibility to track the trajectory of a curved workpiece. The ultrasonic detection module is mainly responsible for transceiving, acquiring and processing ultrasonic signals. The computer control module includes an industrial personal computer (IPC) integrating the ultrasonic acquisition card, manipulator position/attitude acquisition card and other hardware, as well as a software system special for the equipment.

Complex rotary components are widely used in aviation, aerospace, weapons, ships, vehicles and other manufacturing fields. Typical rotary components include bearings, hubs, turbine disks, warheads, projectile bodies, transmission shafts and pipes. To ensure certain strength, toughness and reliability, those components are generally made of metal material

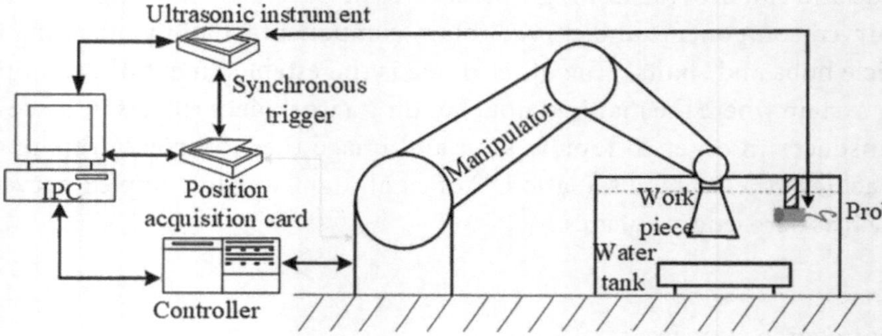

FIGURE 15.1 Structure of a robotic ultrasonic NDT system.

and formed by casting and forging. Take casting as an example. The casting process involves smelting, pouring, curing and other steps. Various process parameters such as pressure, temperature, geometry and chemical composition will affect the product quality. Therefore, it is necessary to check the component quality before the components leave the factory. In addition, most of the rotary components need to bear alternating loads during service. Take wheel hub as an example. The wheel hub in service needs to bear the dynamic load, the impact load caused by road unevenness and the side load caused by cornering. Long-term load will induce crack or even fracture inside the material. Therefore, it is also very necessary to periodically check the parts in service. We specifically studied the fracture of the internal components of wind turbines under long-term cyclic load and random load. The fine cracks in the materials show different growth trends under different loads and finally lead to the material fracture. So far, the ultrasonic NDT system, mainly in the form of frame-type scanning rack driven by a rotating shaft, has been applied to rotary components. The tested components are generally solid shafts or pipes with little diameter change, rather than rotary components with complex shape and structure (such as hubs and turbine disks), for which the combination of manipulator and extension shaft is more suitable.

15.2 AN APPLICATION EXAMPLE OF ROBOTIC NDT TO ROTARY COMPONENTS

15.2.1 Structure of Clamping Device

The turntable and three-jaw chuck in the robotic UT system are used for centering and clamping the rotary components whose diameters are within a certain range. In order to prevent the manipulator from moving to the bottom of a component and interfering with the turntable, the jaws are raised to a certain height to reserve a scanning space for the manipulator. The structural principle of three-jaw chuck is shown in Figure 15.2. The three jaws A, B and C are, respectively, located on the sliding blocks of three ball screw pairs. One end of each lead screw is engaged with the central bevel gear, and the other end of the screw is engaged with one of the three bevel gears A, B and C on the circumference. Thus, it can be seen that when one of the three circumferential bevel gears A, B and C is rotated, the three jaws can move synchronously in the radial direction.

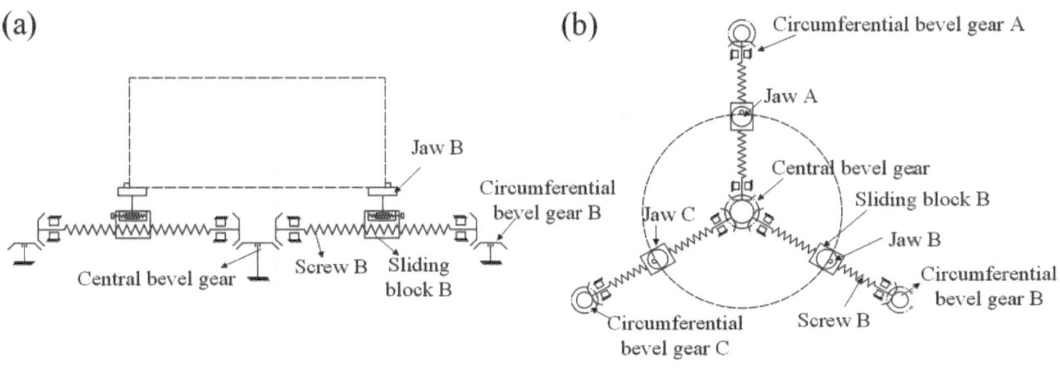

FIGURE 15.2 Structural principle of a three-jaw chuck. (a) Front view of the structure principle of a three-jaw chuck and (b) top view of the structure principle of a three-jaw chuck.

15.2.2 Correction of Perpendicularity and Eccentricity of Principal Axis

According to the Snell theorem of ultrasonic wave conversion at the interface of two different media, the sound beam is very sensitive to the change of incident angle at the interface. In the process of system installation, the positioning error of components will be caused by base tilting, assembly error, clamping deviation and other reasons, so as to affect the defect judgment. In practice, it is found that the errors of perpendicularity and eccentricity of principal axis are the two main causes for the positioning error of a rotary component. The main purpose of error correction is just to ensure that the component axis coincides with the turntable axis. The deflection of principal axis in the vertical direction can be divided into static deflection and dynamic deflection, which should be considered separately. Static deflection is the deflection of the turntable as a whole caused by the base tilting. Dynamic deflection is the non-collinearity of transmission shaft, reducer shaft and motor shaft caused by the assembly error of turntable.

As shown in Figure 15.3, a regular disk is fixed onto the turntable to measure and correct the perpendicularity error of principal axis. When the turntable is static, the manipulator is moved in X and Y directions for the distances l_x and l_y, respectively, in the basic coordinate system {0}. At the same time, the distance changes between the transducer and the disk in X and Y directions, denoted as Δl_x and Δl_y, respectively, are measured in accordance with the ultrasonic ranging principle. Then the deflection angles of the turntable around the Y-axis and the X-axis will be $\theta_y = \arctan\left(\Delta l_x / l_y\right)$ and $\theta_x = \arctan\left(\Delta l_y / l_y\right)$, respectively. The static perpendicularity error of principal axis can be compensated for by modifying the attitude angle of the workpiece coordinate system in the manipulator controller. When the turntable rotates, the manipulator is fixed at a point above the disk and then the sound interval in the vertical direction between the ultrasonic transducer and the disk is measured. Thus, the distance change in the vertical direction, denoted as Δl_z, is calculated to reflect the dynamic deviation caused by the assembly error of turntable, which generally needs to be corrected through reassembly.

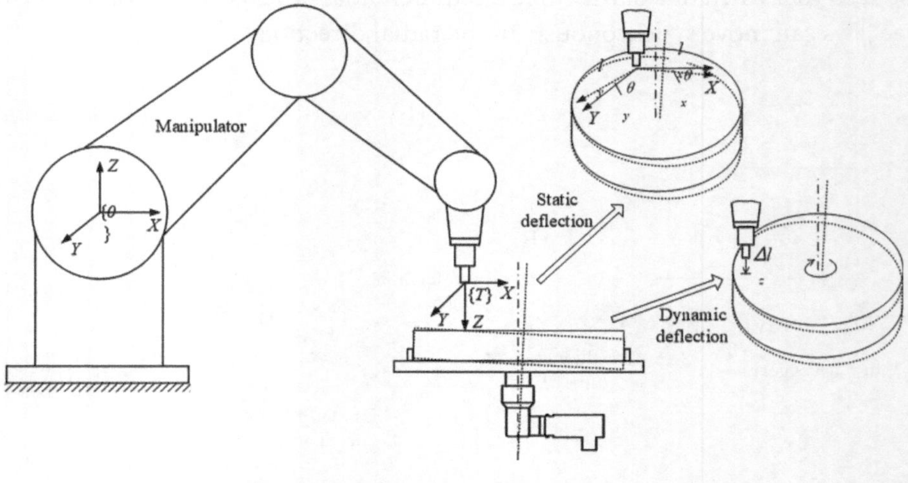

FIGURE 15.3 Measurement of perpendicularity error of a turntable axis.

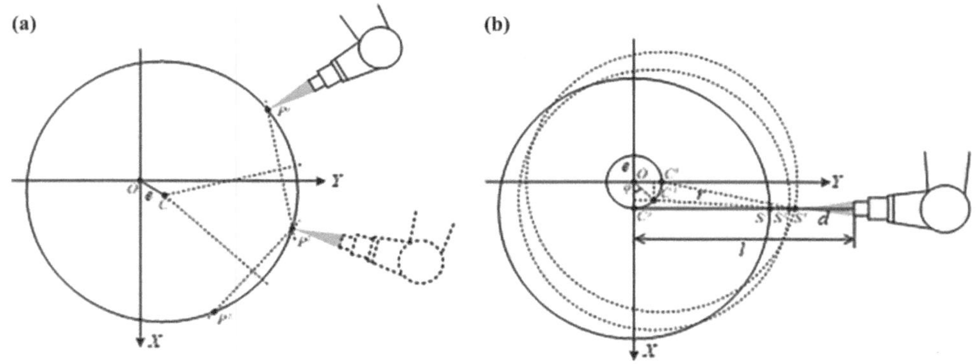

FIGURE 15.4 Measurement of eccentricity error of a turntable axis. (a) Determined the coordinate of the disk axis and (b) determine the position coordinates of the spindle O by rotating the turntable.

Compared with the perpendicularity error of principal axis, the measurement of eccentricity error is more complex. The top sectional view of the disk is shown in Figure 15.4. It is assumed that there is an eccentricity error e between the disk axis C fixed onto a three-jaw chuck and the turntable axis O. Then the coordinates of the disk axis, denoted as $\left(P_x^c, P_y^c\right)$, are determined, as shown in Figure 15.4a. When the disk is fixed, the manipulator is moved in order to measure the coordinates of three points P^1, P^2 and P^3 on the section circumference with respect to the manipulator base coordinate system, denoted as $\left(P_x^1, P_y^1\right)$, $\left(P_x^2, P_y^2\right)$ and $\left(P_x^3, P_y^3\right)$, respectively, in accordance with the sound-interval ranging principle. Then the following equations can be established assuming that the distances from the disc axis to the three points P^1, P^2 and P^3 are equal:

$$\begin{cases} \left(P_x^1 - P_x^c\right)^2 + \left(P_y^1 - P_y^c\right)^2 = \left(P_x^2 - P_x^2\right)^2 + \left(P_y^2 - P_y^c\right)^2 \\ \left(P_x^1 - P_x^c\right)^2 + \left(P_y^1 - P_y^c\right)^2 = \left(P_x^3 - P_x^2\right)^2 + \left(P_y^3 - P_y^c\right)^2 \end{cases} \tag{15.1}$$

The two equations in Eq. (14.1) can be separately expanded and rearranged as follows:

$$\begin{cases} \left(P_x^1\right)^2 - \left(P_x^2\right)^2 + \left(P_y^1\right)^2 - \left(P_y^2\right)^2 = 2\left(P_x^1 - P_x^2\right)P_x^c + 2\left(P_y^1 - P_y^2\right)P_y^c \\ \left(P_x^1\right)^2 - \left(P_x^3\right)^2 + \left(P_y^1\right)^2 - \left(P_y^3\right)^2 = 2\left(P_x^1 - P_x^3\right)P_x^c + 2\left(P_y^1 - P_y^3\right)P_y^c \end{cases} \tag{15.2}$$

The term P_x^c in Eq. (14.2) is transposed:

$$\begin{cases} P_x^c = \dfrac{\left(P_x^1\right)^2 - \left(P_x^2\right)^2 + \left(P_y^1\right)^2 - \left(P_y^2\right)^2}{2\left(P_x^1 - P_x^2\right)} - \dfrac{\left(P_y^1 - P_y^2\right)}{\left(P_x^1 - P_x^2\right)} P_y^c \\ P_x^c = \dfrac{\left(P_x^1\right)^2 - \left(P_x^3\right)^2 + \left(P_y^1\right)^2 - \left(P_y^3\right)^2}{2\left(P_x^1 - P_x^3\right)} - \dfrac{\left(P_y^1 - P_y^3\right)}{\left(P_x^1 - P_x^3\right)} P_y^c \end{cases} \tag{15.3}$$

An equation is established on the basis of Eq. (15.3) to calculate P_y^c, which is then substituted into Eq. (15.3) to calculate P_x^c. The obtained $\left(P_x^c, P_y^c\right)$ are just the coordinates of the disk axis.

Based on the determined disk axis *C*, the position coordinates of turntable axis *O* are deduced by rotating the turntable. In the presence of eccentricity error, the law of turntable motion is similar to the law of motion of eccentric circular CAM mechanism, as shown in Figure 15.4b. At first, the manipulator position/attitude is adjusted so that the sound beam axis is incident on the circumferential surface in the normal direction and then passes through the disk axis point. At this time, the intersection point between the sound beam and the circumferential surface is S^1, the disk axis is at the point C^1 and the rotation angle is defined as $\Phi = 0°$. Then, the manipulator position/attitude is fixed, but the turntable is rotated. A circle with the disk axis *O* as center and the eccentricity *e* as radius will be formed. The sound interval for the ultrasonic transducer to receive the reflection echo will change as the turntable rotates. The change of sound interval reflects the change of the distance *d* between the emitting surface and circumferential surface of the transducer. The distance between the emitting surface of the transducer and the *X*-axis is *l*, and the disc diameter is *r*. The counterclockwise rotation direction of the turntable is taken as positive direction. At $\Phi = 45°$, the intersection point between the sound beam and the circumferential surface is S^2, and the disk axis is at the point C^2. At $\Phi = 90°$, the intersection point between the sound beam and the circumferential surface is S^3, and the disk axis is at the point C^3. Take $\Phi = 45°$ as an example. Perpendiculars are drawn from the point C^2 to the *X*-axis and *Y*-axis respectively. According to the geometric relationship, the motion equation of the distance *d* can be obtained:

$$d = l - e\sin\varphi - \sqrt{R^2 - (e - e\cos\varphi)} \tag{15.4}$$

When the eccentricity error *e* is 0, 5 and 10 mm, the turntable is rotated to measure the sound interval when the transducer receives the reflection echo. Then the distance *d* is deduced from the sound velocity in water. The relationship between *d* and φ is expressed by polar coordinates, as shown in Figure 15.5. It can be seen from the figure that, in a rotation period, even if the eccentricity errors are different, the maximum distance deviation can always be found at two points. The rotation angles at the two points are $\Phi_1 = 90°$ and $\Phi_2 = 270°$, respectively, and the corresponding distance deviations are d_1 and d_2, respectively. Then $\Phi_1 = 90°$ and $\Phi_2 = 270°$ are substituted into Eq. (15.4) to obtain:

$$\begin{aligned} d_1 &= l - e - \sqrt{R^2 - e^2} \\ d_2 &= l + e - \sqrt{R^2 - e^2} \end{aligned} \tag{15.5}$$

It can be seen from Eq. (15.5) that half of the difference between the maximum and minimum distance deviations measured by sound interval during the turntable rotation is just the eccentricity error *e*.

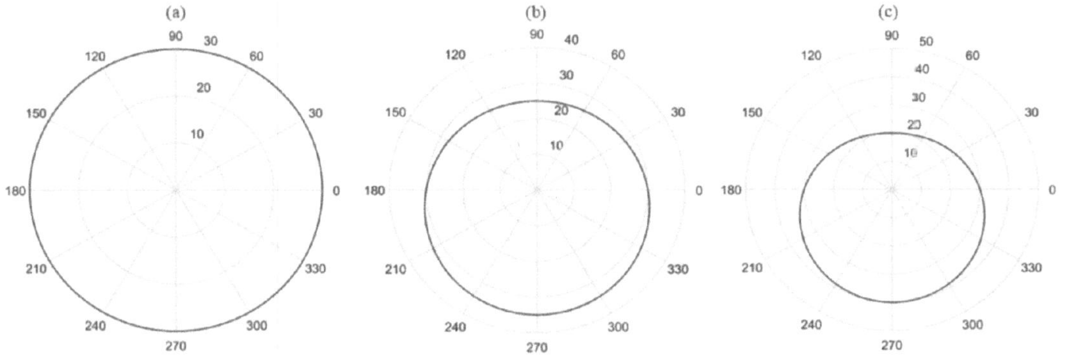

FIGURE 15.5 Relationship between d and φ under different eccentricity errors. (a) Eccentricity error e = 0 mm, (b) eccentricity error e = 5 mm and (c) eccentricity error e = 10 mm.

The eccentricity error needs to be corrected in two steps, namely coarse tuning and fine-tuning, by adjusting the position of the three jaws. Take a single jaw as an example. As shown in Figure 15.6, the lead screw is connected with the bevel gears at both ends through coupling. When the coupling screws close to the central bevel gear are loosened, the rotating lead screw will no longer drive the central bevel gear to rotate. At this time, the position of a single sliding block can be adjusted by rotating the circumferential bevel gears. However, due to the impact of bevel gear backlash, the adjustment accuracy of this method is not high, relying only on coarse tuning. On this basis, a screw is designed between each sliding block and each jaw to fine-tune the jaw position based on the coarse tuning of the slider position. The eccentricity error is finally corrected by coarse and fine-tuning.

15.2.3 Generation and Morphological Analysis of Defects in Rotary Components

The hub studied in this project comes from a heavy-duty vehicle, as shown in Figure 15.7. By analyzing the hub that has fractured and failed, the causes for its cracking and fracture can be classified into the following three categories:

1. **Thermal damage**: During long-term high-speed operation, friction heat will be generated inside the bearing and then transferred to the hub. The most affected area is

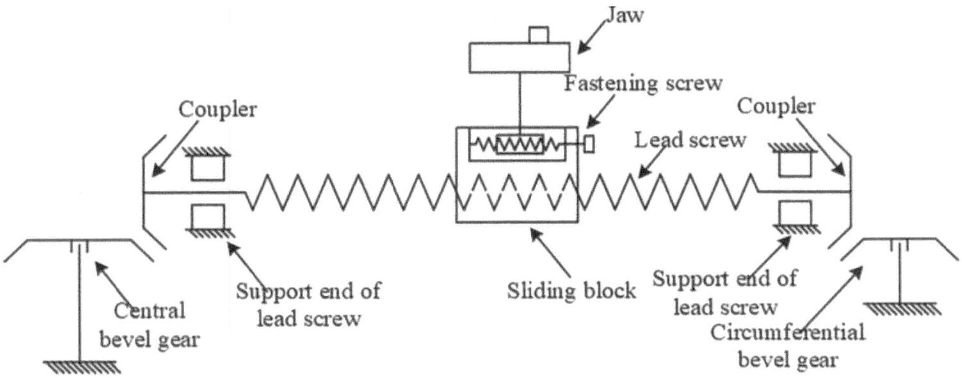

FIGURE 15.6 Driving mechanism of a single jaw.

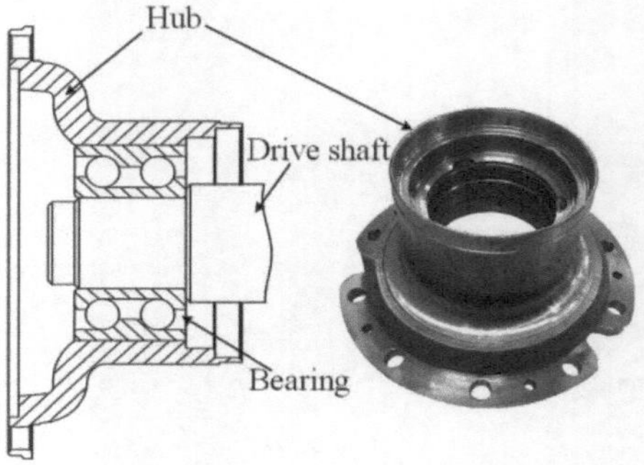

FIGURE 15.7 Wheel hub and the assembly relationship between its components.

the contact area between the hub and the bearing inner race. There are two reasons for cracking and fracture under the influence of friction heat. One reason is that once the heated hub is rapidly cooled, the quenching effect will occur on the bearing surface. Alternated cooling and heating will gradually increase the brittleness of the material and reduce its toughness. The other reason is the expansion caused by heating. When the thermal expansion exceeds a certain limit, a thermal stress will be generated inside the material. Thermal damage will eventually lead to the occurrence of thermal fatigue cracks, which usually grow into a checkerboard pattern and become a source of cracking in other operations.

2. **Fatigue damage**: Under long-term cyclic loading, a large number of microcracks will occur inside the hub but will not cause fatal damage. However, when the cracks grow to a certain size, they can cause an accident. Fatigue cracks tend to appear near the hub surface and then extend to the hub interior. Usually, a fatigue crack develops along a plane. Under cyclic loading, the local stress at the crack tip will change periodically, causing the cracking direction to deviate slightly from the plane. As a result, the crack grows into a "sandbeach" crack. It radiates from the origin to the hub interior, until it develops into a destructive crack with plastic or brittle fracture to expand rapidly. The occurrence of fatigue crack is relatively random and is not easy to be found in the initial stage. So far, fatigue crack has become an important unpreventable cause for vehicle accidents.

3. **Impact damage**: Although a large number of microcracks exist inside the hub, they will not expand in a large scale in short time. However, an impact load will cause the microcracks inside the hub to expand faster than expected and reduce the effective load area of the hub section. The brittle fracture caused by impact damage often occurs suddenly. Some materials may become "brittle" due to the temperature change in the process of machining and operation. For example, the high temperature caused by grinding or bearing friction can increase the content of martensite in steel. From the perspective of microstructure, the increase of martensite will lead to grain boundary brittleness, and the brittle fracture caused thereby will be generally concentrated on

the grain boundary surface. The fractured surface looks like "rock candy". Impact loads will significantly reduce the service time of a hub and thus should be avoided as much as possible during the hub operation.

According to the theory of linear elastic fracture mechanics, crack growth is closely related to external stress loading and crack size. In 1957, G.R. Irwin first proposed the stress intensity factor K_I to estimate the plastic zone size of crack tip. When K_I exceeds a certain critical value, the crack tip will expand suddenly and unstably to form brittle fracture. This critical value represents the resistance of a material against unstable crack propagation and is called fracture toughness K_{mat}. This means that when $K_{I>}K_{\text{mat}}$, the material faces a great risk of fracture caused by unstable crack propagation. Some international standards (for example, British standard BS 748-4:19:97 and BS 7010:2013) have stipulated the method for estimating K_{mat}. A simplified method for calculating the fracture toughness of a material is as follows:

$$a_{pl} = \frac{1}{\pi}\left(\frac{K_{\text{mat}}}{\sigma_Y}\right)^2 \tag{15.6}$$

a_{pi} is the critical size of crack growth and σ_Y is the yield strength of the material. Eq. (15.6) is derived from British standard BS 7010:2013, in which describes in detail the method for judging the fracture resistance of a component by the condition of its internal defects. The analysis of defect generation and component failure fracture shows that cracking is the main cause for fracture after long-term operation, and that the crack size is negatively correlated with fracture resistance. If the crack size exceeds the critical value, it indicates that the component under test is no longer suitable for further service.

15.2.4 Analysis of Error and Uncertainty in the Ultrasonic Detection of Defects inside Rotary Components

It is found in the detection of internal defects by using a focusing ultrasonic transducer that if the reflection surface of a defect is perpendicular to the ultrasonic propagation direction, the reflected wave will be strong enough to correctly judge the shape and size of the defect. However, the defect in a rotary component is mainly the cracking caused by cyclic loading and stress concentration, and the extension direction of the defect deviates from the incident plane by a small angle. This brings some difficulties to the judgment of the reflection echo. When detecting the internal defects of rotary components, the refracted transverse wave generated by an obliquely incident longitudinal wave is preferred. The detection error of the system can be analyzed from the following aspects:

1. Beam diameter

   As shown in Figure 15.8, the longitudinal wave beam emitted by a piezoelectric wafer with the diameter D_l enters the material under test through the coupling medium and gets refracted at the interface between the material and the coupling medium to generate a transverse wave beam. If observed in reverse direction, the transverse wave beam seems to be emitted in the coupling medium by a piezoelectric

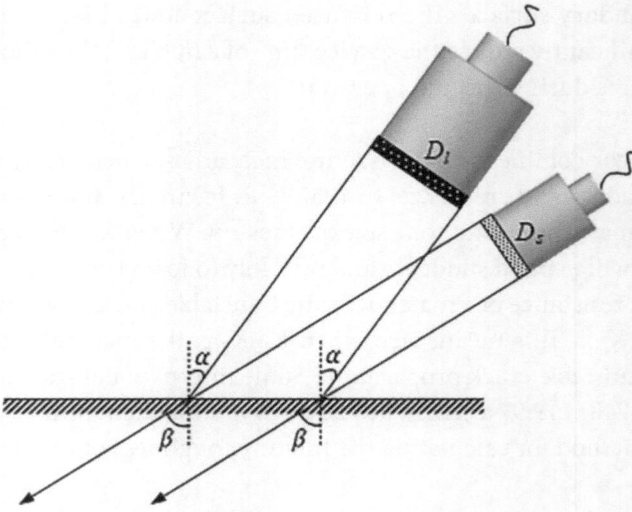

FIGURE 15.8 Effective diameter of a piezoelectric wafer.

wafer with the diameter D_s, whose effective diameter is smaller than the actual wafer. Therefore, the diameter of the refracted sound beam is smaller than that of the longitudinal wave beam before refraction in the following relationship:

$$D_S = \frac{D_l \cos\beta}{\cos\alpha} \tag{15.7}$$

2. Defect resolution

   For a defect that can effectively reflect the ultrasonic wave, its extension length shall be greater than or equal to 1/2 of the wave length and its thickness shall be greater than or equal to 1/4 of the wave length. When the wafer diameter is fixed, the transverse wave velocity at the same frequency is only half of the longitudinal wave velocity and the transverse wave wavelength is also half of the longitudinal wave wavelength. Therefore, when the central frequency of the ultrasonic transducer is fixed, the refracted transverse wave has better defect resolution.

3. Beam directivity

   According to the analysis of sound field parameters of the ultrasonic transducer, the directivity of ultrasonic beam is represented by half diffusion angle $\theta_0 = 70\lambda/D$, where λ is the wavelength and D is the piezoelectric wafer diameter. The transverse wave wavelength is half of the longitudinal wave wavelength and the effective diameter of piezoelectric wafer is small, so the transverse wave has a smaller half diffusion angle and better beam directivity.

   It can be seen from the above three analysis aspects that the transverse wave generated by longitudinal wave refraction can reduce the error in the detection of internal defects of rotary components. In this section, the internal cracks of rotary components are divided into transverse and longitudinal cracks, which are detected by using the refracted transverse wave generated by a longitudinal-wave focusing transducer (see the calculation method below) to improve the detection resolution of internal defects and reduce the influence of uncertain factors.

1. The detection of transverse cracks in the rotary component is shown in Figure 15.9. The focusing length of the transducer is f_1, and the longitudinal wave is incident at the point P_1 with the angle α_1, which is limited between the first critical angle and the second critical angle:

$$\frac{c_w}{c_{ml}} < \sin\alpha_1 < \frac{c_w}{c_{ms}} \tag{15.8}$$

where c_w is the sound velocity in water and c_{ml} and c_{ms} are the longitudinal and transverse wave velocities in the material. According to Snell's law, only transverse wave is refracted in the material between the first and second critical angles, and the refraction angle is:

$$\sin\beta_1 = \sin\alpha_1 \times \frac{c_{ms}}{c_w} \tag{15.9}$$

The propagation distance of sound beam is w_1 in water and m_1 in material. The relationship between them is:

$$\sin\beta_1 = \sin\alpha_1 \times \frac{c_{ms}}{c_w} \tag{15.10}$$

The sound beams refracted by the interface are focused at the point M_1, whose depth from the incident surface is:

$$d_1 = m_1\cos\beta_1 = \frac{c_w}{c_{ms}}\left(f_1 - w_1\right)\cos\beta_1 \tag{15.11}$$

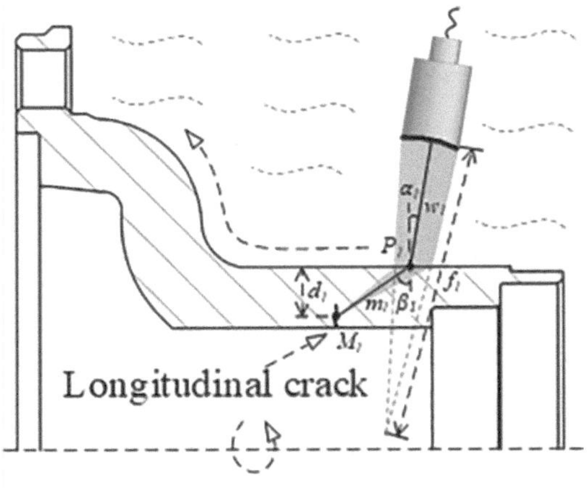

FIGURE 15.9 Detection of transverse cracks in a rotary component.

It can be seen from Eq. (15.11) that by adjusting w_1, the sound beams can be focused at different depths in the material to improve the detection sensitivity and crack position accuracy. In addition, during the adjustment of the focus depth d_1, α_1 should satisfy Eq. (15.8) to ensure that only the refracted transverse waves exist in the material. As shown in Figure 15.9, the ultrasonic transducer moves along the generatrix of the component at a certain incident angle, while the component is driven by the turntable to rotate around the central axis. This spiral motion can be used for the detection of transverse cracks in rotary components.

2. The detection of longitudinal cracks in the rotary component is shown in Figure 15.10. Here, the refracted transverse wave is generated in a different way from that in the transverse crack detection. The ultrasonic transducer is placed vertically. A certain eccentricity l exists between the beam axis and the component center axis, so that the longitudinal wave beam is incident at the point P_2 with the angle α_2. Obviously, the eccentricity l has a trigonometric relationship with the incident angle α_2:

$$\sin\alpha_2 = l/R \tag{15.12}$$

where R represents the outer diameter of the component at the incident point. α_2 increases with l. Therefore, by adjusting l, α_2 can be located between the first and second critical angles, so that only the refracted transverse waves will be generated. The transverse wave beams are focused at the point M_2 in the component. The distance between the component axis and the point M_2 is q. w_2 and m_2, respectively, represent the propagation distances of sound beam in water and in material. Their relationship is similar to that in Eq. (15.10):

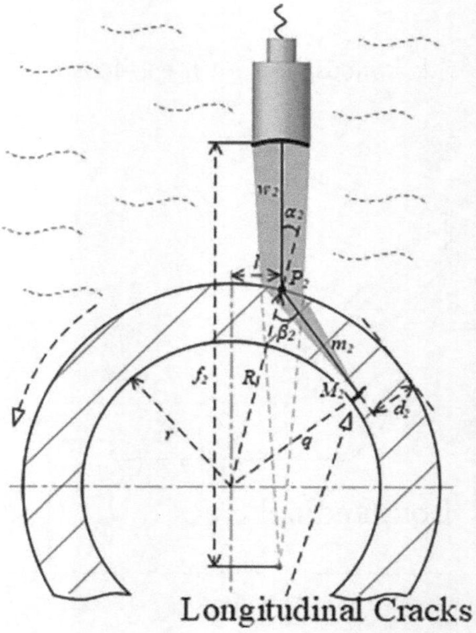

FIGURE 15.10 Detection of longitudinal cracks in a rotary component.

$$\frac{f_2 - w_2}{m_2} = \frac{c_{ms}}{c_w} \tag{15.13}$$

The depth of the focus point M_2 from the incident plane is:

$$d_2 = R - q \tag{15.14}$$

where q can be calculated by the law of cosines in accordance with the following relation:

$$q^2 = R^2 + m_2^2 - 2Rm_2\cos\beta_2 = R^2 + \left[(f_2 - w_2)\frac{c_w}{c_{ms}}\right]^2 - 2R\left[(f_2 - w_2)\frac{c_w}{c_{ms}}\right]\cos\beta_2 \tag{15.15}$$

When $q = r$, the beams are focused on the inner wall. When $q = R$, the beams are focused on the outer wall. By adjusting the water path length w_2, the sound beams can be focused at different depths in the material to ensure the detection sensitivity and positioning accuracy. When the water path length is adjusted, the eccentricity should be controlled within a certain range according to Eq. (14.12) to ensure that only the refracted transverse waves are generated at the interface. As shown in Figure 15.10, the ultrasonic transducer with a certain eccentricity from the component axis moves along the component generatrix, while the component rotates around its central axis. This spiral motion is used for the detection of longitudinal cracks in rotary components.

15.2.5 Application Examples of Robotic NDT of Rotary Components

1. Detection of defects in blade mortise

   The fatigue fracture of the mortise in a turbine disk blade will cause the whole blade or disk to come off, thus endangering the safety of an aircraft and its engine in flight. Therefore, a reasonable method of mortise NDT was studied, and the corresponding testing process was formulated to avoid the occurrence of such accidents. To improve the compactness of material crystal structure, the turbine disk needs to go through the forging process including upsetting and die forging in order to eliminate the original dendrites and form a recrystallized structure. The formation of such fine grains in a uniform structure will substantially reduce the signal noise caused by grain scattering to facilitate ultrasonic detection. Thus, an ultrasonic transducer with higher frequency is suitable for mortise detection. It was found in the experiment that, the 25 MHz water-immersion ultrasonic transducer could detect an artificial defect equivalent to 0.5 mm in a flat bottom hole. The detection probability would be higher for the area-type inclusions or crack defects in a turbine disk greater than 1 mm.

   The surface and mortise of a turbine disk have different characteristics, so they should be considered separately and detected with different techniques. Considering that near-surface resolution and good beam penetration is required for the large thick disk surface containing quite a lot of inclusions, a large-wafer transducer with a diameter of about 15 mm, a large focus diameter and high energy should be selected. The blade mortise is characterized by narrow detection area and small defect size, so the amplitude and signal-to-noise ratio (SNR) of reflected ultrasonic signal should be improved as much as possible and the frequency of the ultrasonic transducer should be appropriately selected to obtain the optimum detection resolution. The turbine disk is fixed on the turntable through a chuck (Figure 15.11). The manipulator end-effector holds the transducer in a specific attitude to ensure that the sound beam is vertically incident into the blade mortise. During the detection process, the manipulator moves along the generatrix of turbine disk, while the turntable rotates.

2. Testing of a pipe with variable diameters

Pipes are widely used in petrochemical and other industrial fields, and their defects are generally dominated by corrosion holes. As shown in Figure 15.12a, a section of aluminum alloy pipe with variable diameters was cut open. Then, a series of artificial flat-bottom holes with different diameters and burial depths were made on the inner wall to simulate the defects caused by corrosion and other reasons. Three groups of defects, each composed of three flat-bottom holes, were distributed in the upper, middle and lower sections of the pipe, respectively. The size and defect distribution of the pipe are shown in Figure 15.12c. The nominal values of workpiece coordinates defined in the trajectory planning as well as the corrected values obtained after the point-cloud matching calculation are shown in Table 15.1. It can be seen that there are deviations between them. During the detection, the turntable is fixed, and only the manipulator moves along the cambered pipe surface to scan the three flat-bottom holes at the maximum pipe diameter. The C-scan images obtained before and after

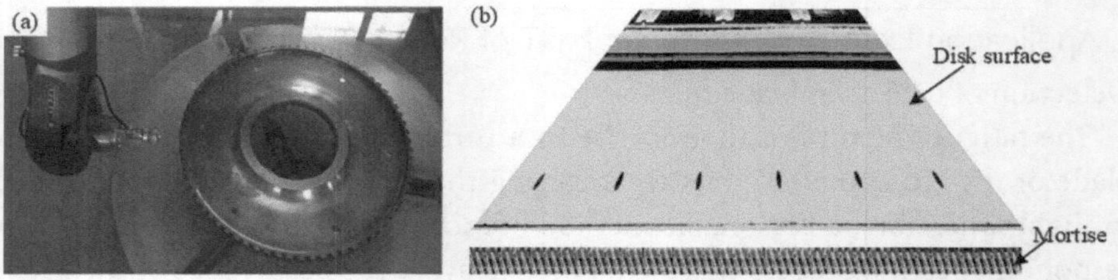

FIGURE 15.11 Detection process and C-scan detection results of a turbine disk. (a) Automate the ultrasonic detection process and (b) the C-scan detection results of a turbine disk.

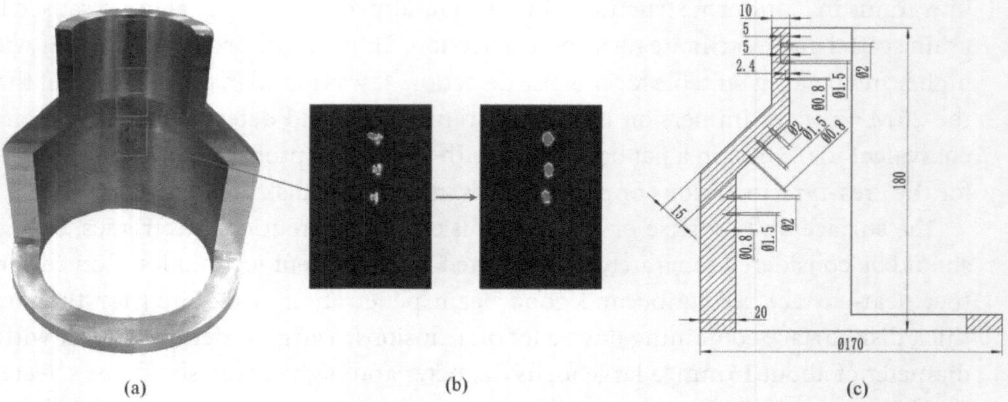

FIGURE 15.12 Pipe with variable diameters and artificial defects. (a) The aluminum alloy pipe with variable diameters, (b) the C-scan images and (c) the size and defect distribution of the pipe.

TABLE 15.1 Comparison of Nominal Coordinates and Corrected Coordinates

Nominal Coordinates (mm)	Corrected Coordinates (mm)	Deviation (mm)
$X = 878.00$	$X = 875.41$	$\Delta X = 2.59$
$Y = 415.00$	$Y = 413.93$	$\Delta Y = 1.07$
$Z = -802.00$	$Z = -802.35$	$\Delta Z = 0.35$
$RX = 1.50$	$RX = 1.03$	$\Delta RX = 0.47$
$RY = 1.50$	$RY = 0.85$	$\Delta RY = 0.65$
$RZ = 0.00$	$RZ = 0.36$	$\Delta RZ = -0.36$

correction are shown in Figure 15.12b. The left image is the C-scan image obtained before the correction of workpiece coordinates, while the right image is the C-scan image obtained after correction. After correction, the three holes can be identified more easily with a higher contrast ratio.

The detection process of variable-diameter pipes is shown in Figure 15.13. During the detection process, the coordinates of beam focus and the sound intervals were simultaneously recorded. As shown in Figure 15.13, the blue solid point cloud is composed of the coordinates of incident points calculated according to the sound intervals. These incident points are on the surface of the component under test and can reflect the actual position/attitude of the component. The red hollow point cloud is composed of the coordinates of beam focus collected through the manipulator controller and can reflect the motion trajectory of the beam focus. It can be seen from the figure that the spatial distributions of the two point clouds coincide with each other very well, indicating that the beam focus is located on the component material. In this case, ideal echo signals with high sensitivity and defect resolution can be received.

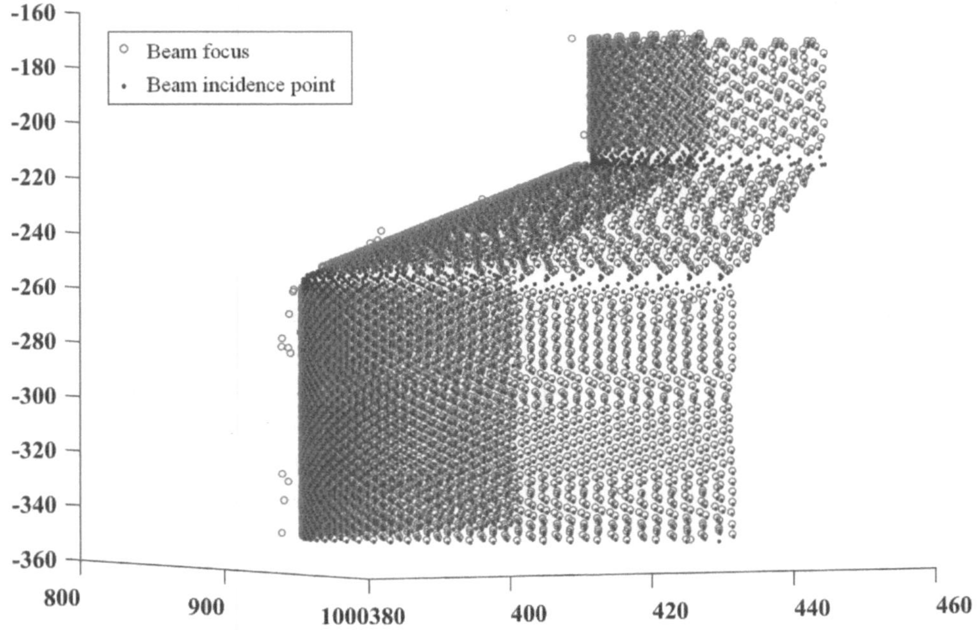

FIGURE 15.13 Beam focus distribution after the correction of coordinates.

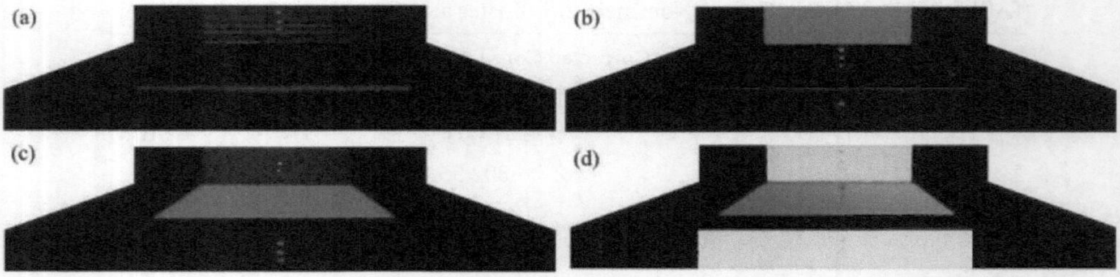

FIGURE 15.14 Detection results of a pipe with variable diameters. (a) Display of defects in the upper section of pipeline, (b) defect display in the middle of the pipeline, (c) defect display in the lower section of the pipeline and (d) display of all defects in the pipeline.

In the experiment, the scanning method featuring the coordinated movement between turntable and manipulator was further introduced to perform the detection, during which the ultrasonic echo signals were collected synchronously through turntable trigger. The obtained C-scan images are shown in Figure 15.14, in which the first three images were obtained through the peak-to-peak imaging of defect echo, and the last image was obtained through the transit-time imaging. In Figure 15.14a, the three defects in the upper section of the pipe are clearly shown. In Figure 15.14b, the three defects in the middle section are mainly detected, and the echo reflected by one lower defect with a relatively small buried depth is also captured by the data gate. In Figure 15.14c, the three defects in the lower section are mainly shown, while the three defects in the upper section are also shown but with a relatively poor contrast ratio. In Figure 15.14d, all the three groups of defects can be displayed in a single C-scan image but each with a poor contrast ratio. By comparing the four images in Figure 15.14, it can be found that the peak-to-peak imaging method has a better contrast – but a smaller detection depth interval – than the transit-time imaging method.

For the first three images in Figure 15.14, three data gates are needed to present all the three groups of defects. For the defects with a large buried depth, the transit-time imaging method can present all the defects in one image but with a poor contrast. Therefore, this method is preferred to locate the defects in the preliminary scanning. On the contrary, the peak-to-peak imaging method applies to further scanning and analyzing the size, location and type of defects.

3. Testing of a shaft

On the shafts under long-term alternating loads, fatigue cracking and fatigue pitting can be easily seen. According to the analysis of defect distribution and stress, the fatigue cracks generally appear in the press-fitting area of a bearing inner race, mostly in the form of dangerous transverse cracks that can cause the shaft fracture easily. In the experiment, a hollow-stepped shaft in the gearbox was tested, as shown in Figure 15.15a. Four flat-bottom holes with the depth of 7.5 mm and the diameters of 0.5, 0.8, 1.0 and 1.5 mm, respectively, were made in the middle section of inner shaft wall, as shown in Figure 15.15b.

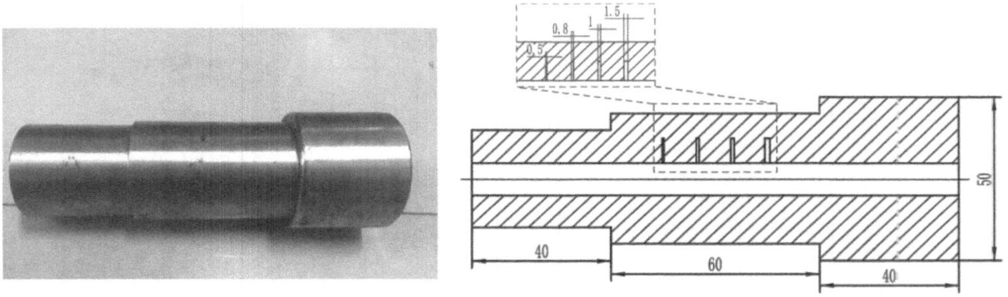

(a) (b)

FIGURE 15.15 Dimensions of a shaft and its artificial defects. (a) Shaft parts and (b) artificial defect distribution map.

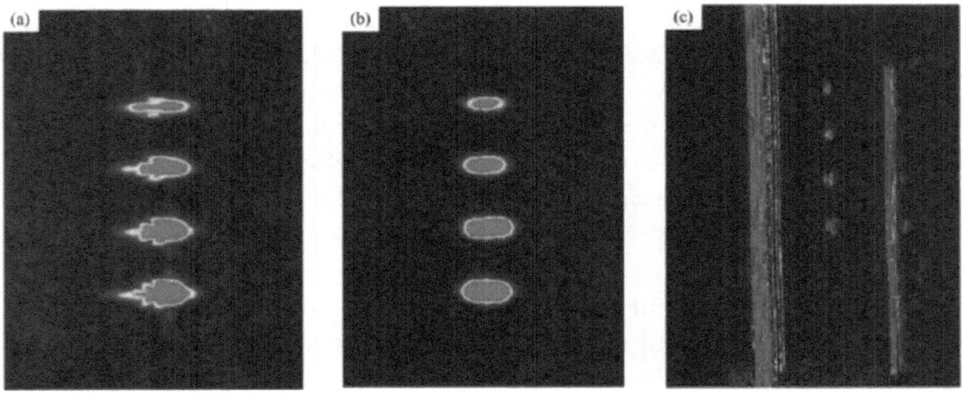

FIGURE 15.16 Detection results of a shaft. (a) Deformed ultrasonic testing C-san, (b) calibrated ultrasonic testing C-scan and (c) Ultrasonic testing B-scan.

The shaft was fixed on the turntable through a three-jaw chuck. The turntable rotated and the manipulator moved along the shaft generatrix. Then the artificial defects on the inner wall, namely four flat-bottom holes, were detected by means of turntable trigger. The detection results obtained by the peak-to-peak imaging method are shown in Figure 15.16. The comparison between Figure 15.16a and b shows that the detected holes are out of shape in the C-scan image of Figure 15.16a due to the eccentricity of the rotating shaft caused by the centering error of the clamping device. By adjusting the clamping device for re-testing, the distortion can be improved, as shown in Figure 15.16b. It can be seen that the clamping deviation of the workpiece will affect the judgment of defect location and size. In addition, the four holes were traversed by B-scan as the manipulator moves along the generatrix. The detection result is shown in Figure 15.16c. The two red lines in the figure are the echoes reflected from the surface and bottom of inner and outer walls. The reflection echoes generated by the four flat-bottom holes are between the inner and outer walls. Through the B-scan image, the depth of a defect in the cross section can be visually displayed.

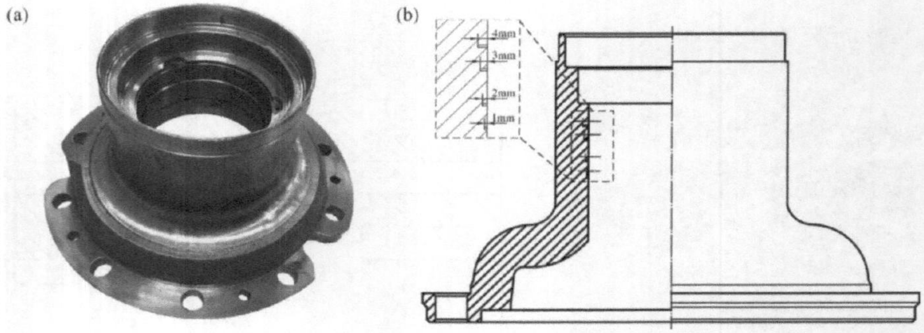

FIGURE 15.17 Dimensions of vehicle hub with artificial defects. (a) vehicle hub requiring nondestructive testing and (b) artificial defects size chart.

4. Testing of a vehicle hub

   Hub is the main load-bearing component of a vehicle. With a high yield in factory, hub is crucial to the driving safety. Every hub needs to be inspected before leaving factory. As shown in Figure 15.17a, the defects in a hub are mainly detected by X-ray due to the complex shape of the hub. The UT of hubs remains in the manual stage with low efficiency and accuracy. Four flat-bottom holes with a diameter of 0.8 mm were drilled on the inner wall of the hub. Their drilling depth are 1, 2, 3 and 4 mm, respectively, and their buried depths are 29, 28, 27 and 26 mm, respectively. The size and distribution of the four hole defects are shown in Figure 15.17b.

   The hub is scanned through the coordinated movement between turntable and manipulator. The detection results obtained by the peak-to-peak imaging method are shown in Figure 15.18a, in which the four artificial defects in the form of flat bottom holes could be clearly identified. In addition, it can also be observed that a large number of blue spots are scattered around the four defects. They are the direct representation of sound wave scattering at coarse grains in the cast iron material. In the detection process, the occurrence of a large number of clutters in the A-scan waveform is caused by the impurities such as grains, flake graphite and composition segregation. With the increase of grain size and scale, the clutter phenomenon will

FIGURE 15.18 Detection results of a hub. (a) C-scan image and (b) the amplified area near the defects.

become more serious. Figure 15.18b shows the amplified area near the defects. The center frequency of the ultrasonic transducer is 10 MHz in the left image and 15 MHz in the right image. The use of the ultrasonic transducer with higher center frequency can improve the detection sensitivity and defect resolution but will aggravate the clutter phenomenon caused by coarse grains. Therefore, the tradeoff between the higher detection resolution of high-frequency transducer and the influence of material characteristics must be weighed carefully.

15.3 ROBOTIC NDT METHOD FOR BLADE DEFECTS

The blades in service may have different types of defects such as porosity, delamination and cracking. All these defects have raised different requirements for the ultrasonic NDT method (immersion testing, gap testing or contact testing), ultrasonic incident angle (vertical or oblique incidence), beam size parameters (focal length and beam diffusion angle) and ultrasound path length. Based on the robotic ultrasonic automatic NDT system presented in the previous chapter, the ultrasonic longitudinal-wave vertical reflection method and the ultrasonic surface-wave detection method are used in this study to detect the internal and surface defects of complex components (such as blades). Next, we will focus on the key technologies and the overall detection scheme.

15.3.1 Robotic Ultrasonic NDT of Blades

The data acquisition card used in this system can set the tracking gates at multiple levels. The setting of two levels of tracking gates can effectively solve the detection problem of near-bottom defects by tracking ultrasonic bottom wave signals but still cannot solve the problem of full-thickness coverage, as shown in Figure 15.19.

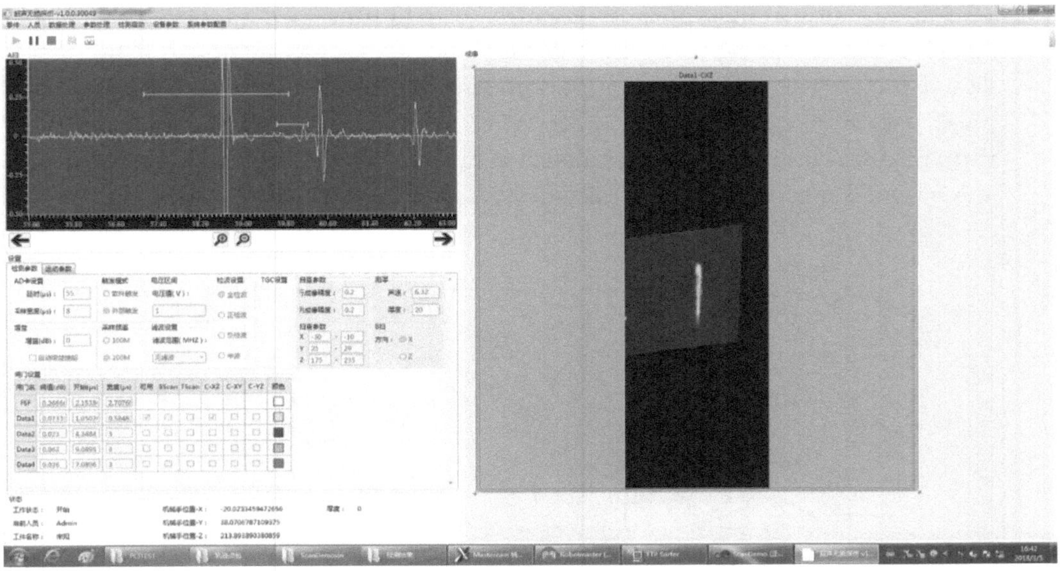

FIGURE 15.19 Multi-level tracking gates.

In order to solve this problem, the length of data gate can be changed by the software program of upper computer in real time to ensure the detection effect of blade defect and thickness. Here, two solutions are provided:

1. At first, collect the A-scan full-wave waveform data of the whole blade. Set two tracking gates at the head wave and bottom wave, respectively, to acquire their real-time positions. After scanning, set the starting position of the imaging data gate according to the head wave position at each scanning point. Set the length of the data gate according to the distance between the bottom wave and the head wave, and then reimage the data to realize the full-thickness scanning.
2. At first, collect the A-scan full-wave waveform data of the whole blade. Determine the variation range of data gate length. Set a fixed time interval or point location spacing. During scanning, change the length of the imaging data gate in real time at a fixed interval to approximately achieve the full-thickness scanning.

15.3.2 Detection by Ultrasonic Vertical Incidence

When using the immersion-type ultrasonic longitudinal-wave vertical reflection method to detect the defects in an aviation blade, the blade is clamped by the manipulator because the rotation velocity of manipulator end joint is faster than that of multi-joint linkage and is easier to control so that the scanning efficiency is higher. The system structure diagram is shown in Figure 15.20. The detection results of blade cracks and bottom holes in this method will be described below.

a. Crack detection experiment

The blade cracks were detected by the vertically incident wave of a SIUI 20M6SJ40Z ultrasonic immersion point-focusing probe with a center frequency of 20 MHz. In this experiment, the water distance was set as 18 mm. A SyncScan1U flaw detector with the following parameters was used: excitation voltage: 200 V; pulse width: 30 ns; damping: 2; receiving bandwidth: 2–20 MHz and receiving gain: 68 dB. A 0.15 × 6 mm crack defect area was made on the blade and scanned. The planned spacing between trajectory points was 1.25 mm, the line spacing was

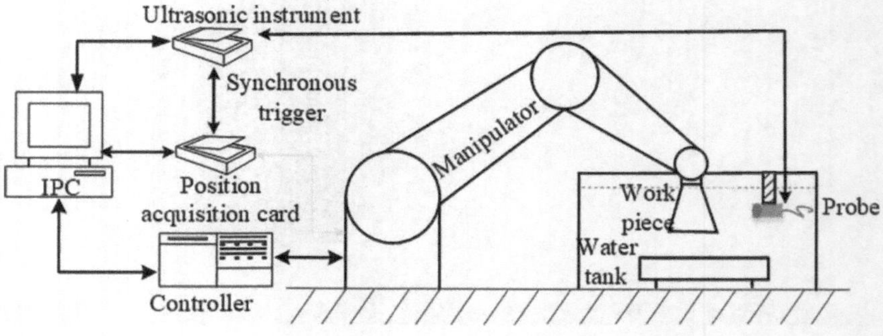

FIGURE 15.20 Structural diagram of an ultrasonic reflection detection system.

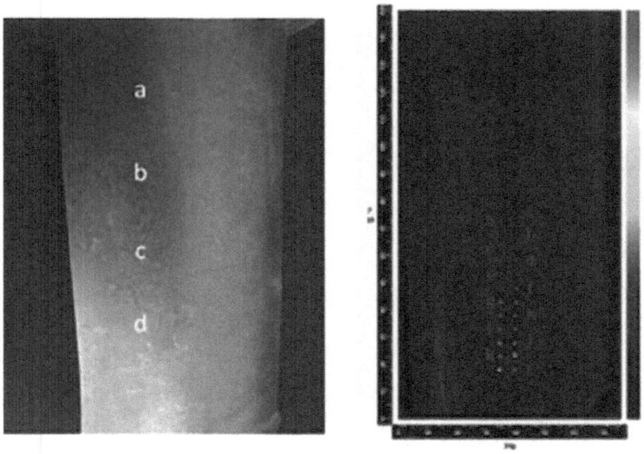

FIGURE 15.21 Physical picture and scanning result of blade cracks.

TABLE 15.2 Designed Dimensions of Crack Defects

No.	Length (mm)	Width (mm)	Depth (mm)
a	6	0.15	0.2
b			0.4
c			0.6
d			0.8

0.10 mm, the distance from the gate to the bottom surface was 0.516 mm and the gate width was 2.579 mm. The physical picture of blade cracks (left) and the scanning result (right) is shown in Figure 15.21. The designed dimensions of crack defects are shown in Table 15.2.

As can be seen from the C-scan image obtained in the experiment, all the designed crack defects have been detected by the vertically reflected ultrasonic longitudinal wave under the experimental conditions. The minimum crack size detected in the blade specimen was 6 mm (length) × 0.15 mm (width) × 0.2 mm (depth).

b. Detection experiment of flat-bottom hole

The flat-bottom holes were detected by the vertically incident wave of a SIUI 20M6SJ40Z ultrasonic immersion point-focusing probe with a center frequency of 20 MHz. In this experiment, the water distance was set as 18 mm. A SyncScan1U flaw detector with the following parameters was used: excitation voltage: 250 V; pulse width: 30 ns; damping: 2; receiving bandwidth: 2–20 MHz and receiving gain: 70 dB. Four flat-bottom holes (Φ0.15, Φ0.2, Φ0.3 and Φ0.4 mm) were made on the blade and scanned. The planned spacing between trajectory points was 1.25 mm, the line spacing was 0.10 mm, the distance from the gate to the bottom surface was 0.449 mm and the gate width was 2.243 mm. The physical picture of flat-bottom holes (left) and the scanning result (right) is shown in Figure 15.22. The designed dimensions of these defects are shown in Table 15.3.

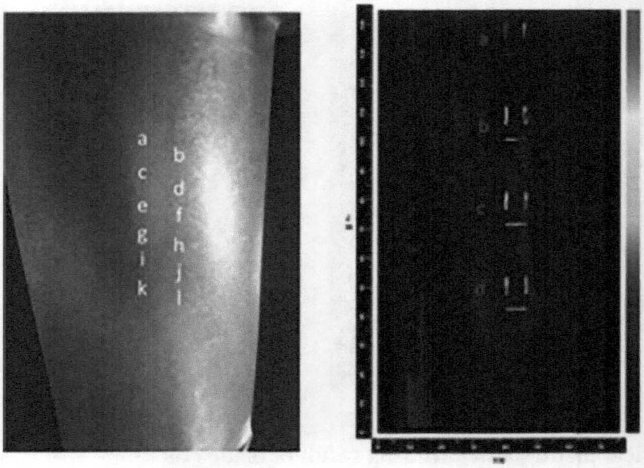

FIGURE 15.22 Physical picture (left) and scanning result of flat-bottom holes (right) in the blade.

TABLE 15.3 Designed Dimensions of Flat-Bottom Holes

No.	Diameter (mm)	Depth (mm)	No.	Diameter (mm)	Depth (mm)
a	0.15	0.5	*g*	0.3	0.5
b		0.8	*h*		0.8
c		1.0	*i*		1.0
d	0.2	0.5	*j*	0.4	0.5
e		0.8	*k*		0.8
f		1.0	*l*		1.0

As can be seen from the C-scan image obtained in the experiment, all the designed hole defects have been detected by the vertically reflected ultrasonic longitudinal wave under the experimental conditions. The minimum hole size detected in the blade specimen was Φ0.15 mm (diameter) × 0.5 mm (depth).

15.3.3 Ultrasonic Surface-Wave Detection Method

1. Simulation of detection process

   To better analyze the process of using a surface wave probe to detect the defects in blade edge surface, a simulation model was establish in Comsol Multiphysics software for numerical calculation. According to the propagation process of surface wave on the surface of titanium alloy blade under the actual detection condition, part of the model was simplified and the detection process was preliminarily simulated on the two-dimensional cross section by using the time-domain finite element method. In this study, another titanium alloy blade, which was convenient for surface wave detection, was selected. A two-dimensional sectional view of the detection system was drawn based on the actual sizes of surface wave probe and blade. According to the material of probe and blade, the model grids were subdivided by a free triangle.

The grid size was set to be less than 1/6 of the sound wavelength in material, in order to ensure enough grids and refine the calculation. A perfectly matched layer was added to the transducer boundary to absorb excess reflection echoes and was divided into five layers of grids through two-dimensional mapping, as shown in Figure 15.23.

The surface wave probe used in the study has a wafer size of 6 × 6 mm and a central frequency of 5 MHz. Therefore, in the pressure-acoustics physical field simulation, a five-peak wave signal with a frequency of 5 MHz was set as the plane-wave radiation condition to simulate the excitation signal emitted by the actual ultrasonic transducer. A small crack defect with a width of 0.1 mm and a depth of 0.3 mm was designed at the blade edge and modeled in the transient solution module. When the plane wave generated by wedge-shaped surface wave probe was incident to a point on the defected blade edge, the time-varying sound pressure amplitude curve of transducer wafer center was simulated, as shown in Figure 15.24.

2. Experimental validation of validity

According to the actual testing requirements and the existing laboratory conditions, a surface wave probe was clamped by the manipulator to scan and detect the blade defects. A fixture and a support frame were designed and fabricated for the curved blade, and another fixture was designed to clamp the probe at the end of the

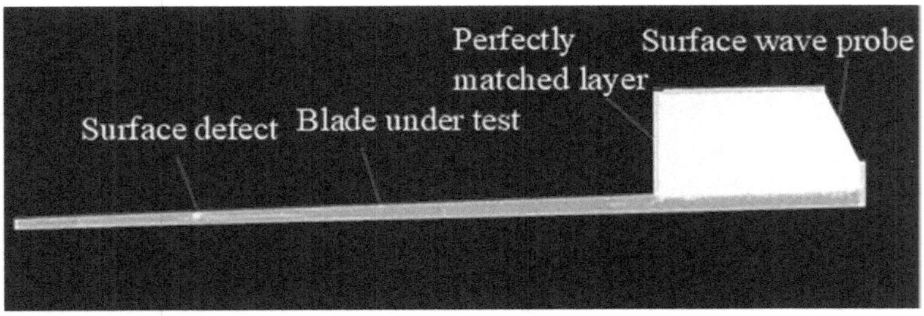

FIGURE 15.23 Grid division of a surface defect model for surface wave detection.

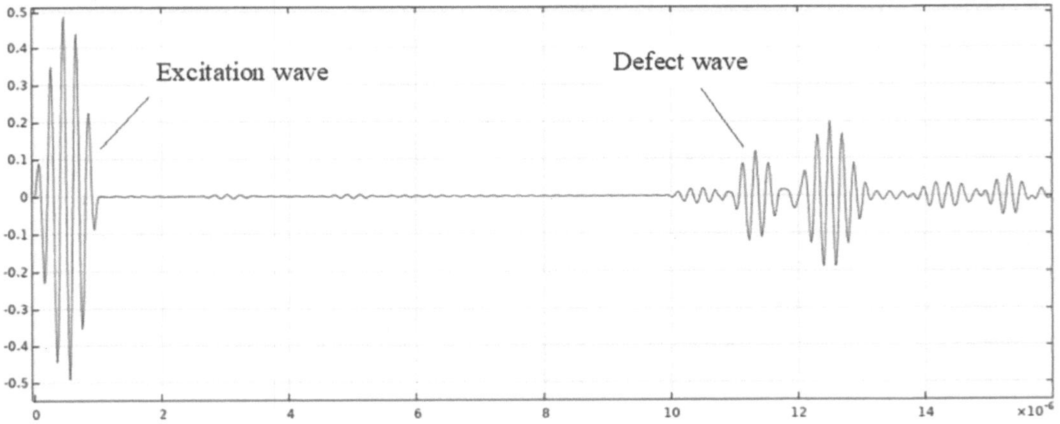

FIGURE 15.24 Simulated amplitude curve of sound pressure at the center of a transducer wafer.

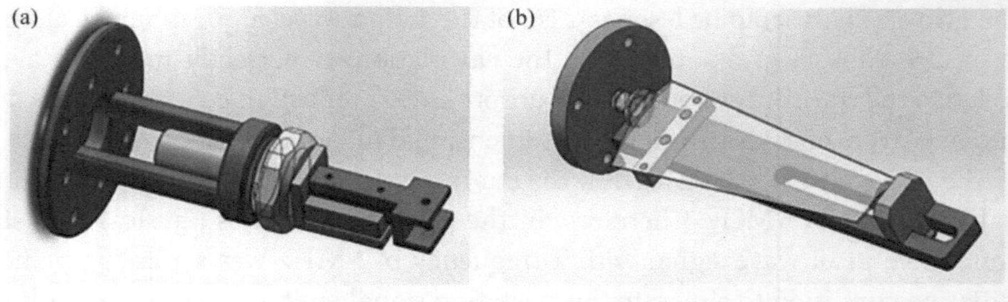

FIGURE 15.25 Surface wave probe fixture (a) and blade fixture (b).

manipulator, so that the system could realize automatic surface wave detection. The 3D structure designs of the two fixtures, both removable, are shown in Figure 15.25. Based on the robotic automatic ultrasonic NDT system, a surface wave probe with the same parameters as the simulated parameters was selected to actually detect the fabricated defects in the blade.

3. Analysis of experimental results

The detection result in the form of oscillogram was obtained by a portable ultrasonic flaw detector connected to the system, as shown in Figure 15.26. It can be seen from the A-scan image that the experimental simulation results of excitation wave and defect wave are consistent. However, the interference curves with high sound pressure amplitude appear in the early and late parts of the experimental curve. This may be caused by the complex transformation of detective surface wave into other longitudinal and transverse waves in practical application.

In addition, the blade edge defects can also be detected by a liquid-immersed ultrasonic longitudinal-wave focusing probe obliquely incident on the blade surface

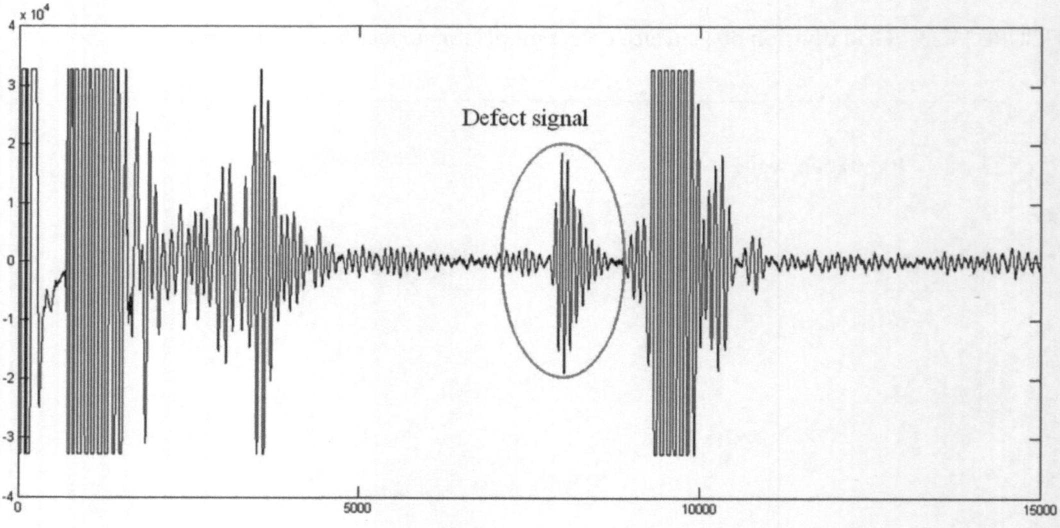

FIGURE 15.26 Detection result in the form of an oscillogram.

(a)

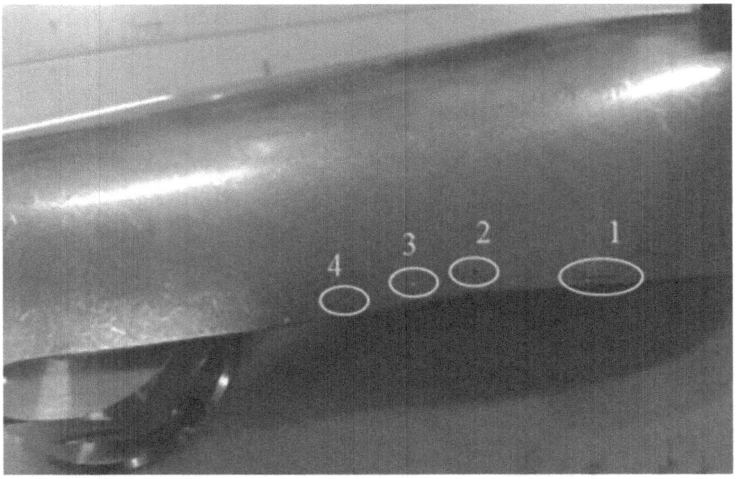

(b)

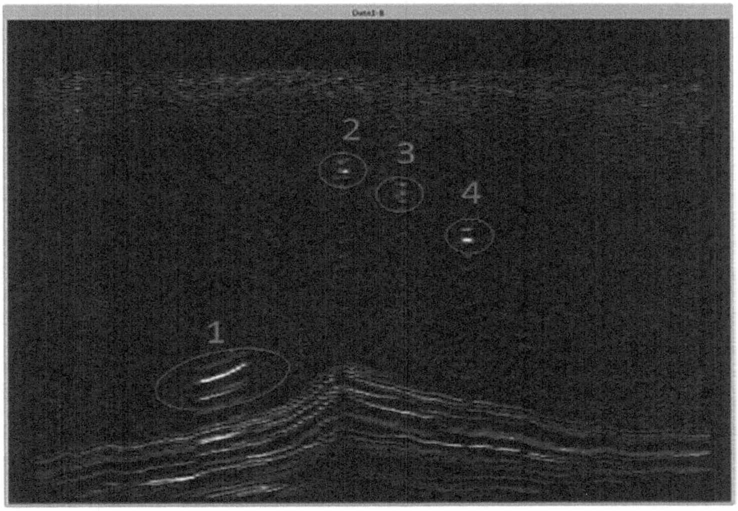

FIGURE 15.27 Physical image and ultrasonic B-scan image of blade defects.

at an angle greater than the second critical angle. This method was used for the ultrasonic B-scan imaging of several flat-bottom holes with a diameter of 0.2 mm and a crack with a width of 0.2 mm at the intake and exhaust edge. The detection results are shown in Figure 15.27. It is not difficult to see that the location and shape of these defects in the B-scan image are in line with reality. Therefore, this method can effectively detect the existence of surface defects at the blade edge.

15.4 ROBOTIC NDT METHOD FOR BLADE DEFECTS

15.4.1 Principle of Ultrasonic Thickness Measurement

In the actual thickness measurement process, the structure and material of tested component, the testing environment and other factors will affect the selection of the ultrasonic probe and thickness measurement method. According to the detection principle, the thickness measurement methods can be divided into three types: the resonance interference method, pulse transmission method and pulse reflection method [1].

1. Resonance interference method

   This method applies to the test pieces with two parallel end faces and a thickness that is an integer (*n*) multiple of half wavelength. An ultrasonic wave with adjustable frequency is vertically incident into the test piece, so that the reflected wave and the incident wave are superposed over each other to form a standing wave and produce resonance. The measured thickness *d* is:

$$d = \frac{\lambda}{2} = \frac{C}{2f_0} = \frac{C}{2\left(f_n - f_{n-1}\right)} \tag{15.16}$$

   where d is the thickness of the tested object; λ is the wavelength of ultrasonic wave; f_0 is the natural frequency of the object; C is the sound velocity in the tested object and f_n and f_{n-1} are adjacent resonance frequencies.

   The voltage-controlled oscillator (VCO) can convert the electrical signals with different amplitudes at the input end into those with variable frequencies at the output end and then load the variable-frequency signals into the ultrasonic transducer, where the ultrasonic waves with variable frequencies can be emitted. Ultrasonic resonance in the specimen will reduce the output impedance of the transducer and maximize the working current of the oscillator. Thus, the frequency of ultrasonic signal can be deduced and the specimen thickness can be calculated. The principle diagram of the resonance interference method is shown in Figure 15.28.

   In the resonance interference method, the accuracy of thickness measurement is high. The measurable thicknesses are mainly 0.1–100 mm and even down to 0.001 mm. However, the end faces of a specimen are required to be parallel and smooth. When its end faces are rough, its thickness is hard to be accurately measured.

2. Pulse transmission method

   In the ultrasonic pulse transmission method, the ultrasonic transducers are placed in the same way as in the resonance interference method. One transducer is placed at either end of the specimen. Of the two transducers, one is responsible for transmitting ultrasonic waves, and the other is responsible for receiving ultrasonic waves. The actual thickness of the specimen can be deduced from the sound interval difference as the ultrasonic wave penetrates the specimen.

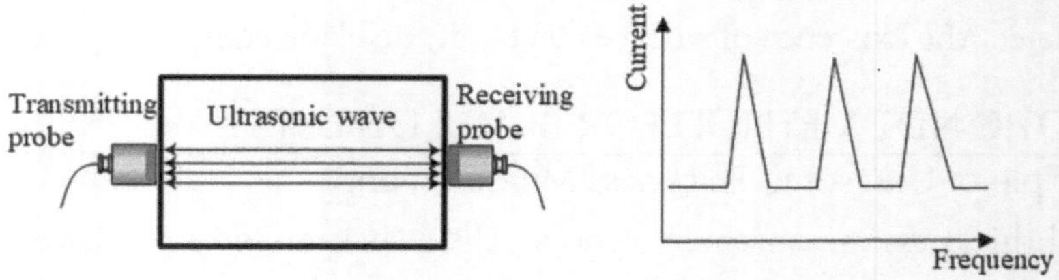

FIGURE 15.28 Principle diagram of a resonance interference method.

As shown in the figure above, the initial excitation wave emitted by the ultrasonic transducer is T, and the subsequent waveform is the transmitted wave R received by the transducer. If the occurrence time of the initial wave and that of the transmitted wave is known to be t_1 and t_2, respectively, the actual thickness of the specimen can be calculated considering the known propagation velocity of ultrasonic wave in the specimen. The disadvantages of this method are complex operation, large space, low detection accuracy and narrow application range.

A pulsed ultrasonic wave with certain energy is emitted by a self-transmitting self-receiving ultrasonic transducer. Then the ultrasonic wave is reflected and transmitted between the surface and bottom of the workpiece under test for many times. Finally, the pulsed reflection echo is received by the transducer, and the workpiece thickness is calculated by using the sound interval difference t between the surface echo and the bottom echo. As long as the sound interval difference is accurately measured, the actual workpiece thickness d can be calculated based on the sound interval difference and the ultrasound velocity c in the workpiece:

$$d = \frac{1}{2} ct \tag{15.17}$$

The thickness measurement principle of the pulse reflection method is shown in Figure 15.29, where T, $B1$ and $B2$ represent the pulsed trigger signal, the first reflection echo signal (head wave) and the second reflection echo signal (bottom wave), respectively, and the subsequent multiple echoes are omitted.

The pulse reflection method can be used to measure the thickness of a specimen with uneven surface or large thickness variation. It has low geometric requirements, high detection accuracy and a wide range of applications.

15.4.2 Calculation Method of Echo Sound Interval Difference

The ultrasonic thickness measurement system based on the principle of the pulse reflection method needs to accurately measure the sound interval difference between two echoes of the ultrasonic wave propagating in the material, in order to calculate the specimen thickness [2]. The calculation method most commonly used in this system is to take the maximum value of surface/bottom echo signals as an eigenvalue and analyze and calculate the sound interval difference between the two echoes by using the eigenvalue. However,

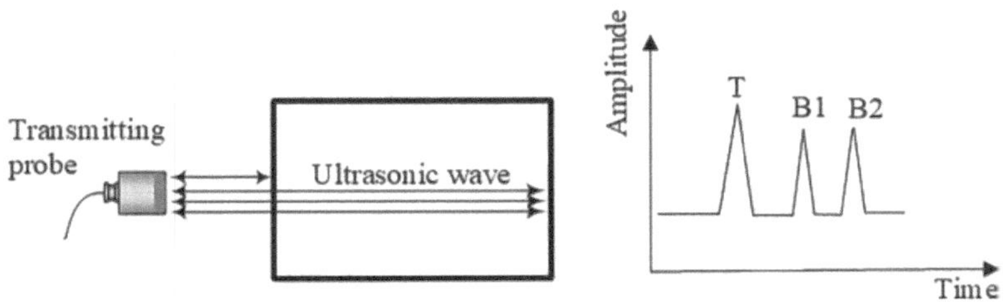

FIGURE 15.29 Principle diagram of a pulse reflection method.

waveform is seriously limited in this method. The complexity of detection process will lead to the position deviation of the maximum or minimum value of ultrasonic echo signals, so the eigenvalue of ultrasonic echo signals can't be accurately extracted and the workpiece thickness can't be accurately measured. Therefore, this method cannot be widely used in the automatic online thickness measurement system where the waveform distortion is easily found [3]. However, the following two methods can effectively solve the above problem.

1. Autocorrelation analysis

The correlation peak analysis using autocorrelation theorem can effectively improve the detection sensitivity and anti-interference ability, improve the accuracy of pulse train identification and save the field-programmable gate array (FPGA) resources. By utilizing the correlation between signals and the mutual independence of noise, the ultrasonic waveform is analyzed and processed by autocorrelation to weaken the noise and strengthen the signal, highlight the useful signal submerged in the interference noise and facilitate the acquisition of detection information. This is exactly the advantage of the detection algorithm based on autocorrelation [3].

The autocorrelation function of the same signal is shown in Figure 15.30. For the specified periodic signal, its autocorrelation function can yield the maximum value only when the time delay τ is 0 or an integer multiple of the period [4]; otherwise, this function will monotonously decrease with the increase of time interval. According to this characteristic, the multiple echoes of ultrasonic repetitive excitation pulse signal in the detection process can be regarded as periodic signals. The autocorrelation calculation result of ultrasonic echo waveform data can show the position point where the correlation function yields the maximum value. Then the sound interval difference between two interface echoes can be identified.

The autocorrelation function is expressed by $Rx(\tau)$, which reflects the correlation of the signal in time shift. The expression of autocorrelation function is [5]:

Periodic signal:

$$R_x(\tau)=\frac{1}{T}\int_0^T x(t)x(t+\tau)dt \tag{15.18}$$

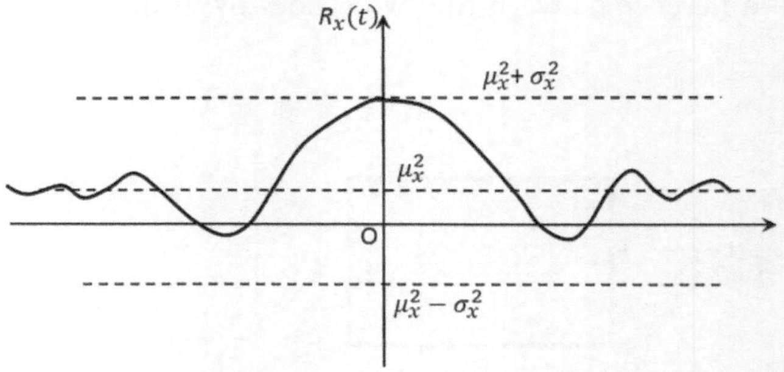

FIGURE 15.30 Autocorrelation function of a signal.

Nonperiodic signal:

$$R_x(\tau)=\frac{1}{T}\int_{-\infty}^{\infty} x(t)x(t+\tau)dt \tag{15.19}$$

The autocorrelation coefficient for describing the autocorrelation degree of a waveform signal is:

$$\rho_x(\tau)=\frac{R_x(\tau)-\mu_x^2}{\sigma_x^2} \tag{15.20}$$

where μ_x is the mean value of the random variable x, and σ_x is the standard deviation of the random variable x. $R_x(\tau)$ will change with τ. When $\tau = 0$, the value of $R_x(\tau)$ is the maximum and is equal to the mean square value of signal, namely $R_x(\tau) = \sigma_x{}^2+\mu_x{}^2$. The value range of $R_x(\tau)$ is $\mu_x^2-\sigma_x^2 \le R_x(\tau)\le \mu_x^2+\sigma_x^2$.

When the time shift τ is discretized, the discrete autocorrelation function of the sequence $x(n)$ can be obtained:

$$_{xx}(m)=\sum_{n=-\infty}^{+\infty} x(n)x(n-m), m=0,\pm1,\pm2... \tag{15.21}$$

2. Sequence similarity detection

   The sequence similarity detection algorithm can also be used in the ultrasonic thickness measurement system based on the pulse reflection method. The basic principle of this algorithm is to analyze the similarity of sequences and find the two sequences with the most similar variation trend from the signal containing two bottom echoes. From the two sequences, the propagation time of ultrasonic wave in the specimen and then the specimen thickness can be derived. The specific steps are as follows. A template sequence $X(N)$ is set in the first bottom echo signal. The sequence $Y(N)$ that matches the template sequence best is found from the second bottom echo signal. Then the ultrasonic propagation time is determined based on the time difference between $X(N)$ and $Y(N)$, as shown in Figure 15.31.

$$\text{MAD}(X,Y)=\sum_{i=1}^{N}\left|X(i)-Y(i)\right| \tag{15.22}$$

   where MAD(X, Y) reflects the degree of matching between $X(N)$ and $Y(N)$. The smaller the MAD(X, Y) is, the higher the matching degree will be. The validity and accuracy of this algorithm have been proved by the results of practical application.

15.4.3 Thickness Measurement Method with Autocorrelation Analysis

For the complex curved workpieces with variable thickness such as aero-engine blades, this chapter presents an ultrasonic thickness measurement method based on the principle of vertical reflection of longitudinal wave pulse. At first, noise reduction and interpolation

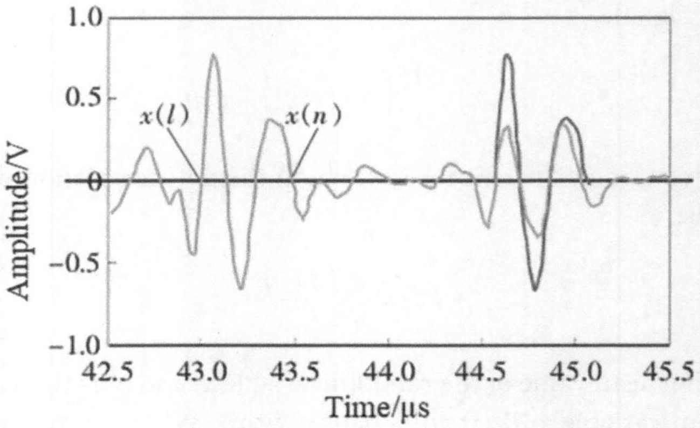

FIGURE 15.31 Principle diagram of sequence similarity detection.

are carried out for the ultrasonic A-scan waveform signal collected by the system. Then, the propagation time of ultrasonic wave in the tested blade is calculated by analyzing the autocorrelation function [6]. Finally, the blade thickness is accurately calculated by using the obtained propagation time and the calibrated sound velocity in the material. This method has no high requirements for the structural characteristics of the specimen and can ensure high measurement accuracy. The specific implementation steps are as follows.

1. Signal acquisition and processing

   The system first needs to collect the ultrasonic echo signal necessary for thickness calculation and then conducts the noise reduction and interpolation for waveform to facilitate subsequent autocorrelation analysis. In the actual detection process, in addition to the use of grounding, shielding wire and other measures to prevent electromagnetic interference, the interference of power-frequency noise signals should be reduced to further improve the SNR of ultrasonic characteristic signals. Periodic power-frequency interference signals can be identified by cross-correlation analysis. Suppose that the sampling signal is $S(t)$. It includes the ultrasonic echo signal $x(t)$ and periodic power-frequency interference signal $x_i(t)$ to be measured, which can be expressed as [7]:

$$\mathbf{x_i}(t)=C\sin(\omega_i t+\phi) \tag{15.23}$$

   According to the trigonometric function formula, Eq. (15.23) can be expanded into:

$$\mathbf{x_i}(t)=C\sin\phi\cos\omega_i t+C\cos\phi\sin\omega_i t \tag{15.24}$$

   Suppose that the sampling time interval is T_0, and the sampling time length is $t_p=NT_0$, an integer multiple of the period of power frequency signal. Then the sampling time series $\mathbf{S}(nT_0)$ can be discretized as

$$\mathbf{S}(nT_0)=x(nT_0)+\hat{A}\cos\hat{\omega}_i nT_0+\hat{B}\sin\hat{\omega}_i nT_0 \tag{15.25}$$

where $\hat{A}, \hat{B}$ and $\hat{\omega}_i$ are the estimated values of A, B and ω_i, respectively, $A = C\sin\phi, B = C\cos\phi$, $n = 0, 1, 2\ldots, N-1$.

The cross-correlation operation of the sampling sequence value $\mathbf{S}(n)$ and $\cos\hat{\omega}_i$ is performed to determine the cross-correlation function value $\hat{R}_A(0)$ at the time delay $\tau = 0$. Then Eq. (15.26) can be obtained:

$$\begin{aligned} R_{xy}(\tau) &= E\left[x(t)y(t+\tau)\right] \\ &= \lim_{T\to\infty}\frac{1}{T}\int_0^T x(t)y(t+\tau)dt \end{aligned} \tag{15.26}$$

$$\mathbf{R}_A(\tau) = \lim_{t_p\to\infty}\frac{1}{t_p}\int_0^{t_p}\mathbf{S}(t)\cos\omega_i(t+\tau)dt \tag{15.27}$$

The values of discrete time series are $S(t)$ and $\cos\omega_i t$ Then Eq. (5.12) can be converted into the summation form:

$$\hat{\mathbf{R}}_A(kT_0) = \frac{1}{N}\sum_{n=0}^{N-1}\mathbf{S}(nT_0)\cos\hat{\omega}_i(n+k)T_0 \tag{15.28}$$

By substituting Eq. (5.10) into Eq. (15.28), the cross-correlation function value $\hat{R}_A(0)$ at the time delay $\tau = kT_0 = 0$ can be obtained:

$$\hat{\mathbf{R}}_A(0) = \frac{1}{N}\sum_{n=0}^{N-1}\left[\mathbf{x}(n) + \hat{A}\cos\hat{\omega}_i n + \hat{B}\sin\hat{\omega}_i n\right]\cos\hat{\omega}_i n \tag{15.29}$$

Since $x(n)$ and $\cos\hat{\omega}_i n$ are independent of each other and the trigonometric functions $\sin\hat{\omega}_i n$ and $\cos\hat{\omega}_i n$ are orthogonal to each other, the above equation can be simplified as

$$\hat{\mathbf{R}}_A(0) = \frac{1}{N}\sum_{n=0}^{N-1}\left[\hat{A}\cos^2\hat{\omega}_i n\right] = \frac{\hat{A}}{2N}\sum_{n=0}^{N-1}\left[1 + \cos 2\hat{\omega}_i n\right] \approx \frac{\hat{A}}{2} \tag{15.30}$$

In the same way, the cross-correlation operation of the sampling sequence value $\mathbf{S}(n)$ and $\sin\hat{\omega}_i n$ is performed to determine the cross-correlation function value $\hat{R}_B(0)$ at the time delay $\tau = 0$:

$$\hat{\mathbf{R}}_B(0) = \frac{1}{N}\sum_{n=0}^{N-1}S(n)\sin\hat{\omega}_i n \tag{15.31}$$

And then we can obtain

$$\begin{cases} \hat{A} = 2\hat{R}_A(0) = \dfrac{2}{N}\displaystyle\sum_{n=0}^{N-1} \mathbf{S}(n)\cos\hat{\omega}_i n \\ \hat{B} = 2\hat{R}_B(0) = \dfrac{2}{N}\displaystyle\sum_{n=0}^{N-1} \mathbf{S}(n)\sin\hat{\omega}_i n \end{cases} \tag{15.32}$$

It can be seen from the above formula that the real-time angular frequency of power-frequency interference signal, denoted as $\hat{\omega}_i$, is the only variable. After the search and iteration of $\hat{\omega}_i$, the actual angular frequency of power-frequency interference signal can be determined according to the peak value of the relation between $\hat{\omega}_i$ and $\hat{C} = \sqrt{\hat{A}^2 + \hat{B}^2}$, where both $\hat{A}$ and $\hat{B}$ have been determined. Then the power-frequency interference signal is:

$$X_i(t) = C\sin\phi\cos\omega_i t + C\cos\phi\sin\omega_i t \tag{15.33}$$

$$\mathbf{x}(n) = \mathbf{S}(n) - \mathbf{x}_i(n) \quad n = 0,1,2,\ldots,N \tag{15.34}$$

The discrete time sequence $X(n)$ of the ultrasonic signal $X(t)$ to be measured can be obtained from the difference between the sampling sequence value $S(n)$ and the sequence value of power-frequency interference signal, denoted as $X_i(n)$. In this way, the useful signal can be separated and extracted, while the power-frequency interference signal can be eliminated.

In addition, when the waveform data within the specified time range is collected by time-domain gate to calculate the blade thickness, the deviation of measurement results is often found due to a small data size. Therefore, linear interpolation is needed for waveform data. Linear interpolation is a type of interpolation using a polynomial as the interpolation function and having zero error on the interpolation node. Compared with other interpolation methods, its operation is simpler and its calculation is more convenient. The geometric significance of linear interpolation is to approximate the function in question by using the line passing through the points A and B in Figure 15.32.

If the waveform data coordinates (x_0, y_0) and (x_1, y_1) are known, the y value corresponding to an interpolation point x within the interval $[x_0, x_1]$ can be obtained by using the following basic calculation procedure:

To determine the interpolation point (x_n, y_n) on the line according to the linear interpolation principle shown in Figure 15.32, two similar triangles can be drawn to obtain $(y_n-y_0)/(y_1-y_0) = (x_n-x_0)/(x_1-x_0)$. The value α on both sides of the equation is called interpolation coefficient, namely the ratio between the distance from x_0 to x_n and the distance from x_0 to x_1. Because the value of x_n is known, the value of α can be derived from formula to be $\alpha = (x_n-x_0)/(x_1-x_0) = (y_n-y_0)/(y_1-y_0)$, and then the value

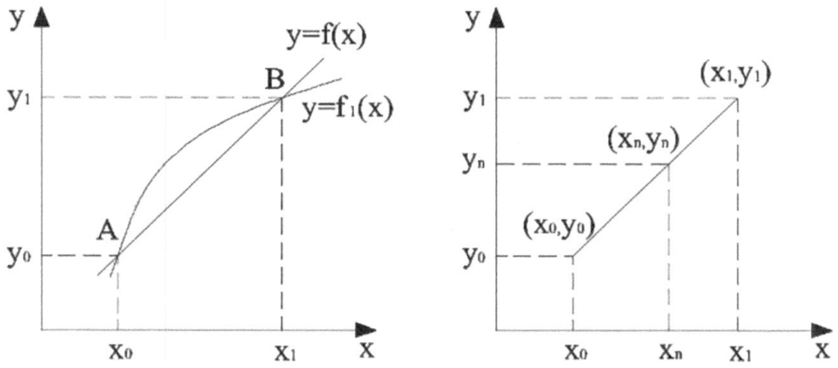

FIGURE 15.32 Principle diagram of linear interpolation.

of y_n can be directly determined as $y_n = y_0 + \alpha(y_1 - y_0)$. In the practical application, if the interpolation multiple is set as m, then the amplitude of the k-th interpolation point is equal to $y_k = y_0 + k/m(y_1 - y_0)$.

Thickness measurement algorithm

The complex components such as variable-thickness blade are usually asymmetric. When measuring the blade thickness, the stability of head wave (the first reflection echo) location in time domain should be ensured. Then the thickness H at each sampling point should be determined on the basis of the sound interval difference Δt between the head wave $B1$ and the bottom wave $B2$ as well as the propagation velocity C_T of ultrasonic wave in the tested material, as shown in Eq. (14.35).

$$T = \frac{C_U \cdot \Delta t}{2} \tag{15.35}$$

According to the principle of ultrasonic detection, the sonic path distance in water should be kept constant by ensuring the manipulator's scanning trajectory precision. By using an ultrasonic transducer with narrow bandwidth and small wavelength (high frequency), the oscillation period of ultrasonic head wave is shorter than the propagation time of ultrasonic wave in the blade. In this way, the accuracy of thickness measurement as well as the SNR and resolution of echo signal can be improved, and the sound interval difference between head wave and bottom wave can be accurately calculated.

The analysis of time-domain sampling signal through the time of flight (TOF) thickness measurement seriously affects the accuracy of thickness measurement. In this case, the gate tracking technique can be used to identify and record the location of interface reflection echo in time domain [8], as shown in Figure 15.33. Only by intercepting the time-domain part of ultrasonic full-wave waveform containing the head wave (primary reflection echo) and the bottom wave (secondary reflection echo), the blade thickness at each sampling point can be deduced from the sound interval difference between the head wave and the bottom wave. Therefore, the

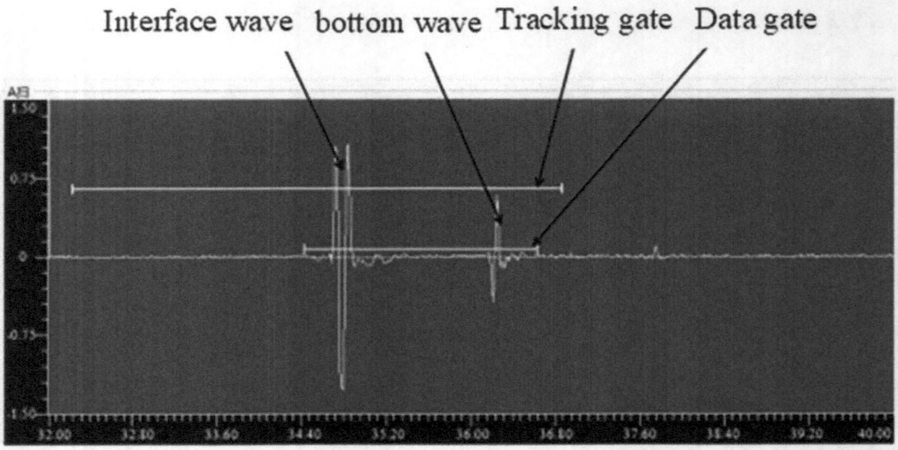

FIGURE 15.33 Tracking of an interface reflection echo gate.

autocorrelation of the intercepted waveform signal should be analyzed in order to determine the correct sound interval difference.

In the actual detection system, the original signal is collected, de-nosed and interpolated to obtain the ultrasonic echo signal $x(m)$ with a length of $2N$. Then the workpiece thickness is calculated with the following method and steps:

1. Before the actual test, the actual thickness of the blade at a measurable point should be measured by standard equipment (such as CMM). Then the actual ultrasonic propagation velocity C_T in the tested material should be calibrated at the measurement point with known thickness.
2. In $x(m)$, the N continuous waveform data points starting from the first data point are taken as the matching signal $y(n)$. The length of data acquisition gate must basically cover the time domain where the head wave and the bottom wave are located.
3. N continuous waveform data points are taken from $x(m)$ to form a new signal $z_i(n)$, where i represents the position of the first waveform data point and satisfies $0 \le i \le N-1$.
4. The discrete autocorrelation function $r(i)$ of $y(n)$ and $z_i(n)$ is calculated:

$$r(i) = \sum_{n=0}^{N-1} y(n) z_i(n) \tag{15.36}$$

5. When i is changing within $[0, N-1]$, step (3) is repeated to calculate all the $r(i)$ values.
6. According to the width of time-domain pulse, the $r(i)$ values near the origin $(i = 0)$ are appropriately eliminated to calculate the maximum $r(i)$ value deviating

from the origin ($i = 0$) by a certain distance and to determine the corresponding data point location i.

7. By using the value of I as well as the sampling frequency f and the interpolation multiple n, the time when the autocorrelation function $r(i)$ reaches its peak can be calculated, so that the sound interval difference Δt between the head wave and the bottom wave can be determined:

$$\Delta t = \frac{i \cdot n}{f} \tag{15.37}$$

8. By using Eq. (5.20), the actual thickness at the sampling point can be calculated.

The location point where the autocorrelation function reaches its maximum value reflects exactly the time-domain position interval in which the ultrasonic signal achieves the maximum likelihood, namely the sound interval difference between the head wave and the bottom wave when the ultrasonic wave propagates in the blade. Therefore, based on this feature of autocorrelation function, the thickness value can be calculated by processing and analyzing the ultrasonic A-scan echo signal collected by the system. The ultrasonic echo signal collected for autocorrelation analysis in the actual detection process and the corresponding autocorrelation function curve are shown in ww. It can be seen that a maximum will appear on the autocorrelation function curve at a large distance from the origin.

3. Implementation of automatic thickness measurement software

According to the actual requirements of thickness measurement, multiple measurement points are selected from the three-dimensional model of the specimen, and then a trajectory point packet is generated by using a series of coordinate

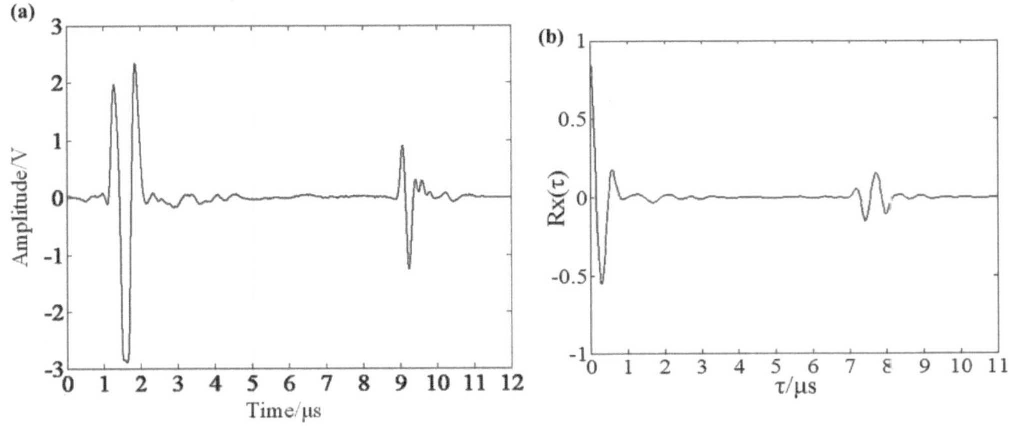

FIGURE 15.34 Ultrasonic echo signal diagram used for autocorrelation analysis, and autocorrelation function curve. (a) Ultrasonic echo signal diagram and (b) autocorrelation function curve.

transformation algorithms. After the point packet is loaded into the automatic thickness measurement program based on the principles presented in the previous sections, single-point thickness measurement and automatic multi-point thickness measurement can be realized. The generated data format is also friendly to later use. The most urgent problem to be solved by thickness measurement software is to operate the manipulator until it accurately moves to a measurement point, where multiple echo data will be selected in accordance with the requirement for average sampling frequency. Then multiple measured values at the same position are obtained with the autocorrelation algorithm, and the average thickness value is calculated to reduce the impact of random errors.

REFERENCES

1. Gu M. *Design of Portable Ultrasonic Thickness Gauge based on FPGA [D]*. Southwest Jiaotong University, 2010.
2. Li F., Fei D., Design of ultrasonic thickness measurement system based on pulse reflection, *Instrument Technique and Sensor*, 2013(4):50–52.
3. Li Z. *Research on Ultrasonic Automatic Detection System for Curved Alloy Components [D]*. Taiyuan: North University of China, 2011.
4. Xia K., He C., Wang W., Dou W., Parameter measurement based on autocorrelation detection [J]. *Digital Technology and Applications*, 2018, 26(5):92–94.
5. Wang X. *Finite Element Method [M]*. Beijing: Tsinghua University Press, 2003.
6. Yang Y. *Testing Technology and Virtual Instrument [M]*. Beijing: China Machine Press, 2010.
7. Ni Y., Tian Y., A waveform recognition algorithm based on fast template matching, *Sensor World*, 2006(4):32–34.
8. Yang Y. *Testing Technology and Virtual Instrument [M]*. Beijing: China Machine Press, 2010.

CHAPTER 16

Typical Applications of Dual-Manipulator NDT Technique

16.1 CONFIGURATION OF A DUAL-MANIPULATOR NDT SYSTEM

According to the position/attitude relationship between the end-effectors of the two moving manipulators, the testing modes of a dual-manipulator ultrasonic nondestructive testing (NDT) system can be divided into two types: the synchronous-motion ultrasonic testing (UT) mode suitable for large components, and the synergic-motion UT mode suitable for small complex components.

16.1.1 NDT Method for Large Components: Dual-Manipulator Synchronous-Motion Ultrasonic Testing

In the dual-manipulator synchronous-motion UT mode, the two manipulators with synchronous motion relationship start the same form of motion at a certain moment, and the relative position/attitude relationship between the two end-effectors remains unchanged or changes by a certain rule during the motion process. In this case, the two six-degree of freedom (DOF) manipulators with end-effectors clamping ultrasonic probes will move synchronously and the UT of the workpiece can be realized under specific constraints. This testing method is mainly oriented to large complex curved composite components. For example, the large composite shell under test in Figure 16.1 is a cavity semi-closed on both sides and composed of a cylinder segment, a spherical segment and a "skirt". In order to achieve the automatic full-coverage scanning of such a component, an acoustic waveguide tube and a special-shaped extension rod tool are introduced to guide ultrasonic wave into the narrow cavity of the component in this testing mode.

A dual-manipulator UT system is mainly composed of a hardware subsystem and a software subsystem. When testing a complex component, the configuration of the hardware subsystem (see the Figure 16.1) mainly includes the manipulator arm and its controller, pulse transceiver, ultrasonic data acquisition card (A/D card), ultrasonic probe (ultrasonic transducer), probe fixture and jet-flow coupling device (water nozzle), special-shaped

DOI: 10.1201/9781003212232-16

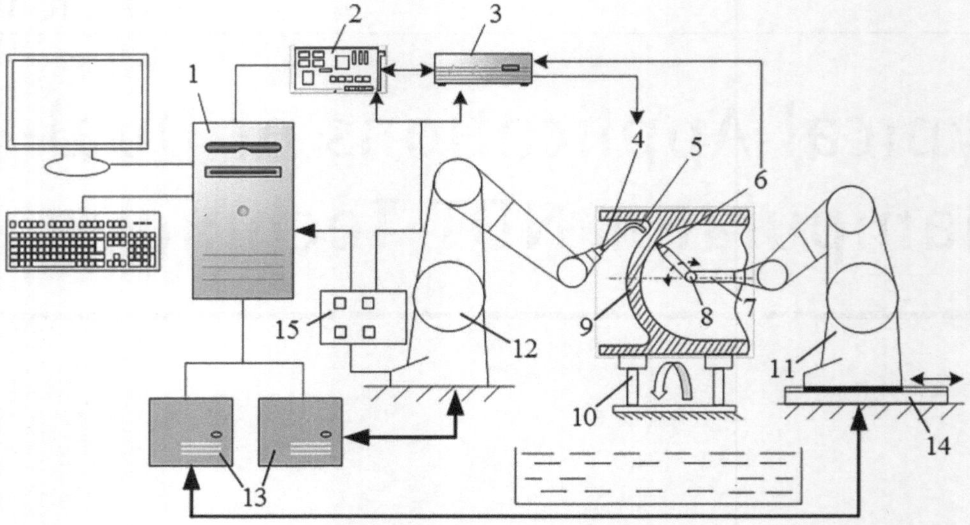

FIGURE 16.1 Dual-manipulator synchronous-motion UT mode. 1, IPC ; 2, ultrasonic acquisition card; 3, pulse transceiver; 4, ultrasonic transmitting probe; 5, acoustic waveguide tube; 6, ultrasonic receiving probe; 7, special-shaped extension rod tool; 8, rotation axis of an extension rod; 9, special-shaped component; 10, auxiliary rotating system; 11, master manipulator; 12, slave manipulator; 13, manipulator controller; 14, manipulator movement platform; 15, manipulator position acquisition card.

extension rod tool (with rotation axis) suitable for the testing of an irregular complex component, acoustic waveguide tube, auxiliary fixation and rotating mechanism (for the component under test), manipulator movement platform, industrial personal computer (IPC) and its peripherals, coupling-fluid circulation systems and other mechanical support structures. The component illustrated in Figure 16.1 is the broken-out section view of a large composite shell. The software subsystem is divided into upper computer software and lower computer software. The hardware provides data, while the software makes decisions. The hardware supports the software. They are organically combined into an automatic UT system for special-shaped complex curved composite components, namely a dual-manipulator ultrasonic NDT system.

16.1.2 NDT Method for Small Complex Components: Dual-Manipulator Synergic-Motion Ultrasonic Testing

In the dual-manipulator synergic-motion UT mode, the two manipulators with synergic motion relationship start to move at a certain moment, and their end-effectors have independent movement trajectories. The end-effector of one manipulator moves relative to the end-effector of the other manipulator. In the motion process, the relative position/attitude relationship between them is changing. In this mode, one manipulator clamps a small component, while the other manipulator clamps the ultrasonic transducer. They move in a synergic manner to realize the comprehensive UT of a small component through one-time clamping under specific constraints. The system configuration in the dual-manipulator synergic-motion testing mode is shown in Figure 16.2.

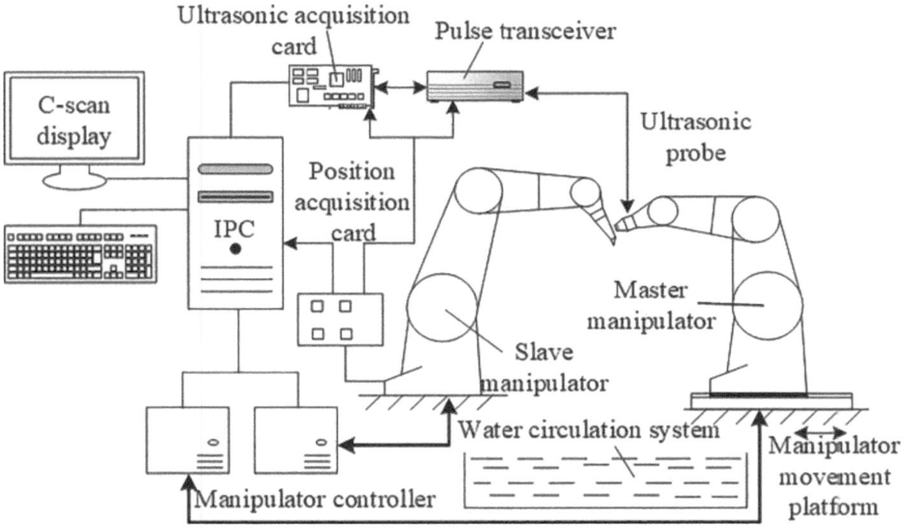

FIGURE 16.2 Dual-manipulator synergic-motion UT mode.

There are mainly two ways for synergic movement in ultrasonic detection:

1. When performing the detection task, one of the two manipulators (usually the manipulator holding the ultrasonic probe) is stationary, while the other manipulator holds the workpiece and moves to complete the task. When the detection task cannot be completed in one operation, the manipulator holding the probe will adjust it to an appropriate position/attitude and the manipulator holding the workpiece will continue the detection movement. This detection mode is called the position/attitude-assisted synergic motion mode.

2. When performing the detection task, both manipulators move at the same time and the ultrasonic transducer and workpiece are in relative motion. Generally speaking, the UT under this motion condition can achieve higher detection efficiency but must deal with more complex kinematic relationship. This detection mode is called linked synergic motion mode. During testing, an appropriate detection mode can be selected depending on the complexity of the tested component.

16.2 AN APPLICATION EXAMPLE OF DUAL-MANIPULATOR ULTRASONIC TRANSMISSION DETECTION

16.2.1 Ultrasonic C-Scan Detection of a Large-Diameter Semi-closed Rotary Component

The dual-manipulator UT system can automatically test a complex component, such as large-diameter semi-closed composite shell [1–3]. This component is an epoxy-resin-based glass fiber workpiece with an outer diameter of Φ600 mm and an inner diameter of Φ550 mm. Some artificial defects were made on the workpiece, as shown in Figure 16.3. These defects, namely flat bottom holes, were arranged into four rows, each row having

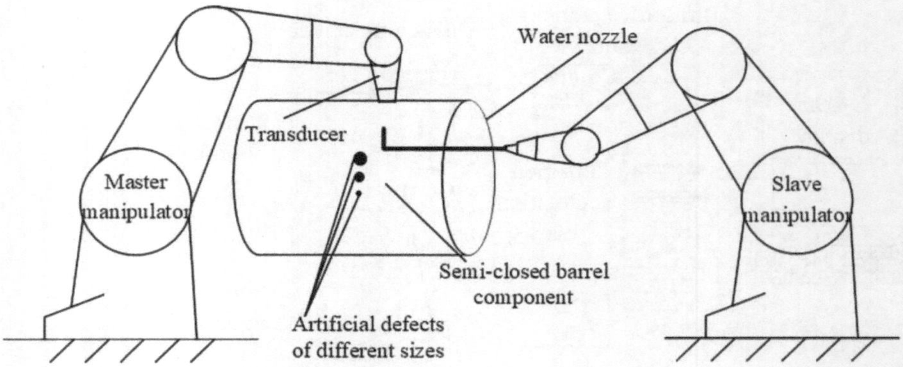

FIGURE 16.3 Dual-manipulator ultrasonic testing of a large-diameter semi-closed composite component.

four holes with the same diameter. The diameters of the holes in the four rows were Φ3, Φ5, Φ7 and Φ10 mm, respectively. The line spacing and sampling spacing on the scanning trajectory were both set as 0.75 mm.

The main components of the UT system are as follows:

1. Two 2.25 MHz water-immersed flat probes, with the wafer size of 0.5 in. and the damping of 200 Ω;
2. An ultrasonic pulse transceiver with the highest excitation voltage of 400 V;
3. A data acquisition card with the highest sampling frequency of 250 MHz.

In the detection experiment, a total of 83, 244 groups of location data [*x*, *y*, *z*] and ultrasonic data [*p*] were collected and arranged in the form of [*x*, *y*, *z*, *p*]. The C-scan image based on these data sets is shown in Figure 16.4, where the red areas are defects.

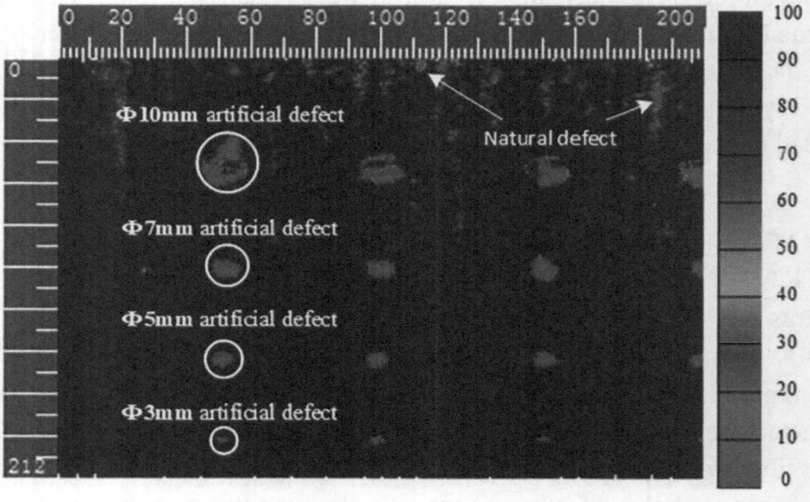

FIGURE 16.4 Ultrasonic C-scan detection result of a large-diameter semi-closed composite component.

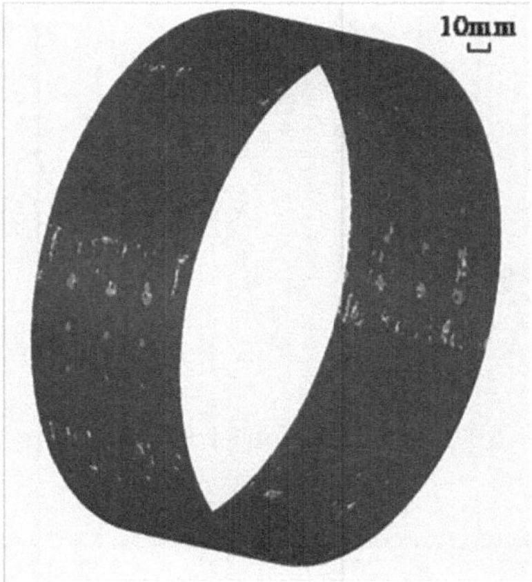

FIGURE 16.5 Ultrasonic three-dimensional C-scan imaging of a segment of a large-diameter semi-closed composite component.

It can be seen from Figure 16.4 that all the Φ3–10 mm artificial defects prefabricated on the workpiece have been detected, and that their size and location on the detection image are consistent with their actual size and location. As can be observed, the image of a Φ10 mm defect is not clear, which is caused by the method of defect fabrication, as described below. At first, the Φ10 mm flat-bottom holes were milled. Then they were filled with resin. During the curing process, the resin penetrated into the bottom of the holes, resulting in poor sound insulation of artificial defects. Therefore, the C-scan image is not clear, especially at the center of the defect. However, the edge of the defect is faintly visible, with a size comparable to that of the artificial defect. In addition, a number of internal natural defects with irregular shapes were detected.

To show the location and size of defects in 3D space more clearly, the 3D C-scan image of the defects in the 200 mm axial annular area is presented in Figure 16.5 based on the detection data. A total of 665,783 sets of location data and ultrasonic data were collected from this area. The 3D C-scan image can describe the size and position of workpiece defects more vividly, intuitively and clearly.

16.2.2 Ultrasonic C-Scan Detection of a Small-Diameter Semi-closed Rotary Component

To further demonstrate the wide application of dual-manipulator UT technique, a semi-closed component made of the same material as the component described in the previous section but with a smaller diameter is used as the detection object [4,5]. This component has an outer diameter of Φ250 mm and an inner diameter of Φ200 mm. Because it has a small internal space, its detection is more difficult. Nine artificial defects were made on the workpiece, as shown in Figure 16.6. The defects were arranged into three rows, each row having three defects of the same size. The diameters of the defects in the three rows were Φ3, Φ4 and Φ5 mm, respectively. The area under UT is just the area containing artificial defects. The scanning mode and parameters here are the same as those set in the previous section.

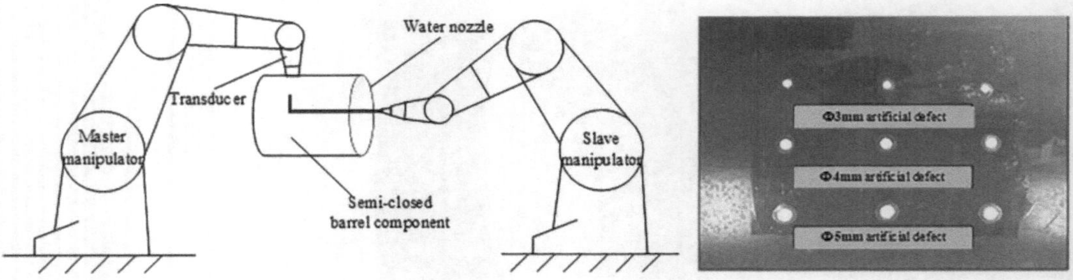

FIGURE 16.6 Dual-manipulator ultrasonic testing of a small-diameter semi-closed component.

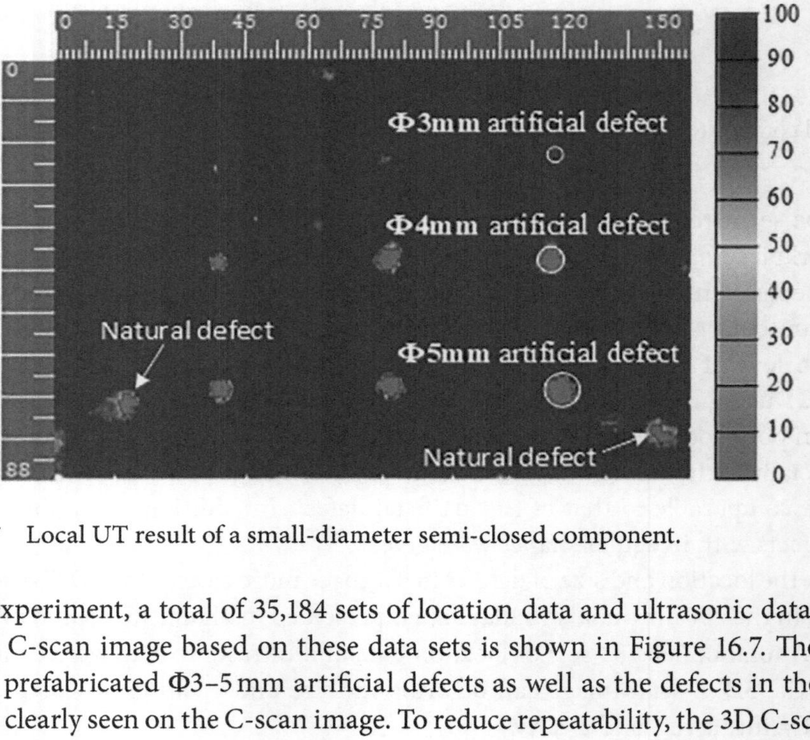

FIGURE 16.7 Local UT result of a small-diameter semi-closed component.

In this experiment, a total of 35,184 sets of location data and ultrasonic data were collected. The C-scan image based on these data sets is shown in Figure 16.7. The size and location of prefabricated Φ3–5 mm artificial defects as well as the defects in the scanned area can be clearly seen on the C-scan image. To reduce repeatability, the 3D C-scan results will not be shown here.

16.2.3 Ultrasonic C-Scan Detection of a Rectangular Semi-closed Box Component

To verify the ability of the dual-manipulator UT system to identify the defects of different shapes, a semi-closed rectangular box was tested through UT. It is made of polymethyl methacrylate (PMMA), a colorless transparent material commonly known as plexiglass. The length (L), width (W), height (H) and thickness of the box are 520, 500, 400 and 12 mm, respectively. As shown in Figure 16.8, three circular defects (Φ3, Φ4 and Φ5 mm) and a triangular defect (10 mm long at the bottom and 25 mm high) were made on the box surface. The step length and the sampling spacing were set as 0.5 mm.

The detection area was set as a 200 × 100 mm rectangular area. In the detection experiment, 78,658 sets of location data and ultrasonic data were collected. The C-scan image based on these data sets is shown in Figure 16.9.

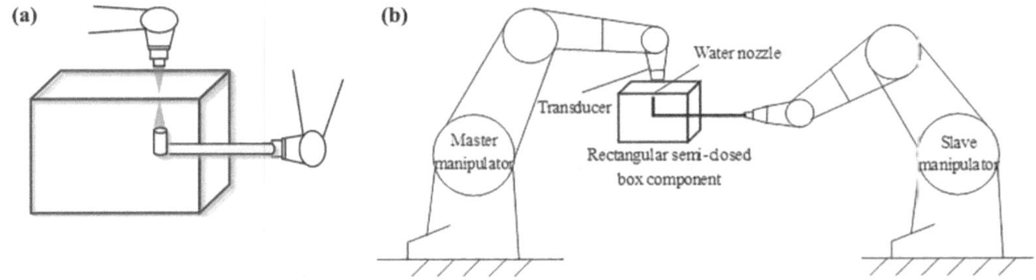

FIGURE 16.8 Dual-manipulator ultrasonic testing of a rectangular semi-closed box component. (a) The rectangular semi-closed box component and (b) the dual-manipulator ultrasonic testing system.

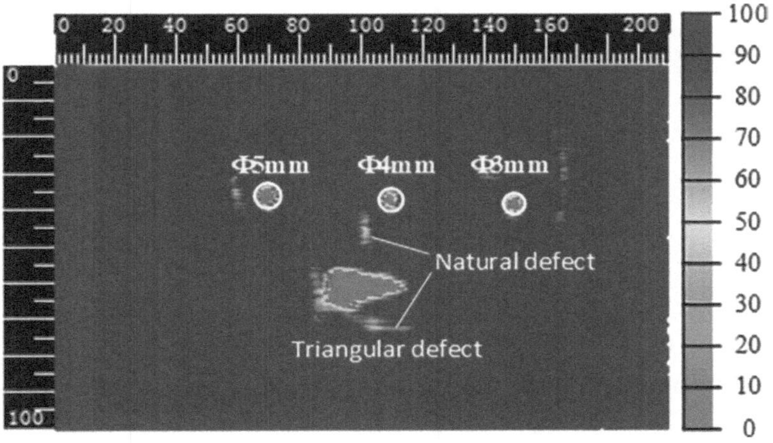

FIGURE 16.9 Local UT result of a rectangular semi-closed box component.

All the Φ3–5 mm artificial defects and the fabricated triangular defects can be clearly seen on the C-scan image. The size and location of the defects can be acquired directly from the image. In addition, some irregular defects also appear on the C-scan image. This is because the workpiece is transparent. The visual inspection found that before the artificial defects were pasted, there were no defects inside the workpiece and on its surface. Therefore, it can be judged that these accidental defects are the pasted artificial defects, which are caused by the bubbles and wrinkles between the tape and the workpiece surface. This indicates that the dual-manipulator UT system has reliable performance, and that the C-scan image has a strong ability to reproduce irregular defects such as triangular defects.

16.2.4 Ultrasonic Testing of an Acoustic Waveguide Tube

In Chapter 2, the UT method and theory of special-shaped components are discussed, and the acoustic characteristics of the waveguide tube and the feasibility of the waveguide UT method are clarified. In this section, an ultrasonic transmitting device where the manipulators clamp an ultrasonic probe, a water nozzle and an acoustic waveguide tube are designed on the basis of double-manipulator detection technique. In this device, ultrasonic waves are led by a waveguide into the inside narrow space of the workpiece, and another manipulator clamps an extension rod tool equipped with an ultrasonic probe and a water nozzle to receive ultrasonic signals from the inside of the workpiece. The detection principle diagram is shown in Figure 16.10.

The sample is a multicomponent-bonded composite sample made of carbon fiber and rubber and is prefabricated with the debonding defects in the sizes of 5, 10 and 15 mm.

The ultrasonic probe studied and analyzed in Chapter 3 and applied to sound field measurement was used as the detection probe. In addition, appropriate detection parameters were set. The excitation voltage was set as 400 V, the receiving gain was set as 16 dB, the ultrasonic scanning range was set as 90 × 120 mm, the step spacing was set as 0.75 mm and the sampling step length was set as 0.5 mm. The C-scan result is shown in Figure 16.11.

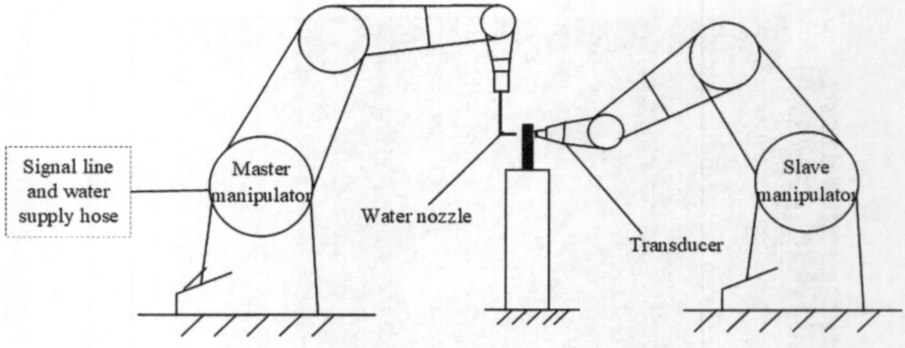

FIGURE 16.10 Verification test of dual-manipulator ultrasonic detection with underwater acoustic waveguide.

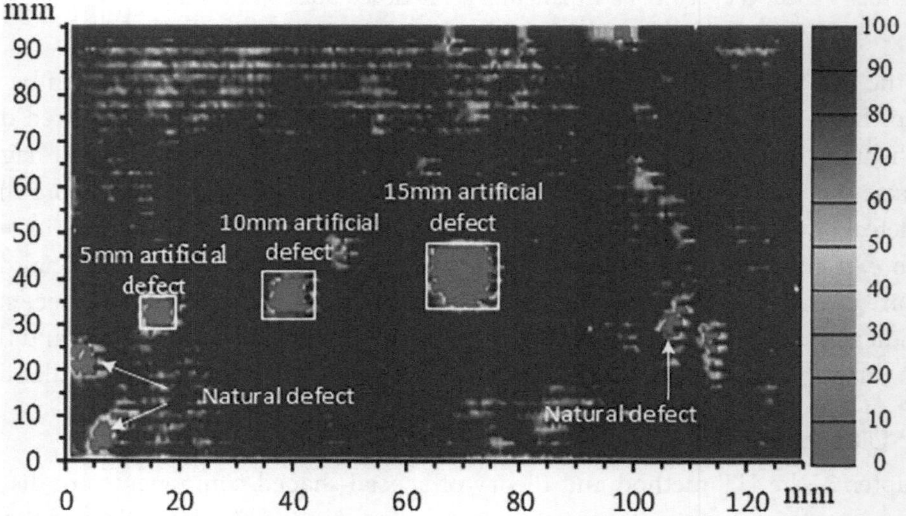

FIGURE 16.11 Experimental result of dual-manipulator ultrasonic detection with underwater acoustic waveguide.

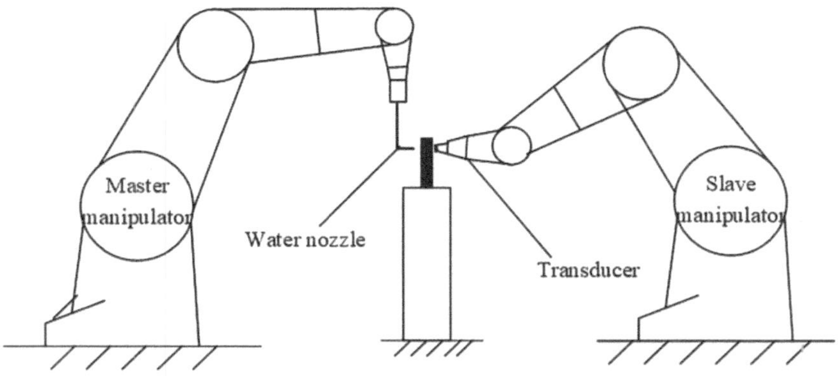

FIGURE 16.12 Comparative experiment of direct ultrasonic detection.

The three artificial defects prefabricated on the workpiece are clearly presented in Figure 16.10. In addition, some internal irregular natural defects have been detected. Although the C-scan result in Figure 16.12 was satisfactory, a comparative experiment was carried out on the same system without acoustic waveguide in order to fully demonstrate the practicability and reliability of the proposed detection concept. As shown in Figure 16.11, the waveguide tube was removed from the water nozzle of the manipulator (5).

The removal of acoustic waveguide will eliminate the ultrasonic attenuation of the system itself. Therefore, in this experiment, the excitation voltage dropped to 300 V, the receiving gain dropped to 4 dB and other parameters were the same as those with the presence of acoustic waveguide. The C-scan result is shown in Figure 16.13.

The comparison between Figures 16.11 and 16.13 shows that the results of the two experiments are almost identical. Although the direction of the ultrasonic wave will change during its propagation in the underwater waveguide, the result of UT will not be affected.

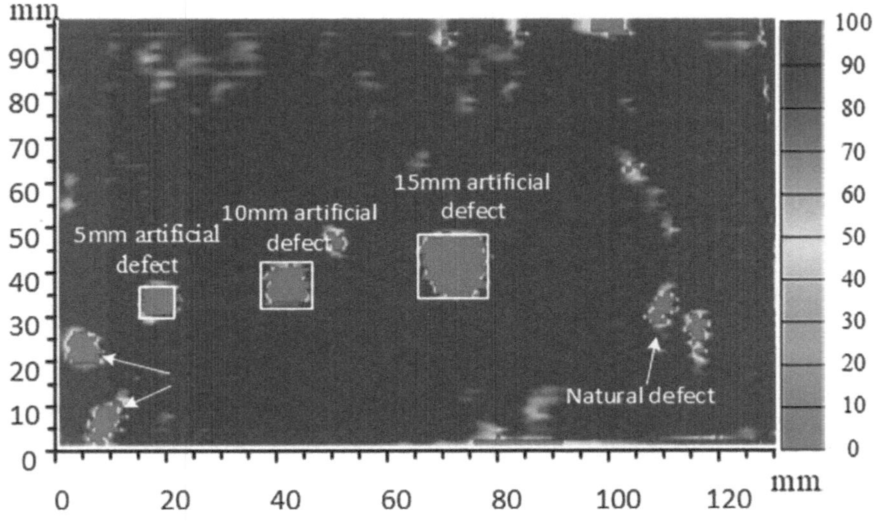

FIGURE 16.13 Result of comparative experiment of direct ultrasonic detection.

Thus, the dual-manipulator ultrasonic NDT with underwater waveguide proves to be reliable. This method can be used in practical engineering to solve the problem that the defects (such as delamination and debonding) in large special-shaped multicomponent-bonded composite components can't be automatically detected with high quality.

REFERENCES

1. Bogue R., The role of robotics in non-destructive testing [J], *Industrial Robot: An International Journal*, 2010, 37(5):421–426.
2. Maurer A., Deodorico W., Huber R., et al., Aerospace composite testing solutions using industrial robots [C]. 18th World Conference on Nondestructive Testing, 2012.
3. Louviot P., Tachattahte A., Gardener D., Robotised UT transmission NDT of composite complex shaped parts [J], NDT in Aerospace 2012- We.3.B.2, 2012, 4th Intern:1–8.
4. Schwabe D.M., Maurer A., Koch R., Ultrasonic testing machines with robot mechanics – A new approach to CFRP component testing [C]. 2nd International Symposium on NDT in Aerospace, 2013.
5. Cooper I., Nicholson I., Liaptsis D., et al., Development of a fast inspection system for complex composite structure [J]. 5th International Symposium on NDT in Aerospace, 2013.